SOUTHEASTERN GRASSLANDS

SOUTHEASTERN GRASSLANDS

Biodiversity, Ecology, and Management

Edited by
JoVonn G. Hill and John A. Barone

The University of Alabama Press Tuscaloosa

The University of Alabama Press
Tuscaloosa, Alabama 35487-0380
uapress.ua.edu

Typeface: Garamond

Manufactured in Korea
Cover image: *Calopogon oklahomensis* (Oklahoma grass pink orchid), a globally rare species restricted to high-quality unplowed prairie remnants, Prairie County, Arkansas.; photo by Michael Warriner/ ANHC
Cover design: Michele Myatt Quinn
Frontispiece: Pulliam Prairie, a Black Belt prairie remnant in Chickasaw County, Mississippi; digital art courtesy of Daniel Otte.

This publication was made possible in part by generous contributions from Wildlife Mississippi and the Mississippi Entomological Museum.

Library of Congress Cataloging-in-Publication Data

Names: Hill, JoVonn Grady, 1978– editor.
Title: Southeastern grasslands : biodiversity, ecology, and management / edited by JoVonn G. Hill and John A. Barone.
Description: Tuscaloosa : The University of Alabama Press, [2018] | Includes bibliographical references and index.
Identifiers: LCCN 2018005491| ISBN 9780817319885 (cloth) | ISBN 9780817391904 (ebook)
Subjects: LCSH: Grassland ecology—Southern States. | Biodiversity conservation—Southern States. | Grassland conservation—Southern States.
Classification: LCC QH104.5.S59 S68 2018 | DDC 577.40975—dc23
LC record available at https://lccn.loc.gov/2018005491

Dedicated to the many students and researchers of southeastern grasslands
who have inspired us.

To our families for their love and support.

In memory of Dr. Sam Riffell.

Contents

Preface

JoVonn G. Hill and John A. Barone

To many residents of the southeastern United States as well as to most visitors, the fact that this forested and agricultural landscape once possessed large areas of natural grasslands comes as a surprise. These grasslands included numerous regions with prairies as beautiful and diverse (in some cases more diverse) as any once found on the Great Plains of Iowa or Kansas. Across the Southeast, these grasslands are home to hundreds of species of plants, and vastly more animals. Indeed, a visitor to one of these prairies in June finds it awash in flowers and brimming with butterflies, beetles, and youthful grasshoppers.

Researchers often use the term "old growth" to distinguish forests that have largely escaped the saw and so are populated by enormous, ancient trees. While intact native grasslands lack this impressive grandeur, they are also old-growth habitats—ancient communities that go back many thousands of years. Unfortunately, their future is much more difficult to visualize. Though there are no precise estimates for the region, it is very likely that close to 99% of the grasslands that existed in the Southeast three years ago have been lost to agricultural development, urban growth, and cattle ranching. Compounding this problem is that many remaining sites are threatened by invasive species and the spread of neighboring forests, no longer held at bay by wildfires.

This book was inspired by the Southeast Prairie Symposium held on the campus of Mississippi State University during May 14–17, 2012. That meeting celebrated a rising tide of interest in the grasslands of the southeastern United States by both researchers and the public alike. The meeting included a variety of talks, visits to both natural and managed Mississippi grasslands, and wide-ranging, informal discussions about the science of these remarkable habitats and their future.

The 20 chapters in this volume arrive at a pivotal moment. It has been over a decade since the University of Alabama Press released Peacock and Schauwecker's *Blackland Prairies of the Gulf Coastal Plain*, a book that inspired the interest of many of the researchers whose work is included in this volume. More recently, Reed Noss's *Forgotten Grasslands of the South* was published, providing a more personal and probing vision of these habitats while introducing them to a broader audience. In the present volume, we have been fortunate to gather talented researchers from across the region to address questions related to the diversity, ecology, and management of southeastern grasslands, as well as to discuss how to restore sites that have been damaged by human activity.

The enthusiasm of these researchers is tempered by the continuing degradation and destruction of existing southeastern grasslands by human development and neglect. Sometimes this occurs simply because of ignorance. To the unaware, a prairie can resemble an abandoned agricultural field, albeit a beautiful one. Thus, our hope is that this book stimulates more interest in these remarkable habitats—including more research—and leads to greater appreciation and conservation.

The publication of this book was supported by a remarkable group of individuals and organizations to whom we are very grateful. We would like to thank the US Fish and Wildlife Service Partners for Fish and Wildlife, Wildlife Mississippi, Mississippi Agricultural and Forestry Experiment Station, Roundstone Native Seed, Wes Burger, Brian Baldwin, and the Department of Biochemistry, Molecular Biology, Entomology, and Plant Pathology at Mississippi State University for providing partial funding for the publication of this book.

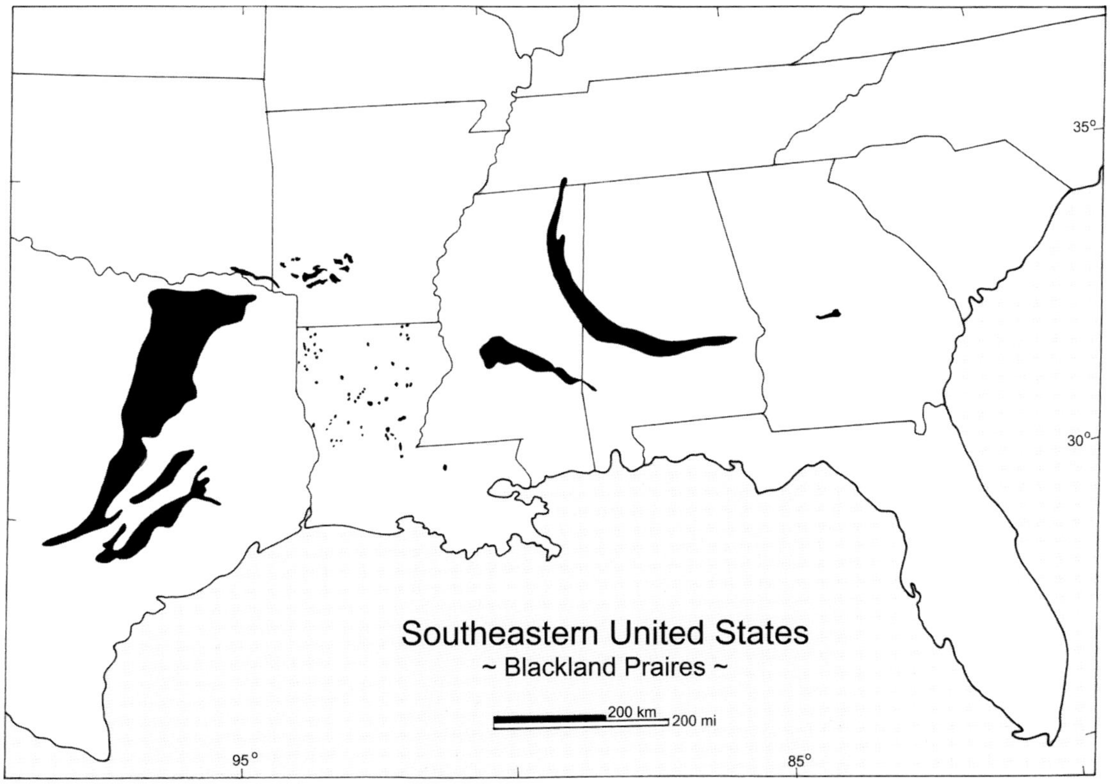

Figure 1.1. Generalized locations of blackland prairies in the southeastern United States. Modified by W. B. Zomlefer from Segars (1987), Griffith et al. (1998), M. MacRoberts and MacRoberts (2002), R. L. Brown (2003), M. MacRoberts et al. (2003), Peacock and Schauwecker (2003), A. Woods et al. (2005), Diggs et al. (2006), and Nature Conservancy (2009).

1

A Synopsis of Southeastern Blackland Prairies

S. Lee Echols and Wendy B. Zomlefer

Introduction

Blackland prairies are among the most prevalent edaphic grasslands across the South but are understudied compared to other North American prairie types (Noss 2013). Blackland ecosystems are characterized by a mosaic of grassland, savanna, shrubland, woodland, and mixed southeastern forest types (NatureServe 2012–2013). Related vegetation is scattered across the Gulf and Atlantic Coastal Plains and farther west: blackland prairies are documented from Alabama, Arkansas, Georgia, Louisiana, Mississippi, Oklahoma, Tennessee, and Texas (Peacock and Schauwecker 2003b; NatureServe 2012–2013) (fig. 1.1). These prairies are associated predominantly with limestone and calcareous clay formations originating during the Upper Cretaceous and Lower Tertiary (ca. 65–30 million years ago [hereafter mya]), when much of the southeastern United States was covered by a shallow sea (Huddlestun and Hetrick 1986; Peacock and Schauwecker 2003b). For this review, the term "blackland prairie" is considered a vernacular vegetation type indicative of southern plant communities dominated by grasses and forbs associated with alkaline clayey soils of high shrink-swell potential and underlain by calcareous parent material of marine origin. This serviceable definition focuses on present-day remnants often representing degraded examples. Mildly acidic soils, as well as more sandy or loamy soils, are also possible (at least historically) within the range of blackland prairie variation (Barone and Hill 2007; Campbell and Seymour 2011a). A variety of soil moisture regimes, from xeric to hydric, are also likely within this historical context (Campbell and Seymour 2011a).

The unique characteristics of the marine sediment–derived soils are the main driver in the formation of blackland prairie vegetation (Noss 2013). Decay of dense root systems from prairie grasses and forbs creates dark organic compounds that persist in the soil and coat mineral particles, creating the coloration responsible for the prairies' name (Diggs et al. 2006). The clays readily absorb water and expand, and then shrink and crack as they dry and the volume of soil decreases. Pettry and Switzer (1993) noted that the size and extent of cracking depends on numerous factors, including absorbed cations, clay content and mineralogy, precipitation, and soil moisture fluctuations. Shrink-swell clays and elevated pH likely decrease invasion of some woody species (Schauwecker 1996). This may explain the extant scattering of small remnants across the southeastern landscape, even in the absence of recent fire. Shallow rooting depth of eroded soil profiles or naturally shallow prairie soils may also play a role in the persistence of degraded prairie remnants (Echols 2007).

Although edaphic conditions are the main driver for the origin of blackland prairies, fire is one of the most important factors in maintaining open prairie vegetation over time (R. Anderson 1990). This is especially true for sites with deeper, less clayey, and slightly acidic soil profiles, where woody invasion may occur more rapidly. Fire benefits to prairie vegetation include decreased woody plant competition, increased nutrient availability, and alteration of microclimatic conditions to favor aboveground plant growth (Moran et al. 2003). Vegetation response to fire depends on factors such as precipitation patterns before and after burning, topography, vegetation structure, depth of soil to water table, frequency, and seasonality (R. Anderson 1990). Diggs and Schulze (2003) recognized the importance of the physical characteristics of clay soils in maintaining prairie vegetation but proposed a soil-dependent fire frequency hypothesis for Texas blackland prairies: larger quantities of high-quality fuel (prairie grasses) on clay soils increase the frequency and intensity of fires, creating a positive feedback loop for fire-adapted grasses that subsequently suppress woody plant growth.

According to Noss et al. (1994), grasslands, including blackland prairies, are the most endangered biome in North America. Blackland prairies are considered critically imperiled or imperiled throughout their range (NatureServe 2012–2013). The fertility of blackland prairie soils has long been exploited (Peacock and Schauwecker 2003a). Extensive cotton farming within the Black Belt of Mississippi and Alabama has played a major role in prairie destruction over the past 200 years (Frost et al. 1986). Overgrazing and deep plowing have caused significant erosion, which has been further exacerbated by the inherent properties of blackland soils (Peacock and Schauwecker 1996). Campbell and Seymour (2011a) note that the upland soils of Black Belt and Jackson Prairies are prone to severe sheet and gully erosion. Other anthropogenic impacts, including urbanization, fire suppression and the subsequent invasion of woody species and exotic invasives, landfill placement, and the construction of catfish ponds, have resulted in range-wide decimation (M. MacRoberts et al. 2003; Peacock and Schauwecker 2003a; Zollner et al. 2003).

Endangered grasslands such as blackland prairies require a systematic approach for effective conservation. The concept of representation—capturing the variation of a system or taxon across its range—is integral to the field of conservation biology (Noss et al. 1994). This chapter summarizes the diversity of blackland prairies across the Southeast on a state-by-state basis, since political boundaries play a large role in the allocation of scant conservation resources. It is hoped that scientists, land managers, and other conservation partners will refer to this chapter for elucidating relationships within and among these endangered grasslands, so that we may better protect this vanishing southern treasure.

Materials and Methods

A literature review was performed for five southeastern states with blackland prairies to compare geology, soils, and vegetation. Blackland prairies—the open, herbaceous, grass- and forb-dominated vegetation types—are the focus of this chapter, but chalk outcrops associated with prairie habitats are also included since they represent open herbaceous communities over shallow chalk or eroded examples of former blackland prairies. Geology discussions focus on formation-level deposits unless otherwise noted. Soil series occurring within blackland ecosystems, but associated with other vegetation types (e.g., forest, savanna, shrubland, woodland), are included here for conservation and restoration planning purposes. Many of these soils occur in complex mosaics with calcareous clays that underlie open prairie vegetation. Others are documented from the greater surrounding blackland landscapes. Some series may have previously supported prairie but are now denuded as a result of various anthropogenic uses. Vegetation descriptions focus on named ecological associations within the International Vegetation Classification framework (NatureServe 2012–2013) (table 1.1), although vegetation descriptions from floristic surveys and other literature sources are also provided. Vascular plant nomenclature follows the USDA PLANTS Database (USDA 2010–2013).

Oklahoma, Tennessee, and Texas were excluded from this review, although these states contain blackland prairies. Tennessee reportedly has degraded remnants of Black Belt Prairie in McNairy County (D. Estes, pers. comm.; NatureServe 2012–2013). The substantial diversity of blackland geology, soils, and vegetation of Texas are beyond the scope of this review. Texas historically contained the largest area of blackland prairies, but the state is generally not considered a part of the Southeast in terms of political boundaries, although eastern Texas shares many of the ecological attributes of the surrounding southeastern states, including a portion of the West Gulf Coastal Plain. Many ecologists consider Texas blackland prairies a southern extension of the "true prairie," or Great Plains prairies stretching northward to Manitoba (O. Collins et al. 1975; Diamond and Smeins 1993; Diggs et al. 2006; Launchbaugh 1955). Additionally, the isolated Fleming prairies of the eastern Texas Piney Woods ecoregion share some floristic and pedologic characteristics with the surrounding southeastern blackland prairies but are underlain by younger, Miocene-age deposits (Echols 2007; Noss 2013). Like Texas, Oklahoma falls outside the southeastern focus area.

Results and Discussion

Alabama

Blackland prairie vegetation occurs in a region known as the "Black Belt." This crescent-shaped area, which is about 40 km long and 500 km wide, extends from northeastern Mississippi into central Alabama (fig. 1.1). The southeastern portion of the Black Belt crescent is in Alabama. Twelve counties historically contained about 700 km^2 of prairies in west-central and central Alabama (Barone 2005b). Montgomery County had about 30,000 ha of prairie, more than any other county in the state. A study by Schotz and Barbour (2009) identified 14,595 prairies collectively covering 6,276 ha, with most prairies in Dallas, Lowndes, and Sumter Counties. They also delineated 10 significant publicly and privately owned Black Belt Prairie complexes. A small portion of the Jackson Prairie region also extends from Mississippi eastward into three Alabama counties (Choctaw, Clarke, and Washington) (NatureServe 2012–2013) (fig. 1.1).

Geology. The Alabama Black Belt Prairie is predominantly underlain by the Demopolis and Mooreville Formations of the Selma Chalk Group. This soft, Upper Cretaceous chalk produces outcrops extending to the south and west (Moran 1995). The Demopolis Formation comprises both pure and impure chalk, as well as marls, including massive-bedded fossiliferous gray marls (Morris et al. 1993). The Mooreville Formation contains a lower percentage of calcareous constituents than the harder, upper Demopolis Formation. The basal Eutaw Formation underlies the Mooreville Formation. The younger Ripley and Prairie Bluff Chalk Formations occur to a lesser extent within the Black Belt region and overlie the Demopolis Formation (J. Gibson 1941). More recent alluvial deposits are present where the Tombigbee River flows through the Black Belt (Barone 2005b).

Soils. Alabama Black Belt soils include many series also found in Mississippi, with the Demopolis and Sumter Series frequently occurring in open prairie vegetation. These alkaline, clay, or silty clay soils have a high shrink-swell potential and are underlain by Selma Chalk. The Trinity and Binnsville Series also underlie prairies to a lesser extent (Schotz and Barbour 2009). Examples of other, more acidic series documented from the Alabama Black Belt (and not necessarily associated with prairie vegetation) include Catalpa, Eutaw, Leeper, Kipling, Oktibbeha, and Vaiden (Soil Service Staff 2013). Table 1.2 includes a comprehensive list of soils of the Alabama Black Belt region.

Vegetation. Open prairie in Alabama is represented mainly by *Schizachyrium scoparium–Sorghastrum nutans–Dalea candida–Liatris squarrosa–(Silphium terebinthinaceum)* Black Belt Herbaceous Vegetation (NatureServe 2012–2013). Dominants include prairie grasses typical of blackland vegetation in other southeastern states, such as *Andropogon glomeratus, A. virginicus, Bouteloua curtipendula, Panicum virgatum*, and *Schizachyrium scoparium* (NatureServe 2012–2013). Jackson Prairie vegetation also occurs within three southwestern Alabama counties, and the following discussion of Mississippi Jackson Prairie vegetation has additional information on this prairie type (NatureServe 2012–2013) (table 1.1).

Schuster and McDaniel (1973) conducted a floristic inventory and vegetation analysis of a 1.2 ha prairie in Greene County, Alabama. *Hypoxis hirsuta* had the highest relative density in early spring, followed by *Schoenolirion croceum*. Campbell and Seymour (2011a) cited these species as evidence for a type of upland prairie on slightly acidic soils within the Black Belt. Schuster and McDaniel (1973) also noted *Schizachyrium scoparium* dominating upper slopes, and *Sporobolus vaginiflorus* dominating lower slopes. A. Jones and Patton (1966) analyzed historical field notes and plats from the Black Belt in Sumter County and found a strong correlation between alkaline clay soils and low tree density. In a similar study of historical Black Belt land-survey data from Montgomery County, Rankin and Davis (1971) reported oaks (*Quercus* spp.) as the dominant trees, followed by pines (*Pinus* spp.).

Arkansas

Blackland prairies occur in seven counties in southwestern Arkansas (Foti 1989). The state has the largest relatively intact holdings within the Southeast, with at least 7,500 ha of blackland prairie and related communities in undisturbed (or near natural) condition (Zollner, pers. comm.; Foti et al. 2003). Approximately 3,500 ha are protected and managed for ecological value (Zollner, pers.

comm.). Important protected prairie sites within Arkansas include Grandview Prairie Wildlife Management Area (Hempstead County, 2,000 ha); Saratoga Blackland Prairie Natural Area (Howard County, 74 ha), Columbus Prairie (Hempstead County, 62 ha), and Terre Noire Natural Area (Clark County, 30 ha). Two tracts (75 ha) are also protected within the Terre Noire landscape (Foti et al. 2003).

Geology. Blackland prairies in Arkansas occur primarily over Cretaceous calcareous substrates, but a narrow strip is underlain by the diverse Midway Group of the Paleogene (Lower Tertiary) period (Foti 1989). Midway Group lithologies include calcareous shale, arenaceous limestone, calcareous glauconitic sandstone, and clay shale (Arkansas Geological Survey 2012). Cretaceous formations and members that underlie prairie and associated vegetation types include Annona Chalk, Arkadelphia Marl, Brownstone Marl, Dierks and DeQueen Limestone (Trinity Group), Ozan, Nacatoch Sands, and Saratoga Chalk (Foti 1989; Hansen et al. 1997; Zollner et al. 2003). Trinity Group deposits date to the Lower Cretaceous, while the others are from the Upper Cretaceous (Foti 1989; Hansen et al. 1997). Most prairie vegetation types occur on shallow beds of soil over chalk, and increased soil depths allow woodland and forest types to flourish.

Soils. Table 1.2 includes a complete list of soil series known from the Arkansas blackland prairie region. Arkansas blackland prairie soils have two to four times the calcium of similar prairie soils in other southeastern states (B. MacRoberts et al. 2011). The most prevalent prairie soil types are the Sumter and Demopolis Series, which may also occur as small inclusions within the more acidic Oktibbeha Series typically associated with surrounding blackland forest types. The Houston Series is another common, typically deeper, alkaline clay soil underlying herbaceous prairie and savanna vegetation. The Terouge and Tuscumbia Series (silty clays) are less common and are associated with mesic to wet, herbaceous and woodland vegetation. Other associated woodland and forest types occur on the more acidic or sandy loams of the Oktibbeha, Kirvin, Kipling, and Sacul Series (Foti 1989; Zollner et al. 2003).

Vegetation. Three association-level open prairie communities have been described within the blackland prairie landscapes of Arkansas, and a fourth has been speculatively designated from degraded remnants (Zollner et al. 2003; NatureServe 2012–2013) (table 1.1). Zollner et al. (2003) found *Sorghastrum nutans–Schizachyrium scoparium–Echinacea pallida–Dalea purpurea* var. *purpurea* Dry-Mesic Blackland Prairie Herbaceous Vegetation to be the matrix type for the state (NatureServe 2012–2013). A rare mesic type has also been described (table 1.1). Foti (1989) documented *Schizachyrium scoparium* as the dominant in 113 prairie plots, and Schauwecker (1996) also found *S. scoparium* with highest frequency in Arkansas prairie plots. Other frequently occurring species included *Berchemia scandens, Chamaecrista fasciculata, Dalea purpurea, Galactia volubilis, Stenaria* spp., and *Rudbeckia hirta. Dalea purpurea* was the most abundant species in an analysis of plant biomass from postburn prairie plots, accounting for about 60% of the biomass (Foti 1990).

Georgia

Georgia blackland prairies are known from Houston, Bleckley, and Peach Counties in central Georgia (Klaus and Patrick 2002). These are located within the Southern Atlantic Coastal Plain ecoregion, whereas the other blackland prairies covered in this review are restricted to the Gulf Coastal Plain (Echols 2007; NatureServe 2012–2013; Peacock and Schauwecker 2003a). There are no historical occurrence estimates for blackland prairie in Georgia. The state recently acquired more than 4,000 ha of the Oaky Woods Wildlife Management Area (WMA), which comprises the majority of high-quality prairie remnants (Echols and Zomlefer 2010; Redmon 2010). Recent efforts by state agencies to restore Georgia chalk prairies have documented over 80 ha of existing prairies and associated communities as well as restorable areas where soils would potentially support this vegetation (Robert Bonds, pers. comm.).

Geology. As with the Jackson Prairies of Mississippi and Louisiana, Georgia blackland prairies are associated with Eocene deposits (Moran 1995; NatureServe 2013). Geologic formations associated with Oaky Woods WMA include the Dry Branch, Ocmulgee, and Tivola Formations, consisting of calcareous clay and limestone deposits of the Upper Eocene. The Twiggs Clay member of the Dry Branch Formation is silty, calcareous, montmorillonite clay that has been mined in central Georgia for its various industrial applications for over 100 years (Eversull 2005). Twiggs Clay is a disjunct eastern extension of Yazoo Clay (Huddlestun and Hetrick 1986; Lynch 2012). The Tivola and Ocmulgee limestone formations are textured, indurated or argillaceous limestone; both formations fall within the Ocala Group, with Tivola Limestone as the basal deposits (Eversull 2005; Huddlestun and Hetrick 1986). The Tobacco Road Sand Formation is an additional Eocene lithosome at Oaky Woods WMA. This formation occurs slightly north of known prairie locations and overlies Twiggs Clay in some areas. It represents a sandy depositional environment when barrier islands were present north of near-shore marine environments responsible for the Ocmulgee Formation (Echols 2007; Lynch 2012).

Soils. Georgia blackland prairies are underlain predominantly by the clayey Sumter Series, a common soil of Arkansas, Mississippi, and Alabama prairies (J. Woods 1967; USDA 2012a). The more acidic Boswell, Oktibbeha, and Susquehanna Series form an intricate mosaic, often associated with Sumter soils. They are composed of sandy loams or sandy clay loams instead of the heavier marly clay soils of open prairie vegetation. The Oktibbeha Series is underlain by chalk, limestone, or marl and is often associated with forest/prairie ecotonal vegetation (Lynch 2012). Other soils occurring in proximity to open prairie sites in Houston County include the deep, acidic fine sandy loams of the Faceville Series, as well as the acidic loamy sands of the Hoffman and Vaucluse Series. The latter two occur in very intricate associations with one another (Lynch 2012; J. Woods 1967).

Vegetation. Georgia blackland prairies are represented by (*Quercus sinuata*) / *Sorghastrum nutans–Ratibida pinnata–Houstonia nigricans* var. *nigricans*–(*Glandularia bipinnatifida*) Georgia Chalk Prairie Herbaceous Vegetation (NatureServe 2012–2013). This ecological association was described as distinct from western blackland prairie counterparts, based on differences in floristic composition and geology (Echols and Zomlefer 2010; Govus 2008). *Sorghastrum nutans* is the most common perennial grass, and *Schizachyrium scoparium* is noticeably absent from all known remnants. Other common and characteristic species of herbaceous prairie vegetation include *Agalinis tenuifolia, Andropogon gerardii, A. virginicus, Aristida purpurascens, Carex cherokeensis, Erigeron strigosus, Glandularia bipinnatifida, Polygala boykinii, Ratibida pinnata, Rudbeckia hirta, Salvia lyrata, Solidago nemoralis, Sporobolus vaginiflorus*, and *Stenaria nigricans* var. *nigricans*. Characteristic woody species include *Berchemia scandens, Celtis tenuifolia, Cercis canadensis, Crataegus aprica, C. crus-galli, C. spathulata, Diospyros virginiana, Fraxinus americana, Quercus muehlenbergii, Rhus copallinum* var. *latifolia, Ulmus alata*, and *Viburnum rufidulum*.

Echols and Zomlefer (2010) provide several local association descriptions for prairie vegetation in Oaky Woods WMA. Two types, one dominated by *Muhlenbergia capillaris* and the other by *Sporobolus clandestinus*, are small patch inclusions within a matrix of other prairie and calcareous woodland types. *Andropogon gerardii* is a codominant with *Sorghastrum nutans* at one location with deeper soil profiles.

Louisiana

Louisiana has four types of isolated calcareous prairies, and two are considered blackland prairies in this review. Jackson Prairies are found in north-central Louisiana within Caldwell Parish near Copenhagen, and there is an additional occurrence in the Catahoula Ranger District of the Kisatchie National Forest (KNF) in Grant Parish. Fleming Calcareous Prairies (western Louisiana) occur in Natchitoches, Rapides, and Vernon Parishes, and occurrences have been documented from Fort Polk and KNF. Cook Mountain Prairies (Keiffer Prairies; west-central Louisiana) are concentrated in Winn Parish within the Winn Ranger District of KNF, and to a lesser extent in Natchitoches and Sabine Parishes. Morse Clay Prairies (northwestern Louisiana) are known from Bossier and Caddo Parishes (Lester et al. 2005; NatureServe 2012–2013). Historical and current estimates of isolated Louisiana prairie habitats vary. Historical estimates range from 800–4,000 ha (Lester et al. 2005) to 16,000–20,000 ha (B. MacRoberts and MacRoberts 1997). B. MacRoberts and MacRoberts (1997) plotted former distributions of prairies in northern and central Louisiana and concluded that fewer than 1% remain. Higher estimates place remaining prairie habitat at 2–10% (Lester et al. 2005).

The two prairie types discussed in the next section, Jackson and Cook Mountain Prairies, share some of the geologic, pedologic, and vegetative characteristics of blackland prairies from surrounding states. Two other prairie types in Louisiana are not treated here. The Morse Clay Prairies have calcareous shrink-swell clayey soils but are restricted to terraces of the Red River. The underlying Pleistocene alluvium is considerably younger (ca. 115,000–130,000 years ago) than geologic formations commonly associated with blackland prairies (NatureServe 2012–2013; USDA 2010–2013). The Fleming Formation represents Upper Miocene stream deposits on a low coastal plain that merge with delta deposits toward the Gulf of Mexico (Diggs et al. 2006). Fleming prairies are therefore associated with younger geology and are also not treated here.

Geology. Within Louisiana, the Cook Mountain and Jackson Prairies are associated with the Cook Mountain Formation and Jackson Group, respectively (Lester et al. 2005). The Cook Mountain Formation underlies portions of the Fayette and San Antonio blackland prairie belts of Texas (Diggs et al. 2006). The Jackson Group of Louisiana is also associated with the Jackson Prairie Belt in Mississippi (Moran 1995). Huner (1939) includes historical descriptions of Louisiana calcareous prairies in his discussion of the geology of Caldwell and Winn Parishes.

The Cook Mountain Formation comprises marly Eocene clays from about 50 mya (Lester et al. 2005). This thick layer of fossil-rich marine muds and mudstones is interbedded with lesser amounts of sand and limestone (Diggs et al. 2006). Jackson Prairies are underlain by the Jackson Group, characterized by Upper Eocene near-shore, beach, and continental deposits associated with an ancient delta system (Diggs et al. 2006). In addition to the Danville Landing and Mosley Hill Formations, central Louisiana Jackson Group sediments contain the Moody's Branch and Yazoo Clay Formations, both of which are associated with Jackson Prairie vegetation in Mississippi (Johnston et al. 2000; Moran 1995). The discussion of Mississippi Jackson Prairie geology provides additional information on the Moody's Branch and Yazoo Clay Formations.

Soils. Louisiana calcareous prairies are associated with several soil series common to blackland prairies of other states, as well as more unique soil types. The Keiffer Series is associated with Cook Mountain Prairies (Lester et al. 2005; USDA 2010–2013). This series is closely related to more widespread Sumter soils; Cook Mountain Prairie soils were tentatively placed within the Sumter Series (L. Smith 1989). Keiffer soils are very deep and are formed in unconsolidated marine sediments, while Sumter soils are moderately deep and are formed in marly clays and chalks (USDA 2010–2013). The Oktibbeha and Hollywood Series underlie calcareous forest associated with the Cook Mountain Prairies (NatureServe 2012–2013).

Louisiana Jackson Prairies are associated with alkaline, calcareous clays containing calcareous concretions and gypsum crystals (NatureServe 2012–2013). Soils associated with this type and with other Louisiana calcareous prairies are in need of further study (Noss 2013). The Maytag Series, a prevalent soil type underlying Mississippi Jackson Prairies (Moran 1995), may also occur in Louisiana (Soil Service Staff 2013).

Vegetation. Vegetation of blackland-related Louisiana prairies includes one prairie type associated with the Cook Mountain Formation, and one prairie type associated with the Jackson Formation (NatureServe 2012–2013; table 1). Floristic inventories of four isolated prairies in central Louisiana found that 98% of the species reported were native (M. MacRoberts and MacRoberts 2002). Species richness in central Louisiana prairies is high, averaging about 100 species/ha (M. MacRoberts et al. 2003).

Cook Mountain Prairies are represented by *Schizachyrium scoparium–Panicum flexile–Carex microdonta* Herbaceous Vegetation (NatureServe 2013). They are the best studied of the two Louisiana blackland prairie types reviewed here. L. Smith et al. (1989) list *Andropogon gerardii*, *Schizachyrium scoparium*, *Sorghastrum nutans*, and *Sporobolus compositus* as typical grasses. Common herbaceous species indicative of relatively undisturbed sites include *Dalea candida*, *D. purpurea*, *Echinacea pallida*, and *Ruellia humilis* (L. Smith et al. 1989). B. MacRoberts and MacRoberts (1996) conducted a vascular plant inventory of two Cook Mountain Prairies, documenting 137 vascular plant species from about 2.8 ha.

Louisiana Jackson Prairies are represented by *Schizachyrium scoparium–Sorghastrum nutans* Jackson Prairie Herbaceous Vegetation (NatureServe 2012–2013). This association, closely related to the Cook Mountain type and to Mississippi's Jackson Prairie vegetation, shares many of the grasses listed for Cook Mountain Prairie, as well as *Bouteloua curtipendula* and *Danthonia spicata* (see NatureServe [2013] for a list of 19 characteristic forbs for this association).

Mississippi

Mississippi has two distinct areas of blackland prairie, the Black Belt and Jackson regions, which have similar vegetation and soils but differ markedly in the underlying geologic formations.

Black Belt Prairie. Black Belt Prairies occur in 11 counties in northeastern and east-central Mississippi (Schauwecker 1996). Barone (2005b) estimated that Black Belt Prairies covered about 700 km^2, or 0.6%, of Mississippi's land area in the 1830s. Extant noteworthy Black Belt Prairie examples and associated communities include the Osborn (Sixteen Section) Prairie in Oktibbeha County (ca. 72 ha) and Pulliam Prairie (ca. 100 ha) in Chickasaw County (Campbell and Seymour 2005b; J. G. Hill, pers. comm.).

Black Belt Prairie—Geology. As in Alabama, Mississippi's Black Belt is underlain predominantly by the Mooreville and Demopolis Formations of the Selma Chalk Group (Moran 1995). Basal sediments consist of the Eutaw Formation, as well as the younger Ripley and Prairie Bluff Chalk Formations to a lesser extent (J. Gibson 1941).

Black Belt Prairie—Soils. Leidolf and McDaniel (1998) reported the Sumter and Binnsville Series underlying a 16.8 ha fragment of remnant upland Black Belt Prairie vegetation within Oktibbeha County. The Houston Series is less common. Other documented Mississippi Black Belt soil series (not necessarily associated with prairie vegetation) include the more acidic Kipling, Oktibbeha, and Vaiden Series, found on old stream terraces, wide ridges, and narrow side slopes; and the Catalpa, Eutaw, and Leeper Series in bottomlands (Soil Service Staff 2013). Table 1.2 includes a comprehensive list of soils known from the Black Belt region of Mississippi.

Black Belt Prairie—Vegetation. Open prairie vegetation is represented mainly by *Schizachyrium scoparium–Sorghastrum nutans–Dalea candida–Liatris squarrosa–*(*Silphium terebinthinaceum*) Black Belt Herbaceous Vegetation (NatureServe 2012–2013). Schauwecker (1996) found *Schizachyrium scoparium* to have the highest frequency in 99 prairie plots in the Mississippi Black Belt. Other high-frequency species included *Aristida dichotoma, Carex cherokeensis, Dalea purpurea, Galactia volubilis, Solidago nemoralis,* and *Symphyotrichum laeve.* Leidolf and McDaniel (1998) conducted a floristic survey at Sixteenth Section Prairie in Oktibbeha County. The dominant open-prairie grasses included *Andropogon glomeratus, A. virginicus, Bouteloua curtipendula, Panicum virgatum,* and *Schizachyrium scoparium.* Common woody species frequently invading this prairie included *Cercis canadensis, Diospyros virginiana, Fraxinus americana, Ilex decidua,* and *Ulmus alata.* In addition to open-prairie types, Leidolf and McDaniel (1998) described another open herbaceous community, chalk outcrops, which are areas of extremely shallow soils or chalk at the surface supporting a limited subset of prairie species as a result of high alkalinity and low fertility. These areas support some species (e.g., *Coreopsis lanceolata, Heterotheca camporum, Mirabilis albida,* and *Silphium laciniatum*) not present in their open-prairie descriptions. Morris et al. (1993) also described a Chalk Outcrops and Barrens community type from the Pontotoc Ridge–Black Prairie transition area. Campbell and Seymour (2011b) conducted a vegetation study of about 100 ha of the Pulliam Prairie complex, a site representing one of the most intact examples of the remnant Black Belt landscape.

Jackson Prairie. The Jackson Prairie Belt occurs in seven counties in central and eastern Mississippi and comprises gently rolling topography approximately 15 to 50 km wide, extending from the eastern edge of the Loess Bluff Region to just across the border of Mississippi and Alabama (Moran 1995; Moran et al. 1997). Barone (2005a) estimated that the Jackson Prairie region covered 19,555 ha in the early nineteenth century. Currently, about 60 prairie remnants are known from the Jackson region, and prairie openings range in size from less than 1 to about 65 ha, totaling about 325 ha for the state (Elsen and Wieland 2003; Moran et al. 2003).

Jackson Prairie—Geology. The Jackson Prairie region is underlain by two formations of the Jackson Group: Yazoo Clay and Moody's Branch, both deposited during the Eocene. The Yazoo Clay Formation is calcareous, contains a high percentage of montmorillonite, and ranges in color from green to gray. The Moody's Branch Formation, the basal formation in the Jackson Group, consists of fossiliferous marls with quartz, sand, and clay (Cooke 1939; Moran 1995; Moran et al. 2003).

Jackson Prairie—Soils. The Maytag Series most commonly underlies Jackson Prairie openings. This series consists of deep, slowly permeable, and well-drained soils derived from chalks or marls (NatureServe 2012–2013; USDA 2012a). Acidic clays are more commonly associated with surrounding mixed pine and hardwood forests (Moran 1995). Additional soil series associated with the Jackson Prairie landscape include Catalpa, Eutaw, Griffith, Houston, Kipling, Leeper, Louin, Okolona, Oktibbeha, Siwell, Sumter, Tuscumbia, and Vaiden (Elsen and Wieland 2003; Moran 1995; Wieland 2000).

Moran et al. (1997) characterized soils from four Jackson Prairie remnants that had slightly to moderately alkaline soils of high natural fertility. Calcium carbonate was moderate to very high, and epipedons (upper soil horizons) were high in organic matter.

Jackson Prairie—Vegetation. Open-prairie vegetation within the Jackson Belt is represented by *Schizachyrium scoparium–Sorghastrum nutans–Dalea purpurea–Silphium integrifolium* Jackson Prairie Herbaceous Vegetation (NatureServe 2012–2013). In addition to these dominant or characteristic species, other common vascular plant species within prairie openings include *Carex cherokeensis* and *Helenium autumnale,* with species such as *Dalea candida, Desmanthus illinoensis, Echinacea purpurea, Manfreda virginica, Muhlenbergia capillaris, Penstemon laxiflorus, Ruellia purshiana, Spiranthes magnicamporum, Sporobolus compositus* var. *macer,* and *Symphyotrichum novae-angliae* occurring within less-disturbed openings (NatureServe 2012–2013). A post oak acid prairie (Moran et al. 1997) has also been described from low slopes within the Jackson Prairie (excluded from table 1.2 because of insufficient information). Moran (1995) characterized the Jackson Prairies as isolated calcareous islands surrounded by mixed pine and hardwood forests and provided a list of common prairie species. S. Jones (1971) briefly described a "virgin prairie" within the Bienville National Forest (ca. 40 ha) known as Harrell Prairie Hill. He included 58 typical or common species for the site. The Mississippi Natural Heritage Program (MNHP 1991) listed these additional indicator species for Jackson Prairie openings in the Bienville National Forest: *Asclepias viridis, Monarda fistulosa, Neptunia lutea,* and *Ruellia humilis.*

Findings across Five States

Geology comparison. Calcareous clays, chalks, and marls of marine origin predominate in the blackland prairies, although diverse groups and formations underlie them. Mud and mudstone, sand, sandstone, shale, and silt constituents are less common. All geologic units originated in the Upper Cretaceous or Lower Tertiary, with the exception of Lower Cretaceous Trinity Group limestone deposits underlying some Arkansas blackland prairies. Arkansas has the largest array of geologic diversity, with at least eight formations of varying lithologies, along with the diverse Midway Group.

The Black Belt of Mississippi and Alabama forms a relatively contiguous geologic unit, underlain primarily by the Demopolis and Mooreville Formations of the Selma Chalk Group. Louisiana Jackson Prairies share the Jackson Group with the Jackson Prairie Belt of Mississippi. Georgia Eocene Chalk Prairies are underlain predominantly by the Dry Branch Formation and members of the Ocala Group. These deposits share an Eocene origin with Louisiana and Mississippi Jackson Prairies. Georgia Twiggs Clay is considered an eastern extension of the Yazoo Clay within the Jackson Group, an intriguing geologic connection between the prairies of Georgia and Louisiana/Mississippi Jackson Prairie Clay (Huddlestun and Hetrick 1986; Lynch 2012).

Soil comparison. Southeastern blackland prairies share numerous soil series (table 1.2). Calcareous clays with high shrink-swell potential dominate blackland prairies across their range. The Sumter Series is a common calcareous clay soil underlying herbaceous prairie vegetation in Alabama, Arkansas, Georgia, and Mississippi. The more acidic Oktibbeha Series, associated with wooded vegetation surrounding prairies, is widespread and documented from all five states. Arkansas has at least 10 soil series in common with the Black Belt region. The Alabama/Mississippi Black Belt shares many soil series, including Binnsville soils for open-prairie vegetation. The Jackson Prairies of Mississippi also have substantial similarities with the nearby Black Belt, but open-prairie vegetation is characterized predominantly by the Maytag Series. This series may occur in related Louisiana Jackson Prairies, where additional soil studies are needed.

Eight soil series are associated with open blackland prairie vegetation (table 1.2). Four are within the Vertisol order, including the Houston, Maytag, Terouge, and Trinity Series. The remaining four series are within three additional orders: Entisols (Demopolis Series); Inceptisols (Keiffer and Sumter Series); and Mollisols (Binnsville Series).

Vegetation comparison. The few prairie remnants scattered across the Southeast often represent degraded examples and do not adequately represent the historical variation. Two of the most commonly reported species across the entire range of blackland prairies are the grasses *Schizachyrium scoparium* and *Sorghastrum nutans,* which are frequently dominant or at least form a component of herbaceous prairie associations. Foti (1989) and Leidolf and McDaniel (1998) note similarities in species composition among Arkansas, Mississippi, and Alabama prairies, such as *Berchemia scandens, Carex cherokeensis, Dalea candida, D. purpurea, Desmanthus illinoensis, Juniperus virginiana,* and *Stenaria* spp. Georgia shares many of these species, especially those of Alabama. Common elements include (but are not limited to) *Erigeron strigosus*, *Polygala boykinii*, *Ratibida pinnata*, *Silphium asteriscus*, *Stenaria nigricans*, *Solidago nemoralis*, and *Sporobolus vaginiflorus*. Noticeably absent in Georgia blackland prairies are *Dalea* species and the *Schizachyrium scoparium*–dominated herbaceous vegetation of Alabama, Arkansas, and Mississippi blackland prairies. *Schizachyrium scoparium* is apparently absent within Georgia blackland prairies (Echols and Zomlefer 2010). The Jackson Prairie openings of Louisiana are considered closely related to those in Mississippi's Jackson Prairie Belt, since they both occur over Jackson Group geology. Additional quantitative vegetation sampling is needed in both prairie types to elucidate relationships (NatureServe 2012–2013). Louisiana Cook Mountain Prairies are unique in that *Carex microdonta* is a characteristic or dominant species (B. MacRoberts and MacRoberts 1996).

Cluster analyses by Barone (2005b) and Echols (2007) placed Mississippi and Alabama Black Belt open-prairie vegetation in different groups, despite their proximity and similarity in geology, soils, and documented prairie types. Since both studies utilized presence-or-absence data, only limited inferences can be made from these results (Barone 2005b). However, these analyses highlight the need for additional vegetation studies within the Black Belt to assess potentially unrecognized variation, especially since this region once covered such a large area. Campbell and Seymour (2011a) provide a model for exploring potential vegetation types across a Black Belt landscape gradient (lowland to upland). This model could also be applied to other blackland systems to reconstruct a more nuanced framework for natural vegetation types surrounding remnant prairie pockets.

Two abiotic factors affecting vegetation (not explicitly addressed in this review) are landscape position and soil moisture gradients (see comments column in table 1.1). Many blackland prairie types across the Southeast have similar or identical soil series, making other factors an important consideration in differentiating ecological associations. Unfortunately, a subset remains of prairie vegetation types that have persisted on the poorest sites not suitable for anthropogenic use. The location and hydrology of prairie remnants within the landscape is important because these factors heavily influence the composition and relative abundances of prairie plant taxa at a site. For example, Arkansas contains at least three types of blackland prairies differentiated in part by moisture regimes (table 1.1).

Conclusions

There is a critical need for additional quantitative vegetation sampling and classification of blackland prairie remnants and associated vegetation types (Peacock and Schauwecker 2003a). These data would help to further assess the blackland prairies in each state and to properly allocate scant resources for preservation and restoration efforts. Attention to the types of geologic formations, soil series, and vascular plant associations is necessary for blackland prairie classification and conservation purposes. The diversity of imperiled and critically imperiled ecological associations (and their underlying soils and lithologies) suggests that blackland prairies differ markedly across their range. Quantitative sampling is critical in the remaining sites where ground cover is still relatively undisturbed, where the most valuable data can be garnered. Sampling is especially critical in the few remaining prairie remnants with characteristics such as deeper intact soil profiles, less clay content, and slightly acidic pH. More-productive sites were lost first because of their suitability for agriculture, pine plantations, and other anthropogenic land uses, and fire suppression further exacerbated the situation. Ongoing restoration efforts and reintroduction of fire back into altered landscapes may also yield more indication of historical vegetation types.

The unique mosaic of associated vegetation makes the diversity of blackland landscapes exceptional (Echols and Zomlefer 2010). Further documentation of surrounding forest, savanna, shrubland, and woodland types is of special concern. These communities also harbor rare and disjunct taxa, and in many cases they are even scarcer and more poorly known than the prairies themselves.

Table 1.1.

Association-level open-prairie communities documented within southeastern blackland prairie regions.

State(s)	Common name of prairie type	IVC association name	Other designation(s)	Global rank	Comments
AL, MS	Black Belt Prairie	*Schizachyrium scoparium–Sorghastrum nutans–Dalea candida–Liatris squarrosa–(Silphium terebinthinaceum)* Black Belt Herbaceous Vegetation	—	G1	Over Sumter and Binnsville Soils in the Black Belt of Alabama and Mississippi; remnants known from McNairy County, southern Tennessee
AL, MS	Mississippi Jackson Calcareous Clay Prairie	*Schizachyrium scoparium–Sorghastrum nutans–Dalea purpurea–Silphium integrifolium* Jackson Prairie Herbaceous Vegetation	Jackson Prairie	G1	Mostly associated with Maytag soils; calcareous clay islands of the Jackson Prairie region in Mississippi; also Washington, Choctaw, and Clarke Counties of Alabama
AR	Upper West Gulf Coastal Plain Dry Calcareous (Blackland) Prairie	*Schizachyrium scoparium–Sporobolus compositus–Fimbristylis puberula* var. *puberula* Wooded Herbaceous Vegetation	—	G2	Dry ridges and chalk outcrops over Sumter and Demopolis soils
AR	Upper West Gulf Coastal Plain Dry-Mesic Blackland Prairie	*Sorghastrum nutans–Schizachyrium scoparium–Echinacea pallida–Dalea purpurea* var. *purpurea* Dry-Mesic Blackland Prairie Herbaceous Vegetation	—	G2G3	Upper and middle slopes, and ridgetops over Sumter and Demopolis clay; matrix type community for AR, responds to sweet clover and redcedar removal
AR	Upper West Gulf Coastal Plain Mesic Blackland Prairie	*Sorghastrum nutans–Andropogon glomeratus–Silphium laciniatum* Mesic Blackland Prairie Herbaceous Vegetation	—	G1?	Along lower slopes, flat tabletops, areas of high soil moisture; Houston and Sumter soils
AR	—	*Lythrum alatum–Panicum anceps–Symphyotrichum lanceolatum* West-Mesic Blackland Prairie Temporarily Flooded Herbaceous Vegetation	—	G1?	Shallow draws, depressions, and level areas where soils are occasionally to seasonally saturated; Houston Clay and Terouge silty clay soils; speculative conceptual community based on degraded sites
GA	Georgia Eocene Chalk Prairie	(*Quercus sinuata*) / *Sorghastrum nutans–Ratibida pinnata–Houstonia nigricans* var. *nigricans–*(*Glandularia bipinnatifida*) Georgia Chalk Prairie Herbaceous Vegetation	Georgia Blackland Prairie	G1	Sumter soils on shoulder and back slopes
LA	Jackson Calcareous Prairie (Louisiana)	*Schizachyrium scoparium–Sorghastrum nutans* Jackson Prairie Herbaceous Vegetation	—	G1	Ridgetops and upper slopes; restricted to north-central Louisiana over the Jackson Formation; closely related to Mississippi Jackson Calcareous Clay Prairie
LA	Cook Mountain Prairie	*Schizachyrium scoparium–Panicum flexile–Carex microdonta* Herbaceous Vegetation	Keiffer Prairie	G1	Silty loams, clays, and silt loams on ridgetops and gentle slopes; west-central Louisiana over the Cook Mountain Formation.

Source: NatureServe (2013); Zollner et al. (2003)

Note: IVC = International Vegetation Classification. Global Rank definitions: G1 = critically imperiled, 5 or fewer occurrences; G2 = imperiled, 6–20 occurrences; G3 = rare or local throughout range, in special habitat, or narrowly endemic, 21–100 occurrences; ? = inexact numeric rank.

Table 1.2.
Geology and soils of blackland prairies by state and prairie type.

State	General prairie type	Geology (formations unless otherwise noted)	Geographically associated soil series
AL	Black Belt Prairie	Eutaw (basal formation), Demopolis, Mooreville, Prairie Bluff Chalk, Ripley	Binnsville*, Brooksville, Catalpa, Demopolis*, Eutaw, Faunsdale, Hannon, Houlka, Houston, Kipling, Leeper, Marietta, Oktibbeha, Searcy, Sucarnoochee, Sumter*, Trinity*, Urbo, Vaiden, Watsonia
	Jackson Prairie	Moody's Branch, Yazoo Clay	Additional information needed; soils likely identical to contiguous Mississippi Jackson Prairie
AR	Blackland Prairie	Annona Chalk, Arkadelphia Marl, Brownstone Marl, Dierks, De Queen, Ozan, Nacatoch Sands, Midway Group, Saratoga Chalk	Demopolis*, Faunsdale, Houston*, Kipling, Kirvin, Leeper, Marietta, Oktibbeha, Sacul, Sumter*, Terouge*, Tuscumbia, Vaiden, Watsonia
GA	Georgia Eocene Chalk Prairie	Dry Branch (Twiggs Clay Member), Ocmulgee, Tivola, Tobacco Road Sand	Boswell, Hoffman, Oktibbeha, Sumter*, Susquehanna, Vaucluse
LA	Cook Mountain Prairie	Cook Mountain	Hollywood, Keiffer*, Oktibbeha
	Jackson Prairie	Jackson Group	Additional information needed
MS	Black Belt Prairie	Eutaw (basal formation), Demopolis, Mooreville, Prairie Bluff Chalk, Ripley	Belden, Binnsville*, Brooksville, Catalpa, Demopolis, Eutaw, Faunsdale, Houlka, Houston, Kipling, Leeper, Marietta, Oktibbeha, Sucarnoochee, Sumter*, Urbo, Vaiden, Watsonia
	Jackson Prairie	Moody's Branch, Yazoo Clay	Catalpa, Griffith, Houston, Kipling, Leeper, Louin, Maytag*, Okolona, Siwell, Sumter, Tuscumbia, Vaiden

Source: Campbell and Seymour (2011a); Elsen and Wieland (2003); Eversull (2005); Foti (1989); J. Gibson (1941); Huddlestun and Hetrick (1986); Launchbaugh (1955); Lester et al. (2005); B. MacRoberts and MacRoberts (1995); Moran (1995); Moran et al. (2003); NatureServe (2013); Schotz and Barbour (2009); Soil Service Staff (2013); Wieland (2000); J. Woods (1967); and Zollner et al. (2003).

* = Series documented underlying open-prairie vegetation.

2

Historical Distribution of Prairies in Arkansas

John A. Barone

Introduction

Of the southeastern states, Arkansas almost certainly had the most rich and diverse set of prairie habitats. In the nineteenth century, prairies were found at sites from the Mississippi River in the east to the Ozark Mountains and Arkansas Valley in the west, and from the banks of the Red River in the southwest corner of the state to the White River in the north. The goal of this chapter is to provide a map of these historical prairies as well as descriptions that expand on information provided by the map.

General Land Office surveys are the most useful source of information on the historical landscape for much of the Southeast in the nineteenth century. In these surveys, conducted initially in the 1840s in Arkansas, surveyors divided the landscape into 6.0 by 6.0 mi (9.7 by 9.7 km) "townships," which were further divided into 1.0 by 1.0 mi (1.6 by 1.6 km) "sections" (Whitney and DeCant 2001). As the surveyors measured out these lines, they made notes on the landscape features, such as streams, swamps, and prairies, that they encountered (Whitney and DeCant 2001; Gray and Schauwecker, chapter 3, this volume). These notes were converted to "plat maps," representing the approximate size and locations of these features. Plat maps have been used many times to describe historical environments in the Southeast (e.g., Nelson 1997; Bragg 2003; Barone 2005b).

While of unparalleled value in understanding historical landscapes, plat maps do have limitations (Whitney and DeCant 2001; Gray and Schauwecker, chapter 3, this volume). For example, plat maps are uneven in quality. One may show many landscape features, including prairies, while the plat map of the neighboring township may be blank, despite surveyors' notes documenting the presence of prairies (Gray and Schauwecker, chapter 3, this volume). Second, smaller prairies that were not close to the section lines would not have been observed by the surveyors and so could not have been included on the plat maps. Third, even though the surveying of the state took many years, the plat maps nevertheless represent the size, location, and position of the prairies over a relatively short period. Over longer periods—hundreds or thousands of years—the prairies likely expanded and contracted with changes in climate, the populations of important herbivores, and levels of natural and anthropogenic fire. Moreover, even at the time of the surveys, the influx of settlers had already altered some parts of the Arkansas landscape and the forces that sculpted that landscape. Certainly by the late nineteenth century, some observers expressed concern that even prairies that were unplowed and apparently undisturbed by people were changing in size and species composition (see the following discussion).

Finally, the diverse landscape of Arkansas includes many types of open habitats besides prairies, including woodlands, glades, and barrens. Historical sources, including plat maps, often fail to distinguish these habitat types. Thus, any map made from plat maps necessarily has a low resolution, showing "prairies" in a very broad sense—largely treeless habitats, dominated by grasses.

Other historical sources can be used to complement plat maps and to draw a fuller picture of what the habitats in the state were once like. For historical Arkansas, these sources include the journals and accounts of explorers and travelers, letters, local histories, and even scientific papers (Edmonds 2001). Such records have their own drawbacks, such as shifts in the meanings of words, difficulties in locating the sites described, and the biases of the authors (Edmonds 2001). While conclusions should be drawn with caution, they are independent sources of information and provide a valuable supplement to prairie maps created from the plat maps.

Thus, in this chapter, maps are presented based on the plat maps, showing the historical distribution of prairies in Arkansas. Additionally, nineteenth- and early-twentieth-century accounts of prairies are quoted and, as much as possible, located to particular prairie areas. The hope is that these resources will provide a historical context to the study of the ecology and natural history of the prairies of Arkansas.

Methods

All of the plat maps for the state of Arkansas were examined on microfilm at the state's archives, maintained by the Arkansas History Commission. (These plat maps are now available online at http://www.glorecords.blm.gov.) When a plat map showed an area of prairie (typically marked with the word "prairie"), the map was printed. These printouts were later scanned, and the resulting file was opened in the geographic information systems software program ArcMap (Esri, Inc., Redland, California). Once in ArcMap, the plat maps were georeferenced, and the outlines of the prairies were traced into shapefiles. (These files are available for download at http://mississippientomologicalmuseum.org.msstate.edu/.)

A search was conducted for written sources describing the historical presence, nature, and extent of prairies in Arkansas. Firsthand descriptions were especially valued, as were earlier sources, particularly from the first half of the nineteenth century, when the authors would have seen the prairies prior to the large-scale alterations that occurred on the landscape with the advent of widespread agriculture. The search for sources describing prairies was extensive but not exhaustive, and for the Grand Prairie region in particular, where the prairie was well known and very frequently described, the sources have been used selectively. All quotations are given verbatim, with brief editorial comments designated with brackets. The approximate locations of the described sites are shown in the figures. Brief quotes that mention prairie only in passing are also presented in the figures but without further comment.

For clarity, the descriptions of prairies are organized by level III Arkansas ecoregions, which unite areas of the state according to their similarities in soils and vegetation (A. Woods et al. 2004). Arkansas has seven level III regions.

Results

According to the plat maps, there were approximately 286,000 ha (707,000 ac) of prairie in Arkansas at the time of the surveys (table 2.1). Of this, about 56%, or 160,000 ha (395,000 ac) were present in the Grand Prairie. Two ecoregions, the Mississippi Valley Loess Plains, with 140 ha, and the Ouachita Mountains, with 220 ha, had between them about 0.1% of the total historical prairie area.

South-Central Plains Ecoregion

A total of 26,150 ha of prairies show up on the plat maps for this ecoregion, spread across three areas: (1) upland prairies of Drew and Ashley Counties, at the eastern end of the region; (2) blackland prairies in Nevada, Hempstead, and Sevier Counties, at the western end; and (3) floodplain prairies of the Red River Valley in the southwestern corner of the state.

Upland prairies of southeastern Arkansas. Bragg (2003) used plat maps and General Land Office survey data to reconstruct the historical plant communities of Ashley County. Not surprisingly, the map of prairies in Bragg (2003) is very similar to the one presented here. A soil survey conducted by Vanatta et al. (1914) for Ashley County confirmed the general size and shape of the larger sites and described prairies as occurring predominantly in the central upland region of the county. They also noted that the prairies were "covered by a growth of wild grasses, chief among which is broom sedge [*Andropogon virginicus*]." Wackerman (1929) wrote that "the vegetation of the prairies is mostly broom sedge, grasses, herbaceous plants, and Crataegus in

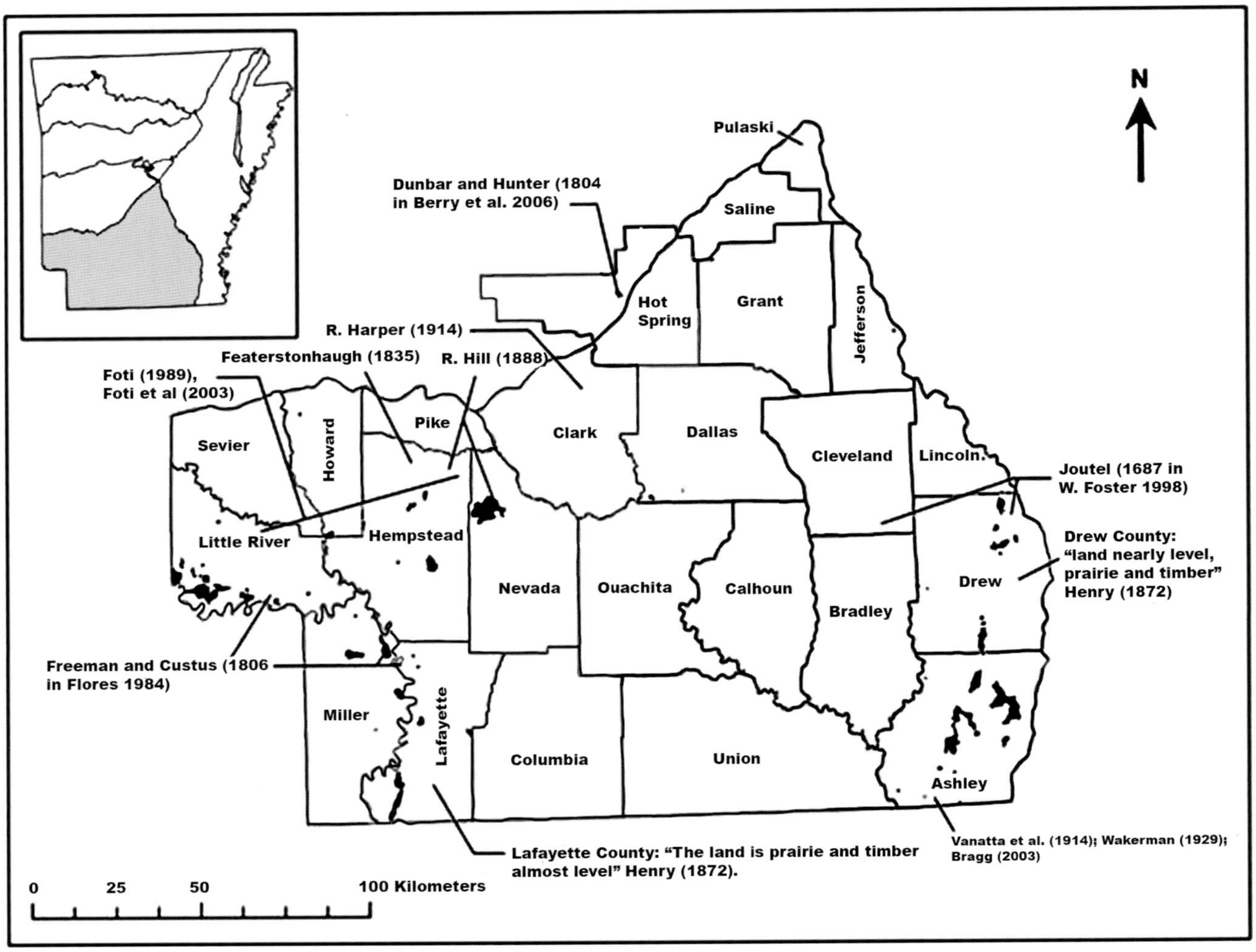

Figure 2.1. Historical distribution of prairies in the South-Central Plains ecoregion of Arkansas. Black areas represent prairies as seen on General Land Office plat maps, mostly from the 1840s. The inset map shows the location of the ecoregion in the state. Lines indicate the approximate locations of prairies described in the citations. Citations without quotes are discussed in the text.

the lowest depressions." He also mentioned the presence of oaks on elevated sites within the prairies. Wackerman (1929) and Bragg (2003) also provided discussions of the soil and other factors leading to the occurrence of prairies in this area. (See also Bragg and Weih 2007.)

The prairies shown on plat maps for northern Drew County may have been observed by Henri Joutel and a party of French explorers, who were part of La Salle's expedition to Texas, when they traveled across Arkansas in the summer of 1687. W. Foster (1998) provided an approximate route for the group. On July 20, Joutel (1687, in W. Foster 1998) made the following comments in his journal: "We came upon rather pretty country with quite lovely plains. The Indians [who were guides] indicated that it stretched very far and that there were many bison there. We marched about five or six [French] leagues [12 to 14.4 miles] toward the northeast that day." On the following day, Joutel wrote that they continued "finding rather beautiful country with quite pretty plains from time to time and very fine trees."

Joutel also observed what may have been prairies in what is now southeastern Cleveland County, though no such prairies show up on the plat maps. Upon crossing the Saline River, Joutel (1687, in W. Foster 1998) noted that "between the clusters of woods are small prairies full of grass but not as tall as in many other places we had passed. Even though this country was very pretty and most agreeable to the sight, the land did not seem very good to me because the soil was sandy and dry and, as a result, grass and trees did not grow too well. After we had gone about a league and a half, we saw a herd of bison grazing." Joutel indicated that they passed through about three (French) leagues (about 11.6 km or 7.2 mi) of this type of landscape. It seems quite possible that these "prairies" were, in fact, barrens, similar to those found at the Hall Creek Barrens Natural Area located about 13 km (or 8 mi) to the northeast of the route laid out by W. Foster (1998). It is also possible that this interpretation of Joutel's route is incorrect and that Joutel, in fact, saw either these barrens or those in the Kingsland Prairie Natural Area, a bit farther to the northeast.

One small area of prairie shows up on the plat maps of Hot Spring County, at the southern edge of the Ouachita Mountains ecoregion. This site was apparently seen by Dunbar and Hunter (in Berry et al. 2006) during their voyage up the Ouachita River in 1804, as they mentioned the presence of "prairie de Champignole" that was "often frequented by Buffalo; some salt licks are to be found near it." Berry et al. (2006) placed the site near Social Hill.

Blackland prairies of southwestern Arkansas. The ecology, historical nature, and extent of blackland prairies in southwestern Arkansas have already been extensively documented (Foti 1989; Foti et al. 2003). Several medium to small prairies show up on the plat maps for Nevada and Hempstead Counties, and Foti (1989) showed that prairies in this region were far more extensive. Using a combination of General Land Office surveyors' notes, soil maps, and numerous field assessments, he determined that a region of prairies stretched from southwestern Clark County across Hempstead and southern Sevier Counties, with an additional area in western Little River County (Foti 1989; Foti et al. 2003). These blackland prairies usually developed on steeper slopes with underlying chalk or marl bedrock and alkaline soils (Foti 1989; Foti et al. 2003; Echols and Zomlefer, chapter 1, this volume).

As noted by Foti (1989), several historical sources described these prairies. For example, the geologist Featherstonhaugh (1835) surveyed parts of Missouri and Arkansas in 1834 and encountered blackland prairies in northern Hempstead County: "The country here appeared to consist of a chain of prairies running westward, and parallel with the Red river for a very great distance. Some of these prairies were mere bald spots, of half an acre and upwards, surrounded by plants, whilst others were said to contain several hundred acres. In every instance they were surrounded with a belt of timber and plants peculiar to the country. I was informed by Judge Cross . . . that these prairies extend probably many hundred miles to the west, and that it is an opinion deserving of being entertained." Later, Featherstonhaugh (1835) described the presence of chalk bedrock and fossils, confirming a description of blackland prairies.

R. Hill (1888) described the geology of southwestern Arkansas and made several references to prairies. For example, on the northern half of Hempstead County, he wrote, "Tradition says that the crest of this Columbus-Marlbrook ridge, which is now all under cultivation, was once a continuous prairie." He also observed, "A dense forest growth, sometimes of pine and hardwood, and sometimes of hardwood alone, covers the whole surface of the country, except a few bare glades, the whole of which do not aggregate 5000 acres [2,023 ha]. These bare spots, with the exception of Prairie d'Ane, at Prescott [in Nevada County], and Bois d'Arc Prairie in southern Hempstead county, and Prairie de Roan at Hope [in Hempstead County], are seldom of any considerable extent, and are scarcely worthy of the name of prairies, for the true meaning of the word prairie not only implies absence of forest growth, but also great extent." He also described these prairies as occurring in sites with white and yellowish clays with lots of siliceous gravel, leading to "glade-like" conditions.

Prairies of the Red River Valley. Some of the extensive prairies seen on plat maps of southwestern Arkansas were witnessed early in the nineteenth century by the Freeman-Custis Expedition, which traveled up the Red River in 1806 (Flores 1984). For example, Freeman and Custis observed a prairie in the northwestern part of Lafayette County that was also the former site of a Caddo village (Flores 1984). Freeman wrote that the site was "a beautiful prairie, on the North-East side of the River," situated at the base of a hill. Later, he added that "in front of this mount lies a beautiful and rich meadow, extending from its base to the river, and downwards for about two miles. It is interspersed with small clumps of trees, and has a small lake or pond in its centre."

Across the Red River in Miller County, several prairies show up on the plat maps. Freeman and Custis appear to have seen some of these as well, as they described "extensive prairies" in the portion of the Red River Valley from present-day Shreveport, Louisiana, northward into Arkansas (Flores 1984). One evening in July, Freeman noted that the expedition camped "in a handsome Prairie, on the left, or west side of the river, elevated 40 feet [12.2 m] above the water. The surface was level and interspersed with clumps of trees, which added much to the beauty of its appearance." Flores (1984) suggested that this is a site just south of Garland City, which corresponds to "Fisher's" prairie on that county's plat maps.

Freeman and Custis (in Flores 1984) saw a prairie in Little River County, as well. "On the 25th [of July] at about 20 miles above the Little river, on the right hand side, ascending, is a prairie, considerably above the water, of a rich soil, and now overgrown with high grass, bushes and briars. The prairie extends back from the river about half a mile, and is bounded by open woods of Oak and Hickory." Flores (1984) placed this prairie (and an associated Caddo village) near present-day Index, though no prairies show up on plat maps at this point. However, several larger prairies were present upriver, so one of these may have been the actual location.

A soil survey of Miller County by Martin and Carr (1903) does not mention prairies. On the plat maps, however, the drawings of the prairies indicate that they were already, in the 1840s, heavily cultivated, with agricultural fields drawn onto the prairies. By 1903, the prairies may have been completely obliterated. An alternative interpretation is provided by Foti et al. (2003), who suggested that these prairies were old fields where Caddo Indians once grew crops.

Mississippi Alluvial Plain

Prairies in this ecoregion can be divided into three large groups geographically.

The Grand Prairie stretched from east-central Lonoke County southeast across Prairie County to southern Arkansas County. Toward the northern end of the ecoregion, a second group ran from western Craighead County south to Cross County, though the southern sites appear to have been smaller and more scattered. A third cluster of prairies was found in the central portion of the alluvial plain, in western St. Francis, southeastern Woodruff, and northwestern Lee Counties.

Grand Prairie. Unlike the prairie areas in some other parts of the state, Grand Prairie has been very well documented. The earliest historical reference to the Grand Prairie area may be found in the chronicles of the de Soto expedition. In 1542, the expedition reached Anilco, a community near the Arkansas River that Hudson (1997) suggested corresponds to the Menard-Hodges archaeological site in Arkansas County. This site sits in a small prairie that was near the southeastern end of the Grand Prairie region. In one description, the chronicler Elvas (Clayton et al. 1993) wrote that de Soto "lodged with all his men in the cacique's town which was located on a level field, and which was all populated for a quarter of a league; while a league and a half-league distant were other very large towns." This level field may have been a prairie.

The botanist Thomas Nuttall (1821), during his 1819 expedition up the Arkansas River, visited the southeastern end of the Grand Prairie in late January and early February, and his narrative includes several descriptions. For example, he wrote, "Here a vast prairie opens to view, like a shorn desert, but well covered with grass and herbaceous plants. Over this vast plain, which proceeds a little to the west of north, computed to be not less than 30 leagues [90 mi or 144.5 km] in length by 10 to 15 [30 to 45 mi or 48.2 to 72.4 km] in breadth, passes the road." Later, he continued, "The great prairie of which we have already spoken, said to be 90 miles in length, contains an invaluable body of land," commenting on its potential value as agricultural land. He also discussed the presence of cattle on the prairie.

McPhail (1837) described prairies in eastern Arkansas, very likely the Grand Prairie: "The prairies are rich in vegetation. In spring, as far as the eye can reach, we behold an almost illimitable carpet of verdue [*sic*], studded with the variegated hues of a thousand flowers, with scarcely a tree or hillock to interrupt the view. Several indigenous species of *mimosae* are found upon these prairies; and I have myself noticed three varieties of the sensitive vine, the *pink, white* and *yellow*—this last without prickles."

In his soil survey of the Stuttgart area, Lapham (1902) made many references to the Grand Prairie. For example, he noted that "Arkansas County consists of nearly equal areas of prairie and of forest land." He also reported that large areas of prairie were used as pasture for cattle and were cut for hay, making use of the native grasses. He also indicated that prairie was found primarily on two soil types within the survey area, Miami Clay Loam and Almyra Silt Loam. These two soil types, as shown on the map, are nearly congruent with areas of prairie shown on plat maps made 60 years earlier.

Even as late as 1911, when R. Harper (1914) visited this part of the state, extensive areas of the Grand Prairie were still present. He noted, for example, that in the 40 km

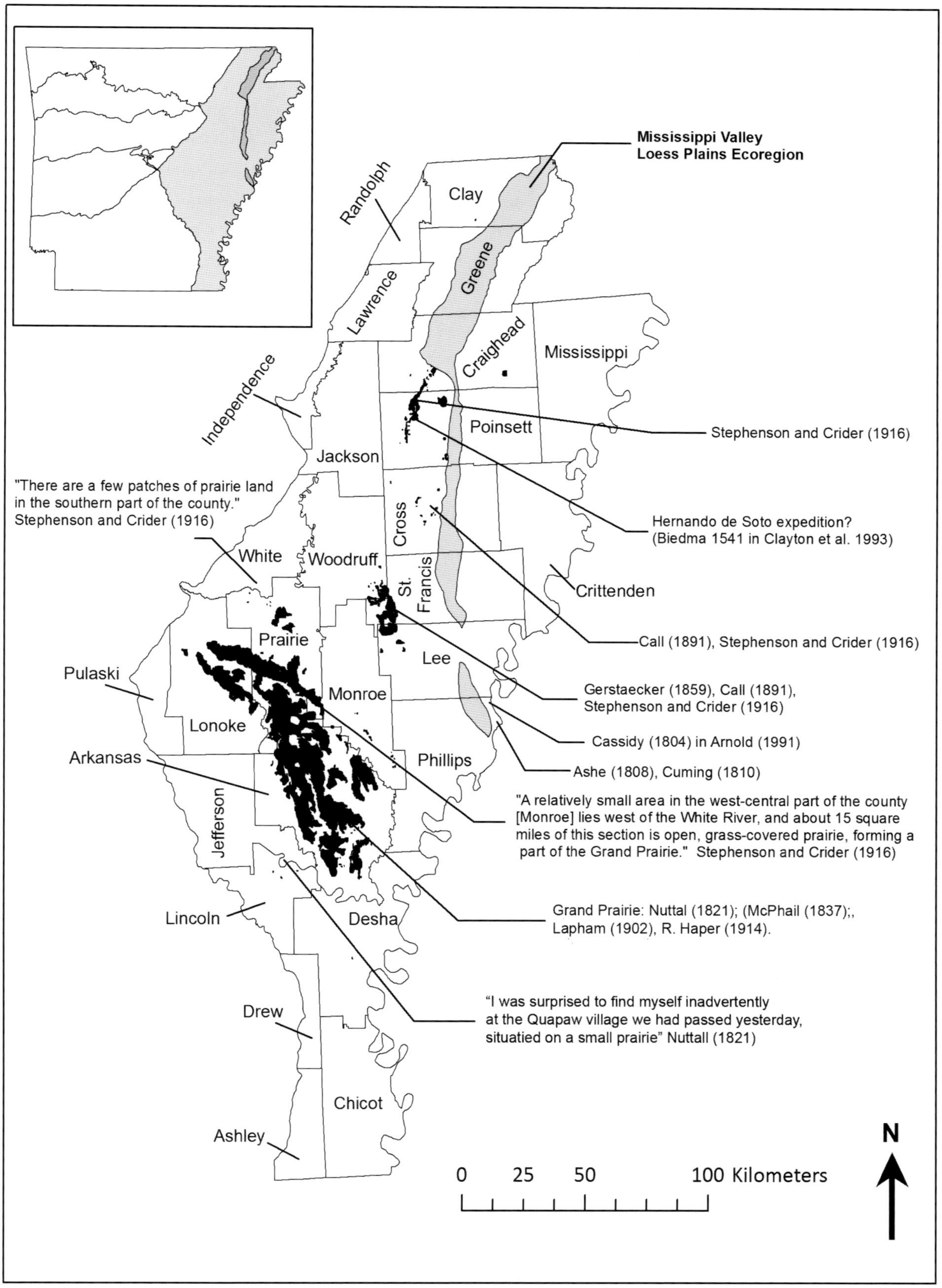

Figure 2.2. Historical distribution of prairies in the Mississippi Alluvial Plain ecoregion of Arkansas. The Mississippi Valley Loess Plains ecoregion is indicated on the map and is found within the Mississippi Alluvial Plain ecoregion. The inset map shows the positions of the two ecoregions within the state.

(25 mi) between Lonoke and DeValls Bluff, "one can look for miles and miles north and south over a vast level grassy plain," with occasional clusters of trees. He also provided a detailed description of the soils, the topography, and a list of the plant species encountered during a walk across the prairie. Despite increasing rice cultivation, he wrote that there were "thousands of acres of undisturbed prairie within easy reach of railroad stations." He did note, however, that hay baling was very common in the Grand Prairie.

Northern Mississippi Valley Alluvial Plain. The second prairie area in this ecoregion included a large site that was 33 km long (20.5 mi) and 1 to 4 km wide (0.6 to 2.5 mi) and ran from south-central Craighead County to western Poinsett County (fig. 2.2). Covering over 3,800 ha, this prairie was located between the L'Anguille River and Bayou De View and is, perhaps, the site with the oldest historical record in the state, as it may have been described by a chronicler with the Hernando de Soto expedition. In late June 1541, the expedition entered a native community of Pacaha, which may have been in northeastern Crittenden County, close to the Mississippi River (Dye 1993). De Soto sent a reconnaissance force northwest from this town for eight days, where, according to the chronicler, the travelers encountered a "very great swampy lake, where we did not even find trees, but rather some great plains, where was grass so tall and so strong that even with the horses we could not force our way through it" (Biedma 1541, in Clayton et al. 1993). While Dye (1993) suggested that this was a description of tallgrass prairies to the northeast in Mississippi County in Arkansas, and in Dunklin County in Missouri, an alternative interpretation is that this group indeed went to the northwest, in the direction of the Cache River (Hudson 1997). If this is the case, then the expedition may have seen the prairies in western Poinsett and Craighead Counties about 300 years before they were mapped by land surveyors.

A much later source, Stephenson and Crider (1916), provides brief descriptions of this prairie area. For example, of western Poinsett County, they wrote that "the interstream areas [between L'Anguille River and Bayou De View] present irregularly distributed tracts of open, grass-covered prairies, separated by slightly lower wood lands." Similarly, in discussing the same prairie area in Craighead County, they remarked that "this lowland is characterized by irregularly distributed patches of grass-covered, gently undulating prairie land and tracts of wooded land."

Farther south, in west-central Cross County, an irregular grouping of about 12 small prairies shows up on the plat maps (fig. 2.2). Call (1891), in a geological survey of the county, supplied a list of townships and sections with prairies present. These correspond exactly to those observed on the plat maps for three townships, but for the fourth, four sections with prairie on the plat maps are absent from Call's list. Call wrote, "There are several small areas of prairie land in Cross county, evidently at one time a single larger prairie which was possibly continuous with that of St. Francis county on the west." The description also notes that the prairies and neighboring forests are poorly drained and that "both rest upon a wet-clayey hard-pan that often forms a great part of the surface soil." Finally, Call said that these prairies "produce several varieties of native grasses in abundance and might be made still more productive than they are now." Stephenson and Crider (1916) saw the same prairies 25 years later: "In the interstream areas west of L'Anguille River the surface is gently rolling and presents irregularly distributed tracts of open grass-covered prairie, separated by slightly lower wooded areas."

Central Mississippi Valley Alluvial Plain. According to the plat maps, another extensive prairie area was located in southeastern Woodruff, eastern St. Francis, and northwestern Lee Counties, west of Big Creek (fig. 2.3). Prairies in this region were briefly described by Friedrich Gerstaecker (1859), a German adventurer who traveled extensively in the United States in the mid-1850s and stayed with a family in western St. Francis County. Gerstaecker (1859) noted, for example, that during a trip to procure some honey from wild bees, "we proceeded to a little prairie about three miles off and soon found the tree which S. had discovered and marked . . . S.'s tree was a decayed red oak, on the verge of the little prairie." Later, describing a work arrangement, he added, "We agreed to undertake the charge of S.'s cattle, to give them salt on the little prairie from time to time, where a tent was to be pitched for us."

Call (1891), in a geological survey, included several short descriptions of prairies in St. Francis County. For example, in describing the streams in the western part of St. Francis County, he wrote, "These creeks drain all the prairie region west of Goodwin." Later, on the same prairies, he noted, "In the open and treeless prairies the surface is quite level, save where the lenticular swells before mentioned form minor topographic features, while the wooded sections are below the level of the treeless plains." He also said that "the drier and more elevated portions of the county form the open prairies while the depressed portions are heavily wooded."

Stephenson and Crider (1916), in their geological survey of the region, provided very similar descriptions for these prairies across all three counties. For St. Francis County: "The interstream areas west of L'Anguille River present irregular patches of open, grass-covered prairie land, separated by slightly depressed patches and belts of timbered land." On the prairies in northeastern Lee County: "The interstream areas are gently undulating, being in part wooded and in part open, grass-covered prairies." And in their description of Woodruff County, they found that "east of the Bayou de Vue the surface presents irregular patches of grass-covered prairie separated by tracts of slightly depressed timbered lands."

Three historical sources mention prairies that were located near where the St. Francis River empties into the Mississippi River, though these prairies are absent from the plat maps for Phillips County. In 1806, Thomas Ashe (1808) explored eastern Arkansas. He wrote, "I passed the mouth of the river of St. Francis, and came to a beautiful prairie a little below it, on which I found one solitary dwelling inhabited by a family who traded with the Indians up the river, and occasionally dealt with the boats descending the Mississippi." Fortescue Cuming (1810) traveled down the Mississippi River and apparently saw the same grasslands about 4.5 mi (7.2 km) south of the mouth of the St. Francis River. He noted, "The Big Prairie is a natural savanna of about sixty acres [24.3 ha] open to the river on the right bank. It is covered with a fine, rich, short herbage, very proper for sheep. Immediately behind it at less than a half a mile [0.8 km] from the river is a small lake eight or nine miles [12.8 to 14.5 km] in circumference, formed in the spring and summer by the Mississippi, which in that season rising, flows up a small canal or (in the language of the country) bayau, and spreads itself over a low prairie." Cuming (1810) suggested that this seasonal flooding produced the grassland. The site was probably located east of Helena in Phillips County, perhaps near Prairie Point Towhead.

Arnold (1991), in a history of colonial Arkansas, related an anecdote about Henry Cassidy, who had title to land on the St. Francis River, possibly in Phillips County. In October 1804, he received an affidavit from a former Spanish commander in the area who confirmed that he "had conceded [to Cassidy] a portion of land in the place called the little prairie."

Arkansas Valley Ecoregion

Plat maps for this region have a total of 35,110 ha of prairie, with a large concentration found in the western end of the Arkansas Valley ecoregion in Sebastian, Franklin, and Logan Counties. However, small areas were also scattered in the eastern and central portions.

The prairies seen on the plat maps for Faulkner County at the eastern end of the ecoregion were apparently observed by Thomas Nuttall during his journey up the Arkansas River in 1819. He stopped at Cadron Creek, and a mile and a half away (2.4 km), he encountered a "small alluvial plain or prairie" (Nuttall 1821). He went on to mention that farmers in the area were "beginning to plough for cotton, an operation here not very laborious, except when breaking up the prairies, as the soil is friable and loamy." He also noted the presence of a small adjacent prairie.

In neighboring Conway County, two small areas of prairie show up on plat maps. Burgess and Ely (1907), in their soil survey of the county, described the "native vegetation" they observed, including "lowland grasses," suggesting that in some instances these sites may have been fairly open, enough to lead to a characterization of "prairie" on the plat maps.

A history of the northwestern portion of Arkansas makes numerous references to prairies in the western end of this ecoregion (*History of Benton* 1889). For example, as part of a description of the movements of the First Arkansas Infantry during the Civil War, this anonymous author wrote that "on [Capt. William C. Parker]'s return, while crossing Haguewood Prairie in Franklin County, he was suddenly confronted by the rebel general Shelby's command, then moving northward on his well-known raiding expedition into Missouri." The same source describes a land title from Franklin County issued in 1816 that mentions the presence of "Aiaita Prairie" or "Choctaw Prairie" on the Arkansas River. The property is "on the river directly southwest of Altus." No prairie is present in this area on the plat

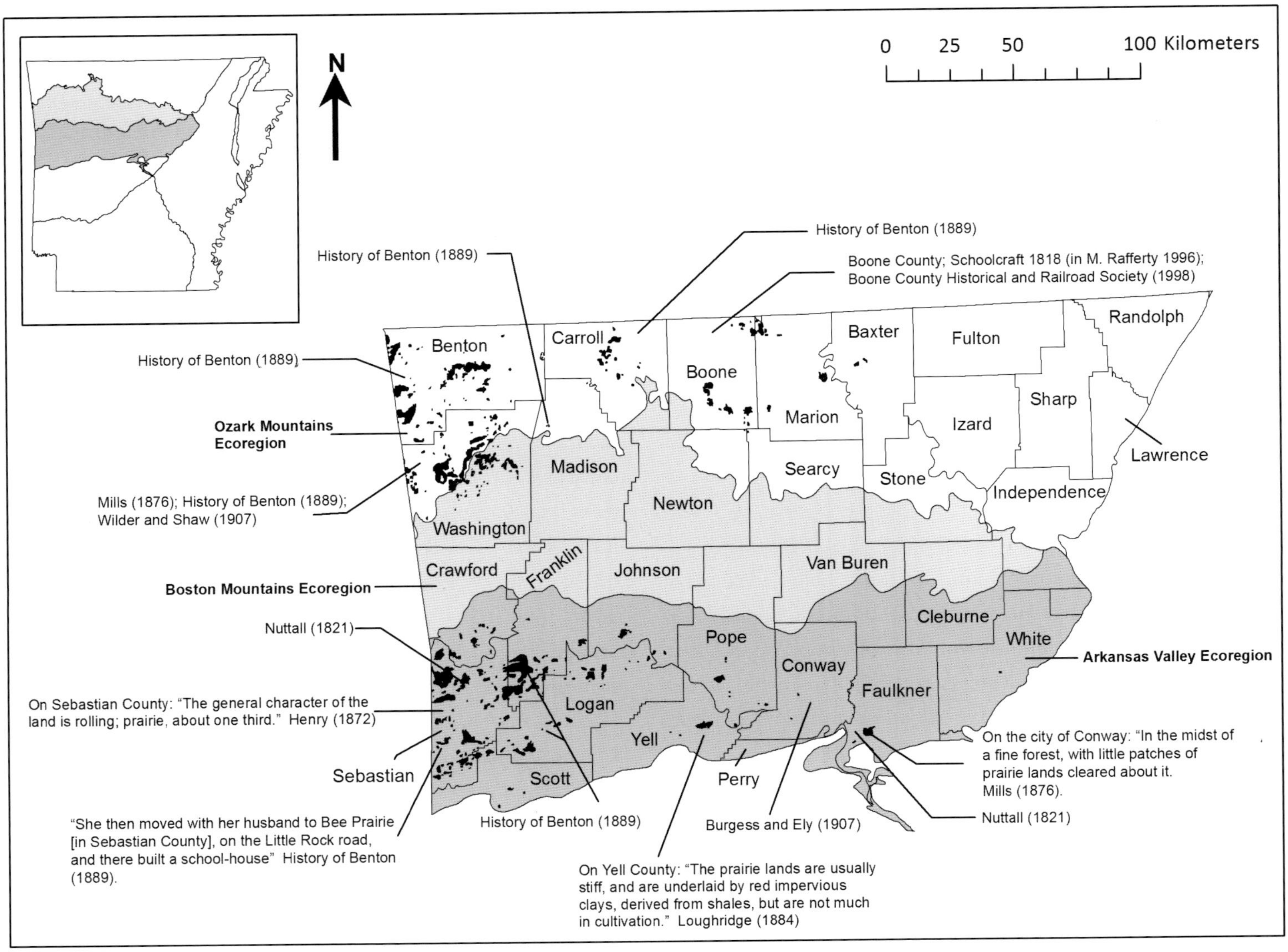

Figure 2.3. Historical distribution of prairies in the Arkansas Valley, Boston Mountains, and Ozark Mountains ecoregions.

maps. The author also said that the town of Charleston, in southwestern Franklin County, "is beautiful, and the broad prairies surrounding it only seem to need a railway to develop them." On plat maps, the location of Charleston is on or near "Potatoe-hill prairie," a prairie of about 2,900 ha (7,166 ac). The author also wrote that "for about five or six miles [8 to 9.6 km], parallel with the river is a series of ridges, beyond which spread the more or less continuous prairies to the Magazine Mountains on the southern boundary." Presumably, these prairies are the ones visible in western Logan County on the plat maps.

Prairies were also carefully observed by Nuttall (1821) during his stay in Fort Smith, in neighboring Sebastian County. While many of his short trips during his visit were to present-day Oklahoma, on May 3, 1819, he visited "cedar prairie, lying about 10 miles south-east of the garrison, and presenting an irregular or undulating surface." Several additional comments follow. For example, he added, "The extensive and verdant meadow, in every direction appeared picturesquely bounded by woody hills of different degrees of elevation."

Mills (1876) also visited Fort Smith and wrote that "the character of the land is rolling prairie, with the most fertile soil, and there is no better land for fruit or for farming purposes in any county in the northwest of the State."

Ozark Highlands and Boston Mountains Ecoregions

The Ozark Highlands ecoregion stretches across the northern tier of the state, and according to the plat maps, 29,930 ha were found from Baxter County westward across Marion, Boone, and Carroll Counties, with the highest concentration in the northwestern corner of the state in Benton and Washington Counties (fig. 2.3). Most of the prairies were on the Springfield Plateau, a flat to rolling area underlain by cherty limestone (A. Woods et al. 2004). Some of the prairies in Washington County extend southward into the Boston Mountains ecoregion, where 5,200 ha of prairies were found.

Northern tier of the Ozark Highlands. The prairies in Boone County are described in a county history published by the Boone County Historical and Railroad Society (1998). This history lists, for example, the names of larger historical prairie sites in the county, including Baker's, Mitchell's, Huzzah, Rolling Marshall, and Sugarloaf. This last site was also encountered by Henry Rowe Schoolcraft (in M. Rafferty 1996) in December 1818, during a trip through southern Missouri and northern Arkansas: "[We] arrived at an early hour in the afternoon at the house of a mr. Coker, at what is called Sugar-Loaf Prairie. This takes its name from a bald hill covered with grass rising on the verge of the river alluvion on the west side of the river, and is discernible at the distance of many miles."

Another late-nineteenth-century history of this part of the state contains numerous accounts of prairies (*History of Benton* 1889). On the prairies in Carroll County, the author wrote that "the principal prairies of the county are the 'Big' prairies in Prairie and Polo Townships, between Big Indian Creek and King's River, six miles [9.6 km] long from north to south and from three to five miles [4.8 to 8.0 km] wide; and Scott's Prairie, in Hickory Township, about three square miles [7.8 km^2] in extent. These tracts, when first known to the pioneer settlers, were covered with the long grass peculiar to similar tracts in the West, and were without any timber growth whatever." These locations match the plat maps very well. A description in the same volume mentions a smaller prairie, also shown on the plats: "When first visited by white settlers this locality was a prairie and hence attracted settlement earlier than the surrounding timbered region" (*History of Benton* 1889).

On the plat maps, only one small area of prairie is present in Madison County, but this site was apparently known to the author (*History of Benton*), who described it thus: "On this same stream, about nine miles [14.5 km] northwest of Huntsville, the road leads, for half a mile [0.8 km], through barrens with a sandy soil, followed by prairie in which sandstone crops out about ten miles [16.1 km] from Huntsville. The prairie is bounded by wooded hills off to the southwest."

Ozark Highlands of northwestern Arkansas. Many prairies were present in the northern half of Washington County, and these were described by a variety of observers. Col. James P. Neal (in *History of Benton* 1889) writes of his life in Washington County: "My first acquaintance with the valley was in 1829. The buffalo had then receded some fifteen or twenty miles to the northwest. Their paths were still numerous, leading mainly from one lick to another. Their heads were scattered all over the prairies, one perhaps to every three acres of land. These licks were depressions in the earth, filled with water a little brackish in wet weather. In the summer, when dry, they were the resorts of buffalo for the salt with which the earth was saturated, and were known as buffalo licks." In a soil survey of the Fayetteville area, Wilder and Shaw (1906) confirmed the presence of prairies in this area, naming Lindsley and Norwood. They also included a discussion of the soils of the prairies, noting, for example, that "the native vegetation on this soil [Gasconade silt loam] was a heavy growth of prairie grass."

Benton County, in the northwestern corner of Arkansas, also had a large amount of prairie. One quote is quite precise: "The area of the county is 900 square miles, or 576,000 acres [2,331 km^2]. The proportion of unmodified prairie is, approximately, 86,000 acres [34,803 ha]; oak barrens or modified prairie, 175,000 acres [70,820 ha]; wooded mountain or ridge territory, 200,000 acres [80,937 ha]; and river and creek valley lands, 86,000 acres [34,803 ha]" (*History of Benton* 1889). The area of "unmodified prairie," 86,000 ac (34,803 ha), is more than twice the area shown on the plat maps (32,377 ac or 13,102 ha), but the source does not indicate how its values were determined. The same source says that when the county was settled, "all of the comparatively level upland was called prairie, while in truth there was but little real prairie. The timber was then very thin, the trees stood far apart, and the country which is now covered with dense growth of young timber was then so open that the wild deer could be seen anywhere at a distance of several hundreds of yards. The entire surface of the earth was then covered with a rank growth of vegetation, consisting of native grasses and wild flowers, which gave to the landscape, especially in the timbered lands, a

Table 2.1.

Area of historical prairie in level III ecoregions of Arkansas, according to General Land Office plat maps.

Level III ecoregion	Prairie area (ha)	Percentage of total
Arkansas Valley	35,110	12.3
Boston Mountains	5,200	1.8
Mississippi Alluvial Plain*	189,060	66.1
Mississippi Valley Loess Plains	140	0.05
Ouachita Mountains	220	0.07
Ozark Highlands	29,930	10.5
South-Central Plains	26,150	9.1
Total	**285,810**	

* Includes the Grand Prairie.

more beautiful appearance than it now has." The source goes on to say that the vegetation was burned annually by local tribes to open up the country and kill off young trees.

Succession of Prairies

A surprising theme in several prairie descriptions, even from the nineteenth century, is how rapidly prairies that had not been obviously disturbed by human activity changed with the arrival of settlers. For example, John Carnall (in *History of Benton* 1889), in a description of Sebastian County in western Arkansas, said, "The other prairies are mostly under a square mile in extent rather than over, though possibly a few years ago they were much larger, for the timber is gradually growing into them from their borders."

Call (1891) made more extensive comments on prairies in St. Francis County: "The gradual encroachment of the forests upon the prairie region of the county, and in fact of the whole eastern portion of the state, cannot escape the attention of even the most casual observer. The advance guard of the encroaching timber is the omnipresent sassafrass bush (*Sassafrass officinale*) [now *Sassafras albidum*] [which] is everywhere engaged in a war of conquest with the typical prairie vegetation. Scrub oak, box elder, and ash follow close in the wake of the sassafrass. The borders of the prairie on every side are distinguished by a close and vigorous growth of this advance guard; unless the skill of man intervene, the Prairie Petite will in a few generations become a reminiscence." He made extensive and similar remarks concerning prairies near DeValls Bluff in Prairie County, where he noted that railroads were introducing weeds.

Vanatta et al. (1914), in their description of prairies of Ashley County, wrote that "old settlers say that the prairies are much smaller than they were 40 years ago, owing to the gradual encroachment of the forest growth." Wackerman (1929) echoed these remarks for the same county, noting that young trees were moving into the prairies along streams.

Thus, well before the end of the nineteenth century, succession was occurring on even relatively undisturbed sites. The historical prairies of Arkansas were a dynamic system, and as settlement and agriculture spread over the state, the factors that may have helped keep prairies open, such as natural and anthropogenic fires and the activities of bison and other herbivores, receded. How early these changes came about is unclear, but they are a reminder that even "intact" prairies seen today may be quite different in their species composition and ecological processes than they were 200 years ago.

Conclusions

Almost all the prairies seen on the plat maps and described by historical sources are now irretrievably gone (Arkansas Natural Heritage Commission 2013). But Arkansas also has a large number of nature preserves spread across the state that protect much of the remaining prairie. For example, as of this writing, the Arkansas Natural Heritage Commission has 16 preserves statewide that include at least some prairie habitat, covering 3,200 ha (8,000 ac). The Nature Conservancy owns or manages six additional protected sites with prairies. The maps and descriptions in this chapter should furnish interested researchers with possible new sites to search for unrecognized remnant prairies.

All habitats have a history that influences their current and future states. The purpose of the maps and descriptions in this chapter is to provide ecologists, land managers, and all prairie lovers with a greater appreciation of the history and diversity of these habitats in Arkansas.

3

The Use of General Land Office Survey Notes to Locate Prairie Patches in the Jackson Prairie Region

Toby Gray and Timothy J. Schauwecker

Introduction

Two regions of blackland prairies—the Jackson Prairie Region, a wedge-shaped area that is widest around the city of Jackson and tapers in a band running southeast into the state of Alabama; and the crescent-shaped Black Belt—exist in Mississippi, stretching from northeastern Mississippi to an area just southeast of Montgomery, Alabama. This chapter explores the importance of using General Land Office (GLO) survey notes and plat maps to understand the pre-European settlement ecosystem of the Jackson Prairie Region. Many researchers have created mosaics of GLO plat maps to represent the spatial characteristics of pre-European settlement landscapes at the state or regional scale (Edwards 1954; A. Jones and Patton 1966; Rankin and Davis 1971; Barone 2005a, 2005b; M. T. Gray 2010). Such map products necessarily convey an image of "islands" of open prairie in a generally forested landscape. The geographical fact of a definite margin between prairie and forest is supported in the GLO survey notes by the many uses of such language as "enter," "cross," and "exit prairie." In many cases, the creators of the plat maps surrounded their prairie polygons with tiny sketched trees to convey graphically the presence of an ecological edge. At many other locations in the Jackson Prairie Region, however, the survey notes suggest additional prairie landscape conditions for areas outside or between the polygons on the plat maps. Observations of areas where prairie patches were too small or too numerous to delineate, or in which the tree density was low enough to support a prairie ecosystem but at no point dense enough to suggest an identifiable margin, are present in GLO notes but not represented on the plat maps. The research presented here suggests that written survey notes, and multiple revised transcriptions thereof, are at least as important as the plat maps (if not more so) in the effort to understand ecological conditions before European settlement. This chapter additionally argues that the Jackson Prairie Region and the Black Belt differ significantly in the size and spatial distribution of the historical patches of prairie (small and scattered in the Jackson Prairie Region, large and dense in the Black Belt). The patches are so scattered in the Jackson Prairie Region, both historically and today, that we propose that "Region" is a more apt term than the commonly used "Belt" when referring to this area. Because of the great distances of presumed "forest" land between patch polygons on the plat maps (significantly greater than in the Black Belt), notes are of particular importance in gaining an understanding of the prairie landscape in this area. Finally, this chapter asserts that multiple copies of the GLO surveys exist and should be preserved. As the deputy surveyors transcribed field notes and prepared official records, some references to prairie conditions were lost. Field notes constitute an eyewitness account of the landscape and as such are an important addition to the later, "cleaner" versions.

The Nature Conservancy (2006) identified temperate

grassland as a major habitat type of global conservation concern. On the North American continent, the loss of natural prairie has been described as a "cataclysmic change" occurring within a single life span (W. Jordan 1997). Less than 1 percent of the pre-European-settlement prairie ecosystem survives today (R. Anderson 2006). In the southeastern United States, blackland prairies occur on calcareous soils of Late Cretaceous to Eocene age as scattered patches, often in belted formations, from Texas to Georgia (Peacock and Schauwecker 2003a; Noss 2013). In the Black Belt, nearly all the prairies featuring the deep, dark, rich soils indicated by the name "blackland" have been converted to agriculture or livestock grazing. The remnant sites today consist mostly of eroded, light gray, chalky or marly clays (Noss 2013).

Blackland prairies in general are physiographically, botanically, and ecologically one of the least-understood systems in North America (Leidolf and McDaniel 1998; Foti 1989; Gordon and Wiseman 1989). Heavy agricultural pressure has reduced the number and size of native prairies to such a degree that very little remains today, mostly in the form of small patches (0.5 to 5.0 ha) in abandoned pastures, along roadsides, in power line rights-of-way, and in public and private natural areas (Peacock and Schauwecker 2003a). W. Jordan (1997) pointed out that by the time Americans began to care about prairie restoration, the old prairies were almost gone. An early challenge for practitioners of prairie restoration was that prairies, in contrast to other kinds of ecosystems, do not come back on their own once they are destroyed. Prairies must be re-created, a complicated process that W. Jordan (1997) characterized as involving ongoing and increasingly sophisticated and conscientious attempts. Locating, identifying, and recording the presence of intact, good-quality prairie relicts is important for the long-term preservation of this rare ecosystem. Methods for doing so may be of particular interest to conservationists in the Jackson Prairie Region, where the patches are known to have been smaller and more sparsely distributed historically than those in the Black Belt.

General Land Office Notes and Plat Maps

The General Land Office (GLO) and the survey system of township and range was established and refined through a series of acts of the US Congress in the early part of the nineteenth century (Fitz 1832). The rectangular survey system subdivided the land into tier and range belts that were intended to be 6 mi (9.6 km) wide. Tier lines, analogous to latitude, run east to west, and range lines, analogous to longitude, run north to south. The townships were divided into 36 sections of 1 mi^2 (2.6 km^2) each. Each section was divided into 160 ac (64.8 ha) quarter sections, which in turn were divided into 40 ac (16.2 ha) quarters. These parcels were the basic units of homestead purchases. Land in the Black Belt and Jackson Prairie Region was surveyed in the 1820s and 1830s. The township and range lines were completed first, with posts set or trees blazed every mile. Working off those landmarks, surveyors then measured and marked the 36 interior sections. In addition to the section corners, the surveyors marked the halfway point of every section line in order to establish quarter sections. Surveyors measured the lines using chains. In most cases, a chain equals 66 ft (20 m), and 80 chains equal 1 mi (1.6 km). Each chain is divided into 100 links. In some of the surveys discussed here, surveyors used a shorter chain (33 ft or 10 m), for which 1 mi equals 160 chains.

In addition to the measurement and marking of the township, range, and section grid, deputy surveyors collected geographical data about the land. Surveyor General Gideon Fitz (1832) instructed the deputy surveyors of Mississippi to record the shape and location of the swamps, lowlands, water courses, ponds, prairies, roads, and paths. The features were then mapped spatially by using the recorded distances from the section corners as map coordinate points. Therefore, the precise location of landscape features such as swamps, prairies, and stream branches on a plat map is increasingly conjectural as distance from the section line increases. Different deputy surveyors approached their task with different levels of attentiveness, resulting in maps with varying levels of detail. Barone (2005b) noted that "some teams of surveyors created plat maps that were highly detailed while others recorded very few features of the landscape, apart from rivers and streams."

Typical examples of the use of GLO records in the research of historical ecologies include a reconstruction of landscape-level tree canopy composition as reference data for restoring tallgrass oak savannas in Iowa (Asbjornsen et al. 2005), a determination of composition and disturbance patterns of pine forests in Michigan (Whitney 1986), and a study that combined GLO data with archaeological evidence to determine forest conditions in the Tombigbee National Forest in the absence of human occupation (Peacock et al. 2008).

GLO Records and Blackland Prairie

A. Jones and Patton (1966) used GLO plat maps and survey notes to delimit open land and forested areas in the Black Belt (defined geologically as the area underlain by the Selma Chalk and Ripley Formations) in Sumter County, Alabama, in 1832 and noted several instances in which prairie patches recorded in the notes were not included on plat maps. Their analysis of tree density (based on notations of distances to witness trees from section corners) concludes that 23.4% of the land in the Black Belt was "unforested" (10 or fewer trees per acre) in 1832, including the 13.4% described as prairie, meaning that 10% of the area was both unforested and not described as prairie. Additionally, 14.5% of the Black Belt was "heavily forested" (120 or more trees per acre), and 62.1% was occupied by forest of intermediate density (between 10 and 120 trees per acre). A. Jones and Patton (1966) displayed the tree density data and prairie polygons in separate figures, so it is difficult to determine how the unforested, nonprairie corners and the prairie patches were related spatially.

Rankin and Davis (1971) also used the boundary of the Selma Chalk as their study area and examined GLO notes and plat maps within that boundary in Montgomery County, Alabama. They did not find any instances in which prairie was mentioned in the survey notes but excluded from the plats. However, if the prairie polygons from the plats and the tree density information from the notes are placed in the same figure, the indefinite boundary between prairie and forest can easily be inferred.

Open land (10 or fewer trees per acre) is generally contiguous with the polygons, while heavily forested land (120 trees or more per acre) generally occupies spaces at the midpoint between polygons. Areas of high tree density within polygons (such as in R17, T15, or, using the coordinates provided, H, 10) are aligned with a "finger" of forest intruding into the polygon, suggesting a stream branch.

Gordon and Wiseman (1989) used GLO records, including original survey notes, along with field observation and high-altitude photography, to document 54 prairie patch locations in the Jackson Prairie Region. These efforts were organized by the Mississippi Natural Heritage Program (MNHP), and its official inventory for Jackson Prairie remnants has since grown to 89 locations (Ken Gordon, pers. comm.). Harrell Prairie Hill, established in the Bienville National Forest as a managed native prairie in 1971, was the only managed prairie in the Jackson Prairie Region until 2008, when the US Forest Service (USFS) collaborated with the Nature Conservancy of Mississippi in the clearing and management of six other sites (Schauwecker and Gray, in prep.). In the case of Harrell Prairie Hill, Gordon and Wiseman (1989) cited the original survey notes as describing "rolling prairie with scattered pine

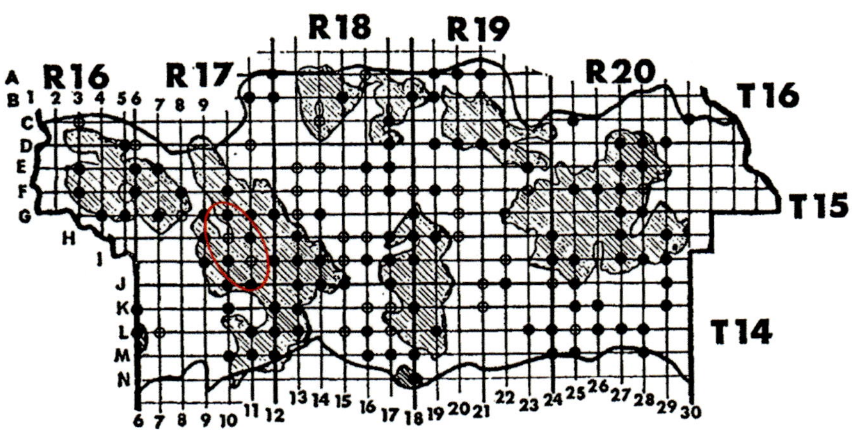

Figure 3.1. A figure from Rankin and Davis (1971). Shaded polygons are prairie patches. Solid dots are unforested section corners. Clear dots are heavily forested section corners. Absence of dots indicates medium tree density. The area outlined in red indicates heavily forested section corners within a prairie polygon, a possible stream bed.

and crabapple thickets." In the set of survey notes used by M. T. Gray (2010), this note is missing. In fact, the survey notes for the entire township lack any mention of prairie. Gordon (pers. comm.) explained that he was working with multiple copies of notes, suggesting that observations of prairies were sometimes lost when notes were transcribed or revised, a point noted in the introduction.

Barone (2005a) used GIS software to create a digital map of prairie patches in the Jackson Prairie Region from original GLO plat maps produced in the 1830s. He discussed evidence that many more prairies existed than were recorded on plat maps, including accounts of prairies documented by other sources that do not appear in the GLO records, polygons cut off by a section line with no corresponding polygon in the adjacent section, and the existence of relict prairies. Barone (2005a) noted that "because of inattentiveness by some teams of surveyors, many prairies, especially smaller ones, are unrecorded and likely lost to history." Our method, described in this chapter, reveals examples of observations of prairies that were recorded in the field notes but lost when the notes were transcribed or when plats were created from them.

In an unpublished master's thesis in landscape architecture at Mississippi State University, M. T. Gray (2010) created a map of historical prairie patches in the Jackson Prairie Region using GLO survey notes alone. These notes were from a set of scanned survey notes formatted as over 64,000 TIFF image files on 14 compact disks provided to Mississippi State University by the office of the Mississippi Secretary of State. Notes associated with townships located within the Jackson Prairie Region were selected from the complete set and read closely for any mention of prairie. These observations were most often rendered as points, measured in chains from section corners, at which the surveyor either entered or exited a prairie. In some cases the point data were not recorded, but the word "prairie" was included in the general note, as in the case of the mile-long boundary between Sections 4 and 5 in Township 8 North, Range 2 East in Madison County: "surface wavy and interspersed with small pieces of prairie." For the cases that included patch entrance and exit points, the chain measurements, the direction traveled, the township and range numbers, and the section numbers defining the line were entered into an Excel database. K. Grala, a geospatial science researcher in the Department of Geosciences at Mississippi State University, developed a tool to automate the process of generating polylines (single or multiple lines generated in a geographic information system [GIS] that graphically represent the surveyor field notes) representing prairie land at the time of the survey from the recorded entrance and exit points. Additional geographic data were added to the GIS.

Prior to visiting any of the sites, M. T. Gray (2010) consulted a team of stakeholders representing the Nature Conservancy of Mississippi, USFS, MNHP, Mississippi Department of Wildlife, Fisheries, and Parks (MDWFP), and Natural Resources Conservation Service (NRCS) to produce criteria regarding the geographic characteristics of highest priority for ranking potential relicts according to conservation potential. After discussion, the stakeholders completed surveys in order to develop weights of importance for eight geographical variables: state park land, USFS land, sixteenth section land, presence of transmission lines, distance to roads, land cover, proximity to other patches, and size of patch. The results of these evaluations were used in a simple map algebra equation to rank the GLO patches in terms of suitability for conservation. In addition, the research team visited 58 of the GLO patches. Sites were selected based on accessibility by public road. Half (29) of the visited sites were so intensively managed (generally either as mown residential areas or as cultivated fields) that prairie plants were absent. Of the half (29) that were left relatively unmanaged, approximately 40% (12) showed no indication of prairie (indistinguishable from the surrounding mixed-forest landscape), 8 featured a few indicator species, and 9 were classified as relict prairies. Of these, 5 were not on the MNHP inventory of Jackson Prairie patches and therefore could be considered newly documented locations of relict prairie (M. T. Gray 2010).

Tabulation of the Number of Prairie Patches Observed

Barone (2005b) used survey notes in his research, but the map he produced was generated entirely from plat maps. M. T. Gray (2010) generated a map entirely from survey notes. Barone graciously shared his GIS product with the authors, who compared the two maps as layers in ArcMap. Within the 129 townships that constituted the study area of M. T. Gray (2010), the number of patches indicated by each of the two methods was tabulated by county. This

Table 3.1.

Comparison of two methods for geolocating historical prairies.

Counties	Gray	Barone	Common to Gray and Barone	Unique to Gray	Unique to Barone
Clarke	10	8	6	4	2
Clarke/Wayne	8	2	2	6	0
Hinds	16	3	3	13	0
Hinds/Rankin	17	15	11	6	4
Holmes/Yazoo	0	1	0	0	1
Jasper	13	7	4	9	3
Jasper/Clarke/Wayne	0	1	0	0	1
Madison	30	14	14	16	0
Madison/Hinds	2	0	0	2	0
Madison/Rankin	6	8	4	2	4
Newton	4	0	0	4	0
Rankin	11	14	9	2	5
Rankin/Scott	5	5	4	1	1
Scott	19	11	9	10	2
Smith	2	2	1	1	1
Yazoo	16	4	4	12	0
Yazoo/Madison	13	12	11	2	1
Totals	172	107	82	90	25

Note: Barone (2005b) mapped and described historical prairie locations for the Jackson Prairie Region using the surveyor-generated plat maps, while M. T. Gray (2010) mapped locations using the survey notes.

tabulation counted each prairie patch as a single observation, regardless of size (table 3.1). The two methods were compared by counting and tabulating, for each township, the number of prairie patches in the plat maps and the number of prairie patches described in the surveyor notes. Prairie patches that were common to each method were counted in order to calculate the difference between the number of common patches and the number of patches that were distinctly counted by each method. This difference represents patches that are unique to each method by township, which were then aggregated by county. Townships that lie across county lines are listed separately, showing all county names.

The results of a direct comparison of the methods illustrate the differences that exist between a plat map–only and a notes-only approach. By count only, Gray mentions 65 prairies in the surveyor notes that were not illustrated on the plat maps observed by Barone. However, a more detailed measure is to compare the prairies that were common to both methods with those that were found independently by each method. This comparison shows that there were 90 mentions of prairie in the surveyor notes that were not shown on plat maps, and 25 prairie patches shown on plat maps that were not mentioned in the surveyor notes. It should be noted that for both methods, a certain number of records were missing, illegible, or other-

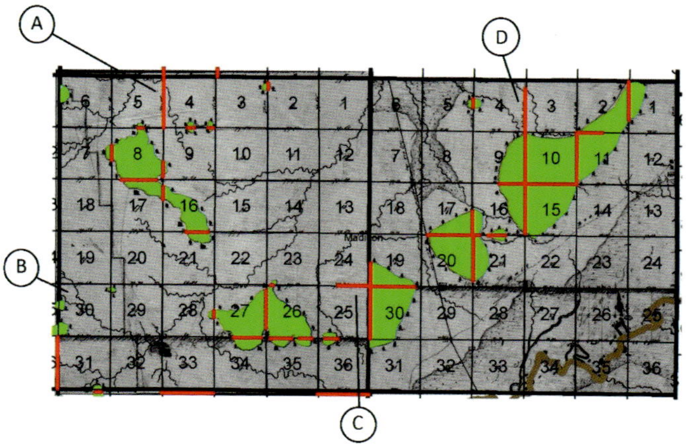

Figure 3.2. Plats for Township 8 North, Ranges 2 (left) and 3 (right) East, in Madison County, with prairie digitized as green polygons. Red lines represent prairie land from transcribed GLO survey notes. The brown line is the county line.

wise unavailable. Neither Gray nor Barone had access to a definitively complete set of GLO records. Neither method is necessarily "better," and both are obviously valuable in reconstructing the habitat that was present at the time of the surveys.

Plats and Notes from the BLM Website

The Bureau of Land Management (BLM) is in the process of scanning GLO survey plats and field notes and making them, along with other GLO records, publicly available on a searchable website (http://www.glorecords.blm.gov/). For this study, we obtained 49 plat maps from the BLM and converted 33 of them into georeferenced raster grids in order to view them as map layers in ArcMap, a geospatial processing program by Esri.

Methods

Researchers searched the BLM website for surveys of the 49 townships in the Jackson Prairie Region for which M. T. Gray (2010) found references to prairie in the survey notes. Plats for all 49 were available for download, but only 18 (37%) included survey notes. All 49 plats were downloaded as PDF files. Plats for 33 townships were converted to JPG files and cropped to the township boundary. These cropped images were converted to georeferenced (Imagine format, file extension .img) raster images using the georeferencing tool set in ArcMap. Georeferencing was accomplished by matching control points on the section corners in the JPG plat image with corresponding corners in a shapefile of section lines obtained from the Mississippi Automated Resource Information System (MARIS). Converting scanned plat maps to georeferenced raster grids in an ArcMap environment allowed the analysts to compare the lines generated by notes alone (M. T. Gray 2010) with patch boundaries (polygons) drawn on plat maps as two separate layers.

Results and Discussion: Discrepancies between Plat Maps and Survey Notes at Three Locations

Prairie polylines derived from notes alone generally match the prairie polygons on the plats exactly. Instances in which the polylines extend beyond polygon boundaries, or appear in the absence of polygons, indicate prairies mentioned in the notes but not featured on the plats. Several such instances at three locations are discussed here.

Madison County

In figure 3.2, the red line between Sections 4 and 5 in Range 2 (label A) was generated because the survey note associated with that mile reads "surface wavy and interspersed with small pieces of prairie." The patches were either too small or too numerous for the deputy surveyor to note the precise margins, so the information was not included on the plat map. In Section 30 (label B), three small patches appear that were not found by M. T. Gray (2010), who was working with digital scans of original notes: in this case the scan is overexposed and the text is illegible. Also in Range 2, between Sections 24 and 25 (label C), the margins of the prairie were noted at 60 and 154 chains traveling east (while most surveys measure a mile as 80 chains, the surveyor here used a 33-foot chain, resulting in a mile

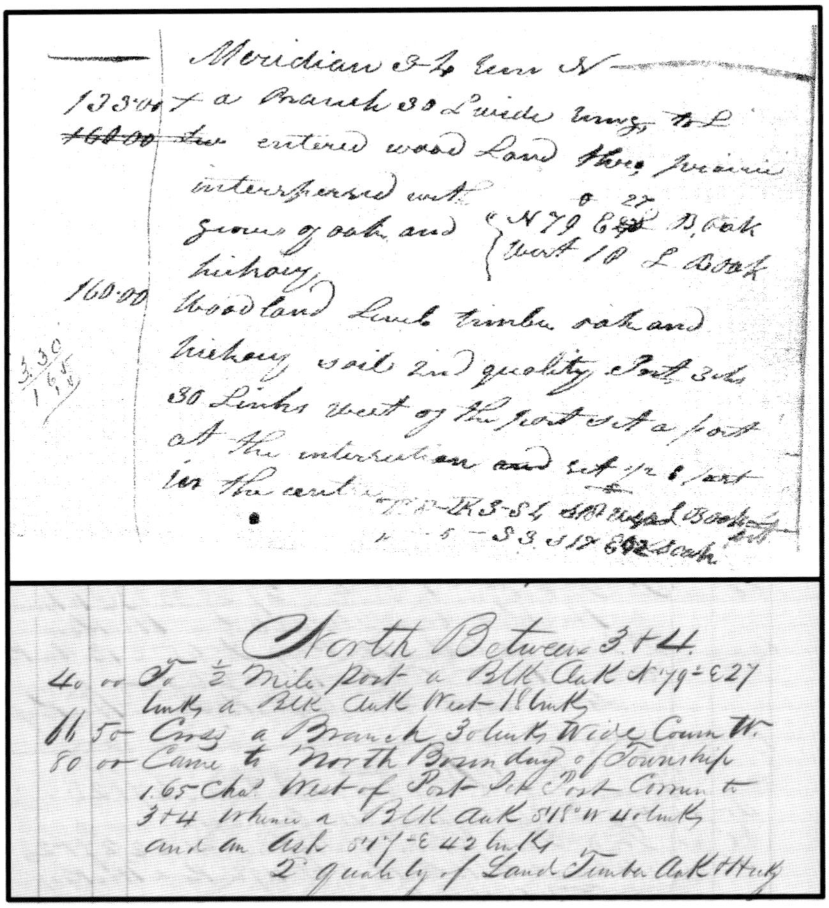

Figure 3.3. Survey notes for the line between Sections 3 and 4 in Township 8 North, Range 3 East, from the set of scanned notes used by M. T. Gray (2010) (top), and scanned notes for the same line from the BLM website (bottom).

of 160 chains). For whatever reason, this prairie was also not shown on the plat, although it is a large patch that extends into the next township, where it was in fact shown on the plat. The general note for this mile reads "thru prairie with some scattering scrubby oak hickory." The note for the line for the mile between Sections 23 and 26 (adjacent on the western side, with the very small prairie patch) reads "surface wavy soil 2nd quality woodland thinly timbered with scrubby oak and hickory." While no prairie margins are noted for the miles between Sections 34/35 and 35/36, the general notes for these lines are almost the same as for the others: "land wavy soil 2nd quality timbered thinly with oak and hickory." Note that the words "woodland" and "scrubby" are dropped. The entire southeastern corner of this section could perhaps be considered a prairie landscape as described by Campbell and Seymour (2011a), with scattered hilltop copses and tree density increasing along streams.

Another line, between Sections 3 and 4 in Range 3 (label D), can be seen in figure 3.2. Survey notes for Township 8 North, Range 3, are available on the BLM website and differ from those used by M. T. Gray (2010) (fig. 3.2); the latter read as follows:

Meridian 3–4 Run N—
133 00 + a branch 30 L wide runs to L [i.e., cross a stream 30 links wide running left, or west]
~~160 00~~ entered wood Land thru prairie

interspersed with groves of
oak and
hickory.
N 79 E ~~28~~ 27 B. oak
West 10 L B. oak

160 00 Woodland level timber oak and
hickory soil 2nd quality. Set 3 chs
30 Links west of the post set a post
at the intersection and set 1/2 post
in the center
T8R3-S4 [illegible: maybe S 18° W
4 L] B oak
T8R3-S3 S 17 E42 B oak

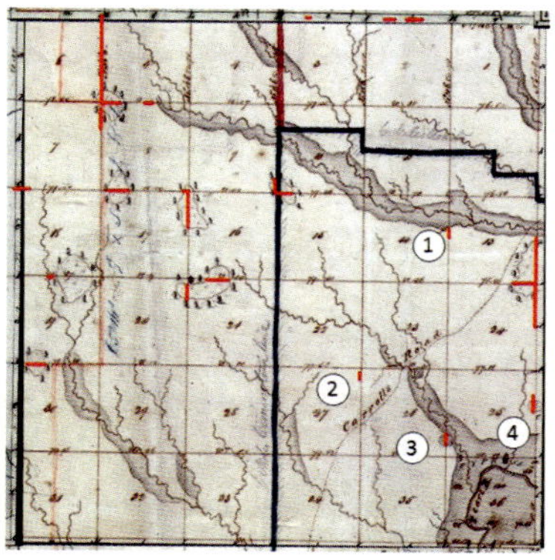

Figure 3.4. Georectified BLM plat map of Township 6 North, Range 1 East, Hinds County, with prairie from survey notes (red lines). Numbers indicate patches discussed in the text.

Figure 3.5. Detail of BLM plat map of Township 6 North, Range 1 East, Hinds County, showing prairie patches with rendered blades of grass within and trees on the margins (Section 17, left), and prairie patches rendered as simple polygons with gray fill (Section 25, right). Numbers refer to patches discussed in the text.

The survey for the same line from the BLM website reads:

North Between 3+4

40 00	To ½ mile post a Blk Oak N 79 + E 27 links a Blk Oak West 18 links
66 50	Cross a branch 30 links Wide Course W.
80 00	Come to North Boundary of Township 1.65 Chains West of Post—Set Post corner to 3+4 Whence a Blk Oak S18° W 40 links and an Ash S17 + E 42 links 2d quality of Land Timber Oak + Hiky [Hickory]

Figure 3.3 (*top*) shows what seems to be an earlier draft written in the field. At 133 chains, the surveyor crosses a branch 30 links wide. This is followed by the observation: "entered woodland through prairie interspersed with groves of oak and hickory." It is unclear at what point along the mile the surveyor entered the prairie, but since the note was made after crossing the branch, this point is likely to have been toward the end of the mile. The note "woodland level" is a description of the general characteristic of the land and applies to the entire mile.

Figure 3.3 (*bottom*) shows what seems to be a "cleaned up" revision of the notes at the top of the figure. The information for the witness trees for the quarter section post is the same: black oak north 79 east 27, black oak west 18; the stream is 30 links wide; and the direction of flow has been clarified as west rather than left. The original chain measurements for the line (in the left margin) have been converted to what they would have been had the surveyor used the more-common 66-foot chain, but oddly, the measurements for the witness trees and the stream branch are left as their original values. The last witness tree in the note, located at south 17 links and east 42 links in both surveys, has changed from a black oak in the first to an ash in the second. Most significant for the research here, the phrase "woodland level" and the reference to prairie have been removed. The "official" note lacks the information about prairies recorded by an eyewitness in the field.

Hinds County

In the survey notes used by M. T. Gray (2010), four small prairie patches were mentioned in the eastern part of Township 6 North, Range 1 East, in Hinds County that, at first glance, seem to be missing from the plat map available on the BLM website.

Two of these four patches do not appear on the plat, and the remaining two are drawn as shaded circles and do not feature the stylized blades of grass in the interior and sketched trees on the margins shown elsewhere on the plat.

Three of these patches, however, are included on a physiographic map of Hinds County by Edwards (1954) and labeled "prairies as mapped in 1820."

Edwards (1954) did not explain how he determined the locations of these patches. The date mentioned (1820) implies that he used GLO survey data, but it is unlikely that he used the plat map now available on the BLM website or that used by Barone (2005b) because the existence of

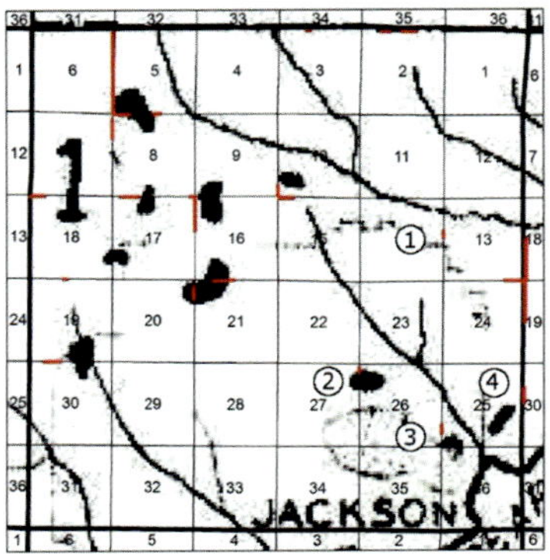

Figure 3.6. Georectified map of Township 6 North, Range 1 East, as presented by Edwards (1954). Three of the four patches found in notes used by M. T. Gray (2010) but not evident on the BLM plat map are shown.

the patches is not directly indicated on those maps. This reinforces the general argument of this chapter that multiple copies of GLO records exist and should be preserved.

The scan of the survey notes used by M. T. Gray (2010) for patch 1 is of poor quality and is not reproduced here. The surveyor traveled north between sections 13 and 14 and passed through 14 chains (462 ft, 141 m) of prairie at about the midpoint of the mile. In the final 30 chains (990 ft, 302 m), he crossed a creek and went through "cane to rich bushy level." The line ends with a monument in a flat swamp. The general note at the end reads "1st half mile wavy, mulatto stiff soil—B Jack and Hky.—Residue red ok, Hky, water ok, gum, ash, elm, [?] Mostly good bottom land, light soil in the Cr bottom. Undergrowth cane and bushes." The present-day location is the corner of Manhattan Drive and Lawrence Road in a residential neighborhood just south of Hanging Moss Creek in Jackson, Mississippi. This patch is not visible at all on the plat map available on the BLM website.

Patches 2 and 3 are featured in the survey manuscript in the following.

On the line between Sections 27 and 26 the surveyor was traveling north "in open thin woods." At 52 chains he crossed a "wagon road." At 80 chains he marked the half-mile point and the witness trees. At 141 chains: "to poor open prairie." At 152 chains he noted "+ [cross] prarie [*sic*] to woods." The general note at the bottom left reads "Poor rolling B jack, post ok & Hky—stiff soil."

On the line between Sections 25 and 26, a mile to the east, the surveyor once again traveled north in "open thin woods." He traveled through "poor open prairie" for 17 chains (561 ft, 171 m) before entering "open poor woods" once again. He crossed a branch three times before setting

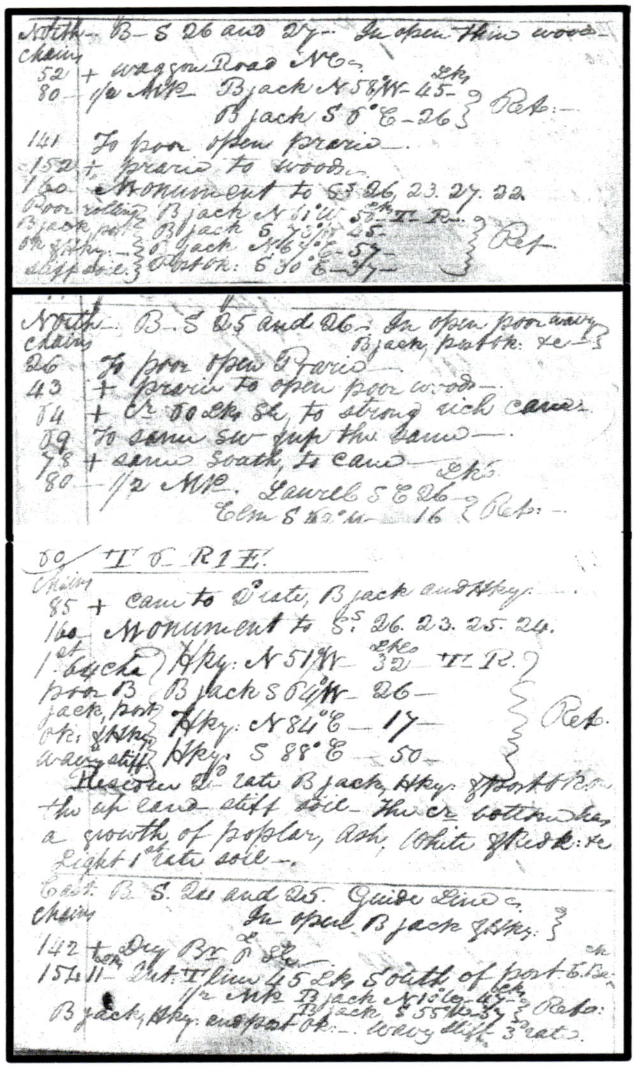

Figure 3.7. Survey notes for the line between Sections 26 and 27 (top) and for the line between Sections 25 and 26 (bottom) in Township 6 North, Range 1 East, Hinds County.

the half-mile post. The general note for the line is "Bjack, Hky and post Ok:- wavy stiff 3rd rate [soil]."

The half-mile post was set at about the point where present-day Highway 25 crosses Eubanks Creek, which is the branch mentioned in the notes and visible on the plat map. The prairie patch on this line existed just north of the parking lot of the Mississippi Museum of Natural Science.

Since the creator of the plat map for this township rendered prairie patches 3 and 4 in figure 3.5 as shaded circles rather than the suggested grass and sketched trees used for the other patches, anyone examining this plat map (including Barone 2005b) would reasonably conclude that these

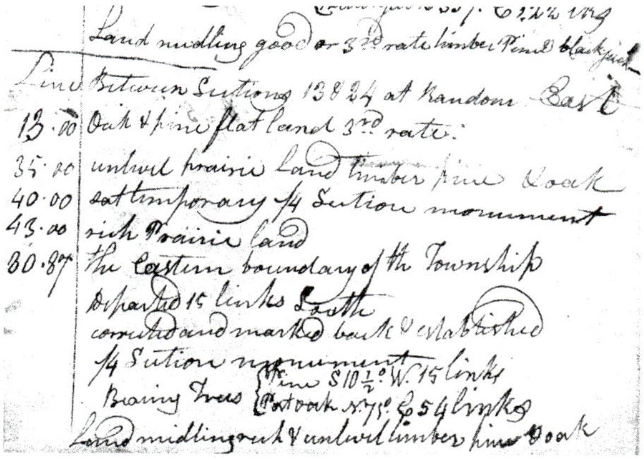
Land midling good or 3rd rate timber Pine & blackjack
Line between Sections 13 & 24 at Random East
13.00 Oak & pine flat land 3rd rate
35.00 unlevel prairie land timber pine & oak
40.00 set temporary ¼ Section monument
43.00 rich Prairie land
80.87 the Eastern boundary of the Township
Departed 15 links South
corrected and marked back & established
¼ Section monument
Bearing trees { Pine S 10½° W. 15 links
Post oak N 75° E 54 links
Land midling rich & unlevel timber pine & oak

Figure 3.8. Survey notes for the line between Sections 13 and 24 in Township 10 North, Range 6 West, Wayne County.

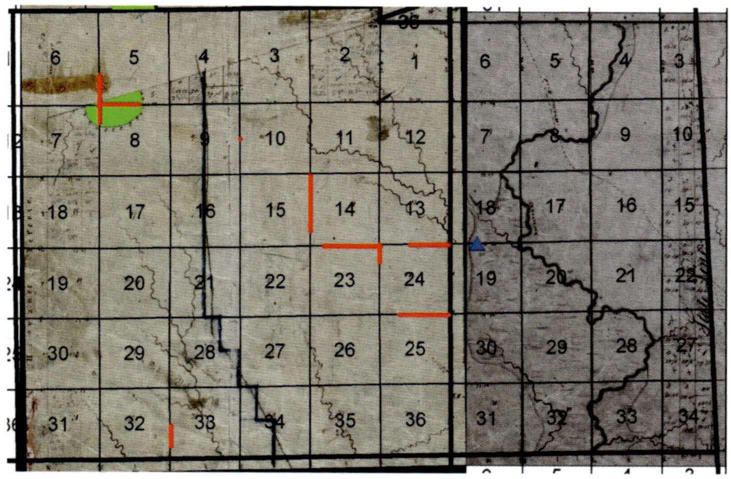

Figure 3.9. Township 10 North, Ranges 5 and 6 West, Wayne County, near the Mississippi-Alabama state line. Raster image of plat from BLM, digitized prairie from plat (green), prairie locations from survey notes (red), and newly documented relict prairie (blue triangle).

polygons represent ponds or some other feature. Only when the notes are digitized and used as a map layer with a georectified plat map in GIS is it obvious that these polygons represent prairie patches. The precise location and extent of pre-European-settlement prairie land is part of the ecological heritage of the city of Jackson and as such is of value to city planners, landscape architects, park managers, and citizens. The actual historical extent of the prairies cannot be obtained from plats alone.

Wayne County

M. T. Gray (2010) found records of prairie in Wayne County in the GLO survey notes that are not recorded on BLM plats and were not included in the plats scanned by Barone (2005a). The deputy surveyor for that area made several observations of prairie along section lines but failed to make clear the locations of the prairie margins, resulting in their exclusion from the plats (fig. 3.7). The surveyor, moving east on the line between Sections 13 and 24, observed "unlevel prairie" at 35 chains and "rich prairie" at 43 chains but failed to note the locations of the entrance and exit points, resulting in insufficient data to generate a polygon on the plat. The location of the polyline is on private land 0.3 mi (483 m) west of Matherville Frost Bridge Road. At a nearby grass roadside opening dominated by *Juniperis virginiana*, M. T. Gray (2010) observed *Ratibida pinnata*, *Rhamnus caroliniana*, *Robinia pseudoacacia*, and *Ruellia humilis*.

This site is one of the five that we consider newly discovered relicts because they are not included in the MNHP inventory. Had this mapping project relied only on plats, the location of this relict and its historical context would have remained unknown.

Conclusions

Reasons that extant, pre-European-settlement prairie patches on the land are missing from nineteenth-century plat maps can be enumerated as follows: (1) the surveyor passed through a prairie but simply neglected to add the observation to the field notes; (2) prairie patches were so small and numerous that recording margins for each would have been inefficient; (3) prairies were not on section lines but rather at unseen locations in the interior of the sections; (4) prairies were recorded in the earlier versions of field notes but excluded from revised copies; and (5) a woodland-prairie gradient in the landscape existed such that the prairie margin could not be precisely determined.

Barone's use of plats (2005a, 2005b) to create a map of the distribution of blackland prairies in the early nineteenth century is a significant achievement. It allows researchers to see for the first time a fairly comprehensive and accurate image of a lost feature at a regional scale. Survey notes provide additional information at a finer level of spatial granularity. GLO plats were generated by surveyors from the information contained in the survey notes

Figure 3.10. Newly documented relict prairie in Wayne County, latitude 31.827 N, longitude 88.525 E. Photograph by Toby Gray, August 2009.

taken in the field. The margins of the prairie openings were drawn according to their recollections from the perspective of individuals engaged in the difficult task of marking a true line in a cardinal direction for one mile at a time. It is unlikely that any deputy surveyor would have taken the trouble to measure the distance from the section line to the prairie margin at any point that did not fall on the line. The georeferenced location of historical prairie margins from plats is therefore most accurate at the point where the prairie margin crosses the section line. That accuracy decays as the distance between the prairie margin and the section line increases.

A significant amount of information regarding the spatial configuration and distribution of prairie patches was lost during the iterative process of creating accurate and useful plats from notes by hand. The iterative process corrected errors but also resulted at times in the loss of information, as exemplified here in the neglected observations of prairie land. Efforts to digitally preserve the survey notes should continue, and they should be expanded to include variant copies of notes, since the earlier copies, while perhaps more prone to error, do constitute the work of eyewitnesses in the field. As stewards of the land, we have failed to preserve the actual prairie patches. Let us not fail to preserve the historical eyewitness records of the locations of those patches.

On the one hand, examining multiple copies of surveys at the Department of State or at various county courthouses is labor intensive. In light of this, we applaud the efforts of the Bureau of Land Management to provide access to GLO records in a searchable website. On the other hand, the use of searchable websites as sources of archival information is a new development in human history. Websites have an ephemeral quality, and the long-term preservation of digital data is arguably best achieved by archiving variant copies of files on multiple servers. State agencies and universities should be encouraged to duplicate the efforts of the BLM and develop specific applications according to regional variations in geographical variables, research priorities, and the quality and condition of their records. Any archivist has to make decisions regarding what is most worth preserving and what should be purged. The research presented here illustrates a case in which the older, messier records are of significant value.

In addition to containing data regarding the historical spatial distribution and configuration of prairie patches, GLO survey notes also contain data about the location of old roads, settlements, fields, and stream branches. The courses of rivers and streams in a region at the time of the survey can be determined from plats, but the notes add a record of the width of each branch wherever it crossed a section line. Variations of stream width across an area are an important parameter for anyone wishing to reconstruct a historical assessment of regional hydrology.

Awareness of the presence of an imperiled ecological habitat, even one nearly lost to history, is important to those who make decisions at site-specific scales. The presence of calcareous soils and prairie indicator species should be considered to present a unique set of opportunities and constraints to any landscape architect, regional planner, or ecologist working in or near any of the southeastern blackland prairie belts. The degree to which any site is proximal or coincident to the pattern of GLO prairies can add value to efforts to conserve or celebrate that part of our ecological heritage. Increasing availability of georeferenced data from the GLO (manuscripts, plat maps, and their digitized equivalents available online from sources such as the BLM) will facilitate communal efforts to preserve what is left and restore what is lost. We hope that an examination of the GLO data, either as notes or as plat maps, will become a standard step in inventory and analysis by actors involved in the multiple uses of land at all scales.

4

Eastern Texas Prairie Landscapes

Jason R. Singhurst and Matt White

Eastern Texas, as defined herein, includes all of the Piney Woods, Oak Woods and Prairies, and the eastern edge of the Blackland Prairie ecoregion of the West Gulf Coastal Plain (Gould 1962; Diggs et al. 1999, 2006; Turner et al. 2003; Nature Conservancy 2003). We have concentrated on five primary prairie landscapes of eastern Texas, including two shared with adjacent Louisiana. We also discuss two ecologically isolated prairie landscapes. We have excluded coastal prairies, as they require extensive discussion and are restricted to their own ecoregion, even though this region is heavily influenced by the southeastern Gulf Coastal Plain.

The Piney Woods occupies 6,038,915 ha (14,922,485 ac) along the eastern edge of Texas and extends into western Louisiana. The southern third of the region, including the Big Thicket (Watson 1979), represents the western part of the lower West Gulf Coastal Plain, while the northern two-thirds is the southwestern portion of the upper West Gulf Coastal Plain. Elevation ranges from about 8 m (25 ft) near Beaumont in the south to over 213 m (700 ft) near Rusk in the north. The climate of the region is humid subtropical, with hot summers and mild winters. Situated at the high end of the Texas east–west rainfall gradient, the Piney Woods receives more rainfall than other regions, ranging from 142 cm (56 in) in Newton County in the east to about 112 cm (44 in) in Walker County in the west. The average length of the growing season ranges from 270 days in Montgomery County to 235 days in Bowie County (Natural Fibers Information Center 1987).

As the name implies, this region is defined by the presence of pine-dominated forests or woodlands. Loblolly pine (*Pinus taeda*) is ubiquitous, occurring in a broad spectrum of habitats throughout the region; it is also widely utilized in commercial pine plantations. Shortleaf pine (*P. echinata*) is most conspicuous on drier sites in the north, while longleaf pine (*P. palustris*) is restricted to the southeastern part of the region. In some areas, pines are mixed with various hardwoods such as oaks (*Quercus falcata*, *Q. stellata*, *Q. alba*, and others), hickories (*Carya* spp.), red maples (*Acer rubrum*), and sweet gums (*Liquidambar styraciflua*). In other areas, particularly in the South where longleaf pine is dominant, natural vegetation is a more open, fire-maintained savanna that varies considerably depending on soils and hydrology. On poorly drained, clayey to loamy soils on upland flats with a fluctuating water table, longleaf pine occurs in the wetland longleaf pine savanna. On well-drained sandy soils of uplands and ridges, it occurs in upland longleaf pine savanna (Bridges and Orzell 1989; Bezanson 2000). The differences between the types are most obvious in the herbaceous layer, which is much more diverse in the wetland longleaf pine savanna than all of the woody layers combined. In good condition, wetland longleaf pine savanna is among the most species-rich communities in Texas.

Other matrix communities include bottomland forests composed of swamp chestnut oak (*Q. michauxii*), overcup oak (*Q. lyrata*), water oak (*Q. nigra*), willow oak (*Q. phellos*), and many other hardwoods. More localized but still large-

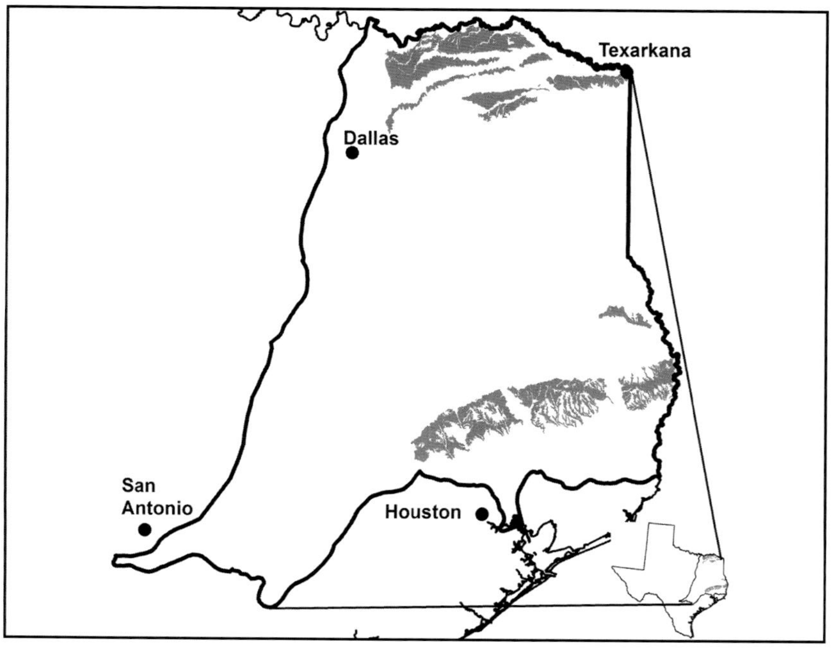

Figure 4.1. Geologic formations that support five primary and two ecologically isolated prairie landscapes in eastern Texas.

scale vegetation types include forests of American beech (*Fagus grandifolia*) and southern magnolia (*Magnolia grandiflora*), which occupy ravine slopes and bottoms, particularly on more clayey soils in the southern part of the region. Similar mesic-slope forests of white oak (*Q. alba*), southern red oak (*Q. falcata*), Shumard red oak (*Q. shumardii*), and other hardwoods occur on sandier sites (Marks and Harcombe 1981; Bezanson 2000). Bald cypress (*Taxodium distichum*)–tupelo (*Nyssa aquatica*) swamp forests occur along some mature streams in various parts of the region.

Oak Woods and Prairies occupies 5,267,193 ha (13,015,517 ac) in a band of varying width stretching from Atascosa County southeast of San Antonio north-northwest to a string of counties along the Red River, and from the northeastern corner of the state westward to Montague County. This main part of the region is often referred to as the Post Oak Savanna (Gould 1975; Bezanson 2000; M. MacRoberts and MacRoberts 2003). Two additional bands, the Eastern and Western Cross Timbers, lie to the west, separated from the main swath by portions of the blackland prairie natural region.

Climate varies considerably along the 644 km (400 mi) north–south axis of this region. In most of the region, the climate is humid subtropical, but in northwestern parts (the Cross Timbers), the climate is merely subtropical. Summers are hot, and winters are generally mild. Average annual rainfall ranges from 74 cm (29 in) in Wilson County in the southwest to 112 cm (44 in) in Walker County along the boundary with the Piney Woods. The growing season is 227 days in Bowie County along the Red River and almost two months longer (280 days) in Wilson County (Natural Fibers Information Center 1987).

The matrix vegetation types in this region are deciduous or mixed evergreen-deciduous woodlands composed of post oak (*Q. stellata*) and various mixtures of blackjack oak (*Q. marilandica*), coastal live oak (*Q. virginiana* or an ecological equivalent), black hickory (*Carya texana*), and eastern redcedar (*Juniperus virginiana*). In the Lost Pines of Bastrop County and in scattered locations to the east, loblolly pine joins this mix. Intermingled with these woodlands, particularly on more clayey substrates, are strips of midgrass or tallgrass prairie dominated by little bluestem (*Schizachyrium scoparium*) and various other species, depending on substrate. At the southern end of the region, mesquite (*Prosopis glandulosa*) invades these prairies and forms tall shrublands. This north–south variation was discussed in detail by McBryde (1933). Riparian woodlands of hackberries (*Celtis* spp.), elms (*Ulmus* spp.), water oak, and other hardwoods line the bottomlands of streams that cross the region on their way to the Gulf of Mexico.

Prairie Landscapes in Eastern Texas

In eastern Texas, five primary prairie landscapes persist, and all are of global conservation concern. All prairie landscapes

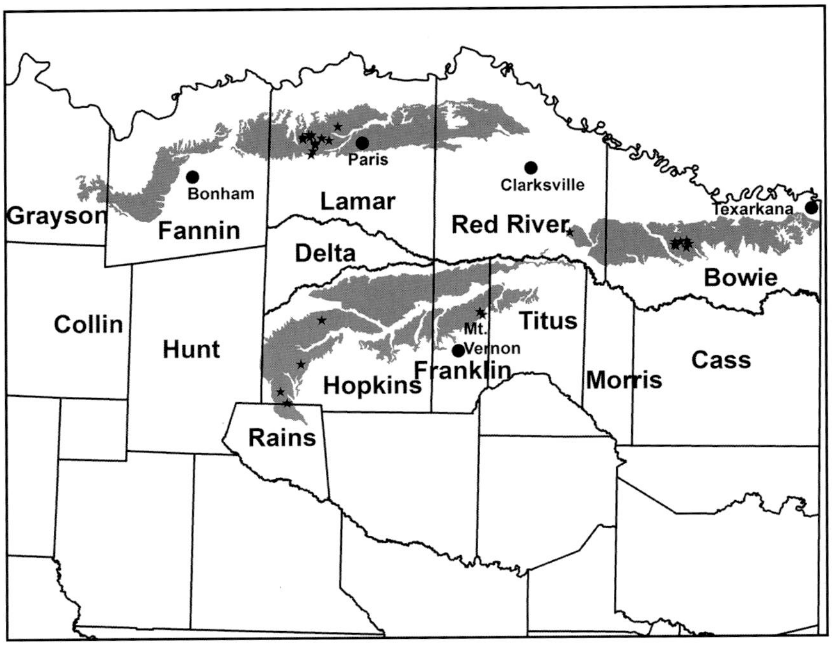

Figure 4.2. Silveus's Dropseed Prairies are ecologically isolated to the Bonham Formation (Late Cretaceous) and Midway Group (Paleocene). Silveus's Dropseed Prairie site locations are labeled with a star.

in eastern Texas are in transition areas between oak-hickory or pine savannas and woodlands. Northeastern Texas is underlain by geological deposits from the Bonham Formation (Late Cretaceous) and Midway Group (Paleocene), which have shaped Silveus's Dropseed Prairies.

These areas are composed of alfisol materials, which include sand, marl, clay, silt, and glauconite. Silveus's Dropseed Prairies have soils of low pH formed mainly on bedrock that is higher in sand content and lower in calcium carbonate. Northeastern Texas is also is underlain by geological deposits from the Eagle Ford Formation, Gober Chalk, Annona Chalk, and Pecan Gap Chalk (all Late Cretaceous), which has led to the development of the Northeastern Texas Blackland Prairies and Chalk Glades.

These areas include chalk with limestone beds, clay, chalky marl, lime sand, marine fossils, and sedimentary shale rock. In deep eastern Texas, centered on the town of San Augustine, the Weches Formation (Eocene) outcrops create small, isolated Weches Prairies and Glades in the form of islands on hilltops.

These areas include a layer of glauconite clay that is exposed at the surface (which is typically rocky and shallow) or covered by a thin layer of calcareous sediment. In southeastern Texas and southwestern Louisiana, Fleming Prairies, which are underlain by the Fleming Formation (Miocene), have developed.

These prairies include calcareous clay soils over sandstone, which contains embedded Cretaceous invertebrate fossils. In the southern portion of eastern Texas, centered on the town of Brenham, the Fleming and Oakville Formations (both Miocene) have formed a more extensive belt of prairie vegetation, the Fayette Prairies (an outlier blackland prairie landscape).

These prairies also occur on calcareous clay soils that are embedded with Cretaceous invertebrate fossils.

Two ecologically isolated prairie landscapes have recently been observed in eastern Texas. Very little is known about the first one, as very few examples persist. We define here for the first time an endemic prairie type currently restricted to Delta and Hopkins Counties, Texas, underlain by the Marlbrook Marl Formation (Late Cretaceous), which has established Marlbrook Prairies.

These prairies occur on clay silt over a limestone bed and glauconite, with marine megafossils. In northeastern Texas, centered on the community of Boxelder in Red River County, the Navarro Group Formation (Late Cretaceous) has formed Morse Clay Prairies. These prairies occur on calcareous clays, silt, and glauconite, with marine megafossils.

Land management practices such as overgrazing, crop production, conversion to pastures or pine plantations, invasive plants, and sand and gravel mining are some of the landscape changes that have played roles in the decline of

Figure 4.3. Silveus's Dropseed Prairie at Daphne Prairie in Franklin County, Texas.

eastern Texas prairie landscapes. A lack of systematic inventories, descriptions, and studies of eastern Texas prairies has led to a poor understanding of their ecology, and the area has taken a back seat with respect to conservation action.

Silveus's Dropseed Prairies

Silveus's Dropseed Prairies (figs. 4.2 and 4.3) occur on distinctive alfisol substrates; they differ drastically in characteristic species from other tallgrass prairie types in Texas and the Midwest. One of the unique features of Silveus's Dropseed Prairies is the low, circular mima mounds or pimple mounds (1 m [3.3 ft] high and 1–14 m [3.3–46 ft] across) (Diamond and Smeins 1993). Soils are usually sandy loams (Diamond 1993), flat to gently rolling, and have a low pH. Dominant grasses include Silveus's Dropseed (*Sporobolus silveanus*), longspike tridens (*Tridens strictus*), Mead's sedge (*Carex meadii*), rosettegrasses (*Dichanthelium acuminatum, D. linearifolium*, and *D. oligosanthes*), and Carolina jointtail (*Coelorachis cylindrica*). Microhabitats are very important in this prairie landscape. Mima mounds that are intact (not reduced in size or flattened) provide an extremely important microhabitat with typically three distinct vegetation zones (top, slope, and base of slope). Mima mounds are typically dominated by little bluestem (*Schizachyrium scoparium*), and the lower slope of the mounds creates seasonal seepage for plants such as

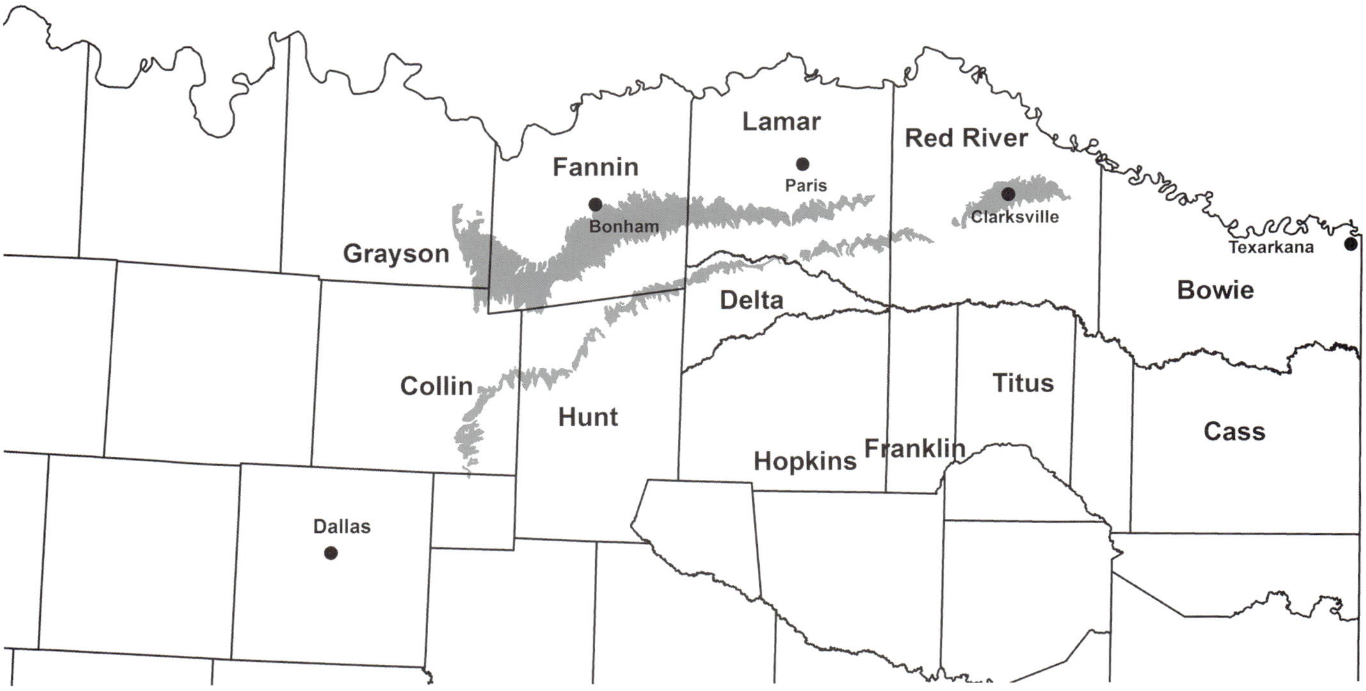

Figure 4.4. Northeastern Texas Blackland Prairie Chalk Glades occur on geological deposits from the Eagle Ford Formation, Gober Chalk, Annona Chalk, and Pecan Gap Chalk (all Late Cretaceous).

Figure 4.5. Blackland prairie on Austin Chalk Formation in Collin County, Texas.

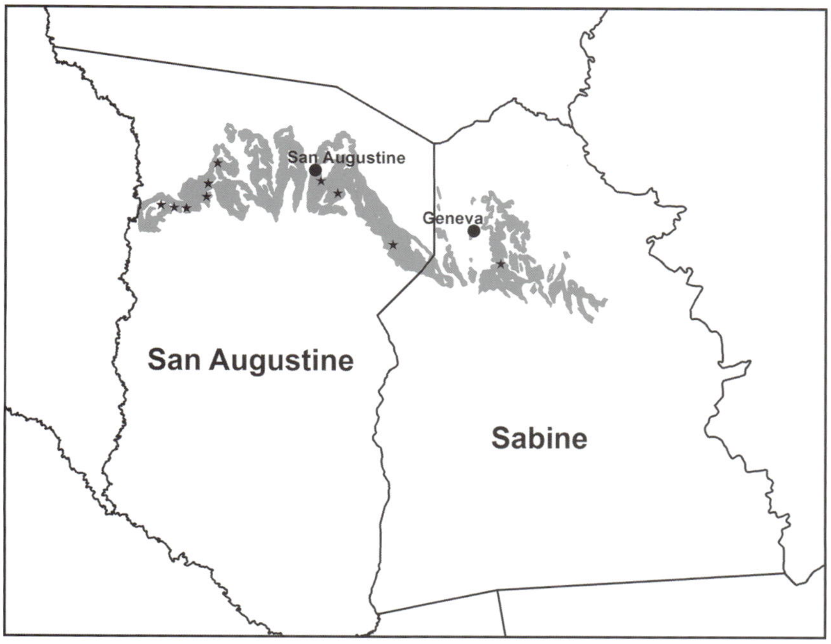

Figure 4.6. Weches Prairies and Glades are restricted to the Weches Formation. Weches Prairies and Glade locations are labeled with a star.

Figure 4.7. Weches Glade vernal pools dominated by pink stonecrop (*Sedum pulchellum*).

Figure 4.8. Texas golden gladecress (*Leavenworthia texana*) in Weches Glade vernal pool.

zigzag bladderwort (*Utricularia subulata*), dwarf sundew (*Drosera brevifolia*), lousewort (*Pedicularis canadensis*), and wild quinine (*Parthenium hispidum*). Seasonally moist swales are frequent between mounds and are dominated by big bluestem (*Andropogon gerardii*), beakrushes (*Rhynchospora glomerata*, *R. caduca*, and *R. capitellata*), nutrushes (*Scleria pauciflora* and *S. georgiana*), and rushes (*Juncus dudleyi*, *J. validus*, and *J. scirpoides*). Maryland meadowbeauty (*Rhexia mariana* var. *interior*), swamp sunflower (*Helianthus angustifolius*), dwarf St. John's-wort (*Hypericum mutilum*), and Texas goldentop (*Euthamia gymnospermoides*) also frequent the moist swales. Intermittent drainageways within these prairies vary between seasonally wet and dry. Seasonally wet drainageways are dominated by gamagrass (*Tripsacum dactyloides*), switchgrass (*Panicum virgatum*), and Florida paspalum (*Paspalum floridanum*). Dry drainageways are often naturally erodible and dominated by sideoats grama (*Bouteloua curtipendula*), ratany (*Krameria lanceolata*), and largeleaf wild indigo (*Baptisia alba*). Small woodlands ($\leq$ 0.1 ha or 0.25 ac) of shortleaf pine (*Pinus echinata*) and oaks (*Quercus stellata*, *Q. marilandica*, *Q. shumardii*, and *Q. falcata*), and easterly limits of mesquite (*Prosopis glandulosa*) woodlands are also scattered infrequently within Silveus's Dropseed Prairies. Silveus's Dropseed Prairies may include isolated saline barrens ("slick spots") around intermittent drainageways. Characteristic saline barren flora is usually dominated by ephemerals, short-lived annuals, or drought-tolerant specialists. Associated flora usually includes pyramid dropseed (*Sporobolus pyramidatus*), narrowleaf sumpweed (*Iva angustifolia*), golden-

Figure 4.9. Weches Prairie in San Augustine County, Texas.

mane tickseed (*Coreopsis basalis*), dwarf plantain (*Plantago hybrida*), rose bluet (*Houstonia rosea*), threadleaf evening primrose (*Oenothera linifolia*), fameflower (*Phemeranthus parviflorus*), and yellow puff (*Neptunia lutea*).

Typical forbs occurring in extant examples of Silveus's Dropseed Prairies include great coneflower (*Rudbeckia maxima*), rough coneflower (*R. grandiflora*), prairie blazing star (*Liatris pycnostachya*), ashy sunflower (*Helianthus mollis*), green milkweed (*Asclepias hirtella*), puffballs (*Marshallia caespitosa*), pale purple coneflower (*Echinacea pallida*), compassplant (*Silphium laciniatum*), roughstem rosinweed (*Silphium radula*), Samson's snakeroot (*Orbexilum pedunculatum*), rattlesnake master (*Eryngium yuccifolium*), prairie camas (*Camassia angusta*), purple prairie clover (*Dalea purpurea*), white prairie clover (*Dalea candida*), and many asters. A few deep sandy soil specialists also occur on the tops of the sandiest mounds, such as silver croton (*Croton argyranthemus*), woolly-white (*Hymenopappus artemisiifolius*), and Texas groundsel (*Senecio ampullaceus*). Silveus's Dropseed Prairies contain regional and narrow endemic plants, rare plants, and peripheral species found only in this landscape in Texas. One highly restricted and endemic plant (Mink et al. 2012), pygmy prairie dawn (*Hymenoxys perpygmaea*), was recently described from saline barrens at Gambill Goose Prairie in Lamar County. To date, it is the only known location globally for this rare plant.

Oklahoma beardtongue (*Penstemon oklahomensis*) (Mink et al. 2010), a narrow endemic known from central and south-central Oklahoma, was recently documented at Gambill Goose Prairie and at a prairie on private land in Hopkins County. A recently proposed candidate for listing, Oklahoma grass pink orchid (*Calopogon oklahomensis*) (USFWS 2011), occurs in a couple of Silveus's Dropseed Prairies, where it is found in moist transitions between mounds and along wet drainageways. Oklahoma grass pink orchid is globally rare (NatureServe 2012–2013); of the 14 states in its range, it is presumed extirpated in Indiana, Iowa, and Wisconsin and is critically imperiled (S1, 0–5 occurrences) in Alabama, Illinois, Kansas, Louisiana, Mississippi, and Texas.

Texas sunnybell (*Schoenolirion wrightii*) (NatureServe 2012–2013) has a narrow range in the Southeast (G3, known from Alabama, Arkansas, Louisiana, and Texas) and occurs in a few Silveus's Dropseed Prairies, where it frequents intermittent swales between mima mounds. Scarlet paintbrush (*Castilleja coccinea*) (Singhurst et al. 2011) is a peripheral that has been documented in three locations in Texas in Silveus's Dropseed Prairies. Other peripheral species worth mentioning that occur in these prairies include green fringed orchid (*Platanthera lacera*) (W. Carr 2009) and scaly blazing star (*Liatris squarrosa* var. *glabrata*) (M. White and Nesom 2012).

Extraordinarily high plant diversity has been documented in Silveus's Dropseed Prairies during recent inventory collections of the flora of Gambill Goose Wildlife Refuge Prairie in Lamar County (40.5 ha or 100 ac, 374 species), Daphne Prairie in Franklin County (364.2 ha or 900 ac, 341 species), Mary Talbot Prairie Preserve in Bowie County (44.5 ha or 110 ac, 304 species), and Tridens Prairie Preserve in Lamar County (39.3 ha or 97 ac, 243 species). The largest remaining Silveus's Dropseed Prairie is the Smiley-Woodfin Prairie (849.8 ha or 2,100 ac), a hay meadow and Texas historical ranch in Lamar County. Tridens Prairie Preserve is a high-quality example owned by the Nature Conservancy. Gambill Goose Wildlife Refuge Prairie is owned by the city of Paris, Texas. Daphne Prairie is owned by a private landowner and includes 48.6 ha (120 ac) of high-quality hay meadow and an additional 315.7 ha (780 ac) of restorable prairie. Mary Talbot Prairie Preserve was recently (2012) purchased by the Native Prairies Association of Texas (NPAT), and arrangements can be made through NPAT to view this high-quality prairie. Silveus's Dropseed Prairies have a very limited global distribution. Very few intact examples exist, and most of the remaining examples have been affected by overgrazing and lack of prescribed burning. Much of the original acreage of this prairie landscape has been converted to cropland and nonnative pasture or has migrated to woodland. This prairie association has not been observed in adjacent Oklahoma or Arkansas and is known only as a species element in Oklahoma. The remaining relatively intact examples that exist are threatened by fire suppression and invasion of exotics. Areas that have been hayed frequently or overgrazed are dominated by early-successional species such as splitbeard bluestem (*Andropogon ternarius*) or woody species such as eastern redcedar (*Juniperus virginiana*).

Northeastern Texas Blackland Prairies with Limestone and Chalk Glades

Blackland prairies with limestone and chalk glades (figs. 4.4 and 4.5) occur in northeastern Texas on chalk and exposed rocky limestone formations (all Late Cretaceous). There are two variations of blackland prairie–associated limestone and chalk glades. One contains limestone pools on terraces just above floodplains of creeks and intermittent drainages, and the other has exposed chalk limestone outcrops on higher points of rolling hills or slopes just below hilltops.

Blackland Prairie Limestone Glades

Blackland Prairie Limestone Glades (Singhurst 2010) accommodate seasonally wet limestone glades of the Blackland Prairie ecoregion of Texas. These glades are typically on the upper rims of gradual to moderately steep slopes, in areas exceeding 1 ha (2.5 ac), and are interdispersed among tallgrass prairie dominated by little bluestem (*Schizachyrium scoparium*), Indiangrass (*Sorghastrum nutans*), and little-tooth sedge (*Carex microdonta*). The lower slopes of the prairie typically grade into woodland dominated by chinkapin oak (*Quercus muehlenbergii*), black oak (*Quercus velutina*), slippery elm (*Ulmus rubra*), eastern redcedar (*Juniperus virginiana*), Ashe's juniper (*Juniperus ashei*), white ash (*Fraxinus americana*), eastern redbud (*Cercis canadensis* var. *canadensis*), Carolina buckthorn (*Frangula caroliniana*), roughleaf dogwood (*Cornus drummondii*), skunkbush sumac (*Rhus trilobata*), and Eve's necklace (*Styphnolobium affine*). Ground and limestone pool moisture is prevalent during late winter. This glade vegetation has developed on the Gober Chalk Formation in Fannin and Lamar Counties, Texas, and on the Lavender Formation in Limestone County, and it characteristically occupies depressions in the limestone that hold water in the winter and early spring. These areas become desiccated and baked in the heat of summer. Characteristic plants on outcrops include the annuals pink stonecrop (*Sedum pulchellum*) and Lady Bird's centaury (*Centaurium texense*) and the perennials Drummond's onion (*Allium drummondii*) and sunbright (*Phemeranthus parviflorus*). The blue-green alga *Nostoc commune* forms mats. Wet pool specialists include small-leaf pygmyweed (*Crassula drummondii*), lowland rotala (*Rotala ramosior*), and hairy cornsalad (*Valerianella amarella*). Characteristic taxa that occupy the pools when they dry out include poverty dropseed (*Sporobolus vaginiflorus*), Dixie stitchwort (*Minuartia muscorum*), southern bluet (*Houstonia micrantha*), purslane (*Portulaca* sp.), southwestern bedstraw (*Galium virgatum*), prairie tea (*Croton monanthogynus*), and eyebane (*Chamaesyce nutans*). Other taxa that are commonly found in crevices of limestone sheets and drier edges include sideoats grama (*Bouteloua curtipendula*), Texas grama (*Bouteloua rigidiseta*), yellow nailwort (*Paronychia virginica*), spreading bladderpod (*Lesquerella gracilis*), fineleaf four-nerved daisy (*Tetraneuris linearifolia*), spider milkweed (*Asclepias asperula*), pasture heliotrope (*Heliotropium tenellum*), prairie bluets (*Stenaria nigricans*), Reverchon's false pennyroyal (*Hedeoma reverchonii*), wedgeleaf draba (*Draba cuneifolia*), tenpetal thimbleweed (*Anemone berlandieri*), pricklypear cactus (*Opuntia* sp.), narrowleaf stoneseed (*Lithospermum incisum*), stiff greenthread (*Thelesperma filifolium*), death camas (*Zigadenus nuttallii*), broomrape (*Orobanche* sp.), foxglove (*Penstemon cobaea*), Great Plains ladies' tresses (*Spiranthes magnicamporum*), and spring pygmycudweed (*Evax verna*).

The Blackland Prairie Limestone Glades community is restricted to the Blackland Prairie ecoregion of Texas. This community has fewer than 10 known occurrences, spread over a relatively narrow geographic area. It is restricted to areas with at least some seepage hydrology. The only publicly protected site is at Fort Parker State Park in Limestone County. A few examples are known from Lamar County, and two in Fannin County. There is a historical occurrence of this glade type in Grayson County, though no recent effort to relocate the site has been made. Its overall coverage in the landscape is limited, and it is threatened by development and land-use conversion. Threats include destruction or degradation by recreational off-road vehicle traffic, gravel and mineral surface mining, and land use related to crop production.

These prairies and glades were surveyed in 1999 by Jason Singhurst and Walter Holmes (plant taxonomist, Baylor University, Waco, Texas). A recent survey conducted by the authors in 2012 documented a few additional glade sites in Fannin and Lamar Counties. Most of the survey work was conducted in spring and early summer, and more examination of the fall flora is needed.

Periodic fire is required to maintain the herbaceous character of the vegetation. In the absence of fire, certain shrubs become much more abundant, especially eastern redcedar (*Juniperus virginiana*). Some sites are subject to invasion by non-native grasses such as Johnsongrass (*Sorghum halepense*), Bermudagrass (*Cynodon dactylon*), and King Ranch bluestem (*Bothriochloa ischaemum*). Other exotic invaders include the woody plants Chinese privet (*Ligustrum sinense*) and Japanese honeysuckle (*Lonicera japonica*).

Blackland Prairie Chalk Glades

The Blackland Prairie Chalk Glades vegetation has developed on the Eagle Ford Formation, Annona Chalk, Austin Chalk, and Pecan Gap Chalk Formations. The following two descriptions are of slight variations of chalk glades found on two different formations. The Annona Chalk Glades surround the community of White Rock northeast of Clarksville in Red River County, Texas. These sites were surveyed in 1999 by Jason Singhurst and Walter Holmes and were revisited in 2005 by Jason Singhurst and Andrew Salywon (Desert Botanical Garden, Phoenix, Arizona). There are no known conservation sites for Annona Chalk Glades.

Annona Chalk Glades are dominated by little bluestem (*Schizachyrium scoparium*), Missouri coneflower (*Rudbeckia missouriensis*), prairie bluets (*Stenaria nigricans*), cusp blazing star (*Liatris mucronata*), and yellow nailwort (*Paronychia virginica*). Characteristic flora includes sideoats grama (*Bouteloua curtipendula*), Mead's sedge (*Carex meadii*), threadleaf bladderpod (*Lesquerella angustifolia*), Topeka purple coneflower (*Echinacea atrorubens*), purple prairie clover (*Dalea purpurea*), white prairie clover (*D. candida*), foxglove (*Penstemon cobaea*), puffballs (*Marshallia caespitosa*), pasture heliotrope (*Heliotropium tenellum*), narrowleaf gumweed (*Grindelia lanceolata*), compassplant (*Silphium laciniatum*), Reverchon's false pennyroyal (*Hedeoma reverchonii*), trailing ratany (*Krameria lanceolata*), spider milkweed (*Asclepias asperula*), green antelopehorn (*A. viridis*), whorled milkweed (*A. verticillata*), azure blue sage (*Salvia azurea*), Carolina larkspur (*Delphinium carolinianum* ssp. *vimineum*), death camas (*Zigadenus nuttallii*), Illinois bundleflower (*Desmanthus illinoensis*), stiff greenthread (*Thelesperma filifolium*), and Lady Bird's centaury (*Centaurium texense*). Seasonal seepage occurs at the base of chalk outcrops and provides microhabitats for limestone spikesedge (*Eleocharis occulta*), Drummond's onion (*Allium drummondii*), and rushes (*Juncus* spp.).

Austin Chalk Glades in northeastern Texas are distributed from Collin County northeast through Hunt, Fannin, Delta, Lamar, and Red River Counties. The following description is based on a two-year detailed plant inventory of Chinquapin Preserve in Collin County by Jason Singhurst and Walter Holmes. Chinquapin Preserve is owned by David Kusin, who has opened this private preserve to a local Montessori school for more than a decade as an outdoor science learning center. Chinquapin Preserve (44 ha or 110 ac) contains high-quality examples of Blackland Prairie and Austin Chalk Glades. The dominant chalk glade flora includes sideoats grama (*Bouteloua curtipendula*), Texas grama (*B. rigidiseta*), hairy grama (*B. hirsuta*), Texas cupgrass (*Eriochloa sericea*), composite dropseed (*Sporobolus compositus*), littletooth sedge (*Carex microdonta*), and silver bluestem (*Bothriochloa laguroides*). Characteristic flora includes little bluestem (*Schizachyrium scoparium*), Indiangrass (*Sorghastrum nutans*), black-eyed Susan (*Rudbeckia hirta*), blacksamson echinacea (*Echinacea angustifolia*), cusp blazing star (*Liatris mucronata*), blue wild indigo (*Baptisia australis*), prairie bluets (*Stenaria nigricans*), pasture heliotrope (*Heliotropium tenellum*), American basketflower (*Centaurea americana*), false boneset (*Brickellia eupatorioides*), Arkansas yucca (*Yucca arkansana*), Leavenworth's eryngo (*Eryngium leavenworthii*), groovestem Indian plantain (*Arnoglossum plantagineum*), prairie bishop (*Bifora americana*), puffballs (*Marshallia caespitosa*), purple prairie clover (*Dalea purpurea*), white prairie clover (*D. candida*), and foxglove (*Penstemon cobaea*). Other associated flora includes Illinois bundleflower (*Desmanthus illinoensis*), Engelmann's daisy (*Engelmannia peristenia*), bigfruit evening primrose (*Oenothera macrocarpa*), stiff greenthread (*Thelesperma filifolium*), stemmy four-nerve daisy (*Tetraneuris scaposa*), narrowleaf gumweed (*Grindelia lanceolata*), compassplant (*Silphium laciniatum*), Reverchon's false pennyroyal (*Hedeoma reverchonii*), Carolina larkspur (*Delphinium carolinianum* ssp. *vimineum*), spreading bladderpod (*Lesquerella gracilis*), standing cypress (*Ipomopsis rubra*), small palafox (*Palafoxia callosa*), azure blue sage (*Salvia azurea*), blue mealy sage (*S. farinacea*), spider milkweed (*Asclepias asperula*), green antelopehorn (*A. viridis*), green comet milkweed (*A. viridiflora*), whorled milkweed (*A. verticillata*), purple poppymallow (*Callirhoe involucrata*), light poppymallow (*C. alcaeoides*), lemon beebalm (*Monarda citriodora*), annual buckwheat (*Eriogonum annuum*), trailing ratany (*Krameria lanceolata*), Jersey tea (*Ceanothus herbaceus*), Texas toothleaf (*Stillingia texana*), tall silverbush (*Argythamnia mercurialina*), Maximilian sunflower (*Helianthus maximiliani*), shiny goldenrod (*Oligoneuron nitidum*), wedgeleaf draba (*Draba cuneifolia*), broadpod draba (*D. platycarpa*), Berlandier's sundrops (*Calylophus berlandieri*), slimflower scurfpea (*Psoralidium tenuiflorum*), Carolina woollywhite (*Hymenopappus scabiosaeus*), Texas bluestar (*Amsonia ciliata* var. *texana*), prairie brazosmint (*Warnockia scutellarioides*), and Lindheimer ironweed (*Vernonia lindheimeri*).

A few rare and Texas endemic plants show up in northeastern Texas glades. These include threadleaf bladderpod (*Lesquerella angustifolia*), which is globally restricted to Annona Chalk Glades in Red River County, Texas, and to Choctaw and McCurtain Counties, Oklahoma (OVPD 2013). The rare summer gayfeather (*Liatris aestivalis*) (O'Kennon and Nesom 2001) is restricted to central and north-central Texas and south-central Oklahoma and appears in a few chalk glades in Collin and Fannin Counties, Texas. Hall's prairie clover (*Dalea hallii*) is a Texas endemic that is periodically encountered in northeastern Texas chalk glades and also occurs in grasslands on eroded limestone or chalk and in oak scrub on rocky hillsides in north-central Texas, south to a few spots on the Edwards Plateau (W. Carr 2009). Lindheimer ironweed (*Vernonia lindheimeri*) is a near endemic to central and north-central Texas and a rare peripheral in adjacent Coahuila, Mexico. Lindheimer ironweed appears from time to time in Austin Chalk Glades. Purple locoweed (*Oxytropis lambertii*) is an outlier from the Rolling Plains ecoregion that drifts into chalk glades in northeastern Texas.

Relatively high plant diversity has been documented in several chalk glades in northeastern Texas. Recent inventory collections of the flora of Chinquapin Preserve have documented 277 plant species. Two significant chalk prairies are publicly accessible, including Oak Point Park Prairie, owned by the city of Plano in Collin County, Texas (124 species in a single fall flora survey); and Bonham State Park in Fannin County, with 212 species.

Native plant invaders include mesquite (*Prosopis glandulosa*), eastern redcedar (*Juniperus virginiana*), and honeylocust (*Gleditsia triacanthos*). Several highly aggressive invasive plants are problematic in chalk glades in northern and northeastern Texas, including King Ranch bluestem (*Bothriochloa ischaemum*), rescuegrass (*Bromus catharticus*), Johnsongrass (*Sorghum halepense*), mourningbride (*Scabiosa atropurpurea*), and St. John's-wort (*Hypericum perforatum*).

Weches Prairies and Glades

Weches Prairies and Glades (figs. 4.6, 4.7, 4.8, and 4.9) occur within the Piney Woods natural region of easternmost Texas. This community occurs on outcrops of marine sediment and glauconitic clays of the Weches Formation in a localized area of eastern Texas (San Augustine, Nacogdoches, and Sabine Counties). Soils are mapped as Trawick Series (Mollic Hapludalfs) and are shallow, rocky, and basic, which inhibits growth of woody vegetation. In the outcrops, soils are seepy and saturated during winter and early spring but become hard and dry in the summer. Large seasonal variations in species dominance can occur. Prairie vegetation is usually dominated by sideoats grama (*Bouteloua curtipendula*), little bluestem (*Schizachyrium scoparium*), cusp blazing star (*Liatris mucronata*), littletooth sedge (*Carex microdonta*), and compact prairie clover (*Dalea compacta*). Other characteristic Weches Prairie species include Cherokee sedge (*Carex cherokeensis*), Texas wintergrass (*Nassella leucotricha*), groovestem Indian plantain (*Arnoglossum plantagineum*), standing cypress (*Ipomopsis rubra*), Texas prairie parsley (*Polytaenia texana*), soft-hair marbleseed (*Onosmodium bejariense*), prairie bluets (*Stenaria nigricans*), roundstem false foxglove (*Agalinis gattingeri*), whorled milkweed (*Asclepias verticillata*), Berlandier's sundrops (*Calylophus berlandieri*), American basketflower (*Centaurea americana*), clasping coneflower (*Dracopis amplexicaulis*), scarlet milkpea (*Indigofera miniata*), stiff greenthread (*Thelesperma filifolium*), pasture heliotrope (*Heliotropium tenellum*), Carolina larkspur (*Delphinium carolinianum* subsp. *vimineum*), basil beebalm (*Monarda clinopodioides*), smooth phacelia (*Phacelia glabra*), white four o'clock (*Mirabilis albida*), and Baldwin's ironweed (*Vernonia baldwinii*).

The federally endangered white bladderpod (*Physaria pallida*) (USFWS 1992) is restricted to Weches Prairies and is often found on edges of freshly disturbed gopher mounds or sparsely covered transitional areas between prairies and glades that maintain very little plant cover. White bladderpod is a globally rare Texas endemic (San Augustine County) and is currently known from only seven remnants.

Dominant species of glade vegetation include pink stonecrop (*Sedum pulchellum*), Arkansas mint (*Clinopodium arkansanum*), and poverty dropseed (*Sporobolus vaginiflorus*) (figs. 4.7 and 4.8). Glades consist of vernal pools over flat rock, rocky outcrops with fine sediments with little or no pooling, and rocky slopes (fig. 4.9). The vernal pools are occupied by highly restricted flora that includes the federally endangered Texas golden gladecress (*Leavenworthia texana*), a Texas endemic restricted to Sabine and San Augustine Counties. Other characteristic vernal pool species include pink stonecrop, limestone quillwort (*Isoetes butleri*) (Holmes at al. 2005), waterthread pondweed (*Potamogeton diversifolius*), twoheaded water-starwort (*Callitriche heterophylla*), low spearwort (*Ranunculus pusillus*), limestone spikesedge (*Eleocharis occulta*), and rushes (*Juncus* spp.). Other characteristic species of outcrops and rocky slopes include sedges (*Carex* spp.), smallflowered milkvetch (*Astragalus nuttallianus* var. *nuttallianus*), corn salad (*Valerianella radiata*), southwestern bedstraw (*Galium virgatum*), pitcher's stitchwort (*Minuartia patula*), Drummond's onion (*Allium drummondii*), Carolina anemone (*Anemone caroliniana*), prairie tea (*Croton monanthogynus*), eyebane (*Chamaesyce nutans* [= *Euphorbia nutans*]), Virginia nailwort (*Paronychia virginica*), wild hyacinth (*Camassia scilloides*), death camas (*Zigadenus nuttallii*), standing cypress (*Ipomopsis rubra*), rosy palafox (*Palafoxia rosea* var. *rosea*), lax hornpod (*Mitreola petiolata*), western daisy (*Astranthium integrifolium*), rough false pennyroyal (*Hedeoma hispida*), Engelmann's adderstongue fern (*Ophioglossum engelmannii*), sandyseed clammyweed (*Polanisia dodecandra* subsp. *trachysperma*), wedgeleaf draba (*Draba cuneifolia*), pricklypear cactus (*Opuntia* spp.), and prairie wedgescale (*Sphenopholis obtusata*). A scattered shrub layer may be present on some sites, including eastern redbud (*Cercis canadensis*), roughleaf dogwood (*Cornus drummondii*), eastern persimmon (*Diospyros virginiana*), eastern redcedar (*Juniperus virginiana*), gum bumelia (*Sideroxylon lanuginosum*), Carolina buckthorn (*Frangula caroliniana*), Eve's necklace (*Styphnolobium affine*), and chinquapin oak (*Quercus muehlenbergii*).

The federally endangered white bladderpod (*Physaria pallida*) (USFWS 1992) and Texas golden gladecress (*Leavenworthia texana*) (USFWS 2012) are found in small-scale plant communities tied to "geologic and hydrologic conditions that are themselves rather rare on the landscape" (Poole et al. 2007). The white bladderpod and Texas golden gladecress are endemic to glade habitats in northern San Augustine and northwestern Sabine Counties and are habitat specialists, occurring only on outcrops of the Weches Formation (Mahler 1987; George and Nixon 1990; Poole et al. 2007). The gladecress grows only in glades on shallow, calcium-rich soils that are wet in winter and spring. These occur on ironstone (glauconite or greenstone) outcrops (Poole et al. 2007). All species within the small genus *Leavenworthia* are adapted to glade habitats that have unique physical characteristics, the most important being a combination of shallow soil and high calcium content (dolomitic limestone or otherwise calcareous soils), where the soil layers have been deposited in such a manner that they maintain temporarily high moisture content at or very near the surface. Typically, only a few inches of soil overlie the bedrock, but in spots, the soil may be almost lacking and the surface barren. The glade habitats that support all *Leavenworthia* species are extremely wet during the late winter and early spring and then dry to the point of being parched in summer (Rollins 1963). These glades can vary in size from as small as a few square meters to larger than 0.37 km^2 (0.14 mi^2) and are characterized as having an open, sunny aspect (lacking canopy) (Quarterman 1950; Rollins 1963). The landscape position of the glades may also play a role in ensuring the cyclic moisture regime required by glade vegetation communities. The Weches Formation consists of bands of ancient marine sediments deposited in a line running roughly parallel to the Gulf of Mexico, from Sabine to Frio Counties. A layer of glauconite clay is either exposed at the surface or covered by a thin layer of calcareous sediment measuring as deep as 50 cm (20 in) (George and Nixon 1990). Glauconite is a characteristic mineral of marine depositional environments, appearing greenish when initially exposed to the atmosphere and later turning red (R. Davis 1966; Nemec 1996). The glauconite is very friable (crumbly) and has low resistance to weathering (Geocaching.com 2010). The soils overlying the clay layer are typically rocky and shallow (George 1987), and all Texas golden gladecress sites are classified within the Nacogdoches, Trawick, or Bub soil series (USDA 2009). Weches outcrops occur in a band averaging 8 km (5 mi) in width that parallels Texas State Highway 21 through northern San Augustine and northwestern Sabine Counties (Sellards et al. 1932, in Diggs et al. 2006). This band has been deeply dissected by erosion, which has created islands of thin, loamy, alkaline soils (pH 7–8) within the normally deep, sandy, acidic soils (pH 4–5) of the Piney Woods region. The glauconite layer of the Weches Formation is fairly impermeable to water, producing saturated, thin upper soils from late fall through spring, which then dry out and harden during summer months (George 1987; Bezanson 2000, in Diggs et al. 2006). Downslope seepage across the Weches terraces may also be important in maintaining the hydrology required by the gladecress. The cyclic moisture regime and the alkalinity of the soils produce conditions unique to the Weches outcrops, allowing certain plants to evolve within these specialized geologic formations (Mahler 1987; George and Nixon 1990).

Another noteworthy plant that occurs in Weches Prairies and Glades is clasping twistflower (*Streptanthus maculatus*), which has a narrow global range of southeastern Oklahoma, southwestern Arkansas, and northeastern Texas. Limestone spikesedge (*Eleocharis occulta*) (S. Smith 2001) is a recently described species restricted to glades in Oklahoma and Texas. Round-stem false foxglove (*Agalinis gattingeri*) is a rare peripheral restricted to a few prairie sites in eastern Texas.

All known sites of Weches Prairies and Glades occur on private land, and several sites have been extirpated. There are no protected examples, although at least one site is under a temporary management agreement established by the US Fish and Wildlife Service. Current threats include open-pit

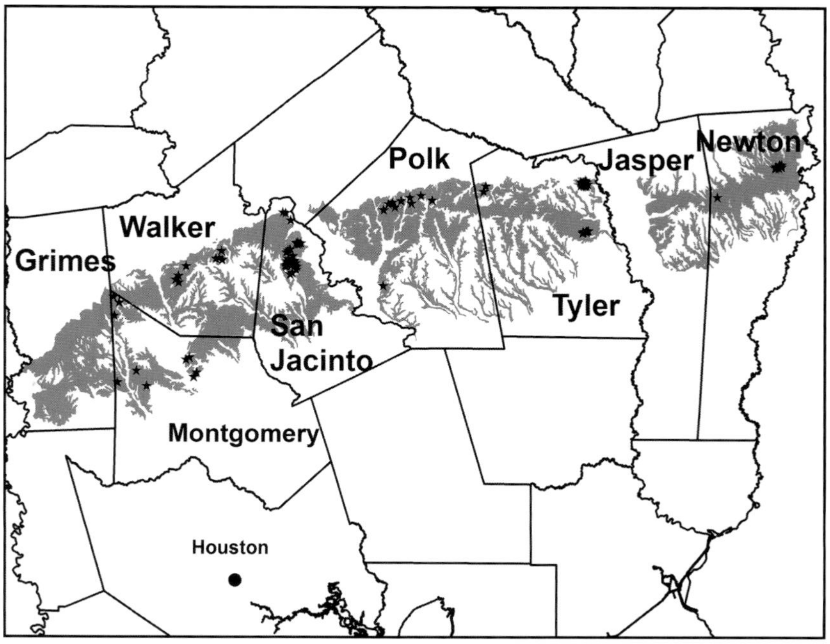

Figure 4.10. Fleming Prairies are restricted to the Fleming Formation. Fleming Prairie locations are labeled with a star.

mining, agricultural land conversion, overgrazing, road construction and maintenance, residential development, herbicide use, and invasive shrub and grass species. Invasive species (exotics and perennial grassy and native woody increasers) include Macartney rose (*Rosa bracteata*), Chinese privet (*Ligustrum sinense*), Japanese honeysuckle (*Lonicera japonica*), Japanese brome (*Bromus japonicus*), Bermudagrass (*Cynodon dactylon*), Japanese clover (*Kummerowia striata*), annual yellow sweetclover (*Melilotus indicus*), suckling clover (*Trifolium dubium*), and others. Heavy encroachment of these species is believed to suppress important herbaceous components of Weches Prairies and Glades, especially white bladderpod and Texas golden gladecress.

Despite the small scale of remaining Weches Prairies (0.04 to 4.0 ha or 0.1 to 10 ac), they still maintain high floral diversity. Floristic documentation has been done for three Weches Prairies in San Augustine County, including Williams Prairie (3.2 ha or 8 ac, 106 species), Blount Prairie (2.8 ha or 7 ac, 124 species), and Hancock Prairie (4 ha or 10 ac, 114 species).

Fleming Prairies

Fleming Prairies (figs. 4.10 and 4.11) are a globally rare prairie type restricted to southwestern Louisiana and southeastern Texas. These calcareous prairies occur in isolated patches and follow a narrow sandstone formation called the Fleming Formation (fig. 4.10), which starts just west of Huntsville, Texas, and follows an arc just south of Livingston through Jasper to just below Toledo Bend Reservoir near Burkeville, Texas, on the Texas and Louisiana border. The soils are deep clays with local sandstone outcroppings, and these prairies are often located on high uplands and at the heads of creeks. They are often dissected by naturally eroded gullies and calcareous ("calciphile") hardwood forests. In general, the flora of these prairies is strongly differentiated from almost all of the mostly sandy, acidic soil communities of southeastern Texas. Very few conservation biologists or naturalists have encountered this rare prairie type because there is limited access; most sites are on private timber company lands.

Fleming Prairies are dominated by little bluestem (*Schizachyrium scoparium*), Missouri coneflower (*Rudbeckia missouriensis*), prairie clovers (*Dalea compacta* and *D. candida*), and cusp blazing star (*Liatris mucronata*). This prairie system is composed of graminoids, composites, and legumes. Characteristic flora varies from one Fleming Prairie to another but usually includes littletooth sedge (*Carex microdonta*), Cherokee sedge (*C. cherokeensis*), Topeka coneflower (*Echinacea atrorubens*), sanguine purple coneflower (*E. pallida*), sideoats grama (*Bouteloua curtipendula*), Texas bluebell (*Eustoma exaltatum*), foxglove (*Penstemon cobaea*), Maximilian sunflower (*Helianthus maximiliani*), asters (*Symphyotrichum* spp.), diamondflowers (*Stenaria nigricans*), soft-hair marbleseed (*Onosmodium bejariense* var. *his-*

Figure 4.11. Fleming Prairie at Russell Grasslands and Forest Preserve in San Jacinto County, Texas.

pidissimum), Indian plantain (*Arnoglossum plantagineum*), narrowleaf gumweed (*Grindelia lanceolata*), prairie acacia (*Acacia angustissima*), purple false foxglove (*Agalinis heterophylla*), rosy palafox (*Palafoxia rosea*), scarlet pea (*Indigofera miniata*), Engelmann daisy (*Engelmannia peristenia*), prairie parsley (*Polytaenia nuttallii*), rattlesnake master (*Eryngium yuccifolium*), false aloe (*Manfreda virginica*), milkweeds (*Asclepias asperula, A. obovata, A. tuberosa, A. viridiflora, A. viridis*), azure blue sage (*Salvia azurea*), basketflower (*Centaurea americana*), snow on the mountain (*Euphorbia bicolor*), plains poppymallow (*Callirhoe alcaeoides*), Kansas gayfeather (*Liatris pycnostachya*), and Appalachian blazing star (*L. squarrulosa*). High-quality examples also include big bluestem (*Andropogon gerardii*) and Indiangrass (*Sorghastrum nutans*).

Some Fleming Prairies include small Catahoula sandstone barrens with rayless goldenrod (*Bigelowia nuttallii*), bluejoint panicgrass (*Panicum tenerum*), Silveus's dropseed (*Sporobolus silveanus*), thin bluestem (*Schizachyrium tenerum*), trailing ratany (*Krameria lanceolata*), yellow stonecrop (*Sedum nuttallianum*), sunbright (*Phemeranthus parviflorus*), and purslane (*Portulaca* spp.). At the privately owned Russell Grasslands and Forest Preserve in San Jacinto County, the Texas endemic bracted gayfeather (*Liatris cymosa*) occurs on a Catahoula barren inclusion within the Fleming Prairie.

Naturally erodible gullies are usually sparsely vegetated and are dominated by sideoats grama (*Bouteloua curtipendula*), Texas grama (*B. rigidiseta*), hairy grama (*B. hirsuta*), and foxglove (*Penstemon cobaea*).

Calcareous woodland drainages are often natural partitions between Fleming Prairie openings. These woodlands are typically narrow (≤ 50 m or 164 ft wide) at the head of drainages and broaden downslope below prairie openings where soils typically change fairly quickly to acidic sand or sandy loams. These woodlands are dominated by an overstory of oaks (*Quercus shumardii*, *Q. muehlenbergii*, and *Q. velutina*), elms (*Ulmus americana*, *U. rubra*, and *U. crassifolia*), sugarberry (*Celtis laevigata*), and gum bumelia (*Sideroxylon lanuginosum*). Subcanopy woody plants include eastern redbud (*Cercis canadensis* var. *canadensis*), roughleaf dogwood (*Cornus drummondii*), lanceleaf buckthorn (*Rhamnus lanceolata* ssp. *glabrata*), Eve's necklacepod (*Styphnolobium affine*), upland swampprivet (*Forestiera ligustrina*), and rusty blackhaw (*Viburnum rufidulum*). The herbaceous layer is dominated by Cherokee sedge (*Carex cherokeensis*), longleaf woodoats (*Chasmanthium sessiliflorum*), inland woodoats (*C. latifolium*), Canada wildrye (*Elymus canadensis*), hairy brome (*Bromus pubescens*), common blue violet (*Viola sororia*), gravelweed (*Verbesina helianthoides*), golden zizia (*Zizia aurea*), Canadian blacksnakeroot (*Sanicula canadensis*), white avens (*Geum canadense*), harvestlice (*Agrimonia parviflora*), purple meadow-rue (*Thalictrum dasycarpum*), green dragon (*Arisaema dracontium*), and meadow garlic (*Allium canadense*). Hairyjoint meadowparsnip (*Thaspium barbinode*), which is disjunct in Texas and rare in the state, has been documented in two woodlands that bisect Fleming Prairies.

Missouri coneflower (*Rudbeckia missouriensis*) is an important component of Fleming Prairies. It is primarily a midwestern species that occurs in Illinois, Missouri, Arkansas, and Oklahoma, but it reaches its southern limits in the Fleming Prairies of southeastern Texas and southwestern Louisiana. Missouri coneflowers are often abundant in early summer (June–July), and observing them at peak flowering in a Fleming Prairie is a breathtaking experience that all plant enthusiasts would enjoy.

Populations of Texas ladies' tresses (*Spiranthes brevilabris* var. *brevilabris*), a globally rare orchid (G1G2S1) in the southeastern United States, have been located in a few Fleming Prairies in Texas (Keith 2007). This is the only spring-flowering (March–April) ladies' tresses orchid that maintains an ovate leaf that lies prostrate on the ground during flowering. Berlandier's groundplum milkvetch (*Astragalus crassicarpus* var. *berlandieri*) is an endemic legume found on eroded gully soils at one Fleming Prairie site in Walker County (*Carr* 12582, TEX) prairies. A disjunct population of lotus milkvetch (*Astragalus lotiflorus*) has been documented at its very southeastern limit at several Fleming Prairies in San Jacinto County. Topeka coneflower (*Echinacea atrorubens*)—restricted to tallgrass prairies in Kansas, Oklahoma, and Texas—and the amazingly showy false gaura (*Stenosiphon linifolius*)—a perennial member of the primrose family that can reach 2 m (6.6 ft) tall—both reach their southernmost range limits in Fleming Prairies. Blue curls (*Phacelia congesta*) occurs along the edges of Fleming Prairie calcareous woodlands at the easternmost extension of its range in Polk and Walker Counties.

Larry Brown (Spring Branch Science Center, Houston) and Eric Keith (Raven Environmental Services, Huntsville, Texas) have conducted several floristic surveys of Fleming Prairies (L. Brown et al. 2002; Keith 2007). Brown studied the flora of Windham Prairie (22 ha or 54.4 ac) in Polk County and recorded 242 plant species and range extensions of several plants. Keith rediscovered Texas ladies' tresses in Texas. The Campbell Group, Hancock Forest Management Group, and Weyerhaeuser Company own and manage high-quality examples of Fleming Prairies in southeastern Texas. These entities have allowed the Texas Parks and Wildlife Department (TPWD) to map these prairies, inventory the flora, and suggest management recommendations to restore the prairies that have been invaded by shrubs and trees, such as eastern redcedar (*Juniperus virginiana*), honeylocust (*Gleditsia triacanthos*), loblolly pine (*Pinus taeda*), and hawthorns (*Crataegus* spp.). All three of these private timber companies have proactively worked toward conserving some portion of their prairies. The largest prairie complex, the Chester Prairie (688 ha or 1,700 ac) in Tyler County, is owned by the Campbell Group and is considered a conservation area. It has recently undergone a prescribed burn (winter 2009) in order to restore the prairie flora. Hancock Forest Management has set aside the Colmesneil Prairie in Tyler County (approximately 40.5 ha or 100 ac) as a distinctive site within its landholdings. The Oakhurst Prairie (82.2 ha or 203 ac) in San Jacinto County is owned by Weyerhaeuser Company and is being cooperatively managed with prescribed fire with the TPWD through funding provided by the Landowner Incentive Program (LIP). The LIP is a state-funded program that is focused on projects aimed at creating, restoring, protecting, and enhancing habitat for rare or at-risk species throughout Texas.

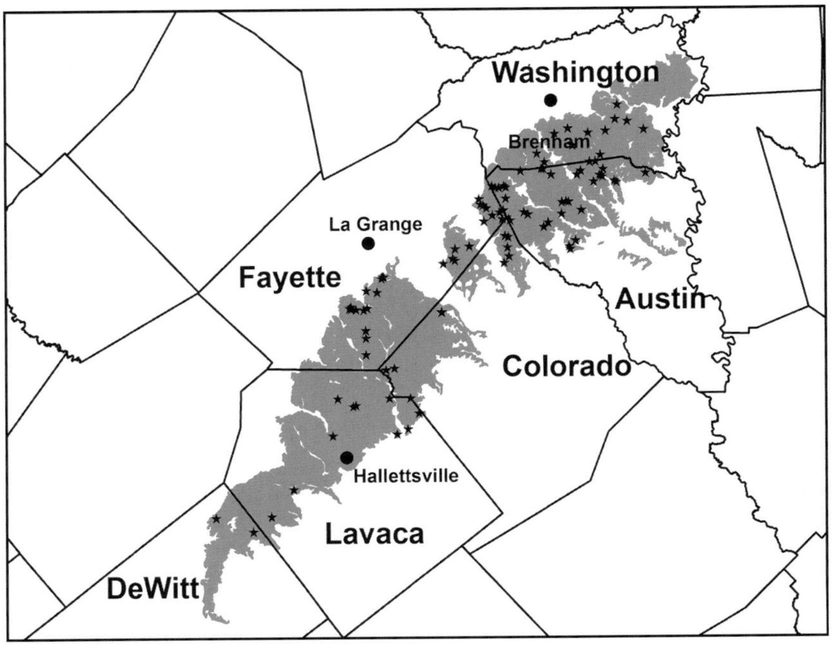

Figure 4.12. Fayette Prairies are restricted to the Fleming and Oakville Formations. Fayette Prairie locations are labeled with a star.

Local efforts by TPWD wildlife biologists Dan Jones and Chris Gregory have included working with the national forests and grasslands of Texas to conduct floristic inventories, restore Fleming Prairie sites on the Sam Houston National Forest (SHNF), and document postrestoration responses. Jones and Gregory have been instrumental in salvaging Fleming Prairie seed and sod from development projects in order to enhance degraded prairie sites in the SHNF. They have recorded 316 of the 386 herbaceous plants documented in Fleming Prairies throughout the SHNF (Montgomery and Walker Counties). The TPWD plans to continue relationships with the private timber corporations in eastern Texas in order to conserve the remaining Fleming Prairies. The agency is willing to assist with botanical surveys and provide management suggestions to any private landowner whose land contains a Fleming Prairie.

The Native Prairies Association of Texas (NPAT) holds a conservation easement on the Russell Grasslands and Forest Preserve in San Jacinto County (122.2 ha or 302 ac), and arrangements can be made through NPAT and the landowner to view this high-quality Fleming Prairie with the globally rare Texas ladies' tresses. The Natural Resources Conservation Service (NRCS) holds a conservation easement on Lone Oak Ranch Prairie (40.5 ha or 100 ac) in Walker County, which also contains the Fleming Prairie type.

Fayette Prairies

Fayette Prairies (figs. 4.12 and 4.13) are globally rare, tallgrass prairie communities that occur on vertisols (Diamond 1980; Smeins and Diamond 1983). The landforms are level to rolling plains, with typically fine-textured, clayey soils that possess a high shrink and swell potential. Fayette Prairies are an island of the blackland prairie (fig. 4.12) east and southeast of the Blackland Prairie ecoregion that follows a southwest to northeast line. Fayette Prairies begin in Lavaca County near the town of Hallettsville, Texas, head northeast through Schulenburg, Fayetteville, and Brenham, and then end at Hempstead, Texas, in Waller County. Fayette Prairies, unlike the contiguous Blackland Prairie ecoregion, are surrounded by the Post Oak Savanna ecoregion to the north, the Coastal Prairie to the south and southeast, the Piney Woods ecoregion to the east, and South Texas Plains to the southwest (Smeins and Diamond 1983). The Fayette Prairies also include a finger of the Edwards Plateau near the town of Schulenburg. Similar to Fleming Prairies, Fayette Prairies contain infrequent inclusions of Catahoula sandstone barrens and outcrops, Oakville sandstone, and local gravel deposits. Consequently, Fayette Prairies have floristic influences from these geological and associated soil inclusions and from adjacent ecoregions. Fayette Prairies historically contained 1.5 million ha (3.7 million ac) of native grassland (Godfrey et al. 1970).

Figure 4.13. Fayette Prairie in Austin County, Texas.

Fayette Prairies are dominated by little bluestem (*Schizachyrium scoparium*), Indiangrass (*Sorghastrum nutans*), big bluestem (*Andropogon gerardii*), and prairie bishop (*Bifora americana*). Secondary grass species include sideoats grama (*Bouteloua curtipendula*), Texas wintergrass (*Nassella leucotricha*), Canada wildrye (*Elymus canadensis*), composite dropseed (*Sporobolus compositus*), and Carolina jointtail grass (*Coelorachis cylindrica*). Associated forbs and legumes include Maximilian sunflower (*Helianthus maximiliani*), cusp blazing star (*Liatris mucronata*), roundhead prairie clover (*Dalea multiflora*), white prairie clover (*D. compacta*), narrow-leaved purple coneflower (*Echinacea angustifolia*), roughstem rosinweed (*Silphium radula*), prairie bluets (*Stenaria nigricans*), heath aster (*Symphyotrichum ericoides*), false boneset (*Brickellia eupatorioides)*, ratany (*Krameria lanceolata*), Drummond phlox (*Phlox pilosa*), light poppymallow (*Callirhoe alcaeoides*), prairie Indian paintbrush (*Castilleja purpurea*), azure blue sage (*Salvia azurea*), rattlesnake master (*Eryngium yuccifolium*), foxglove (*Penstemon cobaea*), green milkweed (*Asclepias viridis*), whorled milkweed (*A. verticillata*), soft-hair marbleseed (*Onosmodium bejariense*), showy bluebell (*Eustoma exaltatum* ssp. *russellianum*), groundplum milkvetch (*Astragalus crassicarpus*), Texas prairie parsley (*Polytaenia texana*), and Illinois bundleflower (*Desmanthus illinoensis*).

Texas chrysopsis (*Chrysopsis texana*) is found in high-quality Fayette Prairies. Texas chrysopsis is mostly restricted to Fayette Prairies, coastal prairies, and sandstone barrens. Its global range is in eastern Texas and the lower coast of Louisiana.

The Fayette Prairie region was settled primarily by German, Czech, and Dutch immigrants who first broke the prairie sod to farm but also maintained hay meadows in native grasses to supply cattle through the winter. This resulted in a multigenerational practice that has allowed many prairie remnants to persist in this region today. However, after more than a century of cultivation and decades of fire suppression, Fayette Prairie remnants are giving way to "improved pastures" and invasion by woody plants like eastern redcedar (*Juniperus virginiana*), huisache (*Acacia farnesiana*), yaupon holly (*Ilex vomitoria*), and, to a lesser extent, mesquite (*Prosopis glandulosa* var. *glandulosa*). These trees have invaded Fayette Prairies, creating thick brush, pushing out the native grasses, and reducing annual and perennial plant diversity. An additional important threat is the spread of non-native grasses, especially Old World bluestems. Fragmentation of the Fayette Prairie by urbanites seeking space and beautiful views in this rolling landscape is threatening some prairie remnants. Some new landowners have different values and less understanding of the precious resources of the Fayette Prairies. A few nontraditional landowners, however, are seeking assistance from wildlife biologists at the TPWD and NRCS regarding better management practices (such as time of haying, prescribed burning, and woody plant removal) to manage these prairies.

Monument Hill and Kreische Brewery State Historical Park in Fayette County maintains the only conserved Fleming Prairie. All other Fleming Prairies are privately owned.

Marlbrook Marl Prairies

Marlbrook Marl Prairies (fig. 4.14), which occur on the Marlbrook Marl Formation and are located on bluffs overlooking the North Sulphur River in Delta County, are some of the rarest in Texas. Recent fieldwork has revealed the presence of a small handful of eroded and somewhat degraded remnants. These tracts are easily recognized in summer by the presence of Missouri coneflower on ledges, glades, and tops of bluffs as well as rock outcrops (Yatskievych 2006a). Missouri coneflower was considered rare in eastern Texas by Correll and Johnston (1979) and was not listed for north-central Texas by Diggs et al. (1999).

These Marlbrook Marl Prairie remnants are found on eroded Ferris Clay, a deep clay that occurs on north-facing slopes and forms long irregular bands, situated atop an underlying geology of Marlbrook Marl and Pecan Gap Chalk (Shelby et al. 1998). A total of five remnants of varying quality were located along a transect approximately 8 km (5 mi) long, generally bordered on the east by State Highway 24 and on the west by Farm to Market 38. With the exception of the highly disturbed Pierce Cemetery, all are on private property.

All sites appear to be eroded, with thin to nonexistent topsoil. The only publicly accessible site is Pierce Cemetery, just west of SH 24 and south of the North Sulphur River. This small, disturbed prairie remnant contains gilgai topography and native grasses, including big bluestem (*Andropogon gerardii*), Indiangrass (*Sorghastrum nutans*), switchgrass (*Panicum virgatum*), little bluestem (*Schizachyrium scoparium*), Texas grama (*Bouteloua rigidiseta*), and the invasive King Ranch bluestem (*Bothriochloa ischaemum*). Forb associates include groovestem Indian plantain (*Arnoglossum plantagineum)*, Carolina woollywhite (*Hymenopappus scabiosaeus*), Carolina larkspur (*Delphinium carolinianum* ssp. *virescens*), Texas thistle (*Cirsium texanum*), fringeleaf wild petunia (*Ruellia humilis*), lemon beebalm (*Monarda citriodora*), Topeka purple coneflower (*Echinacea atrorubens*), false boneset (*Brickellia eupatorioides* var. *texana*), sensitive brier (*Mimosa* sp.), compassplant (*Silphium laciniatum*), whorled milkwort (*Polygala verticillata*), and pasture heliotrope (*Heliotropium tenellum*).

Directly across the road is adjacent grazed, eroded native prairie on private property. By mid to late June, it is a community dominated by forbs, including the following: an extensive population of pasture heliotrope (*Heliotropium tenellum*), prairie bluet (*Stenaria nigricans*), foxglove (*Penstemon cobaea*), purple prairie clover (*Dalea purpurea*), and largebracted plantain (*Plantago aristata*). Infrequent forbs include soft-hair marbleseed (*Onosmodium bejariense*), cusp blazing star (*Liatris mucronata*), and the recently described summer gayfeather (*Liatris aestivalis*). One mile west, on grassland at the edge of an agricultural field, there is also a dominant forb community along both sides of the county road and on private property. In both places, pasture heliotrope is particularly abundant.

Morse Clay Prairie

Morse Clay Prairies (fig. 4.15), sometimes referred to as Morse Clay Calcareous Prairies, are restricted to the Morse

Figure 4.14. Marlbrook Marl Prairie in Delta County, Texas.

Clay soil series in northwestern Louisiana and adjacent Arkansas and Texas (B. MacRoberts et al. 2011). Morse Clay Prairies are surrounded by oak-pine woodlands, and because they are isolated from one another, they are often called "isolated prairies" or "pocket prairies." Morse Clay Prairies are known from Bossier, Caddo, Red River, Rapides, and Natchitoches Parishes in Louisiana and from Lafayette, Columbia, and Little River Counties in Arkansas (Witsell, pers. comm.). In Louisiana, Morse Clay Prairies occur on Pleistocene Red River terraces with reddish, fine-textured, alkaline, calcareous clays and silt loam shrink-swell soils. Fire frequency is different from that of surrounding woodlands; unlike many other prairie landscapes, these prairies maintain openings in the oak-pine woodlands by soil-dependent variations in grass biomass, resulting in differences in fire intensity and extreme soil properties (Diggs and Schulze 2003; Diggs et al. 2006). Morse Clay Prairies have gilgai microtopography and a combination of periodic fire, high soil pH, and extreme physical soil properties that are thought to be important in maintaining this community in a landscape otherwise dominated by forests (at least historically).

Morse Clay Prairies in Texas are dominated by little bluestem (*Schizachyrium scoparium*), Mead's sedge (*Carex meadii*), cusp blazing star (*Liatris mucronata*), and purple prairie clover (*Dalea purpurea*), with scattered shrubby vegetation and occasional mottes of oaks or oak-pine woodlands. Characteristic plant species include scaly blazing star (*Liatris squarrosa*), groovestem Indian plantain (*Arnoglossum plantagineum*), Nuttall's wild indigo (*Baptisia nuttal-*

Figure 4.15. Morse Clay Prairie in Red River County, Texas.

liana), prairie bluets (*Stenaria nigricans*), Barbara's buttons (*Marshallia caespitosa*), prairie pleatleaf (*Nemastylis geminiflora*), false aloe (*Manfreda virginica*), rattlesnake master (*Eryngium yuccifolium*), wild bergamot (*Monarda fistulosa*), yellow puff (*Neptunia lutea*), narrowleaf mountainmint (*Pycnanthemum tenuifolium*), switchgrass (*Panicum virgatum*), wild quinine (*Parthenium integrifolium*), green milkweed (*Asclepias viridis*), Carolina larkspur (*Delphinium carolinianum*), pale purple coneflower (*Echinacea pallida*), and butterfly milkweed (*Asclepias tuberosa*). Typical woody components include post oak (*Quercus stellata*), southern red oak (*Q. falcata*), shortleaf pine (*Pinus echinata*), eastern redcedar (*Juniperus virginiana*), gum bumelia (*Sideroxylon lanuginosum*), Chickasaw plum (*Prunus angustifolia*), deciduous holly (*Ilex decidua*), and eastern redbud (*Cercis canadensis* var. *canadensis*).

Morse Clay Prairies were not known to occur in northeastern Texas until surveys were conducted in the spring of 2012 by Michael Warriner (an invertebrate biologist) and Jason Singhurst with the Wildlife Diversity Program at the TPWD. The focus of the surveys was to visit three isolated prairie pockets (Caughorn Prairie, Shawnee Prairie, and Walker's Prairie) that were mapped in 1850 (T. Jordan 1977) and to document the presence of prairie vegetation, populations of American basketflower (*Centaurea americana*), and bumblebee species utilizing American basketflower. As a result of the survey, two Morse Clay Prairies were documented, one in Shawnee Prairie (~ 4 ha or 10 ac) in Red River County and one in Caughorn Prairie (~ 6.1 ha or 15 ac) in Bowie County. No permanent measures have been taken to conserve these prairies. Further investigation of other potential prairie sites in Bowie and Red River

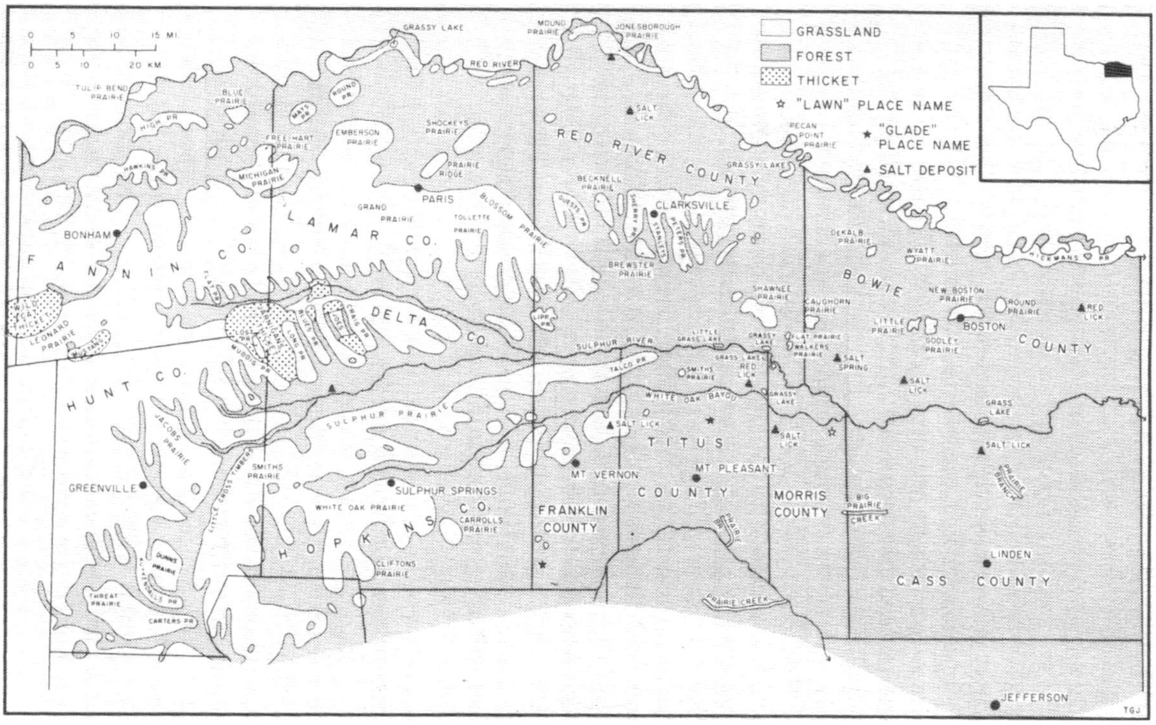

Figure 4.16. Northeastern Texas prairies mapped in 1850 (T. Jordan 1977).

Counties, and possibly in additional counties in northeastern Texas, is needed.

Conclusions

In this chapter, five primary and two ecologically isolated prairie plant communities in eastern Texas are described. The geology and soils on which these prairie communities have developed are discussed, as are differences in the dominant flora, including mention of the rare, endemic, and peripheral plant species associated with the various prairie types. The plant species richness of several of the prairie types is demonstrated. Maps are provided of their historical extent and current range, and a newly described prairie plant community association is introduced. The global rarity of each of these prairie types is emphasized. State, federal, and nongovernmental organizations are encouraged to conserve these precious prairies before the remnants vanish from eastern Texas landscapes. This plant community discussion should support conservation planning, ecological restoration, and future land management decisions relating to these prairie types in eastern Texas and those that extend into Arkansas, Louisiana, and Oklahoma.

5

Floristics of the Louisiana Cajun and Inland Prairies

Charles M. Allen

The Louisiana prairies include the coastal prairie in the southwestern portion of the state (Allain et al. 2000, 2006; Allen and Vidrine 1989; Allen et al. 2001; Post 1969; Vidrine 2010) and the inland prairies in the central and northern parts of the state (Post 1969; L. Smith 1995). Coastal prairie in Louisiana was mostly continuous historically, so it is basically one system, while the inland prairies occur as scattered, isolated small prairies. The objectives of this chapter are to bring together the floristics of Louisiana prairies using published works and to compare and contrast the floras of the different prairie systems.

The coastal prairie, also called the Cajun Prairie in Louisiana, once occupied 1,000,000 ha (2.5 million ac) (Allain et al. 2000, 2006; Allen and Vidrine 1989; Allen et al. 2001; Vidrine 2010; Vidrine et al. 1995). Cajun Prairie occurs in southwestern Louisiana between the Atchafalaya and Sabine Rivers, where most of the European settlers were Cajuns (French Acadians). It was once mostly continuous, interrupted only by the gallery forests that developed along the streams. The soils in the Cajun Prairie are moderately acidic silt loams underlain by a densely packed, hard claypan 20 to 40 cm (8 to 16 in) below the surface (H. Clark et al. 1962; K. Murphy et al. 1986). This prairie developed in an area with 125 cm (50 in) of annual rainfall; forests usually develop in areas with more than 76 cm (30 in) of annual rainfall. The development of prairie in this area could be linked to the hard claypan below the surface, to fires, or to other factors. Most of the prairie has been destroyed, and only a few strips along railroads remain. Studies of Cajun Prairie are based on these railroad remnants, a total of about 81 ha (200 ac). Cajun Prairie grades into coastal marsh to the south, bottomland hardwood forest to the east, and pine savanna forest to the north and west. The herbaceous component of the pine savanna and the Cajun Prairie is very similar, but very few pine trees occur in the prairie. A thorough treatise of many aspects of the Cajun Prairie (history, floristics, faunistics, etc.) can be found in Vidrine (2010).

The inland prairies occur in central and northern Louisiana (Post 1969; L. Smith 1995) in small patches in widely scattered locations. They include calcareous, Morse Clay, and saline soils.

The calcareous system is the largest of the three inland prairies; named prairies include Anacoco in Vernon Parish (Allen et al. 2006); Keiffer in Winn Parish (L. Smith et al. 1989; Allen 1993), and Copenhagen in Caldwell Parish. B. MacRoberts and MacRoberts (1996) studied and reported on the flora of two small calcareous prairies in Natchitoches Parish. Their data are combined with those from the nearby Keiffer Prairie of Winn Parish. A Fleming calcareous forest surrounds most calcareous prairie patches. This forest is characterized by small trees, shrubs, and stunted larger trees. There is a dense canopy and a very limited number of herbaceous species except in openings. The common trees include white ash (*Fraxinus americana*), sweet gum (*Liquidambar styraciflua*), and shortleaf pine (*Pinus echinata*). Small trees and shrubs include parsley hawthorn (*Crataegus marshallii*), cockspur hawthorn (*C. crus-galli*), little-hip hawthorn (*C. spathulata*), gum bumelia (*Sideroxylon lanuginosum*), rusty blackhaw (*Viburnum rufidulum*), and roughleaf dogwood (*Cornus drummondii*). The common vine is rattan (*Berchemia scandens*). The soils of the calcareous prairies are usually moderately alkaline clays (Soil Survey Staff 2003). Apparently, no floristic surveys have been published on the Copenhagen prairie system.

Morse Clay prairies and soils occur on Pleistocene intermediate terraces in northwestern and central Louisiana and southwestern Arkansas (B. MacRoberts et al. 2009b). These prairies typically exist as small grassy openings on side slopes and flats in shortleaf pine with reddish, fine-textured, alkaline, calcareous clay and silt loam soils.

Saline prairies are described in Reid et al. (2010) and B. MacRoberts et al. (2009a). Louisiana saline prairies occur on fluvial terraces adjacent to active floodplains. The soils formed in loamy Late Pleistocene sediments and are one of three soil series: Bonn, Brimstone, and Lafe. The soils are often either extremely dry or waterlogged and are usually strongly alkaline.

Methods

All published floristic works on Louisiana prairies were examined and a list of taxa generated. The nomenclature and family names come from the published work or the USDA PLANTS Database (USDA 2010–2013). The vascular flora of the Cajun Prairie includes 513 taxa in 92 families and 277 genera (Allen et al. 2001). The family with the most taxa is the Asteraceae (80), followed by the Poaceae (78), Cyperaceae (50), Fabaceae (35), Lamiaceae (19), Scrophulariaceae (18), and Onagraceae (15). Allen et al. (2006) reported 177 species from the Anacoco Calcareous Prairie, with 46 species from the Poaceae, 43 from the Asteraceae, and 25 from the Fabaceae. L. Smith et al. (1989) reported 215 taxa from the Keiffer Prairie, including 52 species from the Poaceae, 40 from the Asteraceae, and 19 from the Fabaceae. Allen (1993), adding to the work of L. Smith et al. (1989), reported 135 additional species from the Keiffer Prairie forest system. The number of taxa reported from the Morse Clay prairies is 159 (B. MacRoberts et al. 2009b). This includes 45 families, the largest being the Asteraceae with 40 taxa, Poaceae with 23, and Fabaceae with 17. Reid et al. (2010) reported 219 species from saline prairies in 65 families, with 39 species in the Poaceae, 22 in the Cyperaceae, and 20 in the Asteraceae.

The list of species from Louisiana prairies allowed comparisons to be made between the species from each of the prairies. The number of species unique to each system, plus the number of species that at least two prairies had in common, was determined. The native status of each species was determined using the PLANTS Database (USDA 2010–2013).

Results

A total of 934 species have been reported from all Louisiana prairies (table 5.1). These are in 126 families, with the largest being the Asteraceae with 138 species, followed by the Poaceae with 135, the Cyperaceae with 84, and the Fabaceae with 68. The largest genus is *Carex*, with 30 species, followed by *Cyperus*, *Dichanthelium*, and *Rhynchospora*, each with 14 species.

The Cajun Prairie system had the most unique species, at 212; the Keiffer system had 92; the saline system had 73; the Anacoco system had 73; and the Morse Clay had 16. The two systems with the most species in common were the Anacoco and Keiffer, with 102, and the two with the fewest were Saline and Morse Clay, with no species in common.

The three systems with the most species in common were Cajun, Anacoco, and Keiffer, with 48 species, and those with the fewest were the Anacoco, Saline, and Morse, with no species in common. There were 45 species in common in the Cajun, Anacoco, Keiffer, and Morse Clay systems, but only 2 species in common in the Cajun, Anacoco, Morse Clay, and Saline systems. Seventeen species were common to all five systems—Cajun, Anacoco, Keiffer, Morse Clay, and Saline. A total of 204 species were noted in the forest system adjacent to the prairies. The number of nonnative species was 69.

Four species have been reported in error for Louisiana prairies, including *Elymus canadensis* (= *E. virginicus*), *Euphorbia marginata* (= *E. bicolor*), *Mimosa microphylla* (= *M. nuttallii*), and *Linum rigidum* (= *L. medium* or *L. striatum*). All four errors seem to be based on misidentified specimens, with the probable correct species identification following the name that was in error.

Discussion

As expected, the two calcareous prairies, Anacoco and Keiffer, had the most species in common, at 102. The surprisingly large number of species the Cajun Prairie and Saline Prairie had in common (39) could perhaps be explained by the coastal characteristics (extremely dry or waterlogged soils and strongly alkaline pH) of the inland Saline Prairies. The low number of species the Keiffer, Morse Clay, and Saline Prairies had in common (6) is a mystery since these three are fairly close together geographically. Some of the differences between the numbers of species reported for Louisiana prairie systems could be linked to size of the individual prairies studied, search effort, length of study time, or other factors.

Additional floristic studies should be performed on the other prairies in Louisiana, for instance on the Copenhagen Calcareous Prairie and others. Efforts should also be made to protect all prairie systems in the state from destruction and degradation. Restoration onto appropriate sites and enhancement of current prairies should be pursued, especially the removal and prevention of invasive species.

Table 5.1.

List of species by group and family reported for Louisiana prairies (C = Cajun Prairie; A = Anacoco Calcareous Prairie; K = Keiffer Calcareous Prairie; M = Morse Clay Prairie; S = Saline Prairie; and F = primarily a forest species) (nomenclature and families from USDA 2013).

* = non-native species ** = species misidentified in the literature

Pteridophytes

Aspleniaceae
Asplenium platyneuron (L.) B.S.P. (C, A, S; F)

Dennstaedtiaceae
Pteridium aquilinum (L.) Kuhn (C, A; F)

Dryopteridaceae
Polystichum acrostichoides (Michx.) Schott (A, K; F)

Isoetaceae
Isoetes melanopoda Gay & Durieu ex Durieu (S)

Lygodiaceae
**Lygodium japonicum* (Thunb. ex Murr.) Sw. (C)

Ophioglossaceae
Botrychium biternatum (Sav.) Underwood (A, K; F)
Botrychium lunarioides (Michx.) Sw. (S)
Botrychium virginianum (L.) Sw. (A, K; F)
Ophioglossum engelmanii Prantl (K; F)

Polypodiaceae
Pleopeltis polypodioides (L.) Andrews & Windham (A; F)

Thelypteridaceae
Phegopteris hexagonoptera (Michx.) Fée (A; F)
Thelypteris kunthii (Desv.) Morton (K; F)

Gymnosperms

Cupressaceae
Juniperus virginiana L. (K, M, S; F)

Pinaceae
Pinus echinata P. Mill. (A, K, M, S; F)
Pinus palustris P. Mill. (C, A; F)
Pinus taeda L. (C, A, K, S; F)

Taxodiaceae
Taxodium distichum (L.) L.C. Rich. (C; F)

Monocotyledons

Agavaceae
Manfreda virginica (L.) Salisb. ex Rose (C, A, K, M)
Yucca louisianensis Trel. (K, M, S)

Alismataceae
Sagittaria papillosa Buchenau (S)

Araceae
Arisaema dracontium (L.) Schott (A, K; F)
Arisaema triphyllum (L.) Schott ssp. *triphyllum* (A, K; F)

Arecaceae
Sabal minor (Jacq.) Pers. (A, K, M, S; F)

Bromeliaceae
Tillandsia usneoides (L.) L. (A; F)

Commelinaceae
Commelina erecta L. (C)
Tradescantia hirsutiflora Bush (C, A, K, S)
Tradescantia occidentalis (Britt.) Smyth (S)
Tradescantia reverchonii Bush (A)
Tradescantia virginiana L. (C)

Cyperaceae
Bulbostylis capillaris (L.) Kunth ex C.B. Clarke (C)
Carex abscondita Mack. (M)
Carex alata Torr. (C)
Carex albolutescens Schwein. (C)
Carex amphibola Steud. (A)
Carex arkansana Bailey (S)
Carex blanda Dewey (A, K)
Carex bushii Mack. (S)
Carex caroliniana Schwein. (K)
Carex cherokeensis Schwein. (C, A, K, M; F)
Carex complanata Torr. & Hook. (C, K, S)
Carex crinita Lam. (A)
Carex crus-corvi Shuttleworth ex Kunze (S)
Carex debilis Michx. var. *debilis* (A, K)
Carex digitalis Willd. (A)
Carex flaccosperma Dewey (A, K)
Carex frankii Kunth (C, A)
Carex glaucodea Tuckerman ex Olney (A, K)
Carex granularis Muhl. ex Willd. (K)
Carex lonchocarpa Willd. (A)
Carex longii Mack. (A)
Carex lupulina Muhl. ex Willd. (A, S)
Carex lurida Wahlenb. (A)
Carex meadii Dewey (A, M)
Carex microdonta Torr. & Hook. (C, A, K)
Carex oxylepis Torr. & Hook. var. *oxylepis* (A, K)
Carex planispicata Naczi (K)
Carex retroflexa Muhl. ex Willd. (A, K)
Carex striatula Michx. (A)
Carex styloflexa Buckl. (A)
Carex vulpinoidea Michx. (C)
Cladium mariscus (L.) Pohl ssp. *jamaicense* (Crantz) Kükenth. (C)
Cyperus acuminatus Torr. & Hook. ex Torr. (C, S)
Cyperus croceus Vahl (C)
Cyperus echinatus (L.) Wood (C, S)
Cyperus erythrorhizos Muhl. (C)
Cyperus haspan L. (C)
**Cyperus iria* L. (C)
Cyperus oxylepis Nees ex Steud. (C)
Cyperus pseudovegetus Steud. (C, S)
Cyperus retroflexus Buckley (S)
Cyperus retrorsus Chapman (C, S)
**Cyperus rotundus* L. (C)
Cyperus squarrosus L. (S)
Cyperus strigosus L. (C, A, S)
Cyperus virens Michx. (C)
Eleocharis ambigens Fernald (S)
Eleocharis microcarpa Torr. (C)
Eleocharis montana (Kunth) Roemer & J.A. Schultes (C)
Eleocharis obtusa (Willd.) J.A. Schultes (C)
Eleocharis quadrangulata (Michx.) Roemer & J.A. Schultes (C)
Eleocharis tenuis (Willd.) Schult. var. *verricosa* (Svenson) Svenson (S)
Eleocharis tuberculosa (Michx.) Roemer & J.A. Schultes (C)
Eleocharis wolfii (A. Gray) A. Gray ex Britt. (S)
Fimbristylis annua (All.) Roemer & J.A. Schultes (A)
Fimbristylis autumnalis (L.) Roemer & J.A. Schultes (C)
Fimbristylis castanea (Michx.) Vahl (K)
Fimbristylis miliacea (L.) Vahl (C)

Continued on the next page

Fimbristylis puberula (Michx.) Vahl (C, A, K, M, S)
Fuirena pumila (Torr.) Spreng. (C)
Isolepis carinata Hook. & Arn. ex Torr. (C, S)
Kyllinga brevifolia Rottb. (C)
Kyllinga odorata Vahl (C)
Rhynchospora caduca Ell. (C, A, K, S)
Rhynchospora cephalantha Gray (C)
Rhynchospora chalarocephala Fern. & Gale (C)
Rhynchospora colorata (L.) H. Pfeiffer (C)
Rhynchospora corniculata (Lam.) Gray (C, K, S)
Rhynchospora elliottii A. Dietr. (C)
Rhynchospora globularis (Chapman) Small (C, A, K, S)
Rhynchospora glomerata (L.) Vahl (C)
Rhynchospora harveyi W. Boott (C, S)
Rhynchospora inexpansa (Michx.) Vahl (A)
Rhynchospora microcarpa Baldw. ex Gray (C)
Rhynchospora pusilla Chapman ex M.A. Curtis (C)
Rhynchospora rariflora (Michx.) Ell. (C)
Rhynchospora recognita (Gale) Kral (A)
Scirpus atrovirens Willd. (C, A, K)
Scleria ciliata Michx. (C, M, S; F)
Scleria oligantha Michx. (A, K, M; F)
Scleria pauciflora Muhl. ex Willd. (C)
Scleria reticularis Michx. (C)
Scleria triglomerata Michx. (A; F)
Scleria verticillata Muhl. ex Willd. (C)

Dioscoreaceae
Dioscorea villosa L. (A, K; F)

Iridaceae
Alophia drummondii (Graham) R.C. Foster (S)
Herbertia lahue (Molina) Goldblatt (C, A)
Iris virginica L. (C)
Nemastylis geminiflora Nutt. (M)
Sisyrinchium albidum Raf. (M)
Sisyrinchium angustifolium P. Mill. (C, K)
Sisyrinchium atlanticum Bickn. (C, S)
Sisyrinchium campestre E.P. Bicknell (K)
Sisyrinchium exile Bickn. (C)
Sisyrinchium langloisii Greene (C, M, S)
Sisyrinchium rosulatum E.P. Bicknell (S)

Juncaceae
Juncus acuminatus Michx. (S)
Juncus brachycarpus Engelm. (C, S)
**Juncus capitatus* Weigel (S)
Juncus coriaceus Mack. (K)
Juncus dichotomus Elliott (S)
Juncus dudleyi Wiegand (S)
Juncus effusus L. (C)
Juncus marginatus Rostk. (C, S)
Juncus nodatus Coville (C)
Juncus polycephalus Michx. (C)
Juncus secundus P. Beauv. ex Poir. (S)
Juncus tenuis Willd. (C, A)
Juncus validus Coville (C)
Luzula bulbosa (Wood) Smyth & Smyth (A, S; F)

Liliaceae
Aletris aurea Walt. (C)
Aletris farinosa L. (C)
Allium canadense L. var. *canadense* (C, K, M)
Allium canadense L. var. *mobilense* (Regel) Ownbey (C, A, K, S)
Camassia scilloides (Raf.) Cory (K; F)
Cooperia drummondii Herbert (S)
Habranthus tubispathus (L'Her.) Traub. (S)
Hymenocallis liriosme (Raf.) Shinners (C)
Hymenocallis occidentalis (J. Le Conte) Kunth (A, K; F)
Hypoxis hirsuta (L.) Coville (C, K)
Lilium michauxii Poir. (A, K)
Nothoscordum bivalve (L.) Britt. (C, A, K, M, S)
Polygonatum biflorum (Walt.) Ell. (A, K; F)
Schoenolirion wrightii Sherman (S)
Stenanthium gramineum (Ker-Gawl.) Morong (A, K; F)
Trillium gracile J.D. Freeman (A; F)
Uvularia perfoliata L. (K; F)

Orchidaceae
Calopogon oklahomensis D.H. Goldman (C)
Cypripedium kentuckiense C.F. Reed (A; F)
Listera australis Lindl. (A; F)
Platanthera nivea (Nutt.) Luer (C)
Pteroglossaspis ecristata (Fern.) Rolfe (C)
Spiranthes laciniata (Small) Ames (M)
Spiranthes odorata (Nutt.) Lindl. (K)
Spiranthes praecox (Walter) S. Watson (K)
Spiranthes vernalis Engelm. & Gray (C, S)
Tipularia discolor (Pursh) Nutt. (A, K; F)

Poaceae
Agrostis elliottiana Schult. (S)
Agrostis hyemalis (Walt.) B.S.P. (C, A, S)
Agrostis perennans (Walt.) Tuckerman (A; F)
**Aira caryophyllea* L. (S)
**Aira elegans* Willd. (S)
Alopecurus carolinianus Walt. (C)
Andropogon gerardii Vitman (C, A, K, M)
Andropogon glomeratus (Walt.) B.S.P. (C, A, K, M)
Andropogon gyrans Ashe var. *gyrans* (C, A)
Andropogon ternarius Michx. (C, A, S)
Andropogon virginicus L. (C, A, K, M)
Anthaenantia rufa (Nutt.) J.A. Schultes (C)
Aristida dichotoma Michx. (K)
Aristida longespica Poir. var. *geniculata* (Raf.) Fernald (K)
Aristida longespica Poir. var. *longespica* (C, A, K, S)
Aristida oligantha Michx. (C, A, K, M, S)
Aristida palustris (Chapman) Vasey (A)
Aristida purpurascens Poir. var. *purpurascens* (C, A, K)
Aristida purpurascens Poir. var. *virgata* (Trin.) Allred (K)
Arundinaria gigantea (Walt.) Muhl. ssp. *gigantea* (A, K; F)
**Avena sativa* L. (A)
Axonopus fissifolius (Raddi) Kuhlm. (C, K)
Bothriochloa exaristata (Nash) Henr. (C)
**Bothriochloa ischaemum* (L.) Keng var. *songarica* (Rupr. ex Fisch. & C.A. Mey.) Celarier & Harlan (C, A, K, M)
Bothriochloa longipaniculata (Gould) Allred & Gould (C, M)
Bouteloua rigidiseta (Steud.) Hitchc. (C, A)
Brachyelytrum erectum (Schreb. ex Spreng.) Beauv. (A; F)
Briza minor L. (C, K, S)
**Bromus catharticus* Vahl (C)
**Bromus japonicus* Thunb. (K)
Bromus pubescens Muhl. ex Willd. (A, K; F)
**Bromus racemosus* L. (S)
Chasmanthium latifolium (Michx.) Yates (A, K; F)
Chasmanthium laxum (L.) Yates (A, K; F)
Chasmanthium sessiliflorum (Poir.) Yates (A, K; F)
**Chloris canterai* Arech. (C)
Coelorachis cylindrica (Michx.) Nash (C, K)
Coelorachis rugosa (Nutt.) Nash (C)
Ctenium aromaticum (Walt.) Wood (C)
**Cynodon dactylon* (L.) Pers. (C, K, S)
**Dactylis glomerata* L. (K)
Danthonia spicata (L.) Beauv. ex Roemer & J.A. Schultes (A; F)

TABLE 5.1. *Continued*

Dichanthelium aciculare (Desv. ex Poir.) Gould & C.A. Clark (C, A, K, M, S)
Dichanthelium acuminatum (Sw.) Gould & C.A. Clark (C, A, K, S)
Dichanthelium boscii (Poir.) Gould & C.A. Clark (A, K; F)
Dichanthelium commutatum (J.A. Schultes) Gould (A, K; F)
Dichanthelium consanguineum (Kunth) Gould & C.A. Clark (K)
Dichanthelium depauperatum (Muhl.) Gould (A; F)
Dichanthelium dichotomum (L.) Gould (C, K, S)
Dichanthelium laxiflorum (Lam.) Gould (A, K; F)
Dichanthelium oligosanthes (J.A. Schultes) Gould var. *oligosanthes* (S)
Dichanthelium oligosanthes (J.A. Schultes) Gould var. *scribnerianum* (Nash) Gould (C, A, K)
Dichanthelium ovale (Ell.) Gould & C.A. Clark (C)
Dichanthelium ravenelii (Scribn. & Merr.) Gould (K; F)
Dichanthelium scoparium (Lam.) Gould (C, S)
Dichanthelium sphaerocarpon (Ell.) Gould var. *sphaerocarpon* (C, A, K, S)
Digitaria ciliaris (Retz.) Koel. (C)
Digitaria cognata (J.A. Schultes) Pilger (C, S)
Digitaria filiformis (L.) Koel. (C)
**Digitaria ischaemum* (Schreb.) Schreb. ex Muhl. (C, K)
**Digitaria violascens* Link (C, K)
**Echinochloa crusgalli* (L.) Beauv. (C)
***Elymus canadensis* L. (= *E. virginicus* L.)
Elymus virginicus L. (A, K, M)
**Eragrostis bahiensis* (Schrad. ex J.A. Schultes) J.A. Schultes (C)
Eragrostis elliottii S. Wats. (C, A)
Eragrostis hirsuta (Michx.) Nees (C, A)
Eragrostis lugens Nees (C, A, M, S)
Eragrostis refracta (Muhl.) Scribn. (C, A)
Eragrostis secundiflora J. Presl (K, S)
Eragrostis spectabilis (Pursh) Steud. (C, A, K, M)
**Festuca arundinacea* Schreb. (K)
Festuca paradoxa Desv. (K)
Festuca subverticillata (Pers.) Alexeev (K)
Gymnopogon ambiguus (Michx.) Britton, Sterns & Poggenb. (K)
Gymnopogon brevifolius Trin. (C, A)
Hordeum pusillum Nutt. (A, S)
Koeleria macrantha (Ledeb.) Schult. (K)
Leersia hexandra Sw. (C)
Leersia virginica Willd. (A, K; F)
Limnodea arkansana (Nutt.) L.H. Dewey (C)
**Lolium perenne* L. (C, S)
Melica mutica Walt. (A, K; F)
Muhlenbergia capillaris (Lam.) Trin. (C, A, K)
Muhlenbergia schreberi J.F. Gmel. (K)
**Oplismenus hirtellus* (L.) P. Beauv. (K; F)
Panicum anceps Michx. (C, A, K, M)
Panicum brachyanthum Steud. (C, S)
Panicum flexile (Gattinger) Scribn. (A, K)
Panicum hemitomon J.A. Schultes (C)
Panicum rigidulum Bosc ex Nees var. *rigidulum* (C, K)
Panicum virgatum L. (C, A, K)
**Paspalum dilatatum* Poir. (C, A, K, M)
Paspalum floridanum Michx. (C, A, K, M, S)
Paspalum laeve Michx. (C, A, K, M)
Paspalum plicatulum Michx. (C, A)
Paspalum praecox Walt. (C)
Paspalum pubiflorum Rupr. ex Fourn. (A)
Paspalum setaceum Michx. (C, A, K, S)
**Paspalum urvillei* Steud. (C, A, K)
Phalaris angusta Nees ex Trin. (C)
Phalaris caroliniana Walt. (C, K, S)
Piptochaetium avenaceum (L.) Parodi (A; F)
Poa autumnalis Muhl. ex Ell. (A, K; F)
Saccharum giganteum (Walt.) Pers. (C, A)
Schedonnardus paniculatus (Nutt.) Trel. (S)
Schizachyrium scoparium (Michx.) Nash (C, A, K, M, S)
Schizachyrium tenerum Nees (C, A)
Setaria parviflora (Poir.) Kerguélen (C, A, K, M, S)
**Setaria pumila* (Poir.) Roemer & J.A. Schultes (C, M, S)
Sorghastrum nutans (L.) Nash (C, A, K, M)
**Sorghum halepense* (L.) Pers. (C, A, K)
Spartina spartinae (Trin.) Merr. ex A.S. Hitchc. (C)
Sphenopholis filiformis (Chapm.) Scribn. (S)
Sphenopholis intermedia (Rydb.) Rydb. (K)
Sphenopholis obtusata (Michx.) Scribn. (C, K, M)
Sporobolus clandestinus (Biehler) Hitchc. (K)
Sporobolus compositus (Poir.) Merr. var. *compositus* (C, A, K, M)
Sporobolus compositus (Poir.) Merr. var. *drummondii* (Trin.) Kartesz & Gandhi (K)
Sporobolus compositus (Poir.) Merr. var. *macer* (Trin.) Kartesz & Gandhi (K, S)
**Sporobolus indicus* (L.) R. Br. (K, S)
Sporobolus junceus (Beauv.) Kunth (C, A, K)
Sporobolus neglectus Nash (A)
Sporobolus pyramidatus (Lam.) A.S. Hitchc. (S)
Sporobolus ozarkanus Fernald (K)
Sporobolus silveanus Swallen (C)
Sporobolus vaginiflorus (Torr. ex A. Gray) Alph. Wood (S)
Steinchisma hians (Ell.) Nash (C, S)
Tridens ambiguus (Ell.) J.A. Schultes (C, A)
Tridens flavus (L.) A.S. Hitchc. var. *flavus* (C, A, K, S)
Tridens strictus (Nutt.) Nash (C, A, K, M, S)
Tripsacum dactyloides (L.) L. (C, A, M)
Trisetum interruptum Buckl. (A)
Urochloa platyphylla (Munro ex Wright) R. Webster (C, S)
Vulpia octoflora (Walt.) Rydb. (C, A, K, S)

Pontederiaceae
Pontederia cordata L. (C)

Potamogetonaceae
Potamogeton diversifolius Raf. (S)

Smilacaceae
Smilax bona-nox L. (A, K, M; F)
Smilax glauca Walt. (A, K; F)
Smilax herbacea L. (A, K; F)
Smilax pumila Walt. (A, K; F)
Smilax rotundifolia L. (C, A, K, S; F)
Smilax smallii Morong (A; F)
Smilax tamnoides L. (A, K; F)

Typhaceae
Typha latifolia L. (C)

Xyridaceae
Xyris difformis Chapman var. *difformis* (C)
Xyris laxifolia Mart. (C)
Xyris torta Sm. (C)

Dicotyledons

Acanthaceae
Dicliptera brachiata (Pursh) Spreng. (A; F)
Hygrophila lacustris (Schlecht. & Cham.) Nees (C)

Continued on the next page

TABLE 5.1. *Continued*

Justicia ovata (Walt.) Lindau (C, S)
Ruellia humilis Nutt. (C, A, K, M, S)
Ruellia pedunculata Torr. ex A. Gray (K; F)

Aceraceae
Acer barbatum Michx. (A, K; F)
Acer leucoderme Small (A, K; F)
Acer rubrum L. var. *rubrum* (A, K; F)

Amaranthaceae
**Alternanthera philoxeroides* (Mart.) Griseb. (C)

Anacardiaceae
Rhus copallinum L. (C, A, K, S; F)
Toxicodendron radicans (L.) Kuntze (C, A, K, M; F)

Annonaceae
Asimina triloba (L.) Dunal (A, K; F)

Apiaceae
Centella erecta (L. f.) Fern. (C)
Chaerophyllum tainturieri Hook. (C, A, K)
Cicuta maculata L. (C)
Cryptotaenia canadensis (L.) DC. (K; F)
**Cyclospermum leptophyllum* (Pers.) Sprague ex Britton & P. Wilson (K)
Cynosciadium digitatum DC. (C)
Daucus pusillus Michx. (K, S)
Eryngium integrifolium Walt. (C)
Eryngium yuccifolium Michx. (C, A, K, M)
Limnosciadium pinnatum (DC.) Mathias & Constance (S)
Polytaenia nuttallii DC. (C, K)
Ptilimnium capillaceum (Michx.) Raf. (C)
Ptilimnium costatum (Ell.) Raf. (C)
Ptilimnium nuttallii (DC.) Britt. (S)
Sanicula canadensis L. (A, K; F)
Spermolepis echinata (Nutt. ex DC.) Heller (C, A)
Taenidia integerrima (L.) Drude (K; F)
Trepocarpus aethusae Nutt. ex DC. (K, M, S)
Zizia aurea (L.) W.D.J. Koch (K; F)

Apocynaceae
Amsonia tabernaemontana Walt. (C)
Apocynum cannibinum L. (A, K)
Trachelospermum difforme (Walter) A. Gray (K; F)

Aquifoliaceae
Ilex decidua Walt. (C, A, K, M; F)
Ilex longipes Chapman ex Trel. (A; F)
Ilex opaca Ait. (A, K; F)
Ilex vomitoria Ait. (C, A, K, S; F)

Araliaceae
Aralia spinosa L. (A, K; F)

Aristolochiaceae
Aristolochia serpentaria L. (A, K; F)

Asclepiadaceae
Asclepias lanceolata Walt. (C)
Asclepias longifolia Michx. (C, K)
Asclepias obovata Ell. (C, A, K)
Asclepias stenophylla A. Gray (K)
Asclepias tuberosa L. (C, A, K)
Asclepias variegata L. (A; F)
Asclepias verticillata L. (C, A, K, M)
Asclepias viridiflora Raf. (C, K)
Asclepias viridis Walt. (C, A, K, M)
Cynanchum laeve (Michx.) Pers. (C)
Matelea decipiens (Alexander) Woodson (K; F)
Matelea gonocarpos (Walt.) Shinners (C, A, K; F)

Asteraceae
Ambrosia artemisiifolia L. (C, A, M)
Ambrosia bidentata Michx. (C, A, M)
Ambrosia psilostachya DC. (C, K, M)
Ambrosia trifida L. (C, M)
Antennaria parlinii Fern. ssp. *fallax* (Greene) Bayer & Stebbins (A, K; F)
Arnoglossum ovatum (Walter) H. Rob. (C)
Arnoglossum plantagineum Raf. (K, M)
Baccharis halimifolia L. (C, A, K, M, S)
Bidens aristosa (Michx.) Britt. (C)
Bigelowia nuttallii L.C. Anders. (C, A)
Boltonia asteroides (L.) L'Hér. (C)
Boltonia diffusa Ell. (C, M, S)
Brickellia eupatorioides (L.) Shinners (K; F)
Chromolaena ivifolia (L.) King. & H.E. Robins. (C)
Chrysopsis mariana (L.) Ell. (C)
Chrysopsis pilosa Nutt. (S)
Cirsium carolinianum (Walt.) Fern. & Schub. (A)
Cirsium discolor (Muhl. ex Willd.) Spreng. (K)
Cirsium horridulum Michx. (C, A, K, M)
Conoclinium coelestinum (L.) DC. (A, K)
Conyza canadensis (L.) Cronq. (C)
Coreopsis gladiata Walt. (C)
Coreopsis lanceolata L. (C, A, K, M, S)
Coreopsis palmata Nutt. (M)
Coreopsis pubescens Ell. (C, A)
Coreopsis tinctoria Nutt. (C, A, S)
Coreopsis tripteris L. (C, K)
Echinacea pallida (Nutt.) Nutt. (C, A, K, M)
Echinacea purpurea (L.) Moench (A)
Echinacea sanguinea Nutt. (A, M)
Elephantopus carolinianus Raeusch. (A, K; F)
Elephantopus tomentosus L. (K; F)
Erechtites hieraciifolia (L.) Raf. ex DC. (C, A, K)
Erigeron annuus (L.) Pers. (C)
Erigeron philadelphicus L. (C, K)
Erigeron pulchellus Michx. (A; F)
Erigeron strigosus Muhl. ex Willd. (C, A, K, M)
Erigeron tenuis Torr. & Gray (M, S)
Eupatorium altissimum L. (K)
Eupatorium capillifolium (Lam.) Small (C, A, K)
Eupatorium hyssopifolium L. (C, A)
Eupatorium leucolepis (DC.) Torr. & Gray (C)
Eupatorium perfoliatum L. (C, A)
Eupatorium rotundifolium L. (C, A)
Eupatorium semiserratum DC. (C, K)
Eupatorium serotinum Michx. (C, A, K, M)
Eurybia hemispherica (Alexander) Nesom (C, A, K, M)
Euthamia leptocephala (Torr. & Gray) Greene (C, A, K, S)
Euthamia tenuifolia (Pursh) Nutt. (C)
Evax verna Raf. (S)
Facelis retusa (Lam.) Sch. Dip. (S)
Gaillardia aestivalis (Walt.) H. Rock var. *aestivalis* (C, A, K, M)
Gaillardia aestivalis (Walt.) H. Rock var. *flavovirens* (C. Mohr) Cronquist (K)
Gamochaeta argyrina Nesom (S)
Gamochaeta purpurea (L.) Cabrera (C)
Helenium amarum (Raf.) H. Rock (C, A)
Helenium autumnale L. (K)
Helenium drummondii H. Rock (C)
Helenium flexuosum Raf. (C, K, S)
Helianthus angustifolius L. (C, A, K)
Helianthus hirsutus Raf. (A, K, M)
Helianthus mollis Lam. (C)
Heliopsis gracilis Nutt. (K; F)
Heliopsis helianthoides (L.) Sweet var. *helianthoides* (A; F)
Heterotheca subaxillaris (Lam.) Britt. & Rusby (M)
Hieracium gronovii L. (K; F)
Ionactis linariifolius (L.) Greene (A)
Iva angustifolia Nutt. ex DC. (S)
Iva annua L. (C, K, M, S)
Krigia caespitosa (Raf.) Chambers (C)
Krigia dandelion (L.) Nutt. (C, S)
Krigia occidentalis Nutt. (S)

Table 5.1. *Continued*

Krigia virginica (L.) Willd. (C)
Lactuca canadensis L. (C, K)
Lactuca floridana (L.) Gaertn. (C)
Liatris acidota Engelm. & Gray (C)
Liatris aspera Michx. (A, K, M, S)
Liatris elegans (Walt.) Michx. (C, A)
Liatris pycnostachya Michx. (C, A, K)
Liatris spicata (L.) Willd. (C, A)
Liatris squarrosa (L.) Michx. (C, K, M)
Liatris squarrulosa Michx. (A, K)
Marshallia caespitosa Nutt. ex DC. (M, S)
Mikania scandens (L.) Willd. (C)
Oligoneuron nitidum (Torr. & Gray) Small (C, A, K, M)
Oligoneuron rigidum (L.) Small (K, M)
Packera glabella (Poir) C. Jeffrey (C)
Packera obovata (Muhl. ex Willd.) W.A. Weber & A. Löve (A)
Packera tomentosa (Michx.) C. Jeffery (C)
Parthenium integrifolium L. (M)
Pityopsis graminifolia (Michx.) Nutt. (C, A)
Pluchea camphorata (L.) DC. (C)
Pluchea foetida (L.) DC. (C)
Pluchea rosea Godfrey (C)
Pseudognaphalium helleri (Britt.) A. Anderb. (A)
Pseudognaphalium obtusifolium (L.) Hilliard & Burtt (C, A, K)
Pterocaulon virgatum (L.) DC. (C, S)
Pyrrhopappus carolinianus (Walt.) DC. (C, A, K, S)
Ratibida pinnata (Vent.) Barnh. (A)
Rudbeckia grandiflora (D. Don) J.F. Gmel. ex DC. (C)
Rudbeckia hirta L. (C, A, K, M)
Rudbeckia missouriensis Engelm. ex C.L. Boynt. & Beadle (A)
Rudbeckia subtomentosa Pursh (K; F)
Rudbeckia texana (Perdue) P. Cox & Urbatsch (C)
Silphium asteriscus L. (K)
Silphium gracile Gray (C, A)
Silphium integrifolium Michx. (K, M)
Silphium laciniatum L. (C, A, K, M)
Silphium radula Nutt. (K)
Solidago arguta Aiton (K)
Solidago auriculata Shuttlw. ex. S.F. Blake (K)
Solidago caesia L. var. *caesia* (A, K; F)
Solidago canadensis L. (C, A, K, M)
Solidago nemoralis Aiton (K)
Solidago odora Ait. (C, A, K)
Solidago petiolaris Ait. (K)
Solidago radula Nutt. (K)
Solidago rugosa P. Mill. (C, A, K)
Solidago sempervirens L. var. *mexicana* (L.) Fern. (C)
Solidago ulmifolia Muhl. ex Willd. (K)
**Sonchus asper* (L.) Hill (C)
**Sonchus oleraceus* L. (C)
Symphyotrichum drummondii (Lindl.) Nesom var. *texanum* (Burgess) Nesom (A, K, M)
Symphyotrichum dumosum (L.) Nesom (C, A, K)
Symphyotrichum laeve (L.) Á. Löve & D. Löve (M)
Symphyotrichum lateriflorum (L.) Á. Löve & D. Löve (C, A, K, M)
Symphyotrichum ontarionis (Wieg.) Nesom (M)
Symphyotrichum oolentangiense (Riddell) Nesom (C, K, M)
Symphyotrichum patens (Ait.) Nesom (C, A, K, M)
Symphyotrichum pilosum (Willd.) Nesom (M)
Symphyotrichum praealtum (Poir.) Nesom (M)
Symphyotrichum pratense (Raf.) Nesom (C, A, K, M, S)
Verbesina alternifolia (L.) Britt. ex Kearney (A)
Verbesina helianthoides Michx. (A, K)
Verbesina virginica L. (A, K)
Vernonia gigantea (Walt.) Trel. (C)
Vernonia missurica Raf. (K)
Vernonia texana (Gray) Small (C, K, M)

Berberidaceae
Podophyllum peltatum L. (A, K; F)

Betulaceae
Carpinus caroliniana Walter (A, K; F)
Ostrya virginiana (Mill.) K. Koch. (A, K; F)

Bignoniaceae
Bignonia capreolata L. (A, K; F)
Campsis radicans (L.) Seem. ex Bureau (C, A, K, M; F)

Boraginaceae
Cynoglossum virginianum L. (A, K; F)
Heliotropium tenellum (Nutt.) Torr. (A, K)
Lithospermum tuberosum Rugel ex DC. (A, K)
Myosotis verna Nutt. (C, A, K)
Onosmodium bejariense DC. ex A. DC. (K, M)

Brassicaceae
**Cardamine hirsuta* L. (C)
Cardamine parviflora L. var. *arenicola* (Britt.) O.E. Schulz (C)
**Lepidium densiflorum* Schrad. (S)
Lepidium virginicum L. (C, S)

Buddlejaceae
Polypremum procumbens L. (C, A, K, S)

Cactaceae
Opuntia spp. (S)

Callitrichaceae
Callitriche heterophylla Pursh (C)
Callitriche pedunculosa Nutt. (S)

Campanulaceae
Lobelia appendiculata A. DC. (C, K, M, S)
Lobelia puberula Michx. (C, A)
Triodanis biflora (Ruiz & Pav.) Greene (K)
Triodanis perfoliata (L.) Nieuwl. (C, A, K, S)

Caprifoliaceae
**Lonicera japonica* Thunb. (A, K, M)
Lonicera sempervirens L. (A, K)
Sambucus nigra L. ssp. *canadensis* (L.) R. Bolli (C)
Symphoricarpos orbiculatus Moench. (M)
Viburnum acerifolium L. (K; F)
Viburnum dentatum L. (A, K; F)
Viburnum rufidulum Raf. (A, K, M; F)

Caryophyllaceae
**Cerastium glomeratum* Thuill. (C, S)
Geocarpon minimum Mackenzie (S)
Minuartia drummondii (Shinners) McNeill (S)
Minuartia muscorum (Fassett) Rabeler (S)
Silene antirrhina L. (C)

Celastraceae
Euonymus americanus L. (A; F)

Cistaceae
Helianthemum rosmarinifolium Pursh (S)
Lechea mucronata Raf. (C)
Lechea san-sabeana (Buckley) Hodgdon (S)
Lechea tenuifolia Michx. (C, S)

Clusiaceae
Hypericum cruxandreae (L.) Crantz (C)
Hypericum drummondii (Grev. & Hook.) Torr. & Gray (C, A, S)

Continued on the next page

Hypericum frondosum Michx. (A, K)
Hypericum gentianoides (L.) B.S.P. (C, S)
Hypericum gymnanthum Engelm. & Gray (C)
Hypericum hypericoides (L.) Crantz ssp. *hypericoides* (C, A, K, S)
Hypericum nudiflorum Michx. ex Willd. (C)
Hypericum prolificum L. (K; F)
Hypericum punctatum Lam. (K; F)

Convolvulaceae
Dichondra carolinensis Michx. (C, K, M, S)
Evolvulus sericeus Sw. (C, S)
Ipomoea lacunosa L. (C)
Ipomoea pandurata (L.) G.F.W. Mey. (A, K)
Ipomoea sagittata Poir. (C)
Stylisma aquatica (Walt.) Raf. (C)

Cornaceae
Cornus drummondii C.A. Mey. (C, A, K, M; F)
Cornus florida L. (A, K; F)

Crassulaceae
Crassula aquatica (L.) Schoenl. (S)

Cucurbitaceae
**Cucumis melo* L. (C)
Melothria pendula L. (C)

Cuscutaceae
Cuscuta cuspidata Engel. (K)
Cuscuta gronovii Willd. ex Schult. (S)
Cuscuta indecora Choisy (C)
Cuscuta pentagona Engelm. (M)

Droseraceae
Drosera brevifolia Pursh (C, S)

Ebenaceae
Diospyros virginiana L. (C, A, K, M; F)

Ericaceae
Rhododendron canescens (Michx.) Sweet (A, K; F)
Vaccinium arboreum Marsh. (C, A, K, S; F)
Vaccinium elliottii Chapman (A, K; F)
Vaccinium stamineum L. (K; F)
Vaccinium virgatum Ait. (A, K; F)

Euphorbiaceae
Acalypha gracilens Gray (C, A, K)
**Caperonia palustris* (L.) St.Hil. (C)
Chamaesyce humistrata (Engelm.) Small (C)
Chamaesyce maculata (L.) Small (C, A, K, S)
**Chamaesyce nutans* (Lag.) Small (K, M)
Chamaesyce serpens (Kunth) Small (A)
Croton capitatus Michx. (C, A, K, S)
Croton glandulosus L. (C)
Croton monanthogynus Michx. (A, M)
Croton willdenowii G.L. Webster (C, S)
Euphorbia bicolor Engelm. & Gray (A, K)
Euphorbia corollata L. (C, A, K, M)
***Euphorbia marginata* Pursh (= *E. bicolor* Engelm. & Gray)
Euphorbia spathulata Lam. (C, K, S)
Tragia betonicifolia Nutt. (C, A, K)
Tragia cordata Michx. (A, K; F)
Tragia urticifolia Michx. (K, M)
**Triadica sebifera* (L.) Small (C, A)

Fabaceae
Acacia angustissima (P. Mill.) Kuntze var. *hirta* (Nutt.) B.L. Robins. (A, K, M)
**Aeschynomene indica* L. (C)
Amphicarpa bracteata (L.) Fern. (K; F)
Astragalus crassicarpus Nutt. (K)
Astragalus distortus Torr. & Gray var. *engelmannii* (E. Sheld) M.E. Jones (M, S)
Baptisia alba (L.) Vent. (C, A, K)
Baptisia bracteata Muhl. ex Ell. (C, M)
Baptisia nuttalliana Small (C, A, K)
Baptisia sphaerocarpa Nutt. (C)
Centrosema virginianum (L.) Benth. (C, A, M)
Cercis canadensis L. (A, K, M; F)
Chamaecrista fasciculata (Michx.) Greene (C, A, K, M, S)
Clitoria mariana L. (K; F)
Crotalaria sagittalis L. (C, A, K)
Dalea candida Michx. ex Willd. (C, A, K, M)
Dalea purpurea Vent. (A, K, M)
Desmanthus illinoensis (Michx.) MacM. ex B.L. Robins. & Fern. (K, M)
Desmodium canescens (L.) DC. (K)
Desmodium ciliare (Muhl. ex Willd.) DC. (C, K)
Desmodium cuspidatum (Muhl. ex Willd.) DC. ex D. Don (K)
Desmodium glabellum (Michx.) DC. (K)
Desmodium glutinosum (Muhl. ex Willd.) Wood (A)
Desmodium marilandicum (L.) DC. (K)
Desmodium nudiflorum (L.) DC. (A, K; F)
Desmodium paniculatum (L.) DC. (C, K)
Desmodium pauciflorum (Nutt.) DC. (K; F)
Desmodium rotundifolium DC. (K; F)
Desmodium sessilifolium (Torr.) Torr. & Gray (C)
Erythrina herbacea L. (C, K)
Galactia regularis (L.) B.S.P. (K)
Galactia volubilis (L.) Britt. (C, A, K, M)
Gleditsia triacanthos L. (A, K, M, S; F)
**Glottidium vesicarium* (Jacq.) Harper (C, A, K)
* *Kummerowia striata* (Thunb.) Schindl. (C, S)
Lathyrus pusillus Ell. (S)
Lespedeza capitata Michx. (C)
**Lespedeza cuneata* (Dum.-Cours.) G. Don (C, A, M)
Lespedeza procumbens Michx. (K)
Lespedeza repens (L.) W. Bart. (C, A, K, S)
Lespedeza virginica (L.) Britt. (C, A, K)
Lotus unifoliolatus (Hook.) Benth. (S)
**Medicago lupulina* L. (C, M, S)
**Medicago polymorpha* L. (C)
**Medicago sativa* L. (K)
**Melilotus indicus* (L.) All. (C)
Mimosa hystricina (Small ex Britt. & Rose) B.L. Turner (C)
***Mimosa microphylla* Dryand (= *M. nuttallii* (DC.) B.L. Turner)
Mimosa nuttallii (DC.) B.L. Turner (A, K, M)
Mimosa strigillosa Torr. & Gray (K, M, S)
Neptunia lutea (Leavenworth) Benth. (C, A, K, M, S)
Neptunia pubescens Benth. (C, A)
Orbexilum pedunculatum (P. Mill.) Rydb. var. *psoralioides* (Walt.) Isley (C, A)
Orbexilum simplex (Nutt. ex Torr. & Gray) Rydb. (C, K)
Phaseolus polystachios (L.) B.S.P. (A)
Rhynchosia latifolia Nutt. ex Torr. & A. Gray (K)
Rhynchosia minima (L.) DC. (C)
Senna marilandica (L.) Link (A, K)
Sesbania herbacea (Mill.) McVaugh (K)
Strophostyles leiosperma (Torr. & Gray) Piper (S)
Strophostyles umbellata (Muhl. ex Willd.) Britt. (C, K)
Stylosanthes biflora (L.) B.S.P. (C, K)
Tephrosia onobrychoides Nutt. (C)
Tephrosia virginiana (L.) Pers. (A, K)
Trifolium bejariense Moric. (C)
**Trifolium dubium* Sibth. (C, M)
**Trifolium resupinatum* L. (C)
Vicia ludoviciana Nutt. (C)
Vicia minutiflora F.G. Dietr. (A, K)

Fagaceae
Castanea pumila (L.) P. Mill. var. *pumila* (C)
Fagus grandifolia Ehrh. (A, K; F)
Quercus alba L. (A, K; F)
Quercus falcata Michx. (C, A; F)

Table 5.1. *Continued*

Quercus hemisphaerica Bartr. ex Willd. var. *hemisphaerica* (A; F)
Quercus incana Bartr. (C; F)
Quercus margarettiae Ashe ex Small (S; F)
Quercus marilandica Münchh. (C, A, K; F)
Quercus muehlenbergii Engelm. (A; F)
Quercus nigra L. (C, A, K; F)
Quercus pagoda Raf. (K; F)
Quercus phellos L. (K, S; F)
Quercus shumardii Buckl. var. *shumardii* (A, K; F)
Quercus similis Ashe (S; F)
Quercus stellata Wangenh. (C, A, K, M, S; F)
Quercus velutina Lam. (A; F)
Quercus virginiana P. Mill. (C; F)

Gentianaceae
**Centaurium pulchellum* (Sw.) Druce (C, M)
Sabatia angularis (L.) Pursh (A, K)
Sabatia brachiata Ell. (C)
Sabatia campestris Nutt. (C, K, M, S)
Sabatia gentianoides Ell. (C)
Sabatia stellaris Pursh (C)

Geraniaceae
Geranium carolinianum L. (C, A, K)
Geranium dissectum L. (A, K)

Grossulariaceae
Ribes curvatum Small (S)

Haloragaceae
Myriophyllum pinnatum (Walt.) B.S.P. (S)
Proserpinaca palustris L. (C)

Hamamelidaceae
Hamamelis virginiana L. (A, K; F)
Liquidambar styraciflua L. (C, A, K; F)

Hippocastanaceae
Aesculus pavia L. (A, K; F)

Hydrophyllaceae
Hydrolea ovata Nutt. ex Choisy (C)
Phacelia glabra Nutt. (S)

Juglandaceae
Carya alba (L.) Nutt. ex Ell. (A, K; F)
Carya cordiformis (Wangenh.) K. Koch (A; F)
Carya glabra (P. Mill.) Sweet var. *glabra* (A; F)
Carya glabra (P. Mill.) Sweet var. *hirsuta* (Ashe) Ashe (A; F)
Carya illinoinensis (Wangenh.) K. Koch (C; F)
Carya myristiciformis (Michx. f.) Nutt. (A, K; F)
Carya ovata (P. Mill.) K. Koch (A, K; F)
Carya texana Buckley (K; F)

Lamiaceae
Hedeoma hispida Pursh (A, K, S)
Hyptis alata (Raf.) Shinners (C)
**Lamium amplexicaule* L. (C)
Lycopus americanus Muhl. ex W. Bart. (C)
Monarda fistulosa L. (C, A, K)
Monarda lindheimeri Engelm. & Gray ex Gray (C)
Monarda punctata L. (C)
Physostegia intermedia (Nutt.) Engelm. & Gray (C, S)
Physostegia virginiana (L.) Benth. ssp. *praemorsa* (Shinners) Cantino (C)
Prunella vulgaris L. (K, M, S)
Pycnanthemum albescens Torr. & Gray (C, A, K)
Pycnanthemum muticum (Michx.) Pers. (C)
Pycnanthemum tenuifolium Schrad. (C, A, K, M)
Salvia azurea Michx. ex Lam. (C, A, K, M)
Salvia lyrata L. (C, A, K, M)
Scutellaria elliptica Muhl. ex Spreng. (A)
Scutellaria integrifolia L. (C, A, K, S)
Scutellaria ovata Hill (K)
Scutellaria parvula Michx. (C, K, S)
Stachys crenata Raf. (C)
Stachys floridana Shuttlw. ex Benth. (C)
Teucrium canadense L. (C, K)

Lauraceae
Persea palustris (Raf.) Sarg. (A; F)
Sassafras albidum (Nutt.) Nees (C, A, K)

Lentibulariaceae
Pinguicula pumila Michx. (C)
Utricularia subulata L. (C)

Linaceae
Linum medium (Planch.) Britt. var. *texanum* (Planch.) Fern. (C, K, S)
***Linum rigidum* Pursh, not in Louisiana (= *L. medium* or *L. striatum*)
Linum striatum Walter (K)
Linum sulcatum Riddell (C)

Loganiaceae
Gelsemium sempervirens (L.) St. Hil. (A, K, S; F)
Mitreola petiolata (J.F. Gmel.) Torr. & Gray (C, K)
Mitreola sessilifolia (J.F. Gmel.) G. Don (C)
Spigelia marilandica (L.) L. (A, K; F)

Lythraceae
Ammannia coccinea Rottb. (S)
**Cuphea glutinosa* Cham. & Schlecht. (C)
Lythrum alatum var. *lanceolatum* (Ell.) Torr & Gray ex Rothrock (C, A, K, M)

Magnoliaceae
Magnolia grandiflora L. (A; F)
Magnolia virginiana L. (A; F)

Malvaceae
Callirhoe papaver (Cav.) Gray (C, K)
Hibiscus moscheutos L. ssp. *lasiocarpos* (Cav.) O.J. Blanchard (C)
Modiola caroliniana (L.) G. Don (C)
Sida rhombifolia L. (C)

Melastomataceae
Rhexia mariana L. (C)

Meliaceae
**Melia azedarach* L. (C)

Menispermaceae
Cocculus carolinus (L.) DC. (A, K; F)

Molluginaceae
**Mollugo verticillata* L. (C)

Monotropaceae
Monotropa hypopithys L. (K; F)

Moraceae
Maclura pomifera (Raf.) Schenid. (K, M; F)
Morus rubra L. (A, K; F)

Myricaceae
Morella cerifera (L.) Small (C, A, K; F)

Nyctaginaceae
Mirabilis albida (Walt.) Helmerl (S)

Nyssaceae
Nyssa sylvatica Marsh (C, A, K; F)

Oleaceae
Chionanthus virginicus L. (A, K; F)
Forestiera ligustrina (Michx.) Poir (S)
Fraxinus americana L. (A, K; F)
Fraxinus pennsylvanica Marsh (M, S; F)
**Ligustrum sinense* Lour. (A, S; F)

Onagraceae
Gaura lindheimeri Engelm. & Gray (C)
Gaura longiflora Spach (C, A, K)
Ludwigia alternifolia L. (S)

Continued on the next page

Table 5.1. *Continued*

Ludwigia decurrens Walt. (C)
Ludwigia glandulosa Walt. (C, S)
Ludwigia hirtella Raf. (C)
Ludwigia leptocarpa (Nutt.) Hara (C)
Ludwigia linearis Walt. (C)
Ludwigia palustris (L.) Ell. (C)
**Ludwigia peploides* (Kunth) Raven (S)
Oenothera biennis L. (C)
Oenothera laciniata Hill (C, S)
Oenothera linifolia Nutt. (C, S)
Oenothera pilosella Raf. ssp. *sessilis* (Pennell) Straley (C)
Oenothera spachiana Torr. & Gray (C, S)
Oenothera speciosa Nutt. (C, K, M)

Orobanchaceae
Epifagus virginiana (L.) W. Bart. (A; F)

Oxalidaceae
Oxalis corniculata L. (M, S)
Oxalis dillenii Jacq. (S)
Oxalis stricta L. (C, A, K)
Oxalis violacea L. (C, K)

Passifloraceae
Passiflora incarnata L. (C, K)
Passiflora lutea L. (A, K; F)

Phytolaccaceae
Phytolacca americana L. (C)

Plantaginaceae
Plantago aristata Michx. (C, S)
Plantago heterophylla Nutt. (C)
Plantago pusilla Nutt. (S)
Plantago virginica L. (C, K, S)

Platanaceae
Platanus occidentalis L. (A, K; F)

Polemoniaceae
Phlox divaricata L. (A, K)
Phlox pilosa L. (C, A, K, M)

Polygalaceae
Polygala cruciata L. (C)
Polygala incarnata L. (C)
Polygala leptocaulis Torr. & Gray (C)
Polygala mariana P. Mill. (C)
Polygala nana (Michx.) DC. (C)
Polygala ramosa Ell. (C)
Polygala sanguinea L. (C)
Polygala verticillata L. (C, K)

Polygonaceae
**Polygonum aviculare* L. (S)
Polygonum hydropiperoides Michx. (C, S)
Polygonum ramosissimum Michx. (S)
Rumex hastatulus Baldw. (S)
Rumex verticillatus L. (C)

Portulacaceae
Claytonia virginica L. (C, S)
Phemeranthus parviflorus (Nutt.) Kiger (S)

Primulaceae
**Anagallis arvensis* L. ssp. *arvensis* (C, A)
**Anagallis arvensis* L. ssp. *foemina* (P. Mill.) Schinz & Thellung (A)
Anagallis minima (L.) Krause (C, A, S)
Lysimachia lanceolata Walt. (K)
Samolus valerandi L. ssp. *parviflorus* (Raf.) Hultén (K; F)

Ranunculaceae
Anemone berlandieri Pritz. (K, M)
Anemone caroliniana Walt. (C, S)
Clematis virginiana L. (A)
Delphinium carolinianum Walt. (K, S, M)
Ranunculus fascicularis Muhl. ex Bigelow (C, A, S)
Ranunculus laxicaulis (Torr. & Gray) Darby (C)
**Ranunculus muricatus* L. (C)
**Ranunculus parviflorus* L. (C, A)
Ranunculus pusillus Poir. (C)
**Ranunculus sardous* Crantz. (K, S)

Rhamnaceae
Berchemia scandens (Hill) K. Koch (C, A, K, M; F)
Ceanothus americanus L. (C, A, K)
Frangula caroliniana (Walt.) Gray (A, K; F)

Rosaceae
Agrimonia microcarpa Wallr. (A, K; F)
Crataegus berberifolia Torr. & Gray (K; F)
Crataegus brachyacantha Sarg. & Engelm. (M; F)
Crataegus crusgalli L. (C, A, K; F)
Crataegus marshallii Egglest. (A, K, M; F)
Crataegus spathulata Michx. (A, K, M; F)
Crataegus viridis L. (K; F)
Geum canadense Jacq. (C, K)
Malus angustifolia (Ait.) Michx. (A; F)
Porteranthus stipulatus (Muhl. ex Willd.) Britt. (A; F)
Potentilla simplex Michx. (K; F)
Prunus mexicana S. Wats. (A, K; F)
Prunus serotina Ehrh. (C, A, K; F)
**Rosa bracteata* J.C. Wendl. (A)
Rosa carolina L. (M)
Rosa setigera Michx. (M)
Rubus argutus Link (C, A, K, S)
Rubus flagellaris Willd. (K)
Rubus trivialis Michx. (C, A, K, M, S)

Rubiaceae
Cephalanthus occidentalis L. (C, K; F)
Diodia teres Walt. (C, K, S)
Diodia virginiana L. (C, K)
Galium aparine L. (C, K)
Galium circaezans Michx. (A, K; F)
Galium pilosum Ait. (A, K)
Galium tinctorium L. (C, A, K, S)
Galium uniflorum Michx. (A, K; F)
Galium virgatum Nutt. (C, A)
Houstonia micrantha (Shinners) Terrell (C, A, M, S)
Houstonia nigricans (Lamarck) Fernald var. *nigricans* (C, A, K, M)
Houstonia purpurea L. var. *calycosa* Gray (K, M)
Houstonia pusilla Schoepf. (S)
Houstonia rosea (Raf.) Terrell (S)
Mitchella repens L. (A, K; F)
Oldenlandia boscii (DC.) Chapman (C, A)
**Sherardia arvensis* L. (A)

Rutaceae
Ptelea trifoliata L. (A; F)
Zanthoxylum clavaherculis L. (C, A, S)

Salicaceae
Salix nigra Marsh. (C)

Sapindaceae
**Cardiospermum halicacabum* L. (C)

Sapotaceae
Sideroxylon lanuginosum Michx. ssp. *oblongifolium* (Nutt.) T.D. Pennington (A, K, M, S)

Saxifragaceae
Lepuropetalon spathulatum Ell. (C, S)
Saxifraga texana Buckley (S)

Scrophulariaceae
Agalinis fasciculata (Ell.) Raf. (C)
Agalinis heterophylla (Nutt.) Small ex Britt. (C, M, S)
Agalinis oligophylla Pennell (C, A, K)

Table 5.1. ***Continued***

Agalinis pulchella Pennell (A)
Agalinis purpurea (L.) Pennell (A)
Agalinis skinneriana (Wood) Britt. (C)
Agalinis tenuifolia (Vahl) Raf. (K, M)
Agalinis viridis (Small) Pennell (C, A)
Aureolaria grandiflora (Benth.) Pennell (K, M; F)
Aureolaria virginica (L.) Pennell (A; F)
Bacopa rotundifolia (Michx.) Wettst. (C)
Buchnera americana L. (C, K)
Gratiola flava Leavenworth (S)
Gratiola neglecta Torr. (C, K)
Gratiola virginiana L. (C, S)
Lindernia dubia (L.) Pennell var. *dubia* (C, A)
Mecardonia acuminata (Walt.) Small (C, A, K)
Nuttallanthus canadensis (L.) D.A. Sutton (C, S)
Nuttallanthus texanus (Scheele) D.A. Sutton (C)
Pedicularis canadensis L. (C, A, K, M)
Penstemon digitalis Nutt. ex Sims (C)
Penstemon laxiflorus Pennell (C, K, M)
Penstemon tubiflorus Nutt. (A, K, M)
**Veronica arvensis* L. (C)
Veronica peregrina L. (C)

Solanaceae
Physalis angulata L. (C)
Physalis heterophylla Nees (C, A)
Physalis longifolia Nutt. var. *subglabrata* (Mackenzie & Bush) Cronq. (A)
Physalis pumila Nutt. (A, K)
Solanum americanum P. Mill. (C)
Solanum carolinense L. (C, K, S)
Solanum dimidiatum Raf. (C)
Solanum elaeagnifolium Cav. (C)

Sterculiaceae
**Melochia corchorifolia* L. (C)

Styracaceae
Styrax americanus Lam. (C)
Styrax grandifolius Ait. (A, K; F)

Symplocaceae
Symplocos tinctoria (L.) L'Hér. (A, K; F)

Tiliaceae
Tilia americana L. var. *caroliniana* (P. Mill.) Castigl. (A; F)

Ulmaceae
Celtis laevigata Willd. (C, A, K; F)
Celtis tenuifolia Nutt. (A; F)
Ulmus alata Michx. (A, K, M, S; F)
Ulmus americana L. (C, K; F)
Ulmus crassifolia Nutt. (K, M, S; F)
Ulmus rubra Muhl. (A, K; F)

Urticaceae
Boehmeria cylindrica (L.) Sw. (C, K)

Valerianaceae
Valerianella radiata (L.) Dufr. (C, K, M, S)

Verbenaceae
Callicarpa americana L. (A, K; F)
Glandularia canadensis (L.) Nutt. (K, M)
Glandularia pulchella (Sweet) Troncoso (C)
Phryma leptostachya L. (K; F)
Phyla nodiflora (L.) Greene (C, K, M)
**Verbena bonariensis* L. (C)
**Verbena brasiliensis* Vell. (C, A, K, M)
Verbena halei Small (C, K, M, S)
**Verbena litoralis* Kunth (C)

Violaceae
Viola lanceolata L. (C)
Viola pedata L. (K)
Viola sagittata Ait. (C)
Viola sororia Willd. (A)
Viola walteri House (A, K; F)

Viscaceae
Phoradendron leucarpum (Raf.) Reveal & M.C. Johnston (A; F)

Vitaceae
Ampelopsis arborea (L.) Koehne (C, A, K, M; F)
Parthenocissus quinquefolia (L.) Planch. (C, A, K; F)
Vitis aestivalis Michx. (K; F)
Vitis cinerea (Engelm.) Millard (C, A, K)
Vitis palmata Vahl (K; F)
Vitis rotundifolia Michx. (A, K; F)
Vitis vulpina L. (K; F)

Table 5.2.

Number of plant species in common across the prairie systems in Louisiana.

Systems	Species in common
Cajun–Anacoco	38
Cajun–Keiffer	34
Cajun–Saline	39
Cajun–Morse Clay	5
Anacoco–Keiffer	102
Anacoco–Saline	4
Anacoco–Morse Clay	3
Keiffer–Saline	6
Keiffer–Morse Clay	13
Saline–Morse Clay	0
Cajun–Anacoco–Keiffer	48
Cajun–Anacoco–Morse Clay	5
Cajun–Anacoco–Saline	9
Cajun–Keiffer–Morse Clay	9
Cajun–Keiffer–Saline	13
Cajun–Saline–Morse Clay	0
Anacoco–Keiffer–Morse Clay	14
Anacoco–Keiffer–Saline	2
Anacoco–Saline–Morse Clay	0
Keiffer–Morse Clay–Saline	6
Cajun–Anacoco–Keiffer–Morse Clay	45
Cajun–Anacoco–Keiffer–Saline	31
Cajun–Anacoco–Morse Clay–Saline	2
Cajun–Keiffer–Morse Clay–Saline	6
Anacoco–Keiffer–Morse Clay–Saline	10
Cajun–Anacoco–Keiffer–Morse Clay–Saline	17

6

The Native Flora of Grasslands and Associated Woodlands in the Grand Prairie Ecoregion of Eastern Arkansas

C. Theo Witsell, Thomas L. Foti, and Brent T. Baker

Introduction

The Grand Prairie of eastern Arkansas is one of the most highly altered ecosystems in North America (Heitmeyer et al. 2000). As the region was settled by Euro-Americans, extensive native grasslands were almost entirely converted to farmland or were lost to the encroachment of trees and shrubs. Dramatic changes in ecosystem processes following settlement included changes in the scope and periodicity of fire, extirpation of large native herbivores, interruption of seasonal surface water sheet flow, and changes in the flooding regimes of surface streams. By 1930, almost all presettlement prairie grasslands had been converted to agriculture, and most suitable land was in rice production (Heitmeyer et al. 2000). Unplowed remnants of the original prairie are few, small, and isolated in a highly fragmented landscape. GIS analyses conducted by the authors compared prairies mapped in the General Land Office (GLO) survey prior to 1830 (approximately 161,875 ha [400,000 ac]) with surviving prairie remnants as of 2014 (approximately 182 ha [450 ac]). This loss, approximately 99.9%, is among the highest for any of Arkansas's native habitats.

Recent interest in grassland conservation by several state and federal agencies, private conservation organizations, and the general public has given rise to plans to plant local-genotype prairie vegetation, primarily warm-season grasses, on canal banks, marginal agricultural lands, and other suitable sites. A project to restore prairie vegetation on and around the Stuttgart Airport in Prairie County has been under way for several years, and additional restorations are ongoing at Downs Prairie Natural Area, Raft Creek Wildlife Management Area, and Wattensaw Wildlife Management Area. Significant funding has already been invested in the mass production of local-genotype native grasses for use in these restorations.

Despite this recent interest, however, there has never been a comprehensive floristic inventory of the Grand Prairie grasslands and associated woodlands. In 1914, botanist and ecologist Roland Harper published some observations on a brief trip he made to the Grand Prairie in June 1911. In his account he commented that "to this day no one seems to have attempted to list the plants growing [in the Grand Prairie], or published any photographs of the vegetation." He went on to opine that "the fact that this very interesting vegetation, traversed by a comparatively old and important railroad, has remained undescribed so long seems to indicate a singular apathy on the part of modern phytogeographers" (R. Harper 1914). Sadly, his observation still stands a century later.

The little floristic work that has been done was either sporadic, was focused on only one or a few of the small prairie remnants, or was conducted in the distant past, often on lands that have since been converted to agriculture or other uses. A comprehensive inventory of all remaining prairie remnants has long been needed to determine all the botanical components of the prairie at present and, to the extent possible from sparse historical records, in the past. Results of such an inventory will be useful in determining

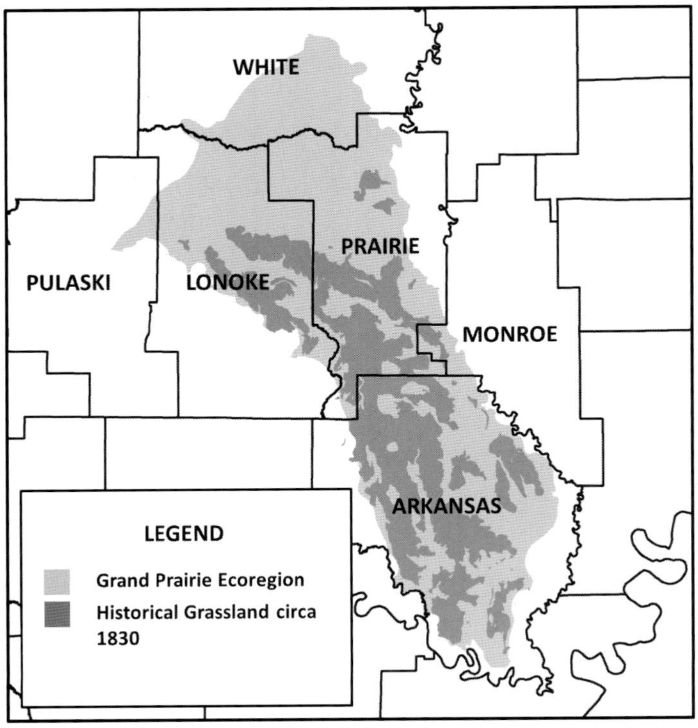

Figure 6.1. Map of the Grand Prairie ecoregion showing major historical open grasslands mapped in the GLO survey (circa 1830).

which plant species are appropriate in restoration sites and in specific microhabitats. We also hope it will be used to guide restoration efforts to replace the full suite of species native to the region. Because of the severe fragmentation of the ecosystem and the ubiquitous non-native invasive species, it is unlikely that these missing native species will return to restoration sites on their own.

The Grand Prairie

The Grand Prairie (level IV) ecoregion covers approximately 364,218 ha (900,000 ac) in Arkansas, Lonoke, Monroe, Prairie, Pulaski, and White Counties in eastern Arkansas (fig. 6.1) and is entirely within the larger Mississippi Alluvial Plain (level III) ecoregion (A. Woods et al. 2004). The Grand Prairie is something of an anomaly within the Mississippi Alluvial Plain in that it is older, higher, and flatter, and much of it was historically covered with different vegetation (grassland) than the rest (Brye and Moreno 2006; Brye and West 2004; Gustafson et al. 1999; R. Harper 1914; Heitmeyer et al. 2000; Irving 1980; Irving et al. 1980). In contrast to the bulk of the Mississippi Alluvial Plain, where alluvium has been deposited within the past 20,000 years or so, and often much more recently, the Grand Prairie has not been reworked by rivers within the past 100,000 years (Saucier 1994).

Geomorphologically, the Grand Prairie is a high, flat terrace underlain by deep deposits of clay laid down by the Arkansas River. Its western border is typically about 3 to 9 m (ca. 10 to 30 ft) higher than the adjacent Bayou Meto bottomlands, but its eastern border is up to about 30 m (100 ft) higher than the adjacent White River bottomlands. Tributary streams, in cutting down to the White River, have in places dissected the terrace into a series of hills up to 16 km (10 mi) wide. In some areas the White River cuts into the Grand Prairie terrace and forms notable bluffs.

Presettlement vegetation in the Grand Prairie consisted of a mosaic of grasslands, wetlands, savannas and open woodlands, and forests. As recently as the nineteenth century, approximately 161,875 ha (400,000 ac) of this region supported prairie grassland and herbaceous wetland, and much of the remainder was prairie-related flatwoods and open upland woodlands. Prairie habitat is confined to high terraces, typically above 64 m (210 ft) in elevation, and is mostly on Calloway, Crowley, and Stuttgart soils (Heitmeyer et al. 2000). These prairies occur on a thin layer of loess topsoil over impervious clay subsoil that lies 30 to 45 cm (12 to 18 in) under the soil surface (Irving 1980; Irving et al. 1980). This subsoil creates the droughty conditions that favor grasses and forbs over woody vegetation (Irving et al. 1980). At the same time, these large contiguous areas of terrace surface covered with grasses ensured that a fire ignited by lightning or by humans could burn thousands to tens of thousands of acres unimpeded. Therefore, the frequency of fire at any given location was increased, and these fires could help maintain the prairie. The prairie probably originated during the hot, dry Hypsithermal period, about 8,000 to 5,000 years ago. The droughty soils and frequent fire allowed the prairie to persist even under the moister conditions of more recent times. The impervious nature of the subsoil also lends itself well to rice farming, a major industry in the area, for which much of the prairie was plowed.

Methods

A combination of extensive field and herbarium work was undertaken by the first author from 2000 to 2014 with the goal of completing a comprehensive floristic inventory of grasslands and related woodlands of the Grand Prairie region. Fieldwork was sporadic, often somewhat opportunistic, and was often done while on the way to do other work elsewhere. Herbarium work was integral to this study. Specimens at ANHC, APCR, BRIT, HDX, HEND/LRU, MO, NLU, STAR, UAM, UARK, UMO, and VDB were examined from the counties in the Grand Prairie ecoregion: Arkansas, Lonoke, Monroe, Prairie, Pulaski, and White. Locality data from specimen labels were analyzed using GIS to determine whether specimens were collected within the boundaries of the Grand Prairie level IV ecoregion (A. Woods et al. 2004). Taxa were included if they were collected from this ecoregion and are considered native; they were collected from one or more of the following habitats: (1) prairies, (2) herbaceous marshes, (3) slash woodlands, (4) prairie-associated flatwoods, (5) prairie-associated upland woodlands, and (6) ditches, roadsides, or other open rights-of-way through these habitats. We recognize noncatastrophic soil disturbance, such as that historically caused by bison and other native animals, as a natural process in prairie ecosystems. As such, we have considered weedy native species, both annual and perennial, found in disturbed grassland habitats to be native and natural elements of the flora. Taxa restricted to bottomland hardwood forests and associated swamps and oxbow lakes along major streams were not considered, nor were taxa restricted to a series of mesic, closed-canopy hardwood forests found in the most protected ravines along the escarpment at the eastern edge of the Grand Prairie (where it drops down to the White River bottomlands). Fire was likely a rare event in these habitats, and they were unlikely to have supported grassland flora historically.

Study Sites

Four state natural areas in the Grand Prairie ecoregion protect prairie remnants: Downs Prairie, Konecny Prairie, Railroad Prairie, and Roth Prairie Natural Areas. These were the primary prairie collection sites for the fieldwork component of this study. Prairie, both disturbed and intact, on the Stuttgart Airport was sampled periodically, and several privately owned prairie remnants were visited where permission for access could be obtained. These include Gray Prairie North, Gray Prairie South, Auersberg Prairie, Weber Prairie, Stewart Prairie, and Fairmont Prairie. Additionally, specimens were examined from a number of other remnants that are no longer extant, and from areas where permission to access was denied by the owners during the present study.

In a region as heavily affected by habitat loss as the Grand Prairie, much of the grassland flora that remains is found in remnants of open woodlands at the edge of areas that were historically prairie. Oak flatwoods and upland woodlands, some extensive and with considerable remnant grassland flora, occur in the Wattensaw Wildlife Management Area, Holland Bottoms Willow Oak Forest Preserve and Wildlife Management Area, Smoke Hole Natural Area, Holloway Memorial Natural Area, Arkansas Post National Memorial, Bayou Des Arc Wildlife Management Area, and upland portions of the Potlatch Conservation Education Center at Cook's Lake. Konecny Grove Natural Area protects a remnant slash woodland and marsh. These were major woodland collection sites. Many other areas along roadsides, ditch banks, and utility rights-of-way hold remnant grassland flora and were sampled opportunistically.

Whenever possible, voucher specimens were collected to document the occurrence of taxa in the region. However, observations by the authors or other knowledgeable staff of the Arkansas Natural Heritage Commission or partner agencies were accepted for some species for which voucher specimens could not be located. These are labeled "observed" in the annotated list of taxa (table 6.1).

History of Botanical Investigations in the Grand Prairie

The English botanist Thomas Nuttall passed through the Grand Prairie in late January 1819 but made few observations on the flora because of the early season (Nuttall 1821). He did remark on several species already beginning to bloom, including what were almost certainly *Claytonia virginica*, *Houstonia pusilla*, and *Nothoscordum bivalve*. He also noted the leaves of *Eryngium yuccifolium* and observed how wet the prairies were at that season, remarking that "the prairie, in consequence of the late rains, appeared almost one continued sheet of water." The earliest plant specimens collected from grasslands in the region are probably those collected by F. L. Harvey in the early 1880s, a number of which survive in the herbaria of the University of Arkansas (UARK), Arkansas Tech University (APCR), Missouri Botanical Garden (MO), and various other national herbaria where Harvey deposited specimens.

Figure 6.2. Prairie remnant in spring with *Castilleja coccinea*, *Orbexilum pedunculatum*, and *Tradescantia occidentalis*. Downs Prairie Natural Area, Prairie County, Arkansas. Photo by Michael Warriner/ANHC.

A brief description and species list were published by R. Harper (1914), but apparently he collected no specimens. E. J. Palmer, working for the Arnold Arboretum, collected from grasslands in the region in the 1920s, Delzie Demaree from the 1940s to the 1970s, Dwight M. Moore from the 1930s to the 1950s, and Sophia McCoy and Marie Locke in the 1970s. Irving (1980) and Irving et al. (1980) provide some of the very few published reports concerning the flora, but these were limited to a few prairie remnants. More recently, collections have been made by the staff of the Arkansas Natural Heritage Commission (1970s to the present, but especially from 2000 to the present by the authors) and by the staff of the University of Arkansas Herbarium (under contract to the Arkansas Natural Heritage Commission) from 2000 to 2005.

Habitats or Plant Communities

Habitats or plant communities considered for this study are those in which graminoid plants are dominant in the herbaceous layer and tree canopy is absent to partially open. These include upland prairie, lowland prairie, disturbed habitat, herbaceous marsh, open oak flatwoods, prairie slash, and open upland woodlands. Each community is described briefly below.

Upland Prairie

Rolling, well-drained uplands on the highest areas of the prairie terrace support dry to dry-mesic tallgrass prairie (fig. 6.2) dominated by *Schizachyrium scoparium* and *Dichanthelium* spp., sometimes with lesser amounts of *Sorghastrum nutans* and *Andropogon gerardii*. Common forbs include *Castilleja coccinea*, *Coreopsis lanceolata*, *Delphinium carolinianum*, *Echinacea pallida*, *Lespedeza virginica*, *Liatris aspera*, *Orbexilum pedunculatum*, *Silphium laciniatum*, *Tephrosia virginiana*, *Tradescantia occidentalis*, and many others. Upland prairie also occurs on scattered nebkhas, or pimple mounds, which are often common in otherwise flat areas of lowland prairie.

Lowland Prairie

Flatter, seasonally wet areas of the prairie terrace support mesic to wet-mesic prairie (fig. 6.3) often dominated by

Figure 6.3. Prairie remnant in fall with *Sorghastrum nutans*, *Panicum virgatum*, and *Helianthus angustifolius*. Konecny Prairie Natural Area, Prairie County, Arkansas. Photo by Michael Warriner/ANHC.

Panicum virgatum, often with *Andropogon gerardii*, *Sorghastrum nutans*, and *Tripsacum dactyloides* codominant (Irving et al. 1980). A variety of sedges (mostly *Carex* and *Eleocharis* spp.) and rushes (*Juncus* spp.) are also common. Common forbs include *Asclepias hirtella*, *A. viridis*, *Eryngium yuccifolium*, *Helianthus angustifolius*, *H. mollis*, *Liatris pycnostachya*, *Rudbeckia grandiflora* var. *alismifolia*, *Tephrosia onobrychoides*, *Tradescantia occidentalis*, and many others.

Disturbed Habitat

While some of the least disturbed prairie remnants retain an assemblage of conservative perennial forbs and grasses, recently disturbed areas often support an entirely different flora. These areas are usually dominated by species not typically found on the less disturbed sites, including a number of weedy annuals and native perennials. Historically, these early-successional species would have had a place in areas impacted by large herbivores and other agents of disturbance, and we regard them as native elements of the grassland ecosystem.

Marsh

Herbaceous marsh habitat (fig. 6.4) has been almost entirely eliminated from the Grand Prairie by the draining or filling of small-depression wetlands for agriculture. However, many marsh species have persisted in agricultural and road ditches (fig. 6.5), especially those adjacent to prairie remnants. Because these habitats serve as "surrogate marshes" in this regard, they were valuable in the present study. Several semi-natural marshes (perhaps with altered hydrology but otherwise more or less intact) remain in the Railroad Prairie Natural Area in Prairie County and in the Konecny Grove Natural Area. Remnant marshes in the Grand Prairie today are dominated by a variety of sedges (mostly *Carex*, *Eleocharis*, *Rhynchospora*, and *Scirpus* spp.) and rushes (*Juncus* spp.), along with *Spartina pectinata*. *Coelorachis rugosa* is common in several marshes on the Railroad Prairie, and *Panicum virgatum* and *Tripsacum dactyloides* are often common around the fringes. Common forbs include *Hibiscus lasiocarpos*, *Ludwigia* spp., *Sagittaria* spp., *Persicaria* spp., and *Hydrolea* spp. In larger marshes, *Typha latifolia* and

Figure 6.4. Herbaceous/shrub marsh dominated by *Persicaria glabra*, *Hibiscus lasiocarpos*, and *Cephalanthus occidentalis*. Lonoke County, Arkansas. Photo by Bill Shepherd/ANHC.

T. domingensis are common. Woody plants commonly become established in these marshes, as they do not burn as readily as surrounding upland prairie. Common species include *Salix nigra*, *Fraxinus pennsylvanica*, *Cephalanthus occidentalis*, and *Amorpha fruticosa*.

The authors visited marshes at several sites repeatedly and found them to contain viable populations of a number of rare plant species, including *Carex opaca*, *Eleocharis wolfii*, *Coelorachis rugosa*, and *Hypericum adpressum*, a species of global conservation concern. Other rare wetland species, such as *Carex arkansana* and *C. fissa* var. *fissa*, are known today only from ditches but were clearly part of the native marsh flora historically.

Marsh habitat was apparently fairly common in the Grand Prairie historically, and a number of collections made by Harvey in the late 1800s appear to have been made from natural prairie marshes. These collections provide some insight into species that occurred in these habitats historically but that are apparently gone from the region today. Examples include a specimen of *Gratiola brevifolia*, collected "about ponds, Grand Prairie, Carlisle" by Harvey in July 1884, apparently on the same trip as he collected *Spiraea tomentosa*, another rare wet-prairie and marsh-edge species no longer found in the region. One year earlier, in July 1883, Harvey made the only known Arkansas collections of *Platanthera nivea* from a wet spot in a prairie in Arkansas County. Two duplicate specimens are known, at UARK and AMES; the label on the AMES specimen notes, "Flowers white, odor like tuber roses, quite plentiful" (G. Johnson 2004). Other rare wetland species apparently persisted into the twentieth century, such as *Sabatia campanulata*, which was collected by Harvey in 1884 and again by Delzie Demaree near Carlisle in 1941 but has not been documented from the Grand Prairie since.

Flatwoods

In extensive areas slightly lower than the highest terrace surface, generally along moderate-sized drainages (often ancient channels of the Arkansas River that formed the terrace), water accumulates by precipitation and sheet flow

Figure 6.5. Ditches in the region, such as this one between a road and a rice field, serve as surrogate habitats for herbaceous marsh species like *Thalia dealbata*, *Rhynchospora corniculata*, and *Leersia oryzoides*. Lonoke County, Arkansas. Photo by Bill Shepherd/ANHC.

from the adjacent higher prairie land and supports flatwoods (fig. 6.6). These flatwoods are in many ways similar to those of the West Gulf Coastal Plain and those elsewhere in the Mississippi Alluvial Plain to the east.

Historically, many prairies in the region transitioned into these open flatwoods communities. These flatwoods occur in the flatter portions of older, higher terraces that rarely if ever flood from adjacent streams. Hydrologic input is strictly from precipitation; the ground is often very wet during the winter and spring (when rainfall is high and transpiration low), and dry during the remainder of the year. This hydroxeric moisture regime is characteristic of both the prairies and other grasslands of the region. Many flatwoods remnants exist in the Grand Prairie, though most are denser today than they were historically because of fire suppression, which may be reflected in the herbaceous flora found in them today. Typical flatwoods in the region are dominated by *Quercus phellos*, *Q. stellata*, *Q. nigra*, and *Fraxinus pennsylvanica* but also contain many other upland and bottomland species present in response to slight differences in elevation (Heitmeyer et al. 2000). The herbaceous layer in more shaded examples is dominated by shade-tolerant sedges such as *Carex tribuloides*, *C. reniformis*, *C. ozarkana*, *C. squarrosa*, *C. typhina*, *C. intumescens*,

Figure 6.6. Oak flatwoods with graminoid understory dominated by *Carex* spp. Arkansas County, Arkansas. Photo by Bill Shepherd/ANHC.

C. flaccosperma, *C. caroliniana*, *C. complanata*, and *Eleocharis tenuis* var. *verrucosa*, along with grasses such as *Chasmanthium laxum*, *Leersia virginica*, and *Agrostis perennans*. More-open examples have a flora more similar to that of the adjacent prairies.

Slash

Prairie slash (fig. 6.7) is a dynamic community that occurs at the upper extremities of small streams in the region, where they just begin to dissect the flat surface of the prairie terrace. These are transitional wetland habitats that fluctuated between prairie and woodland conditions under historical fire regimes. In times of frequent fires, slashes were essentially wet prairies. In times of infrequent fires, they became shrublands or woodlands dominated by early-successional woody species while retaining a diverse herbaceous flora. Konecny Grove Natural Area in Prairie County contains the only protected example of slash woodland in the Grand Prairie. A GLO survey section line forms the western boundary of this area and provides a good illustration of the dynamic nature of this community. While Konecny Grove is mostly woodland today, it was described as being wet prairie during the original survey in 1815, and no woodland was mentioned until one-eighth of a mile south of the southern boundary of the area (Arkansas Land Commissioner 1999).

The canopy at Konecny Grove is dominated by *Celtis laevigata*, *Fraxinus pennsylvanica*, *Ulmus americana*, and *Crataegus viridis*, with a shrub layer of *Ilex decidua*, *Cornus foemina*, and *Sideroxylon lycioides*. The herbaceous layer is extensive and is dominated by a number of sedges, most extensively *Carex hyalinolepis*, which forms dense stands that share dominance with *Elymus virginicus* var. *virginicus* and *Symphyotrichum lanceolatum*. Other characteristic species include *Commelina virginica*, *Cinna arundinacea*, *Iris brevicaulis*, and *Hymenocallis liriosme*. Rare species include *Spiranthes odorata* and *Crataegus* × *canescens*, a single-site endemic shrub.

The term "slash" was not widely used on early herbarium specimens but can sometimes be inferred from label descriptions. For example, a number of collections made by Delzie Demaree with habitat data given as "bottom thickets in

Figure 6.7. Slash woodland dominated by *Celtis laevigata*, *Fraxinus pennsylvanica*, *Ulmus americana*, *Diospyros virginiana*, and *Crataegus viridis*, with an understory of *Carex* spp., *Juncus* spp., and *Spartina pectinata*. Konecny Grove Natural Area, Prairie County, Arkansas. Photo by Michael Warriner/ANHC.

prairie" were likely from slash. R. Harper (1914) described what was likely slash occurring in "very shallow broad elongated depressions or sloughs, which evidently serve as waterways in wet weather." These were "well wooded with oaks, which seem to be mostly *Quercus palustris*." He included a photograph of an open woodland with small trees and a graminoid understory with the caption "*Quercus palustris* in a prairie slough about one and one-half miles south of Hazen, Prairie County."

Upland Woodlands

Some prairies, especially those in more dissected areas, transitioned into open upland oak-hickory woodlands (fig. 6.8). These are better drained, and thus drier, than the flatwoods. Where fires burned from the prairies into the surrounding flatwoods or upland forest, trees would be thinned, allowing the prairie vegetation to occur under a scattering of trees, creating a woodland or post oak savanna. Upland woodlands in the region are typically dominated by *Quercus stellata*, *Q. marilandica*, *Q. falcata*, and *Q. alba*. The shrub layer was likely sparse historically because of frequent fires from adjacent prairies. The herbaceous layer likely ranged from being prairie-like to having extensive areas dominated by *Chasmanthium sessiliflorum* with a diverse mix of woodland forbs.

Results

As a result of this study, 769 taxa of native vascular plants were documented in grasslands and associated woodlands in the Grand Prairie. Table 6.2 lists the number of taxa documented from each habitat (note that some taxa occur in more than one habitat).

The families with the most species were the Asteraceae (112 taxa), Poaceae (103), Cyperaceae (80), Fabaceae (50), Plantaginaceae (20), Lamiaceae (19), Rosaceae (19), Rubiaceae (16), Apiaceae (15), Juncaceae (15), Apocynaceae

Figure 6.8. Upland woodland dominated by *Quercus stellata*. Wattensaw Wildlife Management Area, Prairie County, Arkansas. Photo by Rob Willey/Arkansas Game and Fish Commission.

(14), Fagaceae (14), Onagraceae (13), Euphorbiaceae (12), Polygonaceae (11), Ranunculaceae (11), Hypericaceae (10), and Orobanchaceae (10). The genera with the most species were *Carex* (46 taxa), *Dichanthelium* (19), *Juncus* (14), *Quercus* (14), *Symphyotrichum* (11), *Asclepias* (10), *Cyperus* (10), *Hypericum* (10), *Solidago* (9), *Desmodium* (8), *Eleocharis* (8), *Paspalum* (8), *Helianthus* (7), *Persicaria* (7), *Aristida* (6), *Carya* (6), *Crataegus* (6), *Eragrostis* (6), *Eupatorium* (6), *Houstonia* (6), *Lespedeza* (6), *Ludwigia* (6), *Ranunculus* (6), *Viola* (6), *Vitis* (6), *Agalinis* (5), *Andropogon* (5), *Coreopsis* (5), *Cuscuta* (5), *Euphorbia* (5), *Galium* (5), *Lobelia* (5), *Oenothera* (5), *Panicum* (5), *Rhynchospora* (5), *Sagittaria* (5), *Sisyrinchium* (5), and *Sporobolus* (5).

Plants of Conservation Concern

Forty taxa known from the Grand Prairie are considered rare in Arkansas and are tracked as elements of state conservation concern by the Arkansas Natural Heritage Commission. Specific information on these species is presented in the following pages. They are also noted by the symbol † in table 6.1. Several are considered to be globally rare, while others are rare within the state or region and/or are considerably disjunct from other populations. However, 14 of these are known only from historical specimens, and several have not been documented in the Grand Prairie since the late 1800s or early 1900s. This is no surprise, given the loss of 99.9% of the native prairie in the region. These species are noted by the symbol * in table 6.1.

***Amorpha canescens* (lead-plant)—G5S1**—Known from a single historical (1933) collection from a prairie near Jacksonville in eastern Pulaski County.

***Antennaria neglecta* (prairie pussytoes)—G5S1**—Known from a single historical (1970) collection from Konecny Prairie Natural Area, where it was described as "common." Repeated surveys have failed to relocate it, perhaps because of changes in management since the site was protected (it was managed as a hayfield at the time this species was collected).

***Aristida ramosissima* (S-curve three-awn)—G5S1S2**—Known from a single historical collection (1970) from a roadside in a prairie area.

***Asclepias obovata* (savanna milkweed)—G5?S2**—Rare in high-quality prairie and open flatwoods remnants.

Figure 6.9. *Calopogon oklahomensis* (Oklahoma grass pink orchid), a globally rare species restricted to high-quality unplowed prairie remnants. Prairie County, Arkansas. Photo by Michael Warriner/ANHC.

Callirhoe bushii **(Bush's poppymallow)—G3S3**—Known from a single historical (1970) collection from a roadside in the vicinity of a woodland.

Calopogon oklahomensis **(Oklahoma grass pink orchid)—G3S2**—(fig. 6.9) Rare in high-quality prairie remnants.

Carex arkansana **(Arkansas sedge)—G4S1**—Locally common in at least one roadside ditch in the vicinity of flatwoods and former prairie. Found in both wet, open flatwoods and low prairie elsewhere in Arkansas.

Carex fissa* var. *fissa **(hammock sedge)—G4T3T4S1**—Found in at least one roadside ditch in the vicinity of former prairie.

Carex lupuliformis **(false hop sedge)—G4S1S2**—Found in a wet flatwoods remnant, where it grows with the superficially similar *C. lupulina*.

Carex opaca **(opaque prairie sedge)—G4S2S3**—Locally common in wet prairies, marsh edges, and ditches in former prairie areas.

Coelorachis rugosa **(wrinkled joint-tail)—G5S2**—Rare at edges of marshes within high-quality remnant prairie.

Crataegus* × *canescens **(Stern's medlar)—G1S1**—This single-site endemic (fig. 6.10A and B) was described in 1990 by Phipps (1990) and is known only from 24 individuals in a remnant slash woodland. Lo et al. (2007) hypothesize, based on molecular evidence, that *C.* × *canescens* arose from intergeneric hybridization between introduced *Mespilus germanica* and a native hawthorn, perhaps *C. brachyacantha*. However, neither of these putative parents is known to occur today or to have occurred historically in the region.

Cuscuta glomerata **(rope dodder)—G5S1**—Rare in

Figure 6.10A. *Crataegus* × *canescens* (Stern's medlar) is a single-site endemic known in the wild from just 24 individuals in a remnant slash woodland at Konecny Grove Natural Area, Prairie County, Arkansas. Photo by John Moore.

Figure 6.10B. Close-up of *Crataegus* × *canescens* (Stern's medlar). Konecny Grove Natural Area, Prairie County, Arkansas. Photo by John Moore.

Figure 6.11. *Oenothera pilosella* ssp. *sessilis* (prairie evening-primrose), a globally rare species of prairies and open flatwoods in the region. Railroad Prairie Natural Area, Prairie County, Arkansas. Photo by Devon Holimon.

high-quality remnant prairies, often growing on composites (*Vernonia missurica*, *Helianthus angustifolius*, *Solidago* spp., etc.).

***Dichanthelium dichotomum* ssp. *roanokense* (Roanoke rosette grass)—G5TNRS1**—Rare in low prairie at edge of marshes within high-quality remnant prairie.

***Eleocharis wolfii* (Wolf's spikerush)—G3G4S3**—Locally common in wet prairies, marsh edges, and ditches in former prairie areas.

***Gaura sinuata* (wavy-leaf gaura)—G4G5SH**—Known from a single historical (1968) collection from a remnant prairie along a railroad right-of-way.

***Gratiola brevifolia* (sticky hedge-hyssop)—G4S3**—Known from a single historical (1884) collection labeled "about ponds, Grand Prairie, Carlisle." Plants on this specimen appear similar to ones found in depressions of pine savannas and flatwoods in the Gulf Coastal Plain and different from plants in other Arkansas populations, which are found along mountain streams, mostly in the Ouachita Mountains (D. Estes, pers. comm.).

***Gymnopogon brevifolius* (short-leaf skeleton grass)—G5S2**—Rare in high-quality remnant prairies.

***Hypericum adpressum* (creeping St. John's-wort)—G3S1**—Rare in wet depressions and edges of marshes within remnant prairie.

***Koeleria macrantha* (prairie June grass)—G5S2**—Rare in high-quality upland prairie remnants.

***Lathyrus pusillus* (low vetchling)—G5?S2**—Rare in prairie remnants.

***Muhlenbergia glabrifloris* (inland muhly)—G4?S1**—Rare in prairie remnants.

***Nemastylis geminiflora* (celestial-lily)—G4S3**—Rare in upland prairie remnants.

***Oenothera pilosella* ssp. *sessilis* (prairie evening-primrose)—G5T2S2**—(fig. 6.11) Rare in prairie remnants.

***Paspalum boscianum* (bull paspalum)—G5S1**—Known

Figure 6.12. Specimen of *Platanthera nivea* (snowy orchid) collected from the Grand Prairie in July 1883 by F. L. Harvey. This species has not been documented in Arkansas since then. Photo by Jennifer Ogle/University of Arkansas Herbarium.

from several historical collections (1970 and earlier) from marshes and wet prairies.

Platanthera nivea **(snowy orchid)—G5SH**—(fig. 6.12) Known from two historical (1883) specimens, believed to be duplicates, one at UARK and one at AMES. The UARK specimen gives the locality simply as "Prairies, East Arkansas," but the AMES specimen specifies "Ark. Co. Grand Prairie" and includes the note that the plants were "quite plentiful." It has not been documented in Arkansas since.

Polygala incarnata **(pink milkwort)—G5S1S2**—Known from historical (1955 and earlier) specimens collected from prairie remnants.

Prenanthes aspera **(prairie rattlesnake-root)—G4?S2S3**—Known from historical (1966 and earlier) specimens collected from prairie remnants.

Prunus pumila **var.** ***susquehanae*** **(sand cherry)—G5T4S1**—Rare in two prairie remnants.

Pycnanthemum verticillatum **(whorled mountain-mint)—G5S1**—Rare in high-quality prairie remnants.

Sabatia campanulata **(slender rose-gentian)—G5S1**—Known from historical (1941 and earlier) specimens collected from prairie remnants.

Solidago speciosa **ssp.** ***speciosa*** **var.** ***rigidiuscula*** **(narrow-leaf showy goldenrod)—G5T4S2S3**—Uncommon in high-quality prairie remnants.

Solidago tortifolia **(twist-leaf goldenrod)—G4G5S2—** Uncommon in high-quality prairie remnants.

Spiraea tomentosa **(hardhack)—G5S2**—Known from a single historical (1884) collection from "Prairies N. of Carlisle."

Spiranthes odorata **(fragrant ladies'-tresses orchid)—G5S1**—Locally common in a remnant slash woodland and herbaceous marsh.

Spiranthes praecox **(giant ladies'-tresses orchid)—G5S1S2**—Known from a single site in seasonally wet hardwood flatwoods.

Sporobolus junceus **(pineywoods dropseed)—G5S1S2**—Known from a single site in high-quality upland prairie.

Tradescantia virginiana **(Virginia spiderwort)—G5S1**—Known from a single historical (1968) collection from a prairie remnant.

Trifolium carolinianum **(Carolina clover)—G5S1?**—Known from a single historical (1939) collection from "Des Arc, Prairie Co." No habitat information is given but it is likely prairie or woodland of some kind.

Vernonia lettermannii **(Letterman's ironweed)—G3S3**—This species, which is generally considered endemic to rocky stream banks in the Ouachita Mountains, is known from a single, somewhat surprising historical (1970) collection along a roadside in Prairie County. A second non–Ouachita Mountains collection is also known from an area of former prairie in the Arkansas Valley.

Notable Disjunct, Endemic, and Near-Endemic Species

Grasslands in the Grand Prairie ecoregion share floristic affinities with grassland elements of both the Atlantic and Gulf Coastal Plains and the midwestern prairies. Notable "Coastal Plain taxa" that occur (or once occurred) as disjunct elements in the Grand Prairie include *Aristida purpurascens* var. *virgata*, *Asclepias obovata*, *Coelorachis rugosa*, *Drosera brevifolia*, *Gymnopogon brevifolius*, *Paspalum boscianum*, *Platanthera nivea*, *Sabatia campanulata*, *Solidago tortifolia*, *Spiranthes odorata*, *S. praecox*, *Sporobolus junceus*, and the Coastal Plain form of *Gratiola brevifolia*. Notable "midwestern/Great Plains taxa" that occur as disjunct elements include *Amorpha canescens*, *Antennaria neglecta*, *Cuscuta glomerata*, *Gentiana puberulenta*, *Koeleria macrantha*, *Prenanthes aspera*, *Solidago missouriensis*, *S. speciosa* ssp. *speciosa* var. *rigidiuscula*, and *Sporobolus heterolepis*. Examples of western elements include *Carex fissa* var. *fissa*, *Gaura sinuata*, and *Solanum dimidiatum*, while examples of northern or northeastern elements include *Prunus pumila* var. *susquehanae*, *Pycnanthemum verticillatum*, *Spiraea tomentosa*, and *Tradescantia virginiana*.

The only taxon strictly endemic to the Grand Prairie is *Crataegus* × *canescens*. Two other taxa, *Oenothera pilosella* ssp. *sessilis* and *Helenium campestre*, are somewhat narrow regional endemics locally common in high-quality grassland remnants within the Grand Prairie. Several other species found in the Grand Prairie have their main range within or centered on the Interior Highlands. These include *Callirhoe bushii*, *Calopogon oklahomensis*, *Carex arkansana*, *C. opaca*, *C. ozarkana*, *Corydalis crystallina*, *Hypericum pseudomaculatum*, and *Vernonia lettermannii*.

Acknowledgments

We would like to thank those botanists and collectors who walked the Grand Prairie before us and whose specimens laid the foundation for this research. A number of collectors made especially important collections, including Francis Leroy Harvey, Ernest Jesse Palmer, Delzie Demaree, Dwight Moore, Sophia McCoy, Marie Locke, Bill Shepherd, Sarah Nunn, and Matt Nutt. Likewise, the efforts of the curators and staff of the various herbaria inventoried (ANHC, APCR, BRIT, HDX, HEND/LRU, MO, NLU, STAR, UAM, UARK, UMO, and VDB) were crucial to this work. Special thanks are due to Jennifer Ogle at UARK for her willingness to provide open access to that collection and fulfill numerous requests for data and images. Thanks to Craig Fraiser and Devon Holimon for the use of their photos. JoVonn Hill, Dwayne Estes, and Tanya Miller-Witsell provided critical eleventh-hour support and encouragement, and Samantha Schieman helped with technical editing.

Table 6.1.

Annotated checklist of the native grassland and associated woodland flora of the Grand Prairie ecoregion.

The following list gives information on the native flora in the following format:

Scientific name authority—habitat(s); voucher specimen collector and collection number (acronym of herbarium of deposition); county of voucher specimen [year last documented (for historical specimens only)].

Nomenclature generally follows Gentry et al. (2013), with the exception of the *Dichanthelium acuminatum* complex and *Dichanthelium dichotomum* complex, which partially follow Ladd and Thomas (2015). *Viola affinis* is also recognized, as are infraspecific taxa of *Parthenium integrifolium* and *Lindernia dubia*.

† = taxa listed by the Arkansas Natural Heritage Commission as plants of state conservation concern.

* = known only from historical specimens in the region (older than 35 years). Note: the year last documented in the region is given in [brackets] at the end of the entry.

Observed = taxa observed in identifiable condition by the authors but for which no voucher specimen was collected from the study area or found in herbaria.

HABITAT LEGEND:
UP = Upland Prairie
LP = Lowland Prairie
D = Disturbed
M = Marsh
FW = Flatwoods
SL = Slash
UW = Upland Woodlands

Pteridophytes

Aspleniaceae

Asplenium platyneuron (L.) Britton, Sterns & Poggenb.—UW; Witsell 14-0304 (ANHC); Prairie Co.

Dennstaedtiaceae

Pteridium aquilinum (L.) Kuhn—FW, UW; Observed; Lonoke Co.

Isoetaceae

Isoetes melanopoda Gay & Durieu ex Durieu—LP, M, FW; Witsell 08-0089 (ANHC); Arkansas Co.

Ophioglossaceae

Botrychium dissectum Spreng.—UW; Observed; Lonoke Co.

Ophioglossum pycnostichum (Fernald) Á. Löve & D. Löve—FW; Witsell 02-0112 (HEND); Lonoke Co.

Polypodiaceae

Pleopeltis polypodioides (L.) Andrews & Windham ssp. *michauxiana* (Weath.) Andrews & Windham—FW, UW; Observed; Lonoke Co.

Gymnosperms

Cupressaceae

Juniperus virginiana L. var. *virginiana*—UP, FW, UW; Witsell 02-0078 (HEND); Prairie Co.

Angiosperms—Monocots

Acanthaceae

Justicia ovata (Walter) Lindau var. *lanceolata* (Chapm.) R.W. Long—M, FW, SL; Nunn 7040 (UARK); Prairie Co.

Ruellia caroliniensis (J.F. Gmel.) Steud.—UW; Urban 576 (UARK); Prairie Co.

Ruellia humilis Nutt.—UP; McCoy 158 (UARK); Arkansas Co.

Ruellia pedunculata Torr. ex A. Gray ssp. *pedunculata*—UP, UW; McCoy 202 (UARK); Prairie Co.

Adoxaceae

Sambucus canadensis L.—LP, M, FW, SL; Witsell 14-0260 (ANHC); Prairie Co.

Viburnum rufidulum Raf.—UW; Nutt 62 (UARK); Prairie Co.

Altingiaceae

Liquidambar styraciflua L.—LP, FW, SL, UW; Witsell 14-0308 (ANHC); Prairie Co.

Amaranthaceae

Amaranthus hybridus L.—D; Demaree 37994 (BRIT); Lonoke Co.

Amaranthus spinosus L.—D; Demaree 8414 (HXC); Lonoke Co.

Amaranthus tuberculatus (Moq.) J.D. Sauer—D; Demaree 62895 (MO); Arkansas Co.

Anacardiaceae

Rhus aromatica Aiton var. *aromatica*—UW; Witsell 14-0301 (ANHC); Prairie Co.

Rhus copallinum L.—UP, LP, D, UW; Demaree 15469 (SMU); Lonoke Co.

Rhus glabra L.—UP, LP, D, UW; Nunn 8081 (UARK); Lonoke Co.

Toxicodendron pubescens Mill.—UP, UW; Palmer 25052 (UARK); Lonoke Co.

Toxicodendron radicans (L.) Kuntze—UP, LP, FW, SL, UW; Kellum 79 (UAM); Arkansas Co.

Apiaceae

Chaerophyllum tainturieri Hook.—D; Buchholz s.n. (UARK); Lonoke Co.

Cicuta maculata L.—LP, M, FW, SL; Witsell 02-0434 (ANHC); Prairie Co.

Table 6.1. *Continued*

Cynosciadium digitatum DC.—FW, SL; Witsell 02-0441 (ANHC); Prairie Co.

Daucus pusillus Michx.—UP, D; Locke 4450 (UARK); Arkansas Co.

Eryngium prostratum Nutt. ex DC.—LP, M, FW; Baker 08-0075 (ANHC); Prairie Co.

Eryngium yuccifolium Michx.—UP, LP, FW; McCoy 266 (UARK); Prairie Co.

Limnosciadium pinnatum (DC.) Mathias & Constance—LP, M; Witsell 08-0234 (ANHC); Prairie Co.

Polytaenia nuttallii DC.—UP; Witsell 03-0300 (ANHC); Prairie Co.

Ptilimnium capillaceum (Michx.) Raf.—LP, M, FW, SL; Demaree 55964 (VDB); Lonoke Co.

Ptilimnium nuttallii (DC.) Britton—LP; Nunn 8043 (ANHC); Prairie Co.

Sanicula canadensis L.—SL, UW; Nunn 8170 (UARK); Arkansas Co.

Spermolepis echinata (Nutt. ex DC.) A. Heller—UP; Witsell 08-0031 (ANHC); Prairie Co.

Thaspium trifoliatum (L.) A. Gray—FW; Observed; Lonoke Co.

Trepocarpus aethusae Nutt. ex DC.—FW, SL; Urban 475 (UARK); Lonoke Co.

Zizia aurea (L.) W.D.J. Koch—FW, SL; McCoy 74-4 (UARK); Prairie Co.

Apocynaceae

Amsonia tabernaemontana Walter—LP, M, FW, SL; Witsell 03-0051 (ANHC); Prairie Co.

Apocynum cannabinum L.—UP, LP, FW; Witsell 03-0296 (ANHC); Prairie Co.

Asclepias hirtella (Pennell) Woodson—LP; Nunn 7086 (ANHC); Prairie Co.

† *Asclepias obovata* Elliott—LP, FW; Witsell 02-1522 (ANHC); Lonoke Co.

Asclepias perennis Walter—M, FW; Minteer 319 (UARK); Arkansas Co.

Asclepias purpurascens L.—LP, FW; Butcher s.n. (UAM); Arkansas Co.

Asclepias syriaca L.—D; Butcher s.n. (UAM); Arkansas Co.

Asclepias tuberosa L. ssp. *interior* Woodson—UP, UW; Witsell 08-0236 (ANHC); Prairie Co.

Asclepias variegata L.—FW, UW; DMF 220 (UARK); Arkansas Co.

Asclepias verticillata L.—UP; McCoy 323 (UARK); Prairie Co.

Asclepias viridiflora Raf.—UP; Demaree 15421 (BRIT); Arkansas Co.

Asclepias viridis Walter—LP; Witsell 02-0417 (ANHC); Prairie Co.

Cynanchum laeve (Michx.) Pers.—D, M; Witsell 02-1064 (ANHC); Prairie Co.

Trachelospermum difforme (Walter) A. Gray—LP, FW, SL; Witsell 14-0276 (ANHC); Prairie Co.

Aquifoliaceae

Ilex decidua Walter—LP, FW, SL, UW; Witsell 14-0515 (ANHC); Prairie Co.

Araliaceae

Aralia spinosa L.—FW, UW; Urban 730 (UARK); Lonoke Co.

Aristolochiaceae

Aristolochia serpentaria L.—UW; Observed; Lonoke Co.

Asteraceae

Achillea millefolium L.—UP; Witsell 02-0552 (ANHC); Prairie Co.

Ambrosia artemisiifolia L.—D; Shepherd 418 (ANHC); Arkansas Co.

Ambrosia bidentata Michx.—D; Nunn 7494 (ANHC); Prairie Co.

Ambrosia psilostachya DC. in DC. & A. DC.—UP, D; Witsell 02-1322 (ANHC); Prairie Co.

Ambrosia trifida L.—D, M, FW, SL; Nunn 7523 (ANHC); Prairie Co.

†* *Antennaria neglecta* Greene—UP; McCoy 22 (UARK); Prairie Co.; [1970]

Antennaria parlinii Fernald ssp. *fallax* (Greene) R.J. Bayer & Stebbins—UW; Dillion 6 (UARK); Arkansas Co.

Arnoglossum plantagineum Raf.—LP; Witsell 03-0297 (ANHC); Prairie Co.

Bidens aristosa (Michx.) Britton—LP, D, M, FW, SL; Nunn 7377 (ANHC); Prairie Co.

Bidens bipinnata L.—D, FW, UW; Witsell 14-0302 (ANHC); Prairie Co.

Bidens frondosa L.—M, SL; Nunn 7500 (UARK); Arkansas Co.

Boltonia asteroides (L.) L'Hér.—LP; Locke 4995 (UARK); Lonoke Co.

Boltonia diffusa Elliott—LP, M, FW; Nunn 7079 (ANHC); Prairie Co.

Bradburia pilosa (Nutt.) Semple—UP, D, UW; McCoy 376 (UARK); Prairie Co.

Cirsium altissimum (L.) Spreng.—UP, LP, UW; Nunn 7044 (UARK); Prairie Co.

Conoclinium coelestinum (L.) DC. in DC. & A. DC.—LP, M, FW, SL; Witsell 05-1355 (ANHC); Prairie Co.

Conyza canadensis (L.) Cronquist—D; Nunn 7028 (ANHC); Prairie Co.

Coreopsis grandiflora Hogg ex Sweet var. *harveyana* (A. Gray) Sherff—UP, LP; Palmer 25083 (UARK); Prairie Co.

Coreopsis lanceolata L.—LP; Witsell 02-0538 (ANHC); Lonoke Co.

Coreopsis palmata Nutt.—UP; Dale 6052 (UARK); Lonoke Co.

Coreopsis tinctoria Nutt.—D; Witsell 08-0231 (ANHC); Prairie Co.

Coreopsis tripteris L.—LP, FW; Witsell 14-0251 (ANHC); Prairie Co.

Echinacea pallida (Nutt.) Nutt.—UP; Nunn 7975 (ANHC); Prairie Co.

Echinacea purpurea (L.) Moench—UW; Gray s.n. (NLU); Prairie Co.

Eclipta prostrata (L.) L.—LP, D, M; Witsell 14-0339 (ANHC); Prairie Co.

Elephantopus carolinianus Raeusch.—FW, UW; Observed; Lonoke Co.

Elephantopus tomentosus L.—FW, UW; Observed; Lonoke Co.

Erechtites hieraciifolius (L.) Raf. ex DC. in DC. & A. DC. var. *hieraciifolius*—D; Thomas 130987 (NLU); Lonoke Co.

Erigeron annuus (L.) Pers.—D; McCoy 347D (UARK); Arkansas Co.

Erigeron philadelphicus L. var. *philadelphicus*—D; Witsell 04-0295 (ANHC); Prairie Co.

Continued on the next page

Table 6.1. *Continued*

Erigeron strigosus Muhl. ex Willd.—UP, FW; Witsell 08–0191 (ANHC); Prairie Co.

Erigeron tenuis Torr. & A. Gray—UP, UW; Witsell 03–0294 (ANHC); Prairie Co.

Eupatorium capillifolium (Lam.) Small—D; Observed; Arkansas Co.

Eupatorium hyssopifolium L.—LP; Witsell 05–1362 (ANHC); Prairie Co.

Eupatorium perfoliatum L.—LP, M, FW, SL; Witsell 03–0776 (ANHC); Prairie Co.

Eupatorium rotundifolium L.—LP, FW; Smith 1612 (UARK); Prairie Co.

Eupatorium semiserratum DC. in DC. & A. DC.—LP, FW; Witsell 02–1217 (ANHC); Prairie Co.

Eupatorium serotinum Michx.—D, M; Witsell 02–1413 (ANHC); Prairie Co.

Eurybia hemispherica (Alexander) G.L. Nesom—UP, LP, FW; Witsell 02–1274 (ANHC); Lonoke Co.

Euthamia gymnospermoides Greene—LP; Nunn 7378 (ANHC); Prairie Co.

Euthamia leptocephala (Torr. & A. Gray) Greene ex Porter & Britton—LP, FW, SL; Demaree 54714 (UARK); Arkansas Co.

Fleischmannia incarnata (Walter) R.M. King & H. Rob.—FW, SL; Witsell 14–0307 (ANHC); Prairie Co.

Gaillardia aestivalis (Walter) H. Rock—LP; Demaree 54722 (BRIT); Prairie Co.

Gamochaeta argyrinea G.L. Nesom—UP, D; Buchholz s.n. (UARK); Lonoke Co.

Gamochaeta calviceps (Fernald) Cabrera—D, FW; Thomas 170480J (NLU); Lonoke Co.

Gamochaeta pensylvanica (Willd.) Cabrera—UP, D; McDaniel 51940 (MO); Prairie Co.

Gamochaeta purpurea (L.) Cabrera—UP, D; Baker 08–0091 (ANHC); Arkansas Co.

Helenium amarum (Raf.) H. Rock var. *amarum*—D; Nunn 7059 (UARK); Prairie Co.

Helenium campestre Small—UP, FW; Witsell 08–0188 (ANHC); Prairie Co.

Helenium flexuosum Raf.—LP, M, FW, SL; Witsell 04–0753 (ANHC); Lonoke Co.

Helianthus angustifolius L.—LP; Demaree 54744 (UARK); Arkansas Co.

Helianthus divaricatus L.—FW, UW; McCoy 269 (UARK); Arkansas Co.

Helianthus grosseserratus M. Martens—LP; Nunn 7493 (ANHC); Prairie Co.

Helianthus hirsutus Raf.—UW; Locke 4985 (UARK); Lonoke Co.

Helianthus mollis Lam. in Lam. et al.—UP, LP; McCoy 398 (UARK); Arkansas Co.

Helianthus silphioides Nutt.—UP, LP; Witsell 02–1218 (ANHC); Prairie Co.

Helianthus tuberosus L.—FW, SL; Minteer 331 (UARK); Arkansas Co.

Heterotheca subaxillaris (Lam.) Britton & Rusby ssp. *latifolia* (Buckley) Semple—D; Nunn 7492A (ANHC); Prairie Co.

Hieracium gronovii L.—UW; Reid 980 (UARK); Prairie Co.

Iva annua L.—D, M; Witsell 05–1348 (ANHC); Prairie Co.

Krigia cespitosa (Raf.) K.L. Chambers var. *cespitosa*—LP, D; Witsell 08–0195 (ANHC); Arkansas Co.

Krigia dandelion (L.) Nutt.—UP, UW; Witsell 02–0540 (ANHC); Lonoke Co.

Krigia virginica (L.) Willd.—UP; Observed; Prairie Co.

Lactuca canadensis L.—UP, UW; Witsell 02–1081 (ANHC); Prairie Co.

Lactuca floridana (L.) Gaertn.—FW, SL; Thomas 170576 (NLU); Lonoke Co.

Liatris aspera Michx.—UP, LP, UW; Palmer 24322 (UARK); Prairie Co.

* *Liatris hirsuta* Rydb.—UP; Palmer 24337 (UARK); Prairie Co.; [1923]

Liatris pycnostachya Michx. var. *pycnostachya*—LP, FW; McCoy 272 (UARK); Prairie Co.

Liatris squarrulosa Michx.—UW; Thomas 131029 (NLU); Prairie Co.

Mikania scandens (L.) Willd.—M, FW, SL; Witsell 02–1410 (ANHC); Lonoke Co.

Packera glabella (Poir.) C. Jeffrey—M, FW, SL; Witsell 02–0192 (ANHC); Lonoke Co.

Packera plattensis (Nutt.) W.A. Weber & Á. Löve—UP, LP; Moore 32121B (UARK); Arkansas Co.

Packera tomentosa (Michx.) C. Jeffrey—LP, D; Witsell 02–0539 (ANHC); Lonoke Co.

Parthenium integrifolium L. var. *hispidum* (Raf.) Mears—UP; Witsell 03–0288 (ANHC); Prairie Co.

Parthenium integrifolium L. var. *integrifolium*—UP; Nunn 7976 (UARK); Prairie Co.

Pityopsis graminifolia (Michx.) Nutt. var. *latifolia* (Fernald) Semple & F.D. Bowers—UP; Nunn 7074 (UARK); Prairie Co.

Pityopsis graminifolia (Michx.) Nutt. var. *tenuifolia* (Torr.) Semple & F.D. Bowers—UP; Nunn 7404 (UARK); Prairie Co.

Pluchea camphorata (L.) DC. in DC. & A. DC.—M; Demaree 21645 (MO); Arkansas Co.

†* *Prenanthes aspera* Michx.—UP; Demaree 54724 (UARK); Prairie Co.; [1966]

Pseudognaphalium obtusifolium (L.) Hilliard & B.L. Burtt—UP, UW; McCoy 378 (UARK); Lonoke Co.

Pyrrhopappus carolinianus (Walter) DC. in DC. & A. DC.—UP, LP, D; McCoy 289 (UARK); Prairie Co.

Rudbeckia grandiflora (Sweet) C.C. Gmel. ex DC. in DC. & A. DC.var. *alismifolia* (Torr. & A. Gray) Cronquist—UP, LP; Witsell 03–0779 (ANHC); Prairie Co.

Rudbeckia hirta L.—UP, D; McCoy 377 (UARK); Lonoke Co.

Silphium integrifolium Michx. var. *integrifolium*—LP, FW, UW; Witsell 03–0775 (ANHC); Arkansas Co.

Silphium laciniatum L.—UP; Nunn 7054 (UARK); Prairie Co.
Solidago altissima L. ssp. *altissima*—D; Nunn 7391 (ANHC); Arkansas Co.
Solidago missouriensis Nutt.—UP; Witsell 04–0755 (ANHC); Lonoke Co.
Solidago nemoralis Aiton—UP, UW; Witsell 05–1360 (ANHC); Prairie Co.
Solidago odora Aiton ssp. *odora*—UP, UW; Thomas 130996 (NLU); Lonoke Co.
Solidago rigida L. ssp. *rigida*—UP; Witsell 02–1387 (ANHC); Lonoke Co.
Solidago rugosa Mill.—LP, FW; Witsell 05–1353 (ANHC); Prairie Co.
† *Solidago speciosa* Nutt. ssp. *speciosa* var. *rigidiuscula* Torr. & A. Gray—UP, LP; Witsell 02–1270 (ANHC); Lonoke Co.
† *Solidago tortifolia* Elliott—UP, LP; Witsell 02–1269 (UARK); Lonoke Co.
Solidago ulmifolia Muhl. ex Willd. var. *ulmifolia*—FW, UW; Moore 420355 (UARK); Lonoke Co.
Symphyotrichum dumosum (L.) G.L. Nesom—LP, FW; Witsell 14–0477 (ANHC); Prairie Co.
* *Symphyotrichum laeve* (L.) Á. Löve & D. Löve var. *laeve*—UP; Palmer 24336 (UARK); Prairie Co.; [1923]
Symphyotrichum lanceolatum (Willd.) G.L. Nesom—LP, M, FW, SL; Witsell 02–1406 (ANHC); Prairie Co.
Symphyotrichum lateriflorum (L.) Á. Löve & D. Löve—LP, FW, UW; Demaree 8413 (BRIT); Lonoke Co.
* *Symphyotrichum oolentangiense* (Riddell) G.L. Nesom—UP; Demaree 54900 (BRIT); Prairie Co.; [1966]
Symphyotrichum patens (Aiton) G.L. Nesom—UP, UW; Witsell 05–1347 (ANHC); Lonoke Co.
Symphyotrichum pilosum (Willd.) G.L. Nesom var. *pilosum*—D; Witsell 05–1354 (ANHC); Arkansas Co.
Symphyotrichum praealtum (Poir.) G.L. Nesom—LP, M, FW, SL; Witsell 02–1405 (ANHC); Prairie Co.
Symphyotrichum racemosum (Elliott) G.L. Nesom—LP, D; Nunn 7475 (ANHC); Arkansas Co.
Symphyotrichum subulatum (Michx.) G.L. Nesom var. *ligulatum* (Shinners) S.D. Sundberg—LP, M, SL; Witsell 02–1366 (ANHC); Arkansas Co.
Symphyotrichum turbinellum (Lindl.) G.L. Nesom—FW, UW; Witsell 05–1342 (ANHC); Prairie Co.
Verbesina helianthoides Michx.—UP, FW, UW; Demaree 17594 (BRIT); Lonoke Co.
Verbesina virginica L.—SL, UW; Nunn 7527 (UARK); Prairie Co.
Vernonia baldwinii Torr.—UP, D; Demaree 62817 (BRIT); Arkansas Co.
Vernonia gigantea (Walter) Trel. ex Branner & Coville—FW, SL, UW; McCoy 324 (UARK); Prairie Co.
†* *Vernonia lettermannii* Engelm. ex A. Gray—LP; McCoy 74–2 (UARK); Prairie Co.; [1970]
Vernonia missurica Raf.—UP, LP, SL, UW; Witsell 02–1057 (ANHC); Lonoke Co.
Xanthium strumarium L.—D, M; Witsell 14–0270 (ANHC); Prairie Co.

Berberidaceae

Podophyllum peltatum L.—UW; Witsell 15–0004 (ANHC); Lonoke Co.

Betulaceae

Ostrya virginiana (Mill.) K. Koch—UW; Observed; Lonoke Co.

Bignoniaceae

Bignonia capreolata L.—FW, SL, UW; McCoy 444 (UARK); Arkansas Co.
Campsis radicans (L.) Seem. ex Bureau—LP, D, M, FW, SL, UW; Demaree 15499 (MO); Lonoke Co.

Boraginaceae

Myosotis macrosperma Engelm.—LP, FW, SL; Demaree 57655 (MO); Prairie Co.
Myosotis verna Nutt.—LP; Witsell 02–0531 (ANHC); Lonoke Co.
Phacelia glabra Nutt.—UP; Witsell 04–294 (ANHC); Prairie Co.
Phacelia hirsuta Nutt.—UP, UW; McCoy 114 (UARK); Prairie Co.

Brassicaceae

Cardamine bulbosa (Schreb. ex Muhl.) Britton, Sterns & Poggenb.—FW; McCoy 26 (UARK); Arkansas Co.
Cardamine parviflora L. var. *arenicola* (Britton) O.E. Schulz—LP; Witsell 03–0027.5 (ANHC); Prairie Co.
Cardamine pensylvanica Muhl. ex Willd.—M, FW; Witsell 02–0191 (HEND); Prairie Co.
Draba brachycarpa Nutt. ex Torr. & A. Gray—UP, D; Witsell 07–0003 (ANHC); Prairie Co.
Lepidium densiflorum Schrad.—D; Locke 4980 (UARK); Lonoke Co.
Lepidium virginicum L.—UP, D; Witsell 02–1424 (ANHC); Lonoke Co.
Rorippa palustris (L.) Besser var. *fernaldiana* (Butters & Abbe) Stuckey—M; McCoy 290 (UARK); Prairie Co.

Cactaceae

Opuntia cespitosa Raf.—UW; Nunn 8169 (UARK); Arkansas Co.

Campanulaceae

Lobelia appendiculata A. DC.—UP; Witsell 03–0429 (ANHC); Arkansas Co.
Lobelia cardinalis L.—M, FW; Nunn 7162 (UARK); Prairie Co.
Lobelia inflata L.—D, FW; Observed; Lonoke Co.
Lobelia puberula Michx.—LP, FW; McCoy 395 (UARK); Prairie Co.
Lobelia spicata Lam.—FW, UW; Observed; Prairie Co.
Triodanis perfoliata (L.) Nieuwl. ssp. *biflora* (Ruiz & Pav.) Lammars—UP, D; McCoy 129 (UARK); Prairie Co.
Triodanis perfoliata (L.) Nieuwl. ssp. *perfoliata*—UP, D; Witsell 08–0185 (ANHC); Arkansas Co.

Continued on the next page

TABLE 6.1. *Continued*

Cannabaceae

Celtis laevigata Willd.—FW, SL, UW; Witsell 00-0132 (UARK); Prairie Co.

Caprifoliaceae

Lonicera sempervirens L.—FW, UW; Nunn 8180 (UARK); Arkansas Co.

Caryophyllaceae

Minuartia muscorum (Fassett) Rabeler—LP, FW; Witsell 05-0147 (ANHC); Prairie Co.

Sagina decumbens (Elliott) Torr. & A. Gray ssp. *decumbens*—D; Moore 4728 (BRIT); Prairie Co.

Silene antirrhina L.—UP, LP, D; Witsell 09-0076 (ANHC); Prairie Co.

Chenopodiaceae

Chenopodium album L.—D; Witsell 02-1370 (ANHC); Prairie Co.

Dysphania ambrosioides (L.) Mosyakin & Clemants—D; Observed; Prairie Co.

Cistaceae

Lechea mucronata Raf.—UW; Witsell 02-1319 (ANHC); Prairie Co.

Lechea tenuifolia Michx.—UW; McCoy 215 (UARK); Prairie Co.

Convolvulaceae

Cuscuta campestris Yunck.—M; Smith 4135 (UARK); Prairie Co.

Cuscuta compacta Juss. ex Choisy—M, SL; Witsell 04-1130 (ANHC); Lonoke Co.

Cuscuta cuspidata Engelm.—LP; Demaree 22449 (UARK); Lonoke Co.

† *Cuscuta glomerata* Choisy—LP; Witsell 10-0268 (ANHC); Prairie Co.

Cuscuta indecora Choisy var. *longisepala* Yunck.—LP; Witsell 09-479 (ANHC); Prairie Co.

Dichondra carolinensis Michx.—D; Locke 2555 (UARK); Arkansas Co.

Ipomoea hederacea Jacq.—D, SL; Nutt 87 (UARK); Prairie Co.

Ipomoea lacunosa L.—D, M, SL; Nunn 7413 (UARK); Prairie Co.

Ipomoea pandurata (L.) G. Mey.—LP, D, M; Miller s.n. (UARK); Prairie Co.

Cornaceae

Cornus drummondii C.A. Mey.—FW, UW; Sundell 11610 (UAM); Arkansas Co.

Cornus florida L.—UW; Marsico 5221 (UARK); Arkansas Co.

Cornus foemina Mill.—LP, M, FW, SL; Witsell 08-0196 (ANHC); Arkansas Co.

Cucurbitaceae

Melothria pendula L.—D, M, SL; Witsell 02-1058 (ANHC); Prairie Co.

Droseraceae

Drosera brevifolia Pursh—LP; Witsell 03-0426 (ANHC); Arkansas Co.

Ebenaceae

Diospyros virginiana L.—UP, LP, M, FW, SL, UW; Witsell 14-0271 (ANHC); Prairie Co.

Ericaceae

Lyonia ligustrina (L.) DC.var. *foliosiflora* (Michx.) Fernald—FW; Observed; Lonoke Co.

Vaccinium arboreum Marshall—UW; Witsell 14-0282 (ANHC); Prairie Co.

Vaccinium virgatum Aiton—FW, UW; Nunn 8663 (ANHC); Prairie Co.

Euphorbiaceae

Acalypha gracilens A. Gray—LP, D; Witsell 02-0862 (ANHC); Lonoke Co.

Acalypha rhomboidea Raf.—FW, SL; Observed; Arkansas Co.

Acalypha virginica L.—LP, D, FW, SL; Witsell 02-1407 (ANHC); Prairie Co.

Croton capitatus Michx.—D; Witsell 02-1062 (ANHC); Prairie Co.

Croton glandulosus L. var. *septentrionalis* Müll. Arg.—D; Witsell 14-0244 (ANHC); Prairie Co.

Croton monanthogynus Michx.—D; Witsell 14-0263 (ANHC); Prairie Co.

Croton willdenowii G.L. Webster—LP, D, FW; Witsell 07-0450 (ANHC); Prairie Co.

Euphorbia corollata L.—UP, UW; McCoy 162 (UARK); Arkansas Co.

Euphorbia dentata Michx.—D; Witsell 14-0293 (ANHC); Prairie Co.

Euphorbia maculata L.—LP, D; Witsell 05-1361 (ANHC); Prairie Co.

Euphorbia nutans Lag.—D; Witsell 02-1392 (ANHC); Lonoke Co.

Euphorbia spathulata Lam.—LP, D; Witsell 08-0091 (ANHC); Arkansas Co.

Fabaceae

†* *Amorpha canescens* Pursh—UP; Tillman s.n. (UMO); Pulaski Co.; [1933]

Amorpha fruticosa L.—M, FW, SL; Palmer 24333 (UARK); Prairie Co.

Apios americana Medik.—LP, M, SL; Nunn 7058 (UARK); Prairie Co.

Baptisia alba (L.) Vent. var. *macrophylla* (Larisey) Isely—UP, LP; Witsell 03-0287 (UARK); Arkansas Co.

Baptisia bracteata Muhl. ex Elliott var. *leucophaea* (Nutt.) Kartesz & Gandhi—UP, UW; Harper 27 (UARK); Prairie Co.

Centrosema virginianum (L.) Benth.—FW, UW; Rodgers 135A (UAM); Arkansas Co.

Cercis canadensis L.—UP, D, SL, UW; Observed; Prairie Co.

Chamaecrista fasciculata (Michx.) Greene var. *fasciculata*—D; Witsell 14-0247 (ANHC); Prairie Co.

Chamaecrista nictitans (L.) Moench var. *nictitans*—D; Witsell 14-0249 (ANHC); Prairie Co.

Clitoria mariana L.—FW, UW; Nutt 70 (UARK); Prairie Co.

Crotalaria sagittalis L.—UP, LP; Witsell 03-0781 (ANHC); Prairie Co.

Dalea candida Michx. ex Willd. var. *candida*—UP; Witsell 04-752 (ANHC); Lonoke Co.

Desmanthus illinoensis (Michx.) MacMill. ex B.L. Rob. & Fernald—UP, LP, D; Moore 55-306 (UARK); Arkansas Co.

Table 6.1. *Continued*

Desmodium canescens (L.) DC.—UP, UW; Observed; Prairie Co.
Desmodium ciliare (Muhl. ex Willd.) DC.—UP; Witsell 03-0774 (ANHC); Prairie Co.
Desmodium cuspidatum (Muhl. ex Willd.) DC. ex Loudon—UP, UW; Demaree 22444 (SMU); Lonoke Co.
Desmodium glabellum (Michx.) DC.—UW; Witsell 14-0248 (ANHC); Prairie Co.
Desmodium nudiflorum (L.) DC.—UW; Observed; Lonoke Co.
Desmodium paniculatum (L.) DC.—LP, FW, UW; Witsell 14-0341 (ANHC); Prairie Co.
Desmodium perplexum B.G. Schub.—UW; Nutt 107 (UARK); Prairie Co.
Desmodium sessilifolium (Torr. ex M.A. Curtis) Torr. & A. Gray—UP; Shepherd 319 (ANHC); Lonoke Co.
Dioclea multiflora (Torr. & A. Gray) C. Mohr—FW, UW; Reid 973 (UARK); Arkansas Co.
Galactia volubilis (L.) Britton—D, FW, UW; Witsell 14-0286 (ANHC); Prairie Co.
Gleditsia triacanthos L.—UP, LP, D, M, FW, SL, UW; Observed; Prairie Co.
Glottidium vesicarium (Jacq.) R.M. Harper—D, M; Nunn 7420 (UARK); Prairie Co.
Gymnocladus dioicus (L.) K. Koch—UW; Observed; Lonoke Co.
† *Lathyrus pusillus* Elliott—UP, LP; Witsell 05-153 (ANHC); Prairie Co.
Lespedeza capitata Michx.—UP, LP; Witsell 03-0777 (ANHC); Prairie Co.
Lespedeza frutescens (L.) Hornem.—UW; Witsell 14-0295 (ANHC); Prairie Co.
Lespedeza hirta (L.) Hornem.var. *hirta*—FW, UW; Observed; Prairie Co.
Lespedeza procumbens Michx.—UP, UW; Witsell 14-0305 (ANHC); Prairie Co.
Lespedeza repens (L.) W.P.C. Barton—UP, UW; Witsell 14-0265 (ANHC); Prairie Co.
Lespedeza virginica (L.) Britton—UP, LP; Witsell 02-1321 (ANHC); Prairie Co.
Mimosa quadrivalvis L. var. *nuttallii* (DC. ex Britton & Rose) Beard ex Barneby—UP, LP, UW; Nunn 8071 (UARK); Prairie Co.
Mimosa strigillosa Torr & A. Gray—D, M; Minteer 359 (UARK); Arkansas Co.
Orbexilum pedunculatum (Mill.) Rydb. var. *pedunculatum*—UP, LP; Witsell 02-0545 (ANHC); Prairie Co.
* *Orbexilum simplex* (Nutt. ex Torr. & A. Gray) Rydb.—LP; Tucker 7955 (APCR); Prairie Co.; [1969]
Phaseolus polystachios (L.) Britton, Sterns, & Poggenb.—FW, SL; Witsell 14-0311 (ANHC); Prairie Co.
Senna marilandica (L.) Link—LP, FW; Urban 463 (UARK); Lonoke Co.
Sesbania herbacea (Mill.) McVaugh—D, M; Witsell 02-1063 (ANHC); Arkansas Co.
Strophostyles helvola (L.) Elliott—M, FW, SL; Thomas 170408 (NLU); White Co.
Strophostyles leiosperma (Torr. & A. Gray) Piper—LP; Witsell 14-0245 (ANHC); Prairie Co.
Strophostyles umbellata (Muhl. ex Willd.) Britton in Britton & A. Br.—UP, LP, M; Witsell 14-0267 (ANHC); Prairie Co.
Stylosanthes biflora (L.) Britton, Sterns & Poggenb.—UP; Demaree 37668 (BRIT); Prairie Co.
Tephrosia onobrychoides Nutt.—LP, FW; Witsell 03-0612 (ANHC); Arkansas Co.
Tephrosia virginiana (L.) Pers.—UP; Witsell 03-0298 (ANHC); Prairie Co.
†* *Trifolium carolinianum* Michx.—UP? UW? D?; Brown s.n. (UARK); Prairie Co.; [1939]
* *Trifolium reflexum* L.—UP; Palmer 25041 (UARK); Prairie Co.; [1949]
Vicia minutiflora D. Dietr.—D, UW; Kellum 51 (UAM); Arkansas Co.
Wisteria frutescens (L.) Poir.—M, FW, SL, UW; Observed; Lonoke Co.

Fagaceae
Quercus alba L.—UW; Witsell 14-0298 (ANHC); Prairie Co.
Quercus falcata Michx.—UW; Witsell 14-0291 (ANHC); Prairie Co.
Quercus laurifolia Michx.—FW; Nunn 9936 (UARK); Lonoke Co.
Quercus lyrata Walter—FW, SL; Observed; Prairie Co.
Quercus marilandica Münchh.—UP, UW; Witsell 14-0285 (ANHC); Prairie Co.
Quercus michauxii Nutt.—FW, UW; Observed; Lonoke Co.
Quercus nigra L.—FW; Witsell 14-0292 (ANHC); Prairie Co.
Quercus pagoda Raf.—FW, SL; Urban 460 (UARK); Lonoke Co.
Quercus palustris Münchh.—LP, FW; Witsell 14-0047 (UARK); Prairie Co.
Quercus phellos L.—FW; Witsell 14-0284 (ANHC); Prairie Co.
Quercus shumardii Buckley—FW, UW; Observed; Lonoke Co.
Quercus stellata Wangenh.—UP, FW, UW; Witsell 14-0281 (ANHC); Prairie Co.
Quercus texana Buckley—FW; Nunn 9918 (UARK); Lonoke Co.
Quercus velutina Lam. in Lam. et al.—UW; Witsell 14-0290 (ANHC); Prairie Co.

Gentianaceae
Gentiana puberulenta J.S. Pringle—UP, LP; Witsell 02-1390 (ANHC); Prairie Co.
Sabatia brachiata Elliott—UP; Witsell 03-0503 (ANHC); Prairie Co.
†* *Sabatia campanulata* (L.) Torr.—LP; Demaree 22329 (UMO); Lonoke Co.; [1941]
Sabatia campestris Nutt.—LP; Witsell 08-0364 (ANHC); Prairie Co.

Geraniaceae
Geranium carolinianum L. var. *carolinianum*—D; McCoy 125 (UARK); Prairie Co.

Continued on the next page

Table 6.1. *Continued*

Haloragaceae

Proserpinaca palustris L.—M; Witsell 03–0773 (ANHC); Prairie Co.

Hydroleaceae

Hydrolea ovata Nutt. ex Choisy—M; Nunn 7030 (UARK); Prairie Co.

Hydrolea uniflora Raf.—M; Moore 32815 (UARK); Arkansas Co.

Hypericaceae

† *Hypericum adpressum* Raf. ex W.P.C. Barton—M; Baker 08–101 (ANHC); Prairie Co.

Hypericum drummondii (Grev. & Hook.) Torr. & A. Gray—UP, LP, D; Witsell 05–1356 (ANHC); Prairie Co.

Hypericum gentianoides (L.) Britton, Sterns & Poggenb.—UP; Moore 321050a (UARK); Prairie Co.

Hypericum gymnanthum Engelm. & A. Gray—LP, FW; Witsell 02–0867 (ANHC); Lonoke Co.

Hypericum hypericoides (L.) Crantz ssp. *hypericoides*—FW, UW; Witsell 14–0287 (ANHC); Prairie Co.

Hypericum hypericoides (L.) Crantz ssp. *multicaule* (Michx. ex Willd.) N. Robson—UP, LP; Demaree 22211 (BRIT); Prairie Co.

Hypericum lobocarpum Gatt.—LP; Witsell 14–0315 (ANHC); Prairie Co.

Hypericum mutilum L.—LP, M; McCoy 314 (UARK); Prairie Co.

Hypericum pseudomaculatum Bush ex Britton—UP; Nunn 7994 (UARK); Prairie Co.

Hypericum punctatum Lam.—LP, D; Witsell 14–0250 (ANHC); Prairie Co.

Juglandaceae

Carya alba (L.) Nutt. ex Ell.—UW; Witsell 14–0296 (ANHC); Prairie Co.

Carya cordiformis (Wangenh.) K. Koch—UW; Observed; Lonoke Co.

Carya glabra (Mill.) Sweet—UW; Witsell 14–0283 (ANHC); Prairie Co.

Carya illinoinensis (Wangenh.) K. Koch—LP, FW; Witsell 14–0048 (ANHC); Prairie Co.

Carya laciniosa (Michx. f.) G. Don—FW; Observed; Lonoke Co.

Carya ovata (Mill.) K. Koch—FW, UW; Marsico 5218 (UARK); Arkansas Co.

Lamiaceae

Callicarpa americana L.—FW, UW; Observed; Lonoke Co.

Hedeoma hispida Pursh—LP, D; Eads 10 (UAM); Arkansas Co.

* *Hedeoma pulegioides* (L.) Pers.—SL, UW; Demaree 64063 (BRIT); Arkansas Co.; [1971]

Lycopus americanus Muhl. ex W.P.C. Barton—LP, M; Baker 11–0165 (ANHC); Prairie Co.

Lycopus rubellus Moench—FW; Observed; Prairie Co.

Lycopus virginicus L.—M, FW, SL; Witsell 02–1367 (ANHC); Prairie Co.

Monarda fistulosa L. var. *mollis* (L.) Benth.—UP; Witsell 08–0235 (ANHC); Prairie Co.

* *Physostegia intermedia* (Nutt.) Engelm. & A. Gray—M; Harvey s.n. (UARK); Prairie Co.; [1884]

Physostegia virginiana (L.) Benth. ssp. *praemorsa* (Shinners) P.D. Cantino—LP; Demaree 21672 (BRIT); Arkansas Co.

Prunella vulgaris L. ssp. *lanceolata* (W.P.C. Barton) Hultén—LP, D, FW, UW; Reid 1044 (UARK); Arkansas Co.

Pycnanthemum albescens Torr. & A. Gray—UP, FW, UW; Demaree 15472 (UARK); Prairie Co.

Pycnanthemum muticum (Michx.) Pers.—LP, M, FW; Witsell 07–0449 (OSC); Prairie Co.

Pycnanthemum tenuifolium Schrad.—UP, LP, FW; Witsell 07–0451 (OSC); Prairie Co.

† *Pycnanthemum verticillatum* (Michx.) Pers.—LP; Witsell 07–0452 (ANHC); Arkansas Co.

Salvia lyrata L.—UP, D, FW, UW; McCoy 71 (UARK); Prairie Co.

Scutellaria integrifolia L.—FW; Thomas 170501 (NLU); Lonoke Co.

Scutellaria parvula Michx. var. *australis* Fassett—UP; Witsell 02–0543 (ANHC); Lonoke Co.

Stachys tenuifolia Willd.—FW, SL; Demaree 48878 (BRIT); Arkansas Co.

Teucrium canadense L. var. *canadense*—LP, M, FW, SL; Baker 08–0093 (ANHC); Arkansas Co.

Lauraceae

Sassafras albidum (Nutt.) Nees—UP, LP, UW; Nunn 8066 (UARK); Prairie Co.

Linaceae

Linum medium (Planch.) Britton var. *texanum* (Planch.) Fernald—UP; Witsell 03–0414 (ANHC); Arkansas Co.

Linum striatum Walter—FW; Observed; Lonoke Co.

Lythraceae

Ammannia coccinea Rottb.—LP, M; Witsell 08–0225 (ANHC); Prairie Co.

Lythrum alatum Pursh var. *alatum*—LP, M; Witsell 14–0254 (ANHC); Prairie Co.

Lythrum alatum Pursh var. *lanceolatum* (Elliott) Rothr.—LP, M; Witsell 04–1444 (ANHC); Arkansas Co.

Rotala ramosior (L.) Koehne—LP, M; Moore 32744 (UARK); Arkansas Co.

Malvaceae

†* *Callirhoe bushii* Fernald—FW? UW?; McCoy 277 (UARK); Arkansas Co.; [1970]

Hibiscus laevis All.—M; Moore 48–0667 (UARK); Arkansas Co.

Hibiscus lasiocarpos Cav.—M; Nunn 7042 (UARK); Prairie Co.

Melastomataceae

Rhexia mariana L. var. *mariana*—LP, M, FW; Witsell 14–0236 (ANHC); Prairie Co.

Menispermaceae

Cocculus carolinus (L.) DC.—FW, SL; Locke 5884 (UARK); Prairie Co.

Table 6.1. *Continued*

Molluginaceae

Mollugo verticillata L.—D; Demaree 72162 (MO); Lonoke Co.

Moraceae

Morus rubra L.—FW, SL; Observed; Prairie Co.

Myrsinaceae

Lysimachia lanceolata Walter—LP; Baker 08-102 (ANHC); Prairie Co.

Lysimachia radicans Hook.—FW; Observed; Lonoke Co.

Nyctaginaceae

Mirabilis nyctaginea (Michx.) MacMill.—LP, D; Witsell 05-1357 (ANHC); Lonoke Co.

Nyssaceae

Nyssa sylvatica Marshall—UP, UW; Observed; Lonoke Co.

Oleaceae

Fraxinus americana L.—UP, FW, UW; Nunn 7948 (UARK); Prairie Co.

Fraxinus pennsylvanica Marshall—LP, M, FW, SL; Foti s.n. (UARK); Prairie Co.

Onagraceae

Gaura longiflora Spach—UP, D; Witsell 02-1372 (UARK); Prairie Co.

†* *Gaura sinuata* Nutt. ex Ser.—UP, D; Demaree 57888 (BRIT); Lonoke Co.; [1968]

Ludwigia alternifolia L.—LP, M, FW, SL; Witsell 02-1079 (ANHC); Lonoke Co.

Ludwigia decurrens Walter—M; Witsell 05-1335 (UARK); Arkansas Co.

Ludwigia glandulosa Walter—LP, M; Witsell 14-0478 (ANHC); Prairie Co.

Ludwigia linearis Walter—LP, M; Demaree 38024 (SMU); Lonoke Co.

Ludwigia palustris (L.) Elliott—LP, M; Witsell 14-0241 (ANHC); Prairie Co.

Ludwigia peploides (Kunth) P.H. Raven ssp. *glabrescens* (Kuntze) P.H. Raven—M; Moore 321075 (UARK); Arkansas Co.

Oenothera biennis L.—D; Nunn 7045 (UARK); Prairie Co.

Oenothera laciniata Hill—D; Witsell 02-0550 (ANHC); Arkansas Co.

Oenothera linifolia Nutt.—UP, LP; Witsell 02-0554 (ANHC); Lonoke Co.

† *Oenothera pilosella* Raf. ssp. *sessilis* (Pennell) Straley—UP; Witsell 02-0553 (ANHC); Lonoke Co.

Oenothera villosa Thunb. ssp. *villosa*—D; McCoy 381 (UARK); Arkansas Co.

Orobanchaceae

Agalinis fasciculata (Elliott) Raf.—UP, LP; Witsell 14-0342 (ANHC); Prairie Co.

Agalinis heterophylla (Nutt.) Small ex Britton—LP; Nunn 7157 (UARK); Prairie Co.

Agalinis purpurea (L.) Pennell—LP; Witsell 02-1396 (ANHC); Arkansas Co.

Agalinis tenuifolia (Vahl) Raf.—LP; Demaree 21633 (UMO); Arkansas Co.

Agalinis viridis (Small) Pennell—LP, D; Witsell 14-0340 (ANHC); Prairie Co.

Aureolaria flava (L.) Farw.—FW, UW; Urban 452 (UARK); Lonoke Co.

Aureolaria grandiflora (Benth.) Pennell—SL, UW; Nutt 64 (UARK); Prairie Co.

Buchnera americana L.—UP; McCoy 229 (UARK); Prairie Co.

Castilleja coccinea (L.) Spreng.—UP, LP; Witsell 03-0045 (ANHC); Arkansas Co.

Pedicularis canadensis L.—LP; McCoy 59 (UARK); Prairie Co.

Oxalidaceae

Oxalis dillenii Jacq.—UP, LP, D, UW; Witsell 08-0092 (ANHC); Arkansas Co.

Oxalis stricta L.—LP; Demaree 15496 (BRIT); Prairie Co.

Oxalis violacea L.—UP, LP, FW; Witsell 02-0533 (ANHC); Arkansas Co.

Papaveraceae

Corydalis crystallina (Torr. & A. Gray) Engelm. ex A. Gray—UP, D; Witsell 02-0551 (ANHC); Lonoke Co.

Passifloraceae

Passiflora incarnata L.—LP, D, M, SL; Witsell 03-0780 (ANHC); Prairie Co.

Passiflora lutea L.—LP, FW, UW; Reid 1327 (UARK); Prairie Co.

Penthoraceae

Penthorum sedoides L.—M; Demaree 8404 (SMU); Prairie Co.

Phrymaceae

Mimulus alatus Aiton—M, FW; Nunn 7163 (UARK); Prairie Co.

Phyllanthaceae

Phyllanthus caroliniensis Walter ssp. *caroliniensis*—LP, D, M; Witsell 14-0317 (ANHC); Prairie Co.

Phytolaccaceae

Phytolacca americana L. var. *americana*—D, FW, SL; Witsell 14-0269 (ANHC); Prairie Co.

Plantaginaceae

Bacopa rotundifolia (Michx.) Wettst.—M; Witsell 04-0749 (ANHC); Prairie Co.

Callitriche heterophylla Pursh—M; Moore 330087 (UARK); Lonoke Co.

Callitriche terrestris Raf.—LP, FW; Witsell 08-0086 (ANHC); Arkansas Co.

†* *Gratiola brevifolia* Raf.—M; Harvey s.n. (UARK); Lonoke Co.; [1884]

Gratiola neglecta Torr.—LP, M, FW; Witsell 08-0087 (ANHC); Arkansas Co.

Continued on the next page

Table 6.1. *Continued*

Gratiola virginiana L.—M, FW; Witsell 08-0232 (ANHC); Prairie Co.
Lindernia dubia (L.) Pennell var. *anagallidea* (Michx.) Cooperr.—LP, M; Witsell 14-0264 (ANHC); Prairie Co.
Lindernia dubia (L.) Pennell var. *dubia*—LP, M, FW; Nunn 7116 (UARK); Prairie Co.
Mecardonia acuminata (Walter) Small—LP, FW; Witsell 03-1036 (ANHC); Prairie Co.
Nuttallanthus texanus (Scheele) D.A. Sutton—UP, LP, D; Witsell 02-0536 (ANHC); Prairie Co.
Penstemon digitalis Nutt. ex Sims—UP, LP; Witsell 03-0423 (ANHC); Arkansas Co.
Penstemon laxiflorus Pennell—UP; Witsell 08-0099 (ANHC); Arkansas Co.
Penstemon tenuis Small—LP, M, FW; Moore 350051 (UARK); Lonoke Co.
Penstemon tubiflorus Nutt.—UP, LP; Witsell 02-0541 (ANHC); Lonoke Co.
Plantago aristata Michx.—D; Nunn 7971 (UARK); Prairie Co.
Plantago heterophylla Nutt.—LP, D; Witsell 05-0152 (ANHC); Prairie Co.
Plantago pusilla Nutt.—LP, D; Witsell 08-0032 (ANHC); Prairie Co.
Plantago virginica L.—UP, LP, D; Witsell 02-0558 (ANHC); Lonoke Co.
Veronica peregrina L.—LP, M, FW, SL; Witsell 05-0151 (ANHC); Lonoke Co.
Veronicastrum virginicum (L.) Farw.—UP; Foti s.n. (ANHC); Prairie Co.

Polemoniaceae
Phlox glaberrima L.—LP; Witsell 08-0090 (ANHC); Prairie Co.
Phlox pilosa L. ssp. *ozarkana* (Wherry) Wherry—UP, UW; Witsell 03-0109 (UARK); Prairie Co.
Phlox pilosa L. ssp. *pilosa*—UP, UW; McGraw 56 (UARK); Arkansas Co.

Polygalaceae
†* *Polygala incarnata* L.—UP; Moore 55-266 (UARK); Prairie Co.; [1955]
Polygala mariana Mill.—LP; Witsell 02-0863 (ANHC); Prairie Co.
Polygala sanguinea L.—UP, LP; Witsell 02-0534 (ANHC); Arkansas Co.
Polygala verticillata L.—UP, LP; Witsell 02-0865 (ANHC); Arkansas Co.

Polygonaceae
Brunnichia ovata (Walter) Shinners—D, M, FW, SL; Demaree 70364 (MO); Prairie Co.
Persicaria amphibia (L.) Gray—M; Moore 32749 (UARK); Arkansas Co.
Persicaria bicornis (Raf.) Nieuwl.—LP, M; Witsell 14-0316 (ANHC); Prairie Co.
Persicaria glabra (Willd.) G. Maza—M; Witsell 02-1061 (ANHC); Prairie Co.
Persicaria hydropiperoides (Michx.) Small—M, FW, SL; Witsell 02-1385 (ANHC); Prairie Co.
Persicaria lapathifolia (L.) Gray—D, M; Witsell 02-1364 (ANHC); Prairie Co.
Persicaria pensylvanica (L.) M. Gómez—M; Demaree 72149 (MO); Lonoke Co.
Persicaria punctata (Elliott) Small—LP, M, FW, SL; Witsell 02-1423 (ANHC); Prairie Co.
Rumex altissimus A.W. Wood—LP, M; Witsell 09-0077 (ANHC); Prairie Co.
Rumex hastatulus Baldwin in Elliott—D; Dale 6005 (UARK); Arkansas Co.
Rumex verticillatus L.—M; Moore 510217 (UARK); Arkansas Co.

Portulacaceae
Claytonia virginica L.—UP, LP, FW, UW; Witsell 03-0047 (ANHC); Arkansas Co.

Primulaceae
Anagallis minima (L.) E.H.L. Krause—LP; Witsell 11-0136 (ANHC); Prairie Co.
Primula meadia (L.) Mast & Reveal—UP; Palmer 25047 (UARK); Prairie Co.

Ranunculaceae
Anemone caroliniana Walter—UP; McCoy 34 (UARK); Arkansas Co.
Clematis crispa L.—FW; Buchholz s.n. (UARK); Lonoke Co.
Clematis virginiana L.—LP, SL; Witsell 09-0487 (ANHC); Prairie Co.
Delphinium carolinianum Walter ssp. *carolinianum*—UP; Witsell 03-0299 (ANHC); Prairie Co.
Myosurus minimus L.—FW, SL; Witsell 05-136 (ANHC); Prairie Co.
Ranunculus abortivus L.—LP, SL; Witsell 05-0145 (ANHC); Prairie Co.
Ranunculus fascicularis Muhl. ex J.M. Bigelow—UP; McCoy 6 (UARK); Prairie Co.
Ranunculus hispidus Michx. var. *nitidus* (Chapm.) T. Duncan—FW, SL; Witsell 02-0429 (ANHC); Prairie Co.
Ranunculus laxicaulis Darby—LP, M; Witsell 03-0312 (ANHC); Lonoke Co.
Ranunculus pusillus Poir. in Lam. et al.—M; Palmer 25076 (UARK); Prairie Co.
Ranunculus sceleratus L. var. *sceleratus*—M; Witsell 08-0093 (UARK); Prairie Co.

Rhamnaceae
Berchemia scandens (Hill) K. Koch—LP, FW, UW; Demaree 62921 (BRIT); Arkansas Co.
Ceanothus americanus L.—UP; Witsell 03-0415 (ANHC); Prairie Co.

Rosaceae
Agrimonia rostellata Wallr.—UW; Nutt 95 (UARK); Prairie Co.

Table 6.1. *Continued*

† *Crataegus × canescens* (J.B. Phipps) T.A. Dickinson & E.Y.Y. Lo—SL; Locke 5892 (UARK); Prairie Co.
Crataegus berberifolia Torr. & A. Gray—UW; Shepherd 329 (APCR); Prairie Co.
Crataegus crus-galli L.—UP, UW; Nunn 8168 (UARK); Arkansas Co.
Crataegus marshallii Eggl.—FW, UW; Cottey 55 (APCR); Prairie Co.
Crataegus spathulata Michx.—UW; Shepherd 330 (APCR); Prairie Co.
Crataegus viridis L.—LP, FW, SL; Baker 08-0103 (ANHC); Arkansas Co.
Geum canadense Jacq.—FW, SL, UW; Witsell 02-1070 (ANHC); Prairie Co.
Gillenia stipulata (Muhl. ex Willd.) Nutt. in W.P.C. Barton—FW, UW; Nutt 65 (UARK); Prairie Co.
Potentilla simplex Michx.—UP, LP, FW, UW; Demaree 47648 (MO); Lonoke Co.
Prunus angustifolia Marshall—UP, LP; Witsell 02-0076 (HEND); Prairie Co.
Prunus mexicana S. Watson—FW, SL, UW; Dale 7032 (UARK); Prairie Co.
† *Prunus pumila* L. var. *susquehanae* (Willd.) H. Jaeger—UP, LP; Baker 10-0227 (ANHC); Prairie Co.
Prunus serotina Ehrh.—UP, LP, D, FW, UW; Sundell 8463 (UAM); Prairie Co.
Rosa carolina L.—UP, LP, UW; Witsell 08-0187 (ANHC); Arkansas Co.
Rubus laudatus Berger—UP, LP; Nunn 7967 (UARK); Prairie Co.
Rubus roribaccus (L.H. Bailey) Rydb.—UP, LP; Palmer 25030 (UARK); Prairie Co.
Rubus trivialis Michx.—D; Chenowith 10EX (UARK); Lonoke Co.
†* *Spiraea tomentosa* L.—LP; Harvey s.n. (UARK); Prairie Co.; [1884]

Rubiaceae
Cephalanthus occidentalis L.—M; Nunn 7035 (UARK); Prairie Co.
Diodia teres Walter—UP, D; McCoy 288 (UARK); Prairie Co.
Diodia virginiana L.—LP, M, FW, SL; McCoy 281 (UARK); Prairie Co.
Galium aparine L.—D, FW, SL; Witsell 02-0433 (ANHC); Prairie Co.
Galium circaezans Michx.—FW, UW; Observed; Lonoke Co.
Galium obtusum Bigelow—LP, M, FW, SL; Witsell 03-0290 (ANHC); Prairie Co.
Galium pilosum Aiton—FW, UW; Observed; Prairie Co.
Galium tinctorium L.—FW, SL; Witsell 02-0190 (ANHC); Lonoke Co.
Houstonia caerulea L.—UP, LP, FW; Buchholz s.n. (UARK); Lonoke Co.
Houstonia longifolia Gaertn.—UP; Moore 510206 (UARK); Arkansas Co.
Houstonia micrantha (Shinners) Terrell—LP, D; Gray s.n. (NLU); Prairie Co.
Houstonia purpurea L. var. *purpurea*—UP, FW; Observed; Lonoke Co.
Houstonia pusilla Schoepf—LP, D; Hardcastle 8 (UARK); Lonoke Co.
Houstonia rosea (Raf.) Terrell—LP, D; Witsell 07-0001 (ANHC); Prairie Co.
Mitchella repens L.—FW, UW; Observed; Lonoke Co.
Oldenlandia boscii (DC.) Chapm.—LP, M; Witsell 14-0242 (ANHC); Prairie Co.

Salicaceae
Salix humilis Marshall—UP, LP; McCoy 4 (UARK); Prairie Co.
Salix nigra Marshall—LP, M, SL; Witsell 02-0201 (HEND); Arkansas Co.

Santalaceae
Comandra umbellata (L.) Nutt. ssp. *umbellata*—UP; Witsell 03-0111 (ANHC); Prairie Co.

Sapindaceae
Acer negundo L.—M, FW, SL; Witsell 14-0313 (ANHC); Prairie Co.
Acer rubrum L. var. *drummondii* (Hook. & Arn. ex Nutt.) Sarg.—FW, SL; Urban 723 (UARK); Lonoke Co.
Acer rubrum L. var. *rubrum*—FW, SL, UW; Observed; Lonoke Co.
Aesculus pavia L.—UW; Palmer 24341 (UARK); Prairie Co.

Sapotaceae
Sideroxylon lycioides L.—FW, SL; Witsell 01-0626 (ANHC); Prairie Co.

Saururaceae
Saururus cernuus L.—M, FW, SL; McCoy 220 (UARK); Prairie Co.

Saxifragaceae
* *Micranthes texana* (Buckley) Small—UP; McCoy 1 (UARK); Prairie Co.; [1970]

Solanaceae
Physalis angulata L.—D, M; Witsell 02-1393 (ANHC); Lonoke Co.
Physalis pubescens L.—D; Demaree 64092 (BRIT); Arkansas Co.
Physalis virginiana Mill.—UW; Witsell 14-0306 (ANHC); Prairie Co.
Solanum carolinense L.—D; Nunn 7039 (UARK); Prairie Co.
Solanum dimidiatum Raf.—D; Witsell 02-0422 (ANHC); Prairie Co.
Solanum ptychanthum Dunal—D, SL; Witsell 02-1408 (ANHC); Lonoke Co.

Styracaceae
Styrax americanus Lam. in Lam. et al.—M, FW, SL; Witsell 04-0289 (ANHC); Prairie Co.

Tetrachondraceae
Polypremum procumbens L.—LP, M; Witsell 03-0615 (ANHC); Arkansas Co.

Ulmaceae
Ulmus alata Michx.—UP, SL, UW; Witsell 14-0262 (ANHC); Prairie Co.

Continued on the next page

Ulmus americana L.—LP, FW, SL; Nunn 8055 (UARK); Prairie Co.
Ulmus crassifolia Nutt.—FW; Observed; Prairie Co.
Ulmus rubra Muhl.—LP, FW, SL; Witsell 14-0268 (ANHC); Prairie Co.

Urticaceae
Boehmeria cylindrica (L.) Sw.—M, FW, SL; Nunn 7095 (UARK); Prairie Co.
Parietaria pensylvanica Muhl. ex Willd.—LP, D, FW; Witsell 05-0506 (ANHC); Prairie Co.
Pilea pumila (L.) A. Gray—FW, SL; Witsell 14-0309 (ANHC); Prairie Co.

Valerianaceae
Valerianella radiata (L.) Dufr.—UP, LP, D; Witsell 02-0529 (ANHC); Prairie Co.

Verbenaceae
Phyla lanceolata (Michx.) Greene—LP, M; McCoy 380 (UARK); Arkansas Co.
Verbena bracteata Lag. & Rodr.—D; Demaree 37763 (BRIT); Prairie Co.
Verbena urticifolia L.—LP, D, FW, SL; Witsell 02-1068 (ANHC); Prairie Co.

Violaceae
Viola affinis Leconte—FW, SL; Witsell 02-0194 (HEND); Prairie Co.
Viola bicolor Pursh—D; Kellum 002 (UAM); Arkansas Co.
Viola palmata L.—FW, UW; Observed; Prairie Co.
Viola pedata L.—UP; McCoy 37 (UARK); Arkansas Co.
Viola sagittata Aiton—LP, FW; Witsell 03-0044 (ANHC); Prairie Co.
Viola sororia Willd. var. *sororia*—UW; McCoy 18 (UARK); Prairie Co.

Vitaceae
Ampelopsis arborea (L.) Koehne—LP, M, FW; Minter 340 (UARK); Arkansas Co.
Ampelopsis cordata Michx.—M, FW; Thomas 128841 (MO); Arkansas Co.
Parthenocissus quinquefolia (L.) Planch.—FW, UW; Observed; Lonoke Co.
Vitis aestivalis Michx.—FW, UW; Observed; Prairie Co.
Vitis cinerea (Engelm. in A. Gray) Engelm. ex Millardet—UP, LP, SL; Witsell 02-1422 (UARK); Prairie Co.
Vitis palmata Vahl—M, FW, SL; Witsell 01-0629 (ANHC); Lonoke Co.
Vitis riparia Michx.—M, SL; Witsell 02-0428 (ANHC); Prairie Co.
Vitis rotundifolia Michx.—UW; Witsell 14-0303 (ANHC); Prairie Co.
Vitis vulpina L.—FW, SL; Witsell 02-1060 (ANHC); Prairie Co.

Angiosperms—Monocots

Agavaceae
Camassia scilloides (Raf.) Cory—UP; McCoy 79 (UARK); Arkansas Co.
Manfreda virginica (L.) Salisb. ex Rose—UP, UW; Witsell 03-0778 (ANHC); Prairie Co.

Alismataceae
Alisma subcordatum Raf.—M, FW; Observed; Lonoke Co.
Echinodorus cordifolius (L.) Griseb. ssp. *cordifolius*—M; Toler 24 (UARK); Arkansas Co.
Sagittaria brevirostra Mack. & Bush—M; Moore 420350 (UARK); Lonoke Co.
Sagittaria graminea Michx.—M; Palmer 25042 (UARK); Prairie Co.
Sagittaria montevidensis Cham. & Schtdl. ssp. *calycina* (Engelm.) Bogin—M; Witsell 14-0259 (ANHC); Prairie Co.
Sagittaria papillosa Buchenau—M; McCoy 191 (UARK); Prairie Co.
Sagittaria platyphylla (Engelm.) J.G. Sm.—M; Nunn 8079 (ANHC); Prairie Co.

Alliaceae
Allium canadense L. var. *canadense*—D, FW, SL; McCoy 124 (UARK); Prairie Co.
Allium canadense L. var. *mobilense* (Regel) Ownbey—UP; Hunter 17 (UARK); Prairie Co.
Nothoscordum bivalve (L.) Britton in Britton & A. Br.—UP, LP, FW; Witsell 03-0046 (ANHC); Prairie Co.

Amaryllidaceae
Hymenocallis liriosme (Raf.) Shinners—M, FW, SL; Witsell 03-0112 (ANHC); Prairie Co.
Hymenocallis occidentalis (Leconte) Kunth var. *occidentalis*—FW; McCoy 294 (UARK); Prairie Co.

Araceae
Arisaema dracontium (L.) Schott—FW; Observed; Prairie Co.

Colchicaceae
Uvularia sessilifolia L.—FW; Observed; Lonoke Co.

Commelinaceae
Commelina virginica L.—M, FW, SL; Witsell 00-0126 (UARK); Lonoke Co.
Tradescantia hirsutiflora Bush—UP, LP; Witsell 03-0418 (ANHC); Lonoke Co.
Tradescantia occidentalis (Britton) Smyth var. *occidentalis*—UP, D, FW, SL; Witsell 02-0542 (ANHC); Arkansas Co.
Tradescantia ohiensis Raf.—UW; Moore 350221 (UARK); Prairie Co.
†* *Tradescantia virginiana* L.—LP; Demaree 57470 (BRIT); Arkansas Co.; [1968]

Cyperaceae
Carex abscondita Mack.—FW; Witsell 09-0079 (ANHC); Prairie Co.
Carex albicans Willd. ex Spreng.var. *albicans*—UW; Observed; Prairie Co.
Carex albolutescens Schwein.—LP, FW, SL; Witsell 02-0348 (ANHC); Prairie Co.

Table 6.1. *Continued*

Carex amphibola Steud.—SL; Witsell 02–346 (ANHC); Prairie Co.
Carex annectens (E.P. Bicknell) E.P. Bicknell—M; Witsell 03–0308 (MO); Lonoke Co.
† *Carex arkansana* (L.H. Bailey) L.H. Bailey—M, FW; Hyatt 10,385 (UARK); Prairie Co.
Carex aureolensis Steud.—LP, M, FW, SL; Witsell 02–1083 (ANHC); Arkansas Co.
Carex austrina (Small) Mack.—LP; Witsell 07–0099 (ANHC); Prairie Co.
Carex blanda Dewey—FW, SL, UW; Hyatt 7302 (UARK); Prairie Co.
Carex brevior (Dewey) Mack. ex Lunell—LP; Witsell 08–0193 (ANHC); Arkansas Co.
Carex bushii Mack.—UP; Witsell 04–0292 (ANHC); Lonoke Co.
Carex caroliniana Schwein.—LP, FW; Davis 2239 (VDB); Prairie Co.
Carex cephalophora Muhl ex Willd.—FW, UW; Observed; Prairie Co.
Carex cherokeensis Schwein.—FW, SL; Reznicek 9262 (UARK); Arkansas Co.
Carex complanata Torr. & Hook.—LP, FW, SL; Witsell 02–0425 (ANHC); Prairie Co.
Carex corrugata Fernald—FW, SL; Hyatt 8402 (UARK); Arkansas Co.
Carex crus-corvi Shuttlew. ex Kunze—M, FW; Locke 3042 (UARK); Arkansas Co.
Carex debilis Michx.—FW; Observed; Lonoke Co.
Carex festucacea Schkuhr ex Willd.—LP; Witsell 08–0141 (ANHC); Arkansas Co.
† *Carex fissa* Mack. in Britton et al. var. *fissa*—LP, M; Reznicek 9271 (MICH); Lonoke Co.
Carex flaccosperma Dewey—LP, FW, SL; Witsell 02–0432 (ANHC); Prairie Co.
Carex glaucodea Tuck. ex Olney—LP, FW; Bryson 12310 (UARK); Prairie Co.
Carex grayi Carey—FW; Observed; Lonoke Co.
Carex hirsutella Mack.—UP, UW; Dale 7033 (UARK); Prairie Co.
Carex hyalinolepis Steud.—M, FW; Witsell 07–0132 (ANHC); Arkansas Co.
Carex intumescens Rudge—FW, SL; Nunn 9914 (UARK); Lonoke Co.
Carex joorii L.H. Bailey—M, FW; Miller s.n. (UARK); Prairie Co.
Carex leavenworthii Dewey—LP, D; Moore 330101 (UARK); Arkansas Co.
Carex louisianica L.H. Bailey—FW; Buchholz s.n. (UARK); Lonoke Co.
† *Carex lupuliformis* Sartwell ex Dewey—FW; Baker 14–0126 (ANHC); Lonoke Co.
Carex lupulina Muhl. ex Willd.—M, FW, SL; Baker 08–095 (ANHC); Arkansas Co.
Carex meadii Dewey—UP; Witsell 05–0148 (ANHC); Prairie Co.
Carex muehlenbergii Schkuhr ex Willd.—UP, UW; Moore 51020 (UARK); Arkansas Co.
Carex oklahomensis Mack.—LP, M; Witsell 03–0307 (ANHC); Arkansas Co.
Carex oligocarpa Schkuhr ex Willd.—FW, UW; Witsell 09–0078 (ANHC); Prairie Co.
† *Carex opaca* (F.J. Herm.) P.E. Rothr. & Reznicek—LP, M; Witsell 03–0427 (UARK); Arkansas Co.
Carex oxylepis Torr. & Hook.—FW; Adair 3315 (UARK); Arkansas Co.
Carex ozarkana P.E. Rothr. & Reznicek—LP, M; Davis 2047 (VDB); Prairie Co.
Carex reniformis (L.H. Bailey) Small—M, FW; Nunn 7941 (UARK); Prairie Co.
Carex retroflexa Muhl. ex Willd.—FW, UW; Moore 510204 (UARK); Arkansas Co.
Carex squarrosa L.—LP, FW, SL; Demaree 52137 (VDB); Prairie Co.
Carex triangularis Boeck.—M, FW; Witsell 02–0426 (ANHC); Prairie Co.
Carex tribuloides Wahlenb. var. *sangamonensis* Clokey—LP, FW, SL; Bryson 12317 (UARK); Prairie Co.
Carex typhina Michx.—FW, SL; Nutt 93 (UARK); Prairie Co.
Carex umbellata Schkuhr ex Willd.—LP, FW, UW; Bryson 12314 (UARK); Prairie Co.
Carex vulpinoidea Michx.—M; Witsell 03–0310 (ANHC); Prairie Co.
Cyperus acuminatus Torr. & Hook.—LP, M; Witsell 08–0227 (ANHC); Prairie Co.
Cyperus croceus Vahl—LP, D; Dale 6001 (UARK); Arkansas Co.
Cyperus echinatus (L.) A.W. Wood—UP, UW; Witsell 05–1337 (ANHC); Prairie Co.
Cyperus erythrorhizos Muhl.—M; Witsell 02–1067 (ANHC); Prairie Co.
Cyperus esculentus L.—LP, M, FW, SL; Locke 927 (UARK); Prairie Co.
Cyperus lancastriensis Porter ex A. Gray—LP; Witsell 14–0234; Prairie Co.
Cyperus odoratus L.—M; Kessler 3081 (NLU); Prairie Co.
Cyperus pseudovegetus Steud.—M; Baker 08–0081 (ANHC); Arkansas Co.
Cyperus refractus Engelm. ex Boeck.—LP; Witsell 05–1336 (ANHC); Prairie Co.
Cyperus strigosus L.—LP, M; Nunn 7424 (UARK); Prairie Co.
Eleocharis engelmannii Steud.—M; Toler 27 (UARK); Arkansas Co.
Eleocharis erythropoda Steudel—M; Hyatt 12120 (UARK); Arkansas Co.
Eleocharis lanceolata Fernald—M; Moore 32764 (UARK); Arkansas Co.
Eleocharis obtusa (Willd.) Schult.—M; Witsell 14–0279 (ANHC); Prairie Co.
Eleocharis palustris (L.) Roem. & Schult. in Roem. et al.—M; Witsell 07–0133 (ANHC); Prairie Co.

Continued on the next page

Table 6.1. *Continued*

Eleocharis quadrangulata (Michx.) Roem. & Schult. in Roem. et al.—M; Witsell 09-0472 (ANHC); Prairie Co.

Eleocharis tenuis (Willd.) Schult. var. *verrucosa* (Svenson) Svenson—LP; Witsell 03-0601 (ANHC); Arkansas Co.

† *Eleocharis wolfii* (A. Gray) A. Gray ex Britton in Patt.—LP, M; Witsell 03-0424 (ANHC); Prairie Co.

* *Fimbristylis annua* (All.) Roem. & Schult. in Roem. et al.—M; Moore 32763 (UARK); Arkansas Co.; [1932]

Fimbristylis autumnalis (L.) Roem. & Schult. in Roem. et al.—D, M; Smith 4154 (UARK); Lonoke Co.

Fimbristylis puberula (Michx.) Vahl var. *puberula*—UP, LP; Witsell 03-0413 (UARK); Prairie Co.

Fimbristylis vahlii (Lam.) Link—M; Nunn 7098 (UARK); Prairie Co.

Isolepis carinata Hook. & Arn. ex Torr.—LP, D, M; Witsell 05-0138 (ANHC); Arkansas Co.

Rhynchospora capitellata (Michx.) Vahl—M; Smith 4159 (UARK); Lonoke Co.

Rhynchospora corniculata (Lam.) A. Gray—M; Nunn 7029 (UARK); Prairie Co.

Rhynchospora glomerata (L.) Vahl.—LP, M; Witsell 14-0314 (ANHC); Prairie Co.

Rhynchospora harveyi W. Boott var. *harveyi*—UP, LP; Witsell 08-0223 (ANHC); Prairie Co.

Rhynchospora recognita (Gale) Kral—LP, M; Witsell 03-0419 (ANHC); Prairie Co.

Scirpus cyperinus (L.) Kunth—M; Smith 4107 (UARK); Prairie Co.

Scirpus georgianus R.M. Harper—M; Witsell 08-0237 (ANHC); Prairie Co.

Scleria ciliata Michx. var. *ciliata*—UP, LP; Witsell 08-0222 (ANHC); Prairie Co.

Scleria oligantha Michx.—UW; Witsell 14-0297 (ANHC); Prairie Co.

Scleria pauciflora Muhl. ex Willd. var. *pauciflora*—UP, LP; Witsell 04-758 (ANHC); Lonoke Co.

Scleria triglomerata Michx.—LP; Witsell 14-0319 (ANHC); Prairie Co.

Dioscoreaceae

Dioscorea villosa L.—FW, UW; Observed; Prairie Co.

Hypoxidaceae

Hypoxis hirsuta (L.) Coville—UP, LP; Witsell 02-0535 (ANHC); Prairie Co.

Iridaceae

Iris brevicaulis Raf.—LP, SL; Witsell 02-0435 (ANHC); Prairie Co.

Iris fulva Ker Gawl.—M, FW; Sundell 7118 (UAM); Arkansas Co.

† *Nemastylis geminiflora* Nutt.—UP; Culwell 4887 (UCAC); Prairie Co.

Sisyrinchium albidum Raf.—UP; Witsell 08-0097 (ANHC); Arkansas Co.

Sisyrinchium angustifolium Mill.—LP; Palmer 25061 (UARK); Prairie Co.

Sisyrinchium atlanticum E.P. Bicknell—LP; Witsell 08-0135 (ANHC); Prairie Co.

Sisyrinchium langloisii Greene—LP; Locke 5330 (UARK); Arkansas Co.

Sisyrinchium rosulatum E.P. Bicknell—D; Kellum 81 (UAM); Arkansas Co.

Juncaceae

Juncus acuminatus Michx.—M; Palmer 25066 (UARK); Prairie Co.

Juncus anthelatus (Wiegand) R.E. Brooks—M; Witsell 08-0226 (ANHC); Prairie Co.

Juncus biflorus Elliott—LP, M; Witsell 08-0224 (ANHC); Lonoke Co.

Juncus brachycarpus Engelm. in A. Gray—LP, M; Witsell 03-0420 (UARK); Arkansas Co.

Juncus dichotomus Elliott—LP; Adair 339 (UARK); Arkansas Co.

Juncus diffusissimus Buckley—LP, M; Witsell 08-0233 (ANHC); Lonoke Co.

Juncus dudleyi Wiegand—LP; Dale 6054 (UARK); Arkansas Co.

Juncus effusus L.—M; Witsell 08-0194 (ANHC); Arkansas Co.

Juncus interior Wiegand—LP, M, FW; Witsell 08-0192 (UARK); Arkansas Co.

Juncus marginatus Rostk.—LP, M, FW; Smith 4106 (UARK); Prairie Co.

Juncus nodatus Coville in Britton & A. Br.—M; Witsell 08-221 (ANHC); Prairie Co.

Juncus scirpoides Lam. in Lam. et al.—LP, M; Dale s.n. (UARK); Prairie Co.

Juncus tenuis Willd.—D, FW; Witsell 02-0423 (ANHC); Lonoke Co.

Juncus validus Coville var. *validus*—M; Nunn 9982 (UARK); Lonoke Co.

Luzula bulbosa (A. Wood) Smyth & L.C.R. Smyth—UP; Witsell 03-0028 (ANHC); Prairie Co.

Marantaceae

Thalia dealbata Fraser ex Roscoe—M; Witsell 02-1193 (ANHC); Prairie Co.

Nartheciaceae

Aletris farinosa L.—UP, LP; Witsell 08-0190 (ANHC); Prairie Co.

Orchidaceae

† *Calopogon oklahomensis* D.H. Goldman—UP, LP; Witsell 08-0134 (ANHC); Prairie Co.

Platanthera lacera (Michx.) G. Don in Sweet—UP; Witsell 08-0189 (ANHC); Prairie Co.

†* *Platanthera nivea* (Nutt.) Luer—LP; Harvey s.n. (UARK); Arkansas Co.; [1883]

Spiranthes cernua (L.) Rich.—UP, LP; Witsell 08-0540 (ANHC); Prairie Co.

† *Spiranthes odorata* (Nutt.) Lindl.—M, FW, SL; Witsell 02-1415 (ANHC); Prairie Co.

† *Spiranthes praecox* (Walter) S. Watson in A. Gray et al.—FW; Nunn 9935 (UARK); Lonoke Co.

Spiranthes vernalis Engelm. & A. Gray—UP, LP; Witsell 08-0365 (ANHC); Arkansas Co.

Table 6.1. *Continued*

Poaceae

Agrostis elliottiana Schult.—UP; Witsell 08-0088 (ANHC); Arkansas Co.

Agrostis hyemalis (Walter) Britton, Sterns & Poggenb.—UP, LP, D; Witsell 02-1549 (UARK); Lonoke Co.

Agrostis perennans (Walter) Tuck.—FW, UW; Witsell 05-1341 (ANHC); Prairie Co.

Alopecurus carolinianus Walter—D, M; Witsell 02-0559 (HEND); Lonoke Co.

Andropogon gerardii Vitman—UP, LP, UW; Moore 32820 (UARK); Arkansas Co.

Andropogon glomeratus (Walter) Britton, Sterns & Poggenb.—LP, M; Locke 952A (UARK); Prairie Co.

Andropogon gyrans Ashe var. *gyrans*—UP; Witsell 03-1034 (ANHC); Prairie Co.

Andropogon ternarius Michx. var. *ternarius*—UP; Witsell 02-1368 (ANHC); Prairie Co.

Andropogon virginicus L. var. *virginicus*—UP, LP, D, M, FW, SL, UW; Locke 953A (UARK); Prairie Co.

Aristida dichotoma Michx. var. *dichotoma*—UP, D; Witsell 05-1343 (ANHC); Prairie Co.

* *Aristida longespica* Poir. var. *longespica*—UP, D; Moore 321059 (UARK); Prairie Co.; [1932]

Aristida oligantha Michx.—UP, LP, D; Witsell 02-1384 (ANHC); Prairie Co.

Aristida purpurascens Poir. var. *purpurascens*—UP; Palmer 24327 (UARK); Prairie Co.

Aristida purpurascens Poir. var. *virgata* (Trin.) Allred—LP; Witsell 02-1383 (ANHC); Prairie Co.

†* *Aristida ramosissima* Engelm. ex A. Gray—UP, D; Nielsen 4196 (UARK); Arkansas Co.; [1936]

Arundinaria gigantea (Walter) Muhl.—FW, SL, UW; Witsell 14-0312 (ANHC); Prairie Co.

Axonopus fissifolius (Raddi) Kuhlm.—LP; Baker 08-083 (ANHC); Prairie Co.

Chasmanthium latifolium (Michx.) Yates—FW, SL, UW; Moore 32817 (UARK); Arkansas Co.

Chasmanthium laxum (L.) Yates—FW; Smith 4131 (UARK); Prairie Co.

Chasmanthium sessiliflorum (Poir.) Yates—FW, UW; Hypolite s.n. (UARK); Prairie Co.

Cinna arundinacea L.—LP, FW, SL; Witsell 02-1070 (ANHC); Prairie Co.

Coelorachis cylindrica (Michx.) Nash—UP, LP; Nunn 8092 (UARK); Arkansas Co.

† *Coelorachis rugosa* (Nutt.) Nash—M; Witsell 02-1523 (ANHC); Prairie Co.

Danthonia spicata (L.) P. Beauv. ex Roem. & Schult.—UW; Kral 61826 (VDB); Prairie Co.

Dichanthelium aciculare (Desv. ex Poir.) Gould & C.A. Clark—LP; Witsell 08-0142 (ANHC); Arkansas Co.

Dichanthelium boscii (Poir.) Gould & C.A. Clark—UW; Kellogg 236 (MO); Arkansas Co.

Dichanthelium commutatum (Schult.) Gould—UP, FW, UW; Rodgers 63 (UAM); Arkansas Co.

Dichanthelium depauperatum (Muhl.) Gould—UW; Witsell 14-0300 (ANHC); Prairie Co.

Dichanthelium dichotomum (L.) Gould ssp. *barbulatum* (Michx.) Mohlenbr.—LP, FW; Adair 3354 (UARK); Arkansas Co.

Dichanthelium dichotomum (L.) Gould ssp. *dichotomum*—UP, LP, FW, UW; Reid 977 (UARK); Arkansas Co.

Dichanthelium dichotomum (L.) Gould ssp. *lucidum* (Ashe) Freckmann & Lelong—LP, FW, SL; Witsell 02-1056 (ANHC); Arkansas Co.

† *Dichanthelium dichotomum* (L.) Gould ssp. *roanokense* (Ashe) Freckmann & Lelong—LP; Witsell 03-0314 (ANHC); Prairie Co.

Dichanthelium implicatum (Scribn.) Kerguélen—UP; Moore 510209 (UARK); Arkansas Co.

Dichanthelium lanuginosum (Elliott) Gould—UP, LP, FW; Adair 3355 (UARK); Arkansas Co.

Dichanthelium laxiflorum (Lam.) Gould—FW, UW; Observed; Prairie Co.

Dichanthelium lindheimeri (Nash) Gould—UP, LP, FW; Nielsen 4194 (UARK); Arkansas Co.

Dichanthelium linearifolium (Scribn. ex Nash) Gould—UP, UW; Palmer 25024 (MO); Prairie Co.

Dichanthelium malacophyllum (Nash) Gould—UP, UW; Eads 7 (UAM); Arkansas Co.

* *Dichanthelium oligosanthes* (Schult.) Gould ssp. *oligosanthes*—UP; Kral 61837 (MO); Arkansas Co.; [1978]

Dichanthelium oligosanthes (Schult.) Gould ssp. *scribnerianum*—LP; Witsell 03-0284 (ANHC); Prairie Co.

Dichanthelium polyanthes (Schult.) Mohlenbr.—LP, FW, UW; Nunn 9930 (UARK); Lonoke Co.

Dichanthelium scoparium (Lam.) Gould—LP, D, M; Nunn 7027 (UARK); Prairie Co.

Dichanthelium sphaerocarpon (Elliott) Gould—UP, LP; Palmer 25053 (UARK); Prairie Co.

Digitaria cognata (Schult.) Pilg.—UP; Witsell 04-1356 (ANHC); Prairie Co.

Echinochloa muricata (P. Beauv.) Fernald—D, M; Nunn 7026 (UARK); Prairie Co.

Elymus glabriflorus (Vasey ex L.H. Dewey) Scribn. & C.R. Ball—UP, LP, SL; Witsell 03-0614 (UARK); Prairie Co.

Elymus virginicus L. var. *virginicus*—LP, FW, SL; Baker 11-0112 (ANHC); Prairie Co.

Eragrostis hirsuta (Michx.) Nees—UP; Witsell 09-0478 (ANHC); Prairie Co.

Continued on the next page

Table 6.1. *Continued*

Eragrostis hypnoides (Lam.) Britton, Sterns & Poggenb.—M; Minteer 305 (UARK); Prairie Co.

Eragrostis intermedia Hitchc.—UP, LP, D; Witsell 14-0234 (ANHC); Prairie Co.

Eragrostis pectinacea (Michx.) Nees ex Steud. var. *pectinacea*—D; Observed; Prairie Co.

Eragrostis refracta (Muhl.) Scribn.—LP; Witsell 05-1351 (ANHC); Prairie Co.

Eragrostis spectabilis (Pursh) Steud.—UP, LP; Locke 944 (UARK); Prairie Co.

Eriochloa contracta Hitchc.—D; Oliver s.n. (UARK); Arkansas Co.

Festuca paradoxa Desv.—LP, FW; Witsell 14-0255 (ANHC); Prairie Co.

Glyceria septentrionalis Hitchc. var. *septentrionalis*—M; Witsell 07-0134 (ANHC); Prairie Co.

† *Gymnopogon brevifolius* Trin.—LP; Witsell 02-1320 (ANHC); Prairie Co.

Hordeum pusillum Nutt.—LP; Witsell 02-0347 (ANHC); Lonoke Co.

† *Koeleria macrantha* (Ledeb.) Schult.—UP; Witsell 03-0301 (ANHC); Prairie Co.

Leersia lenticularis Michx.—FW, SL; Observed; Prairie Co.

Leersia oryzoides (L.) Sw.—M; Moore 32772 (UARK); Arkansas Co.

Leersia virginica Willd.—FW; Observed; Prairie Co.

Leptochloa fusca (L.) Kunth ssp. *fascicularis* (Lam.) N. Snow—M; Snow 5811 (MO); Arkansas Co.

Leptochloa panicea (Retz.) Ohwi ssp. *brachiata* (Steud.) N. Snow—M; Demaree 15501 (MO); Prairie Co.

Leptochloa panicoides (J. Presl) Hitchc.—M; Reid 1326 (UARK); Prairie Co.

† *Muhlenbergia glabrifloris* Scribn.—LP; Baker 11-0298 (ANHC); Arkansas Co.

Muhlenbergia schreberi J.F. Gmel.—D; Thomas 131010 (NLU); Lonoke Co.

Panicum anceps Michx. ssp. *anceps*—UP, LP, D, FW, UW; Nielsen 4202 (UARK); Arkansas Co.

Panicum dichotomiflorum Michx. ssp. *dichotomiflorum*—D, M; Nunn 7390 (ANHC); Arkansas Co.

Panicum philadelphicum Bernh. ex Trin. ssp. *philadelphicum*—UP; Witsell 05-1345 (ANHC); Prairie Co.

Panicum rigidulum Bosc ex Nees ssp. *rigidulum*—M, FW; Witsell 09-0476 (ANHC); Prairie Co.

Panicum virgatum L.—LP, M, FW; Nunn 7055 (UARK); Prairie Co.

†* *Paspalum boscianum* Flüggé—LP, M; Locke 908 (UARK); Prairie Co.; [1970]

Paspalum dissectum (L.) L.—M; Adams s.n. (BRIT); Arkansas Co.

Paspalum distichum L.—M; Witsell 08-0229 (ANHC); Prairie Co.

Paspalum floridanum Michx.—LP, M; Witsell 05-1352 (ANHC); Prairie Co.

Paspalum laeve Michx.—LP; Witsell 03-1041 (ANHC); Prairie Co.

Paspalum pubiflorum Rupr. ex E. Fourn.—LP, M; Reid 1050 (UARK); Arkansas Co.

Paspalum repens P.J. Bergius—M; Palmer 24355 (UARK); Prairie Co.

Paspalum setaceum Michx. var. *muhlenbergii* (Nash) D.J. Banks—UP, LP; Witsell 05-1344 (ANHC); Prairie Co.

Phalaris caroliniana Walter—UP, LP; Witsell 02-0424 (ANHC); Arkansas Co.

Saccharum alopecuroides (L.) Nutt.—UP, FW, UW; Locke 943 (UARK); Prairie Co.

Saccharum baldwinii Spreng.—FW; Demaree 8401 (MO); Lonoke Co.

Saccharum brevibarbe (Michx.) Pers. var. *contortum* (Elliott) R.D. Webster—LP; Demaree 72155 (MO); Lonoke Co.

Saccharum giganteum (Walter) Pers.—M; Observed; Arkansas Co.

Schizachyrium scoparium (Michx.) Nash—UP, UW; Moore 321057 (UARK); Prairie Co.

Setaria parviflora (Poir.) Kerguélen—LP; Witsell 02-1052 (ANHC); Prairie Co.

Sorghastrum nutans (L.) Nash—UP, LP; Nunn 7402 (UARK); Prairie Co.

Spartina pectinata Link—LP, M; Moore 32729 (UARK); Prairie Co.

Sphenopholis nitida (Biehler) Scribn.—UW; Marsico 5214 (UARK); Arkansas Co.

Sphenopholis obtusata (Michx.) Scribn.—UP, LP; Witsell 03-0313 (UARK); Prairie Co.

Sporobolus compositus (Poir.) Merr. var. *compositus*—UP; Witsell 03-1038 (ANHC); Prairie Co.

Sporobolus compositus (Poir.) Merr. var. *drummondii* (Trin.) Kartesz & Gandhi—LP; Witsell 05-1359 (ANHC); Prairie Co.

Sporobolus heterolepis (A. Gray) A. Gray—UP; Witsell 04-1446 (ANHC); Prairie Co.

† *Sporobolus junceus* (P. Beauv.) Kunth—UP; Witsell 02-1399 (ANHC); Arkansas Co.

Sporobolus vaginiflorus (Torr. ex A. Gray) Alph. Wood var. *vaginiflorus*—UP, D; Witsell 14-0480 (ANHC); Prairie Co.

Steinchisma hians (Elliott) Nash—LP, M, FW; Witsell 03-0501 (ANHC); Arkansas Co.

Tridens × *oklahomensis* (Feath.) Feath.—UP, LP; Witsell 04-1354 (ANHC); Lonoke Co.

Tridens flavus (L.) Hitchc. var. *flavus*—UP, D, UW; Witsell 14-0294 (ANHC); Prairie Co.

Tridens strictus (Nutt.) Nash—LP, D, FW; Nunn 7053 (UARK); Prairie Co.

Tripsacum dactyloides (L.) L.—LP, M, FW; Witsell 03-0421 (ANHC); Arkansas Co.

Vulpia octoflora (Walter) Rydb. var. *octoflora*—UP; Adair 333 (UARK); Arkansas Co.

Zizaniopsis miliacea (Michx.) Döll. & Asch.—M; Moore 32794 (UARK); Arkansas Co.

Pontederiaceae

Heteranthera limosa (Sw.) Willd.—M; Witsell 04-0750 (ANHC); Lonoke Co.

TABLE 6.1. *Continued*

Heteranthera multiflora (Griseb.) C.N. Horn—M; Witsell 01-0627 (ANHC); Prairie Co.

Potamogetonaceae

Potamogeton diversifolius Raf.—M; Nunn 8035 (UARK); Prairie Co.

Potamogeton foliosus Raf. ssp. *foliosus*—M; Thomforde s.n. (UAM); Lonoke Co.

Potamogeton pusillus L. ssp. *pusillus*—M; Nunn 8036 (UARK); Prairie Co.

Smilacaceae

Smilax bona-nox L.—LP, FW, SL, UW; Witsell 14-0258 (ANHC); Prairie Co.

Smilax glauca Walter—FW, SL, UW; Witsell 14-0299 (ANHC); Prairie Co.

Smilax rotundifolia L.—FW, SL; Witsell 14-0275 (ANHC); Prairie Co.

Typhaceae

Typha domingensis Pers.—M; Minteer 304 (UARK); Prairie Co.

Typha latifolia L.—M; Locke 5883 (UARK); Prairie Co.

Table 6.2. Number of native plant taxa by habitat.

Habitat	Number of species
Upland Prairie	223
Lowland Prairie	299
Disturbed	160
Marsh	213
Flatwoods	263
Slash	129
Upland Woodlands	173

7

Vegetation and Flora on Lowlands in the Central Black Belt of Mississippi—How Low Did the Original Prairie Go?

J. J. N. Campbell and W. R. Seymour Jr.

Introduction

The Black Belt of Alabama, Mississippi, and a small part of western Tennessee is a distinctive calcareous section of the Upper East Gulf Coastal Plain, largely underlain by Upper Cretaceous sediments that are mostly "chalk and marly chalk containing fewer impurities than underlying and overlying formations" (Moore 1985). There has been considerable research on native vegetation of the Black Belt uplands in Alabama and Mississippi before European settlement (Rostlund 1957; Rankin 1974; Peacock and Schauwecker 2003b; Barone 2005a; Barone and Hill 2007; Barone et al. 2008; Schotz and Barbour 2009). We have recently reviewed relevant literature and conducted a detailed survey of flora and vegetation at the Pulliam Prairie, one of the best remnants of original Black Belt grassland in Mississippi (Campbell and Seymour 2011a, 2011b, 2012). This chapter provides further details from lowlands near the Pulliam Prairie and focuses on the following question: What was the nature of the original transitions between the largely open grassy uplands and the mostly wooded lowlands?

Few detailed studies of lowland vegetation in the Black Belt have been published. Only MacDonald (1996) and Leidolf et al. (2002) have conducted complete floristic surveys of counties that lie largely within the Black Belt of Mississippi or Alabama. Botanical information on lowlands typical of peripheral or adjacent regions can be found in the varied reports by Hall and Penfound (1943), Simmons (1968), Beckett and Golden (1982), Crouch and Golden (1997), Wieland (2000), Majure (2007), Majure et al. (2011), and others. There are also important historical notes in the earlier accounts by Mohr (1901), Lowe (1913), and R. Harper (1913, 1943), and descriptions of vegetation in early soil surveys deserve further attention. A good example comes from Chickasaw County (fig. 7.1), where E. Jones and Waldrop (1917) reported: "In the eastern part of the county there are large areas of original prairie land, which supported only clumps of crabapple, honey locust, and wild plum. On uncleared portions of the higher, reddish and brown soils associated with the black prairie there is a growth of post oak, blackjack oak and hickory. The bottomlands in this part of the county support a native growth of sycamore, ash, overcup oak, willow oak, water oak, black locust, and cottonwood."

The existing official classifications for vegetation in the Black Belt have remained somewhat incomplete (NatureServe 2012). For example, NatureServe has not yet listed a seasonally wet prairie type in the region with abundant gamagrass (*Tripsacum dactyloides*), although that species remains one of the most frequent native grasses on lowlands (R. Harper 1913; Campbell and Seymour 2011a). NatureServe (2004) did not list any native grassland in the Tombigbee National Forest, but the southern unit of that forest does lie within the historical area of Black Belt Prairie, and there are significant remnants there (Wiygul et al. 2003). NatureServe did not provide descriptions of forest associations dominated by willow oak (*Quercus phellos*) or water oak (*Q. nigra*) for any national forest in north-

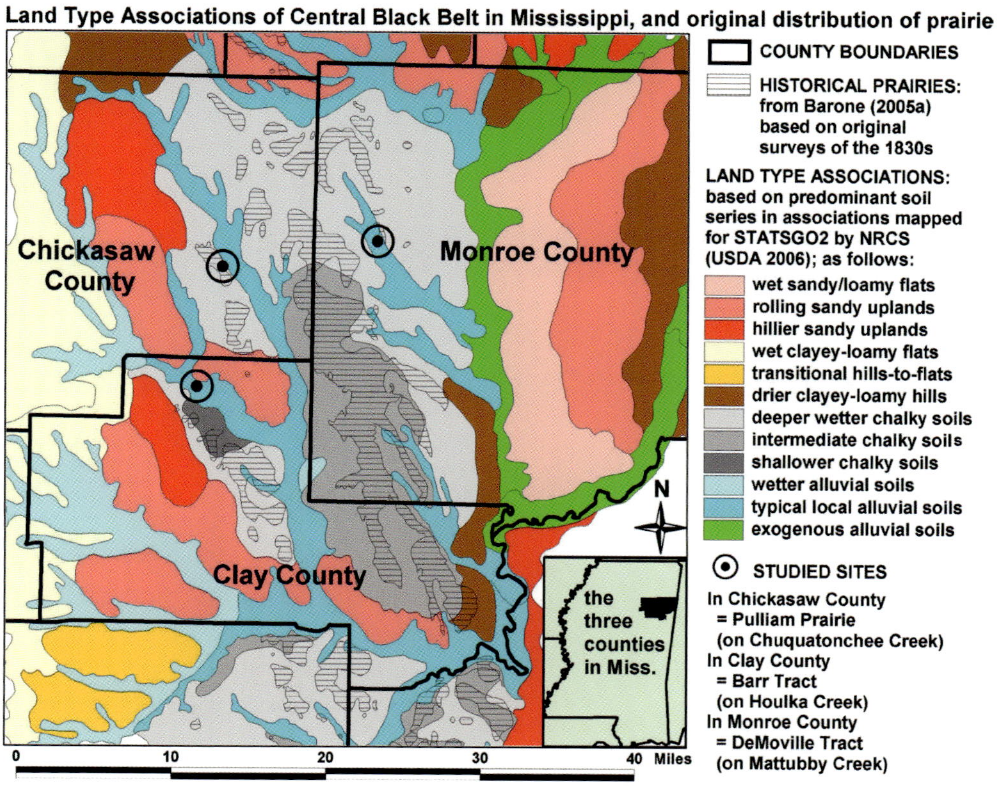

Figure 7.1. Map of central Black Belt in Mississippi, showing locations of study sites overlaid on land-type associations that are based on STATSGO (USDA 2006). Mapping units are groups of soil associations. There is overlap among series, with proportions varying between associations: wet sandy/loamy flats; rolling sandy uplands (locally damp to wet); hillier sandy uplands (mostly well drained); wet clayey-loamy flats; transitions from sandy/loamy uplands to clayey flats; drier clayey-loamy hills (mostly well drained); deeper chalky soils (often damp to wet); intermediate chalky soils; shallower chalky soils (often droughty); wetter alluvial soils (with varied origin and acidity); typical local alluvial soils; exogenous alluvial soils (more loamy-sandy and acid), including old high terraces.

ern Mississippi. However, forest dominated by these oaks is widespread on seasonally drier lowlands and damp upland flats across most of this region, albeit usually much disturbed and in various degrees of conversion from woodland to farmland (Hodges 1997; Campbell and Seymour 2011a, 2011b; Schauwecker et al. 2011; and general observations of the authors).

Uncertainty about the original vegetation is pronounced on topography that is transitional from lowlands to uplands. Most of such land has been converted to farmland, and only scattered woodlots, older pastures, fencerows, and roadsides provide clues to its former condition. Moreover, this transition is not clear in terms of soil trends. There are several poorly drained soil series that are more typical of upland flats than lowlands, and in more gentle topography such series appear to intergrade with soils of high ancient terraces. Similar problems have been noted elsewhere on the Gulf Coastal Plain. One of the most intensive analyses within this whole region was conducted in southern Illinois by P. Robertson et al. (1978), who stated: "Elevation, which at the outset of the study appeared to be an ideal integrating environmental factor, did not adequately represent the gradient complex. This is because topographic form such as shallow depressions on ridge bottoms and low ridges in hardwood bottoms and tip-up mounds confound simple relationships among variables. Also, because of past depositional patterns, sites with similar elevation often have soils which differ markedly in texture, internal drainage and aeration."

This chapter provides a description of vegetation types

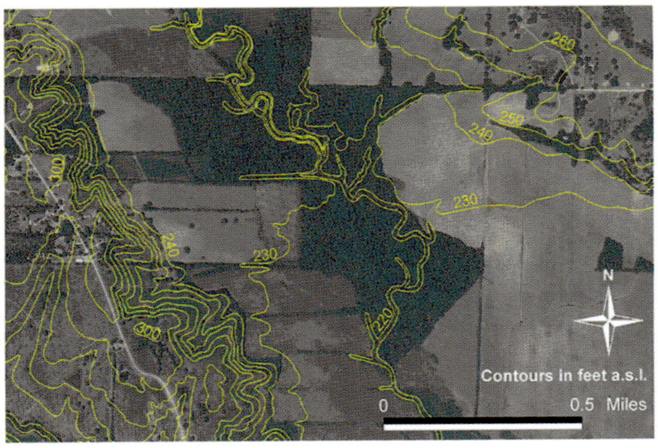

Figure 7.2A. Study area in Monroe County, Mississippi, based on aerial photograph from the National Agriculture Imagery Program (USDA Farm Service Agency). The date of the image is June 30, 2009; the reference code is ortho_1-1_1n_s_ms095_2009_1. Contours in feet are added from MARIS (2009). The DeMoville Tract covers about 200 ac (0.3 mi^2) in the center of this view, where fieldwork was concentrated. Adjacent fields and roadsides were also surveyed.

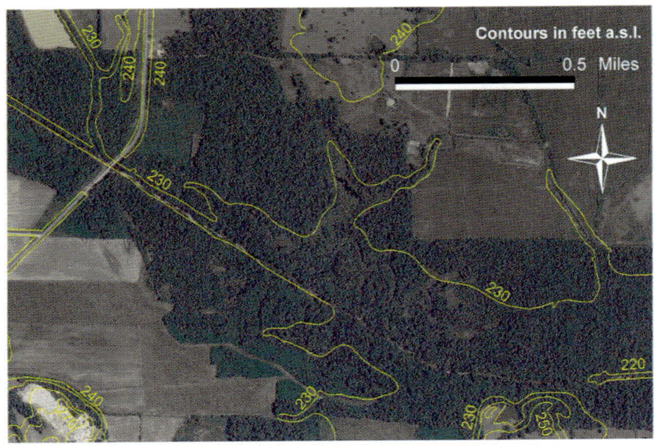

Figure 7.2B. Study area in Clay County, Mississippi, based on aerial photograph from the National Agriculture Imagery Program (USDA Farm Service Agency). The date of the image is June 29, 2009; the reference code is ortho_1-1_1n_s_ms025_2009_1. Contours in feet are added from MARIS (2009). The Barr Tract covers about 300 ac (0.4 mi^2) in the center of this view, where fieldwork was concentrated. Adjacent fields and roadsides were also surveyed.

as well as notes on species of interest at two good remnants of lowland forest. It focuses on indicated associations between soil types and the original vegetation, for which various clues are assembled from our observations and from relevant literature on the East Gulf Coastal Plain. However, more detailed quantitative work at relatively small scales (i.e., 0.1–1 ac plots) will be needed eventually to test ideas from this initial synthesis. The discussion seeks to outline the varied transitions between lowlands and uplands. Overall interpretation is aided by reference to historical accounts of the landscape. Selected rare or "conservative" species from the floristic list (table 7.1) are noted that may indicate the former degree of open conditions in these transitions. Potential applications for conservation are briefly reviewed.

Study Areas and Methods

In order to extend the floristic work at Pulliam Prairie, which occurs mostly on chalky uplands, we selected two sites with relatively well-preserved woods on lowlands of the central Black Belt in Mississippi (figs. 7.1, 7.2a, and 7.2b). The DeMoville Tract is in Monroe County on Mattubby Creek, which drains only from the Black Belt. There has been relatively little channelization of streams within this tract, but herbicide drift from adjacent fields was a problem in some areas. MacDonald (1996) has studied the entire flora of Monroe County, providing a valuable base of information. The Barr Tract is in Clay County on Houlka Creek, which drains from more acid, sandy or clayey uplands on the west side of the Black Belt. The woods were more extensive in this tract, with fewer recent agricultural influences, but Houlka Creek has been channelized. The only previous floristic report from Clay County is a thesis by Simmons (1968) on the more hilly western section.

General geology and soils of the Black Belt have been outlined elsewhere (Campbell and Seymour 2011a, 2011b). It is important to note that the Cretaceous Selma Chalk that defines the most distinctive zones of this region is greatly intermixed with other parent materials at some localities, especially along its western and southern margins. In Monroe County, to the east, the "Selma chalk . . . [lies] unconformably on the Tombigbee sand member of Eutaw Formation. . . . [The] interface is a basal conglomerate consisting of phosphatic nodules, phosphatic molds of fossils, and oyster shells" (Murphree et al. 1966). On uplands in Clay County, Worthen (1909) noted about 70 square miles of prairie: "A rank growth of prairie grasses, with here and there a grove of black-jack and post oak. Only about one half of these prairies is composed of black calcareous soils." To the east in Clay County, there is a more gradual shift from the chalk to less calcareous material, with some soils (Okolona, Brooksville, Kipling) formed in "beds of acid and calcareous clays over thick beds of Selma Chalk." To the southwest, on Pontotoc Ridge, soils (Ora, Prentiss, Stough) are formed in "sediment consisting of non-calcareous sands, silts and clays" (Murphree and Miller 1976). The Barr Tract lies in the transition to Pontotoc Ridge (on "Ripley Sand"), with fine sandy soils on some adjacent uplands and old high terraces (fig. 7.1).

Alluvia of the central Black Belt are complex, based on available mapping of soils by the NRCS (USDA 2006, 2012a, 2012b; see fig. 7.1). In addition to material washed down from local calcareous uplands, several larger creeks have brought in varied exogenous material from adjacent regions, including much fine sand. Alluvium with more influence of chalky uplands usually has a large component of "sticky" (shrink-swell) clay and can fill smaller tributary gullies. Larger floodplains tend to have more loamy soils, especially along natural levees where exogenous silts and sands have been deposited. But in some backwaters of larger creeks, clayey alluvium still prevails (e.g., under the Trinity soil series), presumably with a history of "slack-water" deposition. The alluvial history in this region has been complicated further by recent erosion during the past 200–300 years of intense agricultural development (Campbell and Seymour 2011a, 2011b). Original topsoil on uplands in the central and northern Black Belt in Mississippi contains some glacial loess—at least 1 m thick in some areas (Bettis et al. 2003). It is likely that much of this loess is now in the postsettlement alluvium. Patterns of flooding and drainage have also been greatly modified by artificial channelization and tile drainage during the past 100 years.

Soils of the DeMoville Tract (fig. 7.2a) are mapped as mostly inceptisols (series Leeper, Una, and Tuscumbia) with a silty-clay loam texture (Murphree and Miller 1976; USDA 2012a). In areas with more direct drainage from calcareous uplands, there are vertisols with silty to clayey texture, especially on more poorly drained flats (series Vaiden, Eutaw, etc.) or on better-drained terraces (Marietta, Trinity), where mollisols are also present (Catalpa, narrowly defined). Soils of the Barr Tract (fig. 7.2b) have been mapped mostly as the inceptisol Leeper silty-clay loam (Murphree et al. 1966; USDA 2012a), but the situation is probably more complex. Less mature entisols with more sand and less clay (such as Belden) may actually predominate in the younger alluvial soils, which include much material washed in from uplands to the west. However, Houlka Creek has been channelized, and there has probably been a decline in the input of alluvium onto the original floodplain, allowing more horizon development in some soils. Higher ground in and around the Barr Tract has been mapped mostly as alfisols (Kipling, Ozan, Longview) or ultisols (Stough, Prentiss, Ora) that are silty to sandy loams. More clayey vertisols typical of chalky uplands are absent but do occur within a mile of the site (Griffith, Demopolis).

We made six collecting trips to these two sites in 2010–2011, covering 15 days, as follows: DeMoville Tract (Monroe County) on June 6–7, July 29–30, and September 30, 2010*; Barr Tract (Clay County) on April 18–20, 2010*, and May 23–25, July 26–27, and September 24–25, 2011 [*JC was absent on days with asterisks]. During each trip, we systematically visited all of the major habitat divisions. We used an electric Stealth Predator XR or similar vehicle for transport and for a platform to process herbarium specimens in the field. Based partly on our previous work (Campbell and Seymour 2011a, 2011b, 2012), we initially outlined the major habitats as follows, for application to field notes on each collected species: (A) drier oak woods (with local *Quercus alba*); (B) average oak woods (with *Q. phellos* most common); (C) wetter oak woods (with local *Q. lyrata*); (D) various ponded sites without tree cover; (E/F) riparian zones, from levee crests (E) to water's edge (F). Table 7.1 develops this coding further for typical habitats of each species.

In addition to our fieldwork, we used recent aerial photographs from the National Agriculture Imagery Program of the USDA Farm Service Agency (MARIS 2009), topographic mapping (USGS 7.5-minute topographic quadrangles), and soil mapping by the NRCS (USDA 2006, 2012a) to interpolate our observations of vegetation at these sites. We also conducted a broad regional review of soils and vegetation in order to refine our definitions of habitat types, as initially reported by Campbell and Seymour (2011a). Botanical nomenclature follows primarily the list of vascular flora in Mississippi that is being developed at the Pullen Herbarium in Oxford (McCook and Kartesz 2010), based initially on Kartesz (1999). For a few taxa, we use different names, which we detail and explain in the results. The senior author visited several herbaria in Mississippi and elsewhere in order to improve identifications and clarify taxonomic concepts. Abbreviations for herbaria (APSU, IBE, MISS, MISSA, MO, NCU, TENN, US, etc.) follow Thiers (2012). Charles Bryson (retired from the Agricultural Research Service at Stoneville) generously identified our collections of *Carex*. We assessed the apparent rare-versus-common status for vascular species typical of lowlands in the Black Belt region, with reference to listings of Natural Heritage Programs in Mississippi, Alabama, and elsewhere (MNHP 2006; ANHP 2012; NatureServe 2012–2013). We extended this assessment in several cases to a review of the total reported range of each species (e.g., Kartesz 2012), and to a review of recent taxonomic revisions (e.g., Weakley 2012).

Results and Discussion: Vegetation

Four broadly defined, intergrading classes of native vegetation can be outlined for lowlands in the central Black Belt of Mississippi.

1. A grassland complex, including diverse grasses, sedges, and varied marshy transitions.
2. An oak woodland complex, with willow oak (*Quercus phellos*) often predominating.
3. A ponded complex with more stagnant water, including varied zones with shrubs, marshes, aquatic plants, and seasonally drying shorelines.
4. A riparian complex, ranging from open scoured streambanks to wooded levees over short distances; the woods usually with less oak than the oak woodland complex.

The following notes detail these classes and link them with NatureServe (2012a), if possible. The term "grassland" is used here in a broad sense to indicate vegetation with a large component of Poaceae, Cyperaceae, Juncaceae, or Typhaceae but often with abundant forbs or shrubs. The complementary term "woodland" is used in a broad sense to indicate any vegetation with abundant trees, including moderately open woods as well as the deeper woods or "forest" of NatureServe. The classification below emphasizes floristic composition; more precise subdivision that emphasizes vegetation structure and dynamics is not possible based on this initial survey.

The alphanumeric codes following each heading below (g1 to p3) refer to our evolving sequence of notes on vegetation types that is being developed for the central Black Belt, including uplands. These codes are also used in associated tables and floristic lists. Campbell and Seymour (2011b) employed the same codes, except that "o" and "p" are switched here for more logical sequencing. The symbol ~ implies that the match with previous work is not exact, but there is much overlap of concepts. In the lists of species for each vegetation type, the symbol + indicates that the species occurs widely in several vegetation types and is not strongly concentrated in one type. Binomial names in parentheses have only minor occurrence or uncertain association with the type. The four-number codes used for the vegetation types of NatureServe (2012a) refer to its "Community Element Global Codes" (CEGL) for individual plant associations.

Grassland Complex

These are miscellaneous small areas, narrow zones, or other disturbed sites. They are difficult to classify, but two subdivisions are loosely outlined here.

Drier grassland (~ g1, g2). Clearly intact remnants of the original native grassland do not occur at the two studied lowland sites, but there are several areas that are locally dominated by native species in low roadsides, old fields, fencerows, thin oak woods, and other brushy transitions. These miscellaneous areas are often damp in winter and spring but generally dry out in summer and fall. The more wooded areas have probably been grazed by livestock and burned in past decades. Several typical species listed here were also found mostly as rare to infrequent scatterings along paths through average or drier oak woods (see n1). Similar vegetation was described at the Pulliam Prairie (Campbell and Seymour 2011b; under class "g").

Typical grasses include *Andropogon tenuispatheus* (especially on base-rich soils), *A. virginicus* (local on drier, poorer soils), *Dichanthelium acuminatum* (sensu lato), *D. scoparium* (local in wetter transitions), *Panicum anceps* (local on drier soils), *P. longifolium* (on acid soils), *P. virgatum* (on more base-rich soils), *Paspalum floridanum*, *P. laeve*+, *Setaria parviflora*, *Tridens strictus*, and *Tripsacum dactyloides*. *Tripsacum* is potentially dominant in some areas without mowing or browsing. Sedges include *Carex cherokeensis*+, *C. caroliniana/complanata*, *C. annectans/fissa*, *C. festucacea* (acid soils), *C. glaucodea*, *C. granularis*+, and *C. leaven-*

worthii (and a small amount of *C. striatula*). Additional sedges in section Ovales are expected on richer soils (Campbell and Seymour 2012) but were not confirmed at these sites. Other graminoids include *Cyperus* spp. (especially *C. croceus, C. echinatus, C. retrorsus*), *Juncus* spp. (especially *J. torreyi* on base-rich soils or *J. brachycarpus* on acid soils).

Typical forbs include *Amsonia* sp. nov. (see the notes that follow), *Desmanthus illinoensis*+, (*Euthamia leptocephala*), (*Helianthus angustifolius*), *H. resinosus*+, (*Lobelia puberula*), *Penstemon alluviorum*+, *Phlox carolina, Physostegia angustifolia, Ptilimnium capillaceum, Prunella vulgaris* (sensu lato)+, *Pycnanthemum tenuifolium, Rudbeckia* sp. nov. (see the notes that follow), *Ruellia caroliniensis*+, (*Salvia lyrata*+), (*Solidago altissima* on drier soil), *Symphyotrichum* spp. (especially *dumosum*+, *pilosum*) and *Vernonia missurica*+.

There are no exact matches for such grassland in NatureServe (2012a), but its "poorly defined" 4624 is a reasonable partial match: "*Panicum virgatum - Tripsacum dactyloides* Grand Prairie/Big Barrens Herbaceous Vegetation"—which NatureServe reported from Arkansas, Kentucky, and Tennessee. Other types with some similarity include 7774, "*Sorghastrum nutans - Andropogon glomeratus - Silphium laciniatum*", a drier transition in Arkansas (Zollner et al. 2003); also 3904, "*Baccharis halimifolia - Crataegus berberifolia / Eleocharis* sp. *- Tridens strictus - Euthamia leptocephala*"; and 7963, "*Quercus falcata/pagoda - Carya illinoinensis* (or *myristicaeformius*) */ Silphium integrifolium - Panicum anceps -* (*Carex cherokeensis*)", which are both wooded transitions on the West Gulf Coastal Plain.

Wetter Grassland (~ k1, also similar to c2). These areas usually stay wet or damp for much of the year but dry out in summer to fall. Varied examples often occur in marshy pond margins or old fields, and in disturbed openings such as roadsides within average to wetter oak woods (see n2 to n3).

Typical grasses include *Dichanthelium scoparium, Saccharum giganteum*+ (but probably limited by browsing or mowing), *Panicum* spp. (especially *P. rigidulum*+), *Paspalum* spp. (especially *P. setaceum*+), and *Urochloa platyphylla*+ (a weedy species of bare soil). Sedges include *Carex aureolensis*+, *C. caroliniana, C. hyalinolepis, C. lupulina, C. reniformis, C. triangularis*, and *C. vulpinoidea*. Other graminoids include *Cyperus* spp. (especially *C. erythrorhizos, C. pseudovegetus*, and *C. strigosus*), *Eleocharis* spp. (especially *E. compressa* on basic soil, or *E. tenuis* on acid soil), and *Juncus* spp. (especially *J. biflorus, J. coriaceus*, and *J. effusus*). Cyperaceae are locally dominant, as are Poaceae.

Typical forbs include *Conoclinum coelestinum*+, *Eupatorium serotinum, Hypericum mutilum, Lycopus americanus, Lythrum alatum, Mecardonia acuminata*, (*Persicaria glabra*), *Phlox carolina*+, *Ptilimnium nuttallii*, (*Rhexia mariana* var. *interior*), *Scutellaria integrifolia*+, *Solidago gigantea*, and *Symphyotrichum* spp. (especially *S. dumosum*+). *Solidago rugosa* is notably absent at the studied sites, but it is expected on more acid soils in the region (Leidolf et al. 2002).

The best match in NatureServe (2012a) may be 7962: "*Lythrum alatum - Panicum anceps - Symphyotrichum lanceolatum* Wet-Mesic Blackland Prairie Temporarily Flooded Herbaceous Vegetation." However, that was described only from Arkansas (Zollner et al. 2003), where it has virtually disappeared (with "GH" status). In more northern regions, there is some similarity with NatureServe's rare type 4006 of the Highland Rim: "*Andropogon gerardii - Schizachyrium scoparium - Dichanthelium scoparium - Rhynchospora glomerata*"; and with the more widespread type 2224 of the Great Plains: "*Spartina pectinata - Carex* spp. *- Calamagrostis canadensis - Lythrum alatum -* (*Oxypolis rigidior*)". After frequent mowing or grazing, these types of wet grassland have sometimes become degraded to rush-dominated vegetation, as in the widespread 4112: "*Juncus effusus* Seasonally Flooded Herbaceous Vegetation".

Oak Woodland Complex

This class of vegetation includes relatively mesic, xeric, and hydric variants, which are often combined in general descriptions of "bottomland hardwoods" on the Gulf Coastal Plain. As well as containing much oak (*Quercus*) in more mature stands, especially *Q. phellos* within the Black Belt, the woodland includes locally abundant sugarberry (*Celtis*), ashes (*Fraxinus*), sweetgum (*Liquidambar*), and red maple (*Acer*) in younger stands (e.g., MacDonald 1996; Leidolf et al. 2002; Denley et al. 2002; Hawkins et al. 2009). Abundance of *Q. phellos* in the Black Belt may be associated with the relatively base-rich, clayey soils. This species is much less common in some other regions of Mississippi that tend to have more acid or sandy soils (Leidolf et al. 2002; Majure 2007; Majure et al. 2011; Kartesz 2012).

Compared to most of the "riparian complex" (as defined below), this oak woodland tends to remain wetter through much of the winter and spring but often becomes drier in summer and fall. On the Mississippi River bottomland, oaks are uncommon in the understory, suggesting succession toward ash, elm, and sugarberry in some areas (Oliver et al. 2005). However, P. Robertson et al. (1978) indicated the potential for natural "gap-phase" regeneration of oaks as older stands decline. On the Black Belt bottomlands, there is also relatively little current regeneration of oaks in some woods, but widespread logging and marginal farming during the past 50–150 years, including some use of woodland by livestock, has accompanied the development of much local dominance by oaks and hickories in this region (Murphree et al. 1966, 1974; Murphree and Miller 1976; Hodges 1997)—as it has elsewhere in eastern North America (P. White and White 1996). Before intensive settlement in the early 1800s, it is likely that fires also kept these woods somewhat open (Lowe 1913). The thin-barked maples (*Acer*), beech (*Fagus*), and holly (*Ilex opaca*) are particularly sensitive to fires (USFS 2012c).

Drier Oak Woods (~ n1). These are seasonally dry woods and thickets on high terraces, with "somewhat poor" to "moderate" drainage (USDA 2012a, 2012b). On damper sites, composition grades into the more extensive "average" oak woods described next. Soils exhibit considerable variation, based on available mapping. They include (1) sandy/loamy alfisols in the Barr Tract (Ozan, Longview); or (2) clayey inceptisols (Eutaw, Brooksville) and (3) calcareous vertisols (Houlka, Vaiden) in the DeMoville Tract. Species composition on the more acid sandy soils (1) is locally distinct. Some of these woods may have been transitional to more open shrubby woods and grassland in pre-Columbian times, as suggested by the research of Schauwecker et al. (2011) in the adjacent flatwoods region.

Trees are generally similar to the "average" woods outlined next, with abundant *Quercus phellos* in both, but these drier woods are distinguished by locally frequent *Q. pagoda*, *Carya glabra*, and *Ulmus alata*. Also characteristic, though less frequent, are *Q. alba*, *Q. stellata*, (*Q. velutina*), *C. carolinae-septentrionalis*, *C. ovata*, (*Fraxinus* cf. *smallii*), and (*Morus rubra*+). Some additional species are largely restricted to younger woods and old fields: *Diospyros virginiana*, *Gleditsia triacanthos*, *Juniperus virginiana*, and *Maclura pomifera*. Smaller trees and shrubs, including brushy transitions to fields, contain (*Baccharis halimifolia*), *Callicarpa americana*, (*Chionanthus virginicus*), *Cornus drummondii*+, *Crataegus engelmannii*, *C. marshallii*, *Prunus americana* var. *lanata*, *Rubus argutus*+, *Sideroxylon lycioides*, and *Vaccinium arboreum* (and *V. fuscatum*). Typical vines include *Berchemia scandens*+, *Smilax smallii*+, *S. rotundifolia*+, and *Vitis rotundifolia*+.

Typical grasses include (*Danthonia spicata*), *Dichanthelium dichotomum*, and *Poa autumnalis*+. Sedges include *Carex cherokeensis*+, *C. glaucodea*+, and *C. striatula* (and minor amounts of *C. basiantha*+). *Juncus tenuis*+ is frequent along trails. Typical forbs include *Acalypha rhomboidea*+, *Desmodium paniculatum*+ (sensu lato), *Elephantopus tomentosus*, (*Euphorbia spathulata*), *Galium circaezans*, (*G. virgatum*), *Krigia dandelion*+, *Mitchella repens*, *Podophyllum peltatum*, (*Polygonatum biflorum*+), and *Sanicula canadensis*+. The fern *Asplenium platyneuron*+ is also present.

The following two types of NatureServe (2012a) are not good matches, but they may be the best available for the small patches of mature woods found here on the driest ground: 4775, "*Quercus alba—Quercus nigra—Carya pallida* - (*Quercus pagoda*) / *Magnolia* (*grandiflora*, *macrophylla*)"; and 7226, "*Quercus alba* - *Carya glabra*." These types are described from the East Gulf Coastal Plain. Most of the drier oak woodland at the sites studied here is transitional from such types to the "average" willow oak woods outlined in the subsequent section (especially 4771). However, before settlement and suppression of fire, it is likely that drier lowland oak woods were more extensive and graded into more open areas.

Average Oak Woods (~ n2). This is the most common type of woods at both lowland sites, and it includes the small patches of "swampy woods with wetland oaks" that are mixed into strips of riparian woods through the Pulliam Prairie (Campbell and Seymour 2011b). Typical soil series may generally be inceptisols (especially the silty clay loam Leeper) but grade into entisols near active sloughs (Belden) or into vertisols at upper levels (Houlka). Soils are mostly "somewhat poorly drained" (USDA 2012a, 2012b).

Quercus phellos+ is locally dominant. Also common or locally abundant are *Q. nigra*, *Liquidambar styraciflua*, *Acer rubrum* var. *trilobum*, *Celtis laevigata*+, and *Fraxinus pennsylvanica*+ (especially var. *pennsylvanica*). Other trees include (*Q. michauxii*), *Ulmus americana+*, *U. alata+*, *Carya laciniosa*, (*C. illinoenensis*, perhaps just planted), and (*Diospyros virginiana* var. *pubescens*, rare at field edge). Smaller trees and shrubs include *Amorpha croceolanata*, *Arundinaria gigantea+* (especially near riparian zones), (*Crataegus marshallii*,) *Ilex decidua*, *Rubus trivialis+*, and *Sabal minor+* (locally frequent in thin woods and edges). Vines include *Berchemia scandens+*, (*Dioclea multiflora*), *Vitis cinerea+*, and several other species.

Typical grasses include *Chasmanthium laxum* (locally abundant), *Dichanthelium microcarpon*, *D. polyanthes*, and *Leersia virginica+*. Sedges include *Carex abscondita+*, *C. albolutescens*, *C. corrugata*, *C. debilis*, *C. intumescens*, *C. louisianica*, *C. oxylepis*, and *C. squarrosa*. Typical forbs include *Asclepias perennis*, *Galium obtusum+*, (*Hymenocallis caroliniana+*), *Lysimachia nummularia*, *Phlox carolina+*, (*Ptilimnium nuttallii*), *Ranunculus carolinianus*, *Scutellaria integrifolia+*, (*Stachys hispida*), (*Symphyotrichum lateriflorum*), *Tradescantia ohiensis+*, *Uvularia sessilifolia*, and *Viola* cf. *langloisii+*.

In NatureServe (2012a), the best match may be 7915, "*Quercus phellos - Quercus nigra - Liquidambar styraciflua*," but this type is restricted to the Mississippi Alluvial Plain. In the Black Belt, some of these woods are transitional to drier phases, matching NatureServe's type 4771, "*Quercus phellos - Quercus nigra - Quercus alba / Chasmanthium* (*laxum*, *sessiliflorum*)," which is defined for small stream floodplains of the East Gulf Coastal Plain. There are also frequent transitions to wetter phases (see 2102) or to riparian phases (see 2099). *Sabal minor* appears to be concentrated in marginal areas with more open history at the Barr Tract, but not intensive grazing. Vegetation types with an abundance of this palm are not listed by NatureServe for Mississppi—though it is part of 7869, which was described from the West Gulf Coastal plain "*Quercus phellos - Quercus nigra / Sabal minor - Sebastiania fruticosa*)."

Wetter Oak Woods (~ n3). These areas are similar to the "average" woods outlined previously, but with more shallow pools and sloughs. Typical soil series may be inceptisols (silty clay loams Una and Tuscumbia), or perhaps entisols (silt loams Belden and Mathison) along more active sloughs. These are "somewhat poorly drained" to "poorly drained" soils (USDA 2012a, 2012b).

Trees are generally similar to the "average oak woods" but with local abundance of *Quercus lyrata*. Less frequent characteristic trees include (*Q. texana*), *Fraxinus profunda*, *Acer rubrum* var. *drummondii*, *Crataegus viridis+*, and *Planera aquatica*. *Salix nigra* is locally dominant in transitions to low riparian areas. Though not found at the sites studied here, *Carya aquatica* and *Forestiera acuminata* are known from Monroe County and are expected near the Tombigbee River. Several vines are typically present, with *Brunnichia ovata+* perhaps most frequent; *Vitis palmata+* may also be characteristic but rare.

Typical grasses include *Leersia lenticularis* (or *L. oryzoides* in more open sites), *Panicum rigidulum+*, and *Steinchisma hians+*. Sedges include *Carex aureolensis+*, *C. crus-corvi*, *C. hyalinolepis*, *C. lupuliformis*, *C. lupulina*, *C. sangamonensis*, and *C. socialis*. Typical forbs include *Galium tinctorium+*, *Justicia ovata*, *Lobelia cardinalis*, *Lycopus rubellus*, *Mimulus alatus*, *Penthorum sedoides*, *Saururus cernuus*, and *Trachelospermum difforme+*. *Saururus* is a distinctive local dominant. The fern *Onoclea sensibilis* is locally frequent in the Barr Tract.

In NatureServe (2012a), the best match is 2102, "*Quercus phellos* - (*Q. lyrata*) / *Carex* spp. - *Leersia* spp.," which occurs mostly on lowlands of the central Mississippi Valley. Also, larger patches of woods dominated by *Q. lyrata* may approach 2424, "*Quercus lyrata - Liquidambar styraciflua / Forestiera acuminate*," or 7397, "*Quercus lyrata - Carya aquatic*." For more permanently ponded or riparian areas, see subsequent sections.

Ponded Complex

These are ponds of varied type, with thinly wooded, shrubby, marshy, and aquatic zones. Oaks are generally not common, except for local overcup oak (*Quercus lyrata*). Included here are dried shores and beds of seasonally ponded areas, such as the semiartificial ponds developed for waterfowl on the northeast side of the Barr Tract. These are mostly small areas of only 1–10 acres, which are typically mapped as "water" on soil maps by the USDA (2012a, 2012b) within a general matrix of poorly drained inceptisols (Una-Tuscumbia).

Woody Phases (~ p1, p2). These would be true "swamps" in the strict sense of Hall and Penfound (1943), Basinger et al. (1997), Wieland (2000), Denley et al. (2002), and others. However, the examples studied here are depauperate compared to some other regions of Mississippi and Alabama, with little *Taxodium* and without any *Nyssa* spp., *Forestiera acuminata*, *Itea virginica*, or *Rosa palustris*. Two intergrading variants can be outlined.

1. Cypress ponds (p1): Such vegetation is better developed in some regions adjacent to the Black Belt (e.g., Hodges 1997; Ervin et al. 2006). At the two sites studied here, bald cypress (*Taxodium distichum*) is rare, being restricted to a few small ponds within oak-dominated woods. *Acer rubrum* var. *drummondii* and *Fraxinus profunda* are associated with *Taxodium* elsewhere in Mississippi and do occur at the sites studied here, but they are generally infrequent and do not form a distinct vegetation type, such as 2422 of NatureServe (2012a).

2. Shrubby ponds (p2): These are mostly around areas with more permanent ponding, and buttonbush (*Cephalanthus occidentalis*) is generally dominant. Associated shrubs at these sites include *Cornus foemina* plus occasional *Styrax americanus* and *Planera aquatica*. Typical forbs include *Bidens discoidea* (frequent on stumps and bases of shrubs and trees) and *Lycopus rubellus*+, together with much intermixed marshy shoreline (see the following).

In NatureServe (2012a), there are generic *Taxodium* and *Cephalanthus* types for southeastern states: 2420, "*Taxodium distichum / Lemna minor* Forest"; and 2191, "*Cephalanthus occidentalis / Carex* spp. - *Lemna* spp. Southern Shrubland". For transitions to *Salix nigra*, see 4773 and notes under "lower riparian level." Wieland (2000) has described more developed examples of these types farther west in Mississippi, with additional characteristic species. Farther east within the Black Belt, Hall and Penfound (1943) documented cypress-gum swamp on relatively acid alluvium (pH 5.3–6) near the Alabama River, with abundant *Nyssa biflora* (sensu lato) as well as *N. aquatica*.

Herbaceous Phases (~ k2, p3, p4). Such vegetation has highly varied composition, probably reflecting the varied hydrologies and disturbance histories. There is much intermixing of these phases. In NatureServe (2012a), only partial matches can be found for these vegetation types. More detailed descriptions are needed at local and regional levels.

1. **Seasonally Dry Shores and Pond Beds** (k2). This vegetation occurs mostly in shallow ponds managed for waterfowl, with annuals dominating. Some of these species are also common weeds of wet swales in farmland and might well have occurred around animal licks and wallows in pre-Columbian times. Typical grasses include *Echinochloa* spp. (*E. muricata* and aliens), *Eragrostis* spp. (especially *E. hypnoides*), and *Panicum* spp. (especially *P. dichotomifolium*). Other graminoids include *Eleocharis* spp. (especially *E. obtusa*), *Fimbristylis* spp. (*F. annua*, *F. autumnalis*), *Cyperus* spp. (especially *C. erythrorhizos*, *C. strigosus*), and *Juncus* spp. (especially *J. acuminatus*, *J. diffusissimus*). Typical forbs include *Ammannia* spp. (especially *A. coccinea*), *Hydrolea uniflora*+, *Lindernia dubia*+ (sensu lato), *Ludwigia* spp. (especially *L. glandulosa*), *Persicaria* spp. (especially *P. pensylvanica*, *P. setacea*), *Richardia scabra*, and *Scleranthus annuus*+; the latter, at least, is clearly alien.

In NatureServe (2012a), there are no clear matches. However, there is much ecological parallel with 2039: "*Polygonum* [*Persicaria*] spp. - *Echinochloa* spp. - *Distichlis spicata* Playa Lake Herbaceous Vegetation," which occurs on claypans in the central Great Plains. There are similarities, as well, to types with abundant *Eragrostis hypnoides* on shorelines in Atlantic states (4341, 6483, 6608); see also drier phases of Majure's "river channel" community (2007).

Similar areas occur nearby on the Noxubee Refuge with "moist-soil management" for waterfowl, with drier (as in 1, here) and wetter (as in 2, in the next section) phases (M. J. Gray et al. 1999). In these areas, Brock et al. (1994) measured seed production of *Persicaria* spp. (*P. densiflora* and *P. hydropiperoides* sensu lato), *Ludwigia glandulosa*, and *Rhynchospora corniculata*. The *Ludwigia* had the highest production but was not selected for food. *Persicaria hydropiperoides* (probably including *P. setacea*) had much less seed production. Disking on drier ground is often used to reduce the development of more brushy vegetation, but that can also lead to weedy native species like *Sesbania herbacea* (= *S. exaltata*), which has seeds that are toxic to vertebrates.

2. **Wet to Slightly Dry Marshy Shores** (p3). Typical grasses are *Saccharum giganteum* (especially on higher ground), *Leersia oryzoides* (on wetter ground,

or *L. lenticularis* in areas with more shade), *Phanopyrum gymnocarpon* (locally dominant) and *Steinchisma hians+*. Typical sedges and allies include *Carex lupuliformis, C. lupulina, Cyperus* spp. (especially *C. pseudovegetus+*), *Eleocharis* spp. (especially *E. quadrangulata* in deeper water), *Rhynchospora corniculata*, and *Scirpus cyperinus*. Other graminoids include *Juncus* spp. (especially *J. dudleyi, J. biflorus*), and locally *Sparganium americanum*. Typical forbs include *Hibiscus laevis*, (*Hydrolea uniflora+*), *Ludwigia decurrens+, Mikania scandens+, Mitreola petiolata, Oldenlandia boscii*, (*Penthorum sedoides+*), *Persicaria* spp. (especially *P. setacea* on drier soil, or *P. glabra = densiflora* on damper/richer soil), *Pluchea camphorata*, and *Rumex verticillatus*.

In NatureServe (2012a), there is some similarity to two types described mostly from the Gulf Coastal Plain: 4966, "*Polygonum densiflorum* - (*Saccharum giganteum*) Herbaceous Vegetation"; and 8555, "*Polygonum* spp. - *Phanopyrum gymnocarpon* Seasonally Flooded Herbaceous Vegetation," which may be associated with relatively base-rich soils. Denser and taller perennial marsh, with more graminoids and less *Polygonum* (*Persicaria*) or other sprawling low or annual plants, can be matched by two other types: 3866, "*Scirpus cyperinus* Seasonally Flooded Southern Herbaceous Vegetation," which is widespread in southeastern states, perhaps on relatively acid soils; and 4769, "*Sparganium americanum - Saccharum giganteum* Herbaceous Vegetation," which is mostly on the Gulf Coastal Plain, perhaps on relatively base-rich soils. Partial matches in other literature include the "*Eleocharis* sp.–*Scirpus* sp.–*Ludwigia* sp.–*Penthorum sedoides*–*Cyperus*" marsh of Wieland (2000); the "*Sparganium americanum* marsh" of Hall and Penfound (1943); and the wetter phases of Brock et al. (1994) and the "river channel" community of Majure (2007).

3. **Subaquatic Vegetation around More Permament Ponds** (p4 in part). The aquatic flora at the two studied sites is remarkably depauperate, with virtually no Alismatales (Araceae-Lemnaceae, Alismataceae, Hydrocharitaceae-Najadaceae, Potamogetonaceae) except for a few Lemnaceae and *Potamogeton* cf. *pusillus*. Graminoids are generally less frequent than on drier sites, except for local *Leersia* spp. (especially *L. oryzoides*) and *Carex* spp. (especially *C. lupulina*). Typical subaquatic species—emergent plants rooted in shallow water—include *Ludwigia palustris, L. peploides* (more eutrophic sites), *Persicaria hydropiperoides*, and *Proserpinaca palustris*. The rare "lake-cress"—*Neobeckia aquatica*—was found around one small semi-permanent pond with more stable wooded banks.

In NatureServe (2012a), the more permanently wet pond borders outlined here are partly matched by the broadly defined 4290: "*Polygonum* (*hydropiperoides, punctatum*) - *Leersia* spp. Herbaceous Vegetation." The local dominance of *Ludwigia* spp. is matched by 7835, "*Ludwigia peploides* Herbaceous Vegetation," which is most extensive on the Gulf Coastal Plain (see also 6483, with *L. palustris* in Atlantic states). Majure (2007) indicated similar vegetation in "back-water sloughs" south of the Black Belt region, but with less eutrophic conditions and a more diverse wetland flora.

4. **Aquatic Vegetation within More Permanent Ponds** (p4 in part). Floating aquatics are mostly Lemnaceae, especially *Spirodela punctata* and *Wolffia columbiana*; also, a few plants of *Hottonia inflata*. Submersed aquatics are *Ceratophyllum echinatum, Myriophyllum heterophyllum*, and *Potamogeton* cf. *pusillus*.

In NatureServe (2012a), these purely aquatic phases are part of the generic floating type, 3059, "*Lemna* spp. Permanently Flooded Herbaceous Vegetation," and the generic submersed type, 4725, "*Potamogeton* spp. - *Ceratophyllum* spp. - *Elodea* spp. Permanently Flooded Herbaceous Vegetation." As above, Majure (2007) indicated similar vegetation south of the Black Belt region.

Riparian Complex

These are zones along streams that are directly influenced by flooding, with much physical scouring or deposition of alluvium. Oaks are less common that in the "oak complex," but some species are locally common at upper levels, especially *Quercus michauxii* and *Q. shumardii*. Sycamore (*Platanus*), boxelder (*Acer negundo*), ashes (*Fraxinus*), and elms (*Ulmus*) are among the most characteristic riparian trees. However, in more open floodplains or more intensely farmed land, more weedy riparian trees can become predominant, especially cottonwood (*Populus deltoides*) and black willow (*Salix nigra*). R. Harper (1913) listed the latter two species as the most common trees on lowlands in the Black Belt region. Today, trees typical of more mature riparian woods often prevail, presumably because of succession and reforestation along major streams (Murphree et al. 1966, 1974; Murphree and Miller 1976; Hodges 1997).

Upper-Level Riparian Woods (~ m1). This habitat occurs on relatively well-drained ground with occasional flooding by flowing streams, especially along natural or artificial levees. Typical soil series may be mostly loamy to clayey vertisols (Marietta, Trinity?), or perhaps loamy to sandy entisols on more exogenous alluvium (Kirkville?). However, soil surveys do not generally provide detailed mapping for this zone. The vegetation here intergrades considerably with more widespread oak woodland types (see preceding sections).

Composition is highly variable, but typical trees that are locally frequent include *Acer floridanum, Carya cordiformis, Fagus grandifolia, Juglans nigra, Liriodendron tulipifera, Quercus shumardii, Q. michauxii*, and *Q. nigra*. Less frequent species include *Ilex opaca, Morus rubra, Prunus serotina, Ulmus rubra*, and *Fraxinus* spp. (including *americana*, sensu lato, on drier sites). Smaller trees and shrubs include (*Amelanchier arborea*), *Arundinaria gigantea, Asimina triloba, Carpinus caroliniana, Cercis canadensis* (on drier ground), *Crataegus mollis, Ostrya virginiana* (drier), and *Viburnum rufidulum* (drier). Diverse woody vines are present, including locally abundant *Bignonia capreolata, Parthenocissus quinquefolius, Smilax* spp. (especially *S. hispida* and locally *S. smallii*), and *Vitis* spp. (especially *V. vulpina*). Herbaceous vines include *Cayaponia quinqueloba* and *Cocculus carolinus* +.

There are few typical grasses, but *Dichanthelium joorii* and *Melica mutica*+ may be characteristic. Sedges include *Carex amphibola, C. basiantha*+, *C. crebriflora*, and *C. oxylepis*. Other monocots include *Allium canadense, Arisaema dracontium*, and *Smilax lasioneura*. Typical forbs include *Ageratina altissima, Aristolochia serpentaria*, (*Blephilia ciliata*), (*Campanulastrum americanum*), (*Dasistoma macrophylla*), *Elephantopus carolinianus*+, *Fleischmannia incarnata, Galium aparine*+, *Geum canadense*, (*Heliopsis gracilis*), (*Helianthus resinosus*+), *Lactuca floridana, Melanthera nivea*, (*Myosotis macrosperma*+), *Persicaria virginiana*+, *Phlox divaricata, Phytolacca americana*+, (*Rudbeckia triloba* var. nov.), *Ruellia strepens*+, *Sanicula odorata, Solidago caesia* var. *zedia*, (*Spigelia marilandica*+), *Symphyotrichum ontarionis*+, (*Teucrium canadense*+), *Trepocarpus aethusae*+, and *Vernonia gigantea*+.

In NatureServe (2012a), the best match may be 2099, "*Quercus michauxii - Quercus shumardii - Liquidambar styraciflua / Arundinaria gigantea*," a widespread subhydric type in southeastern states. Sections with more acid sandy soils, frequent *Q. nigra*, and local *Q. pagoda* may also approach 7354; see also Wieland (2000) and Majure (2007). However, the woods described here appear to be relatively mesic and base rich, with more frequent *Acer floridanum*. The most mesic sections suggest some similarity to 4671 (more northern) and to 7458 (more southern), types, with much *Acer floridanum* on calcareous bluffs of the Coastal Plain.

Medium-Level Riparian Woods (m2, m3). These woods are typical of stream banks and adjacent areas that are more frequently flooded than levees, with soils that are well drained (m2) to moderately well drained (m3). These woods grade into somewhat poorly drained oak woods on terraces and along sloughs (n2, n3). Soils are probably similar to those in the upper riparian level but have more organic matter and consistent high fertility (including the mollisol Catalpa).

Typical trees include *Acer negundo* (locally dominant near stream banks), *Fraxinus pennsylvanica* var. *subintegerrima* (locally dominant near sloughs), *Ulmus americana*+, *Celtis laevigata*+, *Platanus occidentalis*+, and *Populus deltoides*+. *Populus deltoides* was probably much more common on fresh alluvium in the Black Belt region before channelization of streams (R. Harper 1913). Smaller trees and shrubs include *Sambucus canadensis*+ (and perhaps *Amorpha fruticosa*). There are diverse abundant vines, as are generally found in riparian zones; *Ampelopsis cordata*+ may be particularly characteristic.

Typical grasses include *Chasmanthium latifolium, Cinna arundinacea*+, and *Elymus virginicus*+. Sedges include *Carex blanda* (clayey soils), and *C.* cf. *gracilescens* (sandy soils); *grayi* may also be expected.

Typical forbs include *Commelina virginica*, (*Bidens* cf. *aristosa*), *Cardamine bulbosa, Cryptotaenia canadensis*, (*Dicliptera brachiata*), *Glechoma hederacea*+, (*Helianthus tuberosus*), *Packera aurea*+, *Phlox paniculata, Pilea pumila*, (*Silphium perfoliatum*), *Symphyotrichum lanceolatum, Verbesina alternifolia, Viola* cf. *langloisii*+, and *Zizia aurea*.

In NatureServe (2012a), 5033, "*Acer negundo*," is the strictly riparian type of stream banks. Its 2427, "*Fraxinus*

Figure 7.3. Diagram showing the concept of major gradients among native vegetation types in the central Black Belt of Mississippi. Drier upland types are above, with more acid in the top row; damper lowland types are below. More mature or mesic woodland (deeply shaded) is mostly to the left; more open, stressed or disturbed grassland (lightly shaded) is mostly to the right. Alphanumeric codes in parentheses refer to the sequence of notes on vegetation types in the text (where codes are provided after vegetation names); these codes are modified and extended from Campbell and Seymour (2011b). Note that different kinds of disturbance are not distinguished; fire may have been the major driver before settlement, but large animals and, after settlement, mowing or plowing have also been important. Underlining indicates the types with the most degradation from plowing; for greater detail, see Campbell and Seymour (2015). A third dimension, related to calcareous versus acid soils, could be added to enhance this scheme and to further subdivide some of the vegetation types.

pennsylvanica - Ulmus americana - Celtis laevigata / Ilex decidua," is the less well-drained type of backwaters; see also Wieland (2000). These are widespread types in eastern states, but NatureServe also outlines several segregates associated with calcareous soils that might be developed here (as in 7778 to the west or 7730 to the east). In other regions of Mississippi (e.g., Wieland 2000), *Acer saccharinum* is locally dominant in similar forest on low wet banks, and *Betula nigra* is locally dominant on relatively infertile fresh sands and gravels. Both of those species are completely absent at the two sites studied here, as in most of the Black Belt region (Kartesz 2012).

Lower-Level Riparian Woods (o1). These are thin, swampy woods on fresh alluvium, grading into more stable ponded sloughs farther from active streams. Typical soils are probably poorly drained inceptisols (especially Tuscumbia). Only a few areas were observed in this study.

Salix nigra is locally dominant. Other characteristic trees may include *Fraxinus pennsylvanica* and *Acer negundo* in riparian areas. *Taxodium distichum* and *Quercus lyrata* may be present locally in transitions to more stable ponded sloughs, where *Cephalanthus occidentalis* is a frequent shrub (see previous discussion). Vines are diverse, with *Brunnichia*+, *Ampelopsis cordata*+, and *Vitis* spp. (*cinerea*+, *palmata*?) among the most frequent.

Typical grasses include *Leersia oryzoides* (in relatively open areas along sloughs) and *Glyceria striata* (in more shady transitions to slightly higher ground). Typical forbs include *Apios americana, Boehmeria cylindrica, Packera glabella*+ (also a common weed on damp uplands), *Samolus floribundus*+, *Helenium autumnale, Tradescantia ohiensis*+, and *Xanthium strumarium* (locally abundant and also a common weed on damp uplands). Several common species are annuals typical of freshly exposed alluvium in sunny areas.

In NatureServe (2012a), this vegetation matches two widespread types: 2103, "*Salix nigra*," or 4773 in more open ponded areas, "*Salix nigra* / (*Cephalanthus occidentalis*)."

Riparian Shorelines of Varied Type (drier o2, wetter o3). These are miscellaneous exposed areas along streams, often with extensions into adjacent fields and roadsides. Annuals are generally dominant. Typical grasses on better-drained deposits (o2) include *Leptochloa panicea* (locally dominant) and *Paspalum* spp. (especially *laeve*). Associated forbs include *Sesbania herbacea*+ (locally dominant), *Amaranthus tuberculatus*+, *Iva annua*+, *Xanthium strumarium*+, *Chamaesyce* spp.+ (at least on roadsides, trails, and other disturbed areas nearby), and *Mollugo verticillata*+ (expected based on MacDonald 1996). Plants on wetter ground (o3) include *Ludwigia decurrens, Penthorum sedoides, Samolus floribundus, Mimulus alatus, Rumex altissimus*, and (*Tradescantia ohiensis*+); *Hottonia inflata* is a rare aquatic plant in pools.

There has been little consistent published description of such vegetation. Wieland (2000) described a structurally similar type in his study area on the Pearl River to the west in Mississippi: "*Leptochloa mucronata–Ammannia coccinea–Strophostyles helvula*." Wieland's *L. mucronata* may be included within *L. panicea*, sensu lato; he also listed *Paspalum laeve*, in common with our sites. Majure (2007) indicated a somewhat similar "river channel" community along the Chunky River to the south, but with little overlap in species and grading more into swampy sloughs. There is no clear match for such vegetation in NatureServe (2012a).

Results and Discussion: Flora

Before summarizing the flora of these two lowland sites, notes are provided here on taxa that have obscure status or that have been largely unknown in Mississippi before now.

Notes on Taxonomy and Nomenclature

The 23 taxa listed below have been assigned names in this chapter that have had little or no prior usage in Mississippi, have not been listed by McCook and Kartesz (2010), or have had difficult or inconsistent identification across southeastern states, as noted by Weakley (2012). Most of these taxa have remained somewhat controversial segregates of more broadly interpreted taxa, or there has been confusion in nomenclature that is now resolvable. In a few cases distinction is clear, but the appropriate name remains uncertain or the plant is undescribed. Preferred names are listed first, and common synonyms are provided in brackets after the equal sign = if there is reasonably clear equivalence. In the complete floristic list (see table 7.1), some additional common synonyms are listed for selected taxa.

Amorpha croceolanata Wats. [= *A. fruticosa* var. *croceolanata* (P.W. Wats.) P.W. Wats. ex Mouille; or often included with typical *A. fruticosa*]. See Campbell and Seymour (2012) for details.

Amsonia sp. nov. [= *A. "ornata"* ined. of Sidney McDaniel, previously included within *A. rigida* Shuttleworth ex Small or *A. tabernaemontanum* Walt.]. This undescribed species has been known to McDaniel and associates for at least 20 years (MacDonald 1996; Leidolf et al. 2002). It has been found only in the Black Belt plus adjacent lowlands of Mississippi (DSC, IBE, MISS, MISSA) and Alabama (A. Schotz, pers. comm.), where it usually occurs in thin swampy woods, thickets, and adjacent roadsides on lowlands. These plants resemble *A. salicifolia* Pursh, a distinct narrow-leaved segregate of *A. tabernaemontana*, but they may be closer to *A. rigida* because of their small inflorescences, glabrous flowers, and smaller fruits. Their flowers are paler than others in this complex—pale lavender to white, with a yellowish central ring. *A. rigida* occurs mostly in depressions of pinelands farther south on the Gulf Coastal Plain, but the name has also been applied to these plants of the Black Belt (MISS; Kartesz 2012). Chris Doffitt (pers. comm.) is completing research on this complex at Mississippi State University.

Andropogon tenuispatheus (Nash) Nash [= *A. glomeratus* (Walt.) B.S.P. var. *pumilus* (Vasey) L.H. Dewey; or often included with *A. glomeratus*]. See Campbell and Seymour (2012) for details.

Aristolochia serpentaria L. var. *hastata* (Nutt.) Duchartre [= *A. hastata* Nutt.]. An approproate combination in the segregated genus *Endodeca* is not yet available (Weakley 2012). These plants appear distinct from typical *serpentaria* in their linear to lance-deltoid shape (versus ovate to ovate-oblong), with divergent basal lobes (versus just cordate), ca. 1–2 cm wide (versus 2–5 cm). Also, leaves tend to be less pubescent and relatively thin textured. Var. *hastata* occurs mostly on the southeastern Coastal Plain (from eastern Texas to Virginia), generally in subhydric bottomland forests. Typical var. *serpentaria* has a more widespread south-central range, typically in mesic to subxeric upland forests. Narrow leaves may be an adaptation to reduce egg laying by the eastern pipe-vine swallowtail butterfly, which uses leaf shape as a search image (Rausher and Odendaal 1986).

Boltonia spp. Identifications here remain tentative. Most or all observed plants are referable to *B. diffusa* Ell. var. *interior* Fern. & Grisc., but taxonomic boundaries within this genus need deeper research (Karaman-Castro and Urbatsch 2006; Weakley 2012). A few collections (SC-1554, SC-2298) suggest transitions to *B. asteroides* (L.) L'Her., which has been reported from the region, mostly along larger rivers (Leidolf et al. 2002) but also along upland roads (Campbell and Seymour 2012). Variation within *B. asteroides* is complex, and plants in this region may be referrable to var. *glastifolia* (Hill) Fern.

Carex cf. *gracilescens* Steud. [? = *C. laxiflora* Lam. var. *gracillima* F. Boott]. Distinction of this species from *C. blanda* Dewey has been problematic in some regions, with intergradation sometimes postulated. There appears to be significant variation within each traditionally recognized species, as reflected in reported chromosome numbers: 2n = 33, 38, and 40 in *C. gracilescens*; 2n = 30, 32, 34, 36, and 40 in *C. blanda* (Gleason and Cronquist 1991; Yatskievytch 2006a, 2006b; P. Ball and Reznicek 2002). The best key character for separating these two species has been considered the color of basal sheaths: "purple, reddish purple, or purple tinged" in *C. gracilescens* versus "brownish" in *C. blanda*. However, this character can often be indistinct (see also Yatskievytch), and Fernald (1950) previously used a different set of characters. Some collections here from submesic to subhydric terraces at both sites have been provisionally assigned to *C. gracilescens*, using published treatments, but C. T. Bryson (pers. comm.) considers them referable to Boott's "*minor*" morph of *C. blanda*: "[This] narrow leaf form is more common on shaded heavy clay soils often over chalk in the Black Belt, and on shaded wet heavy clay soils in the Delta Region often with *Sabal minor* as an associate."

Carex cf. *fissa* MacKenzie. Distinction of these plants from *C. annectens* (Bickn.) Bickn. and *C. triangularis* Boeckl. was difficult in some cases. C. T. Bryson (pers. comm.) considers that our material of "*C.* cf. *fissa*" is referable to *C. annectens*, but further study of variation within both species is warranted. *C. fissa* is a more robust species distinguished by its relatively large perigynia (3.2–4 × 2–2.6 mm versus 2.2–3 × 1.5–2.2 mm), with paler color (dull yellow green to pale brown versus yellow brown to golden). It has two recognized varieties, disjunct to the west and south of northern Mississippi (P. Ball and Reznicek 2002). More fieldwork is needed to determine whether *C. annectens* fills a geographical or ecological gap between these two segregates of *C. fissa*. Both *C. annectens* and *C. fissa* appear uncommon to rare in the Black Belt region (MNHP 2006; ANHP 2012; NatureServe 2012–2013; Kartesz 2012). However, lack of records may reflect lack of botanical survey.

Carex section *Ovales*: *C. sangamonensis* (Clokey) Mohlenbr. [= *C. tribuloides* Wahl. var. *sangamonensis* Clokey; or often included with the species]. This taxon is centered in the lower Mississippi Valley and has recently been elevated to species status (P. Ball and Reznicek 2002; Mohlenbrock 2011). Compared to typical *C. tribuloides*, its perigynia are less elongated (with l/w = 2.2–3 mm versus 3–5 mm). Also, it tends to have a more elongated inflorescence (with separated versus clustered spikes) and narrower leaves (ca. 2–4.5 mm versus 3–7 mm). It is locally abundant in wooded seeps and swamps (versus more open wetlands). Other members of this section at the studied sites include the following (see table 7.1 for details): *C. reniformis* (wetter-tending open woods), *C. albolutescens* (better-drained woods), and *C. festucacea* (seasonally wet open areas). Also expected, but known mostly from more open or seasonally drier sites in the region (table 7.2; Campbell and Seymour 2012; C. T.

Bryson, pers. comm.), are *Carex longii* Mack., *C.* cf. *normalis* Mack., *C. molesta* Mack. ex Bright, and *C. brevior* (Dewey) Mack. ex Lunell.

Dichanthelium acuminatum (Sw.) Gould & C.A. Clark and allies [= *Panicum lanuginosum* Lam., sensu lato]. The treatment here follows R. J. LeBlond in Weakley (2012). Members of this problematic complex that were found at these sites include the following taxa. Further research is needed to determine how distinct these are in morphology and ecology.

(1) *D. acuminatum* (Sw.) Gould & C.A. Clark var. *fasciculatum* (Tuck.) Freckmann; this is the most widespread, weedy member of the complex.

(2) *D. longiligulatum* (Nash) Freckmann, perhaps grading into *D. acuminatum* var. *lindheimeri* (Nash) Gould & Clark; this is widespread in southeastern states and farther south on seasonally drying wetlands. Other botanists in the region have recognized *D. acuminatum* var. *lindheimeri*, but not *D. longiligulatum* (MacDonald 1996; Leidolf et al. 2002). These two taxa may not be distinguishable in the lower Mississippi Valley.

(3) *D.* cf. *leucothrix* (Nash) Freckmann, perhaps grading into (1); this is a southeastern taxon typical of seasonally wet pinelands. It may also grade into *D. meridionale* (Ashe) Freckmann of more northern regions, which was reported from the Pulliam Prairie (Campbell and Seymour 2012).

Dichanthelium joorii (Vasey) Mohlenbr. [= *Panicum joorii* Vasey; *D. commutatum* (Schultes) Gould ssp. *joorii* (Vasey) Freckmann & Lelong]. Treatment here follows Mohlenbrock (1986). See Campbell and Seymour (2012) for details.

Dichanthelium microcarpon (Muhl. ex Ell.) Mohlenbr. [= *D. dichotomum* (Linnaeus) Gould var. *ramulosum* (Torrey) LeBlond; or sometimes combined completely with *D. dichotomum*]. Treatment here follows Mohlenbrock (1986). This is sometimes confused with *D. dichotomum* var. *nitidum* (Lam.) LeBlond, which has been reported as a subspecies or species in this region (Leidolf et al. 2002). The latter is generally a distinct southern taxon that may be more restricted to acid soils, but it may intergrade with *D. microcarpon*.

Dichanthelium polyanthes (Schultes) Mohlenbr. [= *D. sphaerocarpon* (Ell.) Gould var. *isophyllum* (Scribn.) Gould & C.A. Clark]. Treatment here follows Mohlenbrock (1986), Freckmann and Lelong (2003), and Weakley (2012). This taxon is distinct from typical *D. sphaerocarpon*, consistently growing on wetter ground.

Diospyros virginiana L. var. *pubescens* (Pursh.) Dippel [usually included with the species]. This is recognizable mostly along the lower Mississippi River and farther west, where it is most distinct in wetlands (Fernald 1950; Weakley 2012; D. Estes, pers. comm.). The leaves of var. *pubescens* are thinly but persistently covered with relatively long hairs below (versus glabrous or glabrescent), subcordate (versus cuneate to rounded), and often relatively large; also, twigs are hirsutulous (versus glabrous). These characters may be associated with the higher chromosome number (2n = 90 versus 60) and the unusually large fruits that have occasionally been found in this species. However, variation often appears to be more complex or continuous, perhaps involving past hybridization with *D. texana* Scheele (2n = 30) to the southwest, and more analysis is needed (Yatskievytch 2006a, 2006b, Eckenwalder 2009).

Eupatorium cf. *scabridum* Ell. [= *E. rotundifolium* var. *scabridum* (Ell.) Gray]. Treatment of this species here follows Weakley (2012). These plants may have resulted from hybridization of *E. rotundifolium* with *E. semiserratum* DC. and are concentrated locally on the Coastal Plain from Texas to South Carolina (NatureServe 2012–2013).

Gonolobus suberosus (L.) R. Brown var. *granulatus* (Scheele) Krings & Q.-Y. Xiang [previously included in *Matelea suberosa* (L.) Shinners, sensu lato; or in *G. gonocarpus* (Walt.) Perry, sensu lato]. See Campbell and Seymour (2012) for details.

Heliopsis gracilis Nutt. [= *H. minor* (Hook.) C. Mohr of Small (1930); = *H. helianthoides* (L.) var. *gracilis* (Nutt.) Gandhi & R.D. Thomas; or often included with *H. helianthoides*]. See Campbell and Seymour (2012) for details.

Justicia lanceolata (Chapm.) Small [= *J. ovata* (Walt.) Lindau var. *lanceolata* (Chapm.) R. W. Long]. Treatment here follows Small (1933) and Weakley (2012). This distinct taxon occurs mostly on the Gulf Coastal Plain, while typical *J. ovata* occurs on the Atlantic Coastal Plain.

Panicum longifolium Torr. [= *P. rigidulum* Bosc ex Nees var. *pubescens* (Vasey) Lelong]. Treatment here follows Weakley (2012). Although often treated as a variety of *P. rigidu-*

lum, this taxon is distinct, with virtually no documented intermediates, and it generally occurs in habitats with less ponding and more seasonal drying. With ongoing revision of *Panicum* (sensu lato), the accepted name for this species may come to be *Coleataenia longifolia* (Torr.) Soreng; see Weakley (2012) for details.

Ranunculus carolinianus DC. [= *R. hispidus* Michx. var. *nitidus* (Chapm.) T. Duncan]. Treatment here follows Weakley (2012). This widespread southeastern and midwestern diploid (2n = 32) species has often been confused with the northern tetraploid species (2n = 64), *R. caricetorum* Greene (Fernald 1950; Whittemore 1997). Both are wetland plants, with distinctively long recumbent to stoloniferous branches often produced after flowering (especially precocious in *R. carolinianus*). Those two species are usually less hairy than their upland relatives, *R. hispidus* and *R. fascicularis* (both diploids). *R. carolinianus* differs from all of these species in its relatively short (ca. 3.5–5 mm long versus 5–11 mm), reflexed sepals (versus spreading or tardily reflexed); and in its typically less numerous, larger achenes (ca. 3.7–5 mm long versus 2–4.3 mm), with a wider marginal wing (0.4–1.2 mm versus 0.1–0.2 mm).

Rudbeckia sp. nov., aff. *tenax* C.L. Boynt. & Beadle [*R. tenax* itself is often included in *R. fulgida* Ait., sensu lato]. See Campbell and Seymour (2012) for details.

Rudbeckia triloba L. variant. Collections of *R. triloba* were made only at the DeMoville Tract in riparian woods on or near a natural levee. These plants were relatively short and hairy, with small heads produced in June. Typical *R. triloba* flowers mostly in August, occasionally in July or September. Research is under way to determine whether these collections represent an undescribed variety. At NCU, similar plants were noted from Texas, Louisiana, Oklahoma, Arkansas, Mississippi, and Alabama.

Symphyotrichum dumosum (L.) G.L. Nesom variant [= *Aster dumosus* L., sensu lato]. All collections here may be referable to var. *coridifolium*, a combination not yet formally made. This widespread southeastern segregate, as *A. dumosus* var. *coridifolius* (Michx.) Torr. & Gray, has been combined with typical *dumosum* in most recent treatments (e.g., Brouillet et al. 2006). In its strict sense, var. *dumosum* is centered in northern Appalachian regions and is unknown in Mississippi (Fernald 1950; J. C. Semple, pers. comm.). Plants of the *S. dumosum* complex can also be confused with *S. racemosum* (Ell.) Nesom and *S. lateriflorum* (L.) Á. Löve & D. Löve—the collection referred to *S. lateriflorum* here may in fact be hybridized.

Viola cf. *langloisii* Greene [often included with the *V. sororia-papilionacea* complex, as in *V. sororia* Willd. var. *missouriensis* (Greene) McKinney]. A satisfactory treatment is still not available for stemless blue violets of Mississippi, but some guidance comes from the varied treatments of McKinney and Russell (2002), Ward (2006), B. A. Sorrie in Weakley (2012), and earlier manuals. The Black Belt lies in or near areas of potential contact between the largely southeastern *V. langloisii*, the midwestern *V. missouriensis* Greene, the widespread eastern *V. papilionacea* Pursh (sensu Fernald 1950), and the Appalachian-centered *V. affinis* Le Conte. Concepts of these species have often been blurred by lack of morphological characters and by apparent intergradation. The provisional "edge" given here to *V. langloisii* might be changed. Most or all plants observed in the Black Belt—including the "*V.* cf. *missouriensis*" of Campbell and Seymour (2012)—do have a thin scattering of hairs on the upper surfaces of leaf lobes, which might suggest affinity with *V. affinis*. Instead, an argument can be made for combining all of these taxa, plus others, into one complex species—*V. sororia* Willd. Yet the lowland plants observed here do appear somewhat distinct from the plants in attenuated fingers of riparian woods at Pulliam Prairie (Campbell and Seymour 2012).

Species with Little or No Prior Record in Mississippi

Some of the following eight taxa appear to be uncommon or rare, but others are common and have been confused with related taxa. Plants with uncertain identification or undescribed status are excluded from this list, but see the preceding section, "Notes on Taxonomy and Nomenclature" (especially under *Amsonia* and *Rudbeckia*). For details of collections, see table 7.1.

Amorpha croceolanata Wats. See notes of Campbell and Seymour (2012).

Carex aureolensis Steud. See notes of Campbell and Seymour (2012), who did not list their one record of the closely related species *C. frankii* Kunth from the Pulliam Prairie: SC-0156. The latter appears to be much less common in Mississippi than *C. aureolensis*, with which it has been confused until recent years (P. Ball and Reznicek 2002; Majure 2007; Majure et al. 2011).

Crataegus reverchonii Sarg., sensu lato; including *C. discolor* Sarg., *C. palmeri* Sarg., and *C. mohrii* Beadle as varieties, based on work of J. B. Phipps (Chester et al. 2009) and R. Lance in Weakley (2012). This member of the *C. crus-gallii* group, with three to five styles per flower and a tendency to suborbicular leaves, is known mostly from the Edwards Plateau to the Ozark region in Texas, Oklahoma, Arkansas, and Missouri (Kartesz 2012). However, with the inclusion of *C. discolor*, *C. palmeri*, and *C. mohrii*, this species does range into states east of the Mississippi River. Sargent (1905) noted the range of *C. mohrii* as follows: "Western Georgia to central Alabama and Mississippi, and northward to middle Tennessee; abundant and of its largest size in the low flat woods of central Alabama, and ascending into the poorer and drier soils of the neighboring hillsides and low mountain slopes." Without varietal determination, *C. reverchonii* (sensu lato) is reported here from Monroe County, Mississippi, based on a vegetative collection.

Diospyros virginiana L. var. *pubescens* (Pursh) Dippel. See taxonomic notes in the preceding section. Fernald (1950) indicated the following range: "Fla. to Ark., n. to Va., s. Ill. and s. Ia." This variety was listed for Mississippi by McCook and Kartesz (2010), but no collection appears to have been previously detailed in print or recorded in herbarium databases. It is reported here from Clay County, based on a vegetative collection.

Eleocharis compressa Sullivant. See notes of Campbell and Seymour (2012). There have been only one or two prior published reports of this species from Mississippi (C. T. Bryson in Leidolf et al. 2002; S. Smith et al. 2002; Kartesz 2012). Further work on variation in this species and its relatives is needed across southeastern states. *E. elliptica* Kunth is a northeastern species that may intergrade locally with *E. compressa*, and it has also been found rarely in the Black Belt region (C. T. Bryson in Leidolf et al. 2002).

Galium obtusum Bigelow var. *filifolium* (Wiegand) Fernald. This taxon has been reported from New Jersey to Alabama, mostly on the Atlantic Coastal Plain and adjacent Piedmont (USDA 2010–2013; Weakley 2012). It reported here from Clay County, Mississippi, apparently the first published state record.

Geranium pusillum L. This European alien has been reported from most states north of Mississippi (USDA 2010–2013), but it is rare to absent on most of the southeastern Coastal Plain (Kartesz 2012; Weakley 2012). It is reported here from Clay County, apparently the first published record from Mississippi. The closely related species *G. molle* L. was first collected from the state in 2007 (Majure et al. 2011).

Solidago caesia L. var. *zedia* R.E. Cook & Semple. This was recently described (R. Cook and Semple 2004) and has been reported from scattered sites across the Gulf Coastal Plain, including Mississippi (Semple and Cook 2006). However, there have been virtually no published details of its distribution within the state, and available lists or databases for the state do not segregate this taxon from typical *caesia* (e.g., Kartesz 2012; USDA 2010–2013). During this study, it was found at only one site in Clay County, along the old artificial levee along Houlka Creek.

Notes on Rare Species

The following 16 species found at the lowland sites appear to be rare within the Black Belt region, based on a broad floristic review including Flora of North America Editorial Committee (1993–2010), MNHP (2006), ANHP (2012), Kartesz (2012), USDA (2010–2013), NatureServe (2012b), Weakley (2012), and other recent literature. Under each species, the likely degree of rarity in Mississippi is indicated by standard codes: S1 = endangered; S2 = threatened; S3 = moderately rare; S4 = somewhat rare to uncommon. In a few cases, these ranks are adjusted from the recent listing of the Mississippi Natural Heritage Program (MNHP 2006). Most of these species are not imperiled but do have significant interest as "watch-list" or "conservative" plants that can provide useful indications of ecological context. Excluded here are some taxa that have few records but that tend to have uncertain identifications or that deserve much further collection: for example, *Carex* spp., such as *C. annectens*, *C. gracilescens*, *C. sangamonensis*, and *C. socialis* (S1? in AL); and *Juncus* spp., such as *J. interior* and *J. secundus* (S1 in AL).

Ammannia auriculata Willd. This is an annual of ephemeral ponds and shores, with a range centered in the southern Great Plains. There are only a few records from Mississippi, mostly near the Mississippi River, and it is considered rare in the state (S2S3?). The species is tentatively reported here, based on a depauperate specimen collected at the edge of a shallow drying pond at the back of the Barr Tract.

Amorpha croceolanata P.W. Wats. See the preceding notes

on taxonomy, also in Campbell and Seymour (2012). The status of this shrub is hard to determine without further revision. It may be somewhat uncommon locally (S3S4?) and globally (G3G4?).

Amsonia sp. nov. [*A.* "*ornata*" ined.] See the preceding notes on taxonomy. This undescribed species may be considered moderately rare locally (S3?) and globally (G3?).

Campanulastrum americanum (L.) Small. This palatable forb of submesic eutrophic woodland (often browsed out by deer) is widespread across east-central states, but rare to absent on most of the southeastern Coastal Plain, where it is largely restricted to calcareous regions. It is moderately rare in Mississippi as a whole (S3S4), but it is occasional to locally frequent in wooded habitats of the Black Belt, as found here (table 7.1) and elsewhere (MacDonald 1996; Leidolf et al. 2002; Campbell and Seymour 2012).

Carya laciniosa (Michx. f.) G. Don. This tall tree (with unpalatable foliage but much sought-after nuts) is widespread across east-central states on damp fertile soils. It is rare in Mississippi (S3), with some local concentrations in the Black Belt, but even in this region it is generally uncommon to rare (Little 1971; Leidolf et al. 2002); see table 7.1 for our data. It has recently been discovered farther west in the Delta region of the state (Skojac et al. 2007).

Crataegus reverchonii Sarg. This lower midwestern hawthorn has not been reported from Mississippi before (see preceding notes), and its potential for endangered or threatened status here should be investigated further. There may be some biogeographic parallel with *Maclura pomifera*—another species from Texas to Missouri that appears to have had disjunct pre-Columbian occurrences in the Black Belt (J. L. Selzer and J. G. Hill, Mississippi State University, pers. comm.). Habitats are similar, generally in disturbed submesic woods on base-rich soils.

Dasistoma macrophylla (Nutt.) Raf. This semiparasitic forb occurs mostly in thin woods on calcareous soils across east-central states, and it is rare to absent on most of the Coastal Plain. It is moderately rare in Mississippi (S3S4), but somewhat concentrated in the Black Belt. It has been found in small numbers at both lowland sites studied here, the Pulliam Prairie, and other scattered sites in the region, especially at edges of thin submesic woods (MacDonald 1996; Leidolf et al. 2002).

Eleocharis compressa Sullivant. This spike-rush is widespread on damp calcareous soils in midwestern states, but it is rare to absent across most of the southeastern Coastal Plain. Its status in Mississippi is hard to determine, but it may be threatened (S2?). It was found here only around margins of shallow ponds at the back of the Barr Tract in Clay County.

Fraxinus cf. *profunda* (Bush) Bush. This hexaploid tree (2n = 138) is widely distributed over eastern states, but concentrated around more permanent wetlands and with a fragmented range similar to that of *Populus heterophylla* L. (Little 1971; Nesom 2010). In Mississippi, it is moderately rare (S3S4), with most records concentrated on lowlands near the Mississippi River. There is at least one previous record from the Black Belt—Lowndes County (Kartesz 2012)—and it is reported here from both Clay and Monroe Counties. However, our records are based only on vegetative collections. A more thorough revision of *Fraxinus* collections from this region is needed, considering Nesom's recent work on the genus—*F. biltmoreana* Beadle [? = *F. smallii* Beadle] is to be expected on these lowlands as well.

Heliopsis gracilis Nutt. See notes of Campbell and Seymour (2012). This woodland forb has a somewhat fragmented range centered on the Gulf Coastal Plain, and it is generally infrequent to rare (G3G4?). It may be threatened to moderately rare in Mississippi (S2S3?). A few scattered plants were found at both lowland sites, as well as in riparian zones at the Pulliam Prairie (Campbell and Seymour 2012). MacDonald (1996) and Leidolf et al. (2002) reported it as "uncommon" or "occasional," and only from regions adjacent to the Black Belt.

Hottonia inflata Ell. This aquatic plant has a widespread but somewhat fragmentary southeastern range, and it is considered rare to endangered in most states, especially Atlantic states (NatureServe 2012–2013). It may be threatened to rare in Mississippi (S2S3?), but its status is hard to assess. It has recently been found in the Delta region (Skojac et al. 2007), and there is a curious cluster of records in or near the Black Belt (Kartesz 2012). During this study, only a few plants were found—in the Barr Tract, along shorelines of the larger natural pond and along a pool of Houlka Creek.

Neobeckia aquatica (Eat.) Greene [= *Armoracia lacustris* (Eat.) Wieg.]. This subaquatic plant is concentrated in a broad zone from the Great Lakes to the central Mississippi Valley to the Gulf Coastal Plain. It is listed as rare or imperiled by Natural Heritage Programs in most states across the range, and it may be globally rare (G3?); it is endangered across most southeastern states (NatureServe 2012–2013). In Mississippi, *Neobeckia* may be threatened (S2?), with a curious cluster of records in or near the Black Belt as well as on bottomlands of the Mississippi River. During this study, it was found in only one small natural pond in the Barr Tract, near the transition from wooded lowlands to former farmland. Single records were also listed by MacDonald (1996)—from "rare, low wet area near Tombigbee River"—and by Leidolf et al. (2002)—from "grass/forb meadows."

Packera aurea (L.) Á. Löve & D. Löve. This forb is widespread in thin damp woods across eastern North America, except on most of the southeastern Coastal Plain. In Mississippi, it is known from only a few northeastern counties (S2S3?). MacDonald (1996) provided the first state record—from "alluvial woods along Mattubby Creek"—and during this study it was also found along the same creek.

Phlox carolina L. This southeastern species is often confused with *P. glaberrima* L. (centered in east-central states) or *P. triflora* Michx. (centered in Appalachian regions), and it may contain distinct varieties (Wherry 1955; Weakley 2012; P. Zale, Ohio State University, pers. comm.). Its status in Mississippi is hard to determine without further revision, but it may be rare (S3S4). There appears to be some concentration of records in or near the Black Belt region. Typical *carolina* was found locally in drier lowland oak woods in the Barr Tract. MacDonald (1996) reported the species as occasional in "low mesic to wet-mesic open areas"; and Leidolf et al. (2002) reported ssp. *angusta* Wherry as locally common in the "Interior Flatwoods" region on the western side of the Black Belt. Typical *P. carolina* was also reported from the Pulliam Prairie (Campbell and Seymour 2012), but that now appears to be based on misidentification of *P. pilosa* L. Both species are highly variable in pubescence at these sites.

Ptelea trifoliata L. In its typical variety, this unusual, unpalatable small tree with running roots ranges widely from southeastern states to the Great Lakes region. It is rare to absent in more extensive lowlands of the lower Ohio and lower Mississippi Valley, and in much of the central Appalachians. On the Coastal Plain in Mississippi, it is generally rare (S3S4) and concentrated in or near the Black Belt, Buhrstone/Lime Hills, and Jackson Prairie regions (Majure 2007; Kartesz 2012). It was found here only in the DeMoville Tract, with a few plants in thin oak woods. *Ptelea* was also listed as "uncommon" by MacDonald (1996) in "low wet woods along Tombigbee River," and by Leidolf et al. (2002) in bottomland hardwoods and riparian sites.

Scirpus pendulus Muhl. Although widespread on seasonally wet base-rich soils across eastern states, this bulrush appears to be uncommon in Mississippi based on current mapping (S3S4?). It was locally common at the Pulliam Prairie (Campbell and Seymour 2012) but was noted only once at the lowland sites. Further work may be needed in the state to distinguish *S. pendulus* from the closely related southern species *S. lineatus* Michx.

Notes on Alien Species

Of the 513 taxa documented in this study, 57 (11%) are clearly alien, with origins outside North America, as indicated by double asterisks in the left margin of table 7.1. There is uncertainty about the native or alien status of about 15 additional species (3%), as indicated by single asterisks. These aliens—or potential aliens—are concentrated in more open, disturbed sites on the drier soils of the study sites: 44 (including 6 with uncertain or adventive status) are most common in the "Grassland Complex" outlined previously, comprising 28% of all species in that habitat class (table 7.2). In thickets, edges, and other transitions between grassland and woodland, this proportion is only 8%. Low percentages also occur among species typical of the Pond Complex (15%), Riparian Complex (6%), and Oak Woodland Complex (4%). Similar percentages (7–15%) are found in varied habitats on the Pulliam Prairie (Campbell and Seymour 2012).

The 20 species noted below are locally frequent or present a significant invasive threat in the future. Only a few are locally common and persistent aliens in deeper woods, as opposed to brushy edges, old fields, or roadsides: *Glechoma hederacea*, *Lonicera japonica*, *Microstegium vimineum*, and *Persicaria longiseta*. Most of the other aliens are uncommon to rare and do not pose significant problems for native vegetation. Excluded from this list are several minor species that are locally frequent but largely restricted to highly disturbed roadsides or other frequently mowed or plowed sites. Also excluded is *Urochloa platyphylla* (Munro ex C. Wright) R.D. Webster. This "signal-grass" is a locally abundant weedy annual on bare ground in seasonally wet fields and along shores of drying ponds. Its native or alien status remains unclear, based on somewhat conflicting statements in recent manuals; for example, compare Weakley (2012) with Wipff and Thompson (2003). It has been considered native during recent restoration of the Cajun Prairie in Louisiana (Allen and Thames 2007).

Cynodon dactylon (L.) Pers. This creeping perennial warm-season grass from warm regions of Eurasia ("Bermudagrass") is locally abundant in old fields and along roadsides across the region. It can be an initial problem for plantings of native plants, but it is relatively short statured and sun demanding, so it does not become a persistent problem in restoration of woodland or grassland.

Echinochloa: *E. colonum* (L.) Link, *E. crus-galli* (L.) Beauv., and *E. esculenta* (A. Braun) H. Scholtz. These alien "barnyard" grasses are locally frequent in wetlands of the region because of spontaneous spread or, especially in the case of *E. esculenta*, active plantings for waterfowl. The native *E. muricata* (Beauv.) Fern. is also locally frequent, but there has been virtually no attempt to increase its proportion of seed in the marketplace for wildlife food plots. These grasses are particularly successful in seasonal swales, shallow drying ponds, and gradually receding shores.

Festuca arundinacea Schreb. [= *F. elatior* L. var. *arundinacea* (Schreb.) Wimmer; = *Schedonorus phoenix* (Scop.) Holub]. This perennial European cool-season grass ("tall fescue") is locally dominant along roads and in mowed fields. Like *Cynodon*, it can interfere with native plantings, but it does not present a long-term problem for restoration.

Glechoma hederacea L. This Eurasian weed is a locally common creeping plant in damp riparian woods, especially where ground is disturbed or fresh alluvium deposited. It is not a major problem for recovery of taller native plants.

Landoltia punctata (G. Meyer) D.H. Les & D.J. Crawford [= *Spirodela oligorhiza* (Kurz) Hegelm.]; and other alien aquatic species. Special problems resulting from the invasion of alien aquatic plants are not detailed here, except to note the general dominance of *Landoltia* in more permanent ponds in the Barr Tract—this species from the Southern Hemisphere is now infamous because of the appearance of a herbicide-resistant strain (Koschnick et al. 2006). Another potentially problematic alien in these ponds is *Myriophyllum spicatum* L., which needs further documentation on these lowlands and elsewhere in region (Kartesz 2012).

Lespedeza cuneata (Dum. Cours.) G. Don. This East Asian perennial ("sericea lespedeza") is locally dominant in some drier old fields on these lowlands, where it has often been planted during initial conversion from farmland. The species would be a problem if open grassy conditions were desired, but it disappears within a decade or so if succession is allowed.

Ligustrum sinense Lour. This East Asian shrub is locally frequent in drier woods on these lowlands, especially near edges or after old-field succession, but it does not appear to become as abundant in the Black Belt region as in adjacent hills. This species can largely prevent tree regeneration and change ecosystem function, especially on less eutrophic soils elsewhere in southeastern states (e.g., Mitchell et al. 2011). Yet it can also become a major forage species for deer (Stromayer et al. 1998).

Lonicera japonica Thunb. This woody vine from East Asia ("Japanese honeysuckle") is widespread and locally abundant in drier woods on these lowlands, especially in younger woods or near edges. However, it is generally absent from the grasslands here, and burning is well known to keep this species from invading grasslands.

Lysimachia nummularia L. This low Eurasian weed spreads locally over bare ground in riparian woods and wetlands, usually on wetter sites than *Glechoma*. It is not a major problem for recovery of taller native plants.

Microstegium vimineum (Tren.) A. Camus. This unusual warm-season annual grass ("Japanese/stilt-grass") is locally common in well-drained woods on these lowlands, especially where sandy soil is exposed. In Mississippi, *Microstegium* can become locally dominant and influence native species, especially after disturbances (Brewer 2011). But there is evidence that native perennials generally can recover to outcompete most of this alien during succession (Menon and Holland 2012). The potential role of fire in reducing or increasing *Microstegium* remains uncertain (Emery et al. 2011).

Paspalum: *P. dilatatum* Poir. and *P. urvillei* Steud. These alien grasses from South America are locally abundant along mowed roads (especially *P. dilatatum*) and in old fields (especially *urvillei*). As with *Cynodon*, they do not present long-term problems for restoration; see also Allen and Thames (2007).

Persicaria: *P. longiseta* (de Bruijn) Kitagawa and *maculosa* [see synonyms in table 7.1]. These annual smartweeds are locally common, especially the East Asian *longiseta*, which extends more into partial shade; *maculosa* is restricted to open areas. There is probably a general ecological balance with the native smartweeds, especially *punctata*, which often shares space at the same sites as *longiseta*. Two other alien Polygonaceae are locally frequent, but only in fields or along roads, especially where there are ditches: *Rumex crispus* L. and *R. pulcher* L.

Ranunculus sardous Crantz. This annual is locally abundant on seasonally damp disturbed ground in fields and thin woods. It can be a persistent problem in small areas of native vegetation, but the species is generally outcompeted with succession.

Rosa multiflora Thunb. ex Murr. This is a locally common shrub along woodland edges and roadsides, but it is not generally common or invasive within native vegetation on these lowlands.

Sorghum halepense (L.) Pers. This aggressive perennial grass from the Mediterranean region ("Johnsongrass") has become widespread and locally dominant in fields and along roads. It can be a significant problem for restoration of native grassland (Campbell and Seymour 2011a).

Verbena brasiliensis Vell. This South American weed is widely scattered in open areas on drier parts of these lowlands. While not abundant, it can be a locally confusing volunteer in grassland restoration; see also Allen and Thames (2007).

General Discussion

Ecological Gradients from Lowlands to Uplands

If a reliable classification of native vegetation in this region is to be developed, more attention to ecological gradients is needed. The initial outline of gradients developed here in diagramatic form (fig. 7.3) can be tested with more systematic analysis of data, ideally from a comprehensive series of plots plus details of disturbance history and soils. An additional dimension for this diagram, related to calcareous versus acid soils, should also be generated. NatureServe (2012a) has provided invaluable descriptions of reported vegetation types from varied sources, but without a model of ecological gradients, the fundamental science is eluded—environmental factors that have led to differences in species composition remain largely guessed at rather than demonstrated. Also, it is often difficult to know at what scale the NatureServe descriptions were derived and should be applied. For some purposes, such as mapping a county, details of vegetational patterns within areas of less than 1–10 acres are irrelevant, while for other purposes, such as understanding dynamic processes, such small-scale variation is often the primary interest.

Our model of gradients among soil series (Campbell and Seymour 2015) provides an interpretation of the transition from lowlands to uplands. Soils that are mapped frequently on lowlands (fig. 7.1) include some that overlap with the chalky uplands in their characteristics, especially Leeper (a vertic epiaquept) and Griffith (a hapludert), and both of these also occur widely on uplands (fig. 7.1). These two soil series are quite similar to the strictly lowland soils, Tuscambia (a vertic epiaquept) and Trinity (a hapludert), respectively. Moreover, the detailed mapping of our three study sites (USDA 2012a) indicates that other inceptisols and vertisols with moderate to poor drainage occur in the transitions from lowlands to uplands at these localities. At the wetter end of the gradients, especially on the gentlest slopes and flats, these poorly drained soils link lowlands directly to uplands without any mesic or subxeric transition.

The information on vegetation assembled here and elsewhere (Campbell and Seymour 2011a, 2011b) includes notes on types centered between uplands and lowlands—types that have rather little documentation from previous research in the region. Their descriptions and suggested associations with soil types still deserve further development before they can be considered reliable, but the following provisional summaries are possible (see fig. 7.3 for alphanumeric codes to segregates).

1. **Relatively mesic wooded levees** (m1, j2, j3): with local occurrence of swamp chestnut oak (*Quercus michauxii*), black walnut (*Juglans nigra*), sugar maple (*Acer floridanum*), beech (*Fagus*), yellow poplar (*Liriodendron*), and other associates. It would be interesting to know whether the disturbance-sensitive maple and beech are widely scattered along levees in the region, or whether the sites studied here are unusual and oaks generally predominate instead (*Quercus alba*, *Q. michauxii*, *Q. nigra*, *Q. pagoda*, *Q. shumardii*). On flats at the southwestern edge of the Black Belt, Ervin et al. (2006) found that these oaks were relatively common on drier ground in the Noxubee National Wildlife Refuge, together with much sweetgum (*Liquidambar*) but little beech and no sugar maple.

2. **Canebrakes and associated thin woods or thickets** (l, h3): cane (*Arundinaria gigantea*) is locally frequent in better-drained woodland at both

lowland sites, from mesic levees to outer edges that transition into farmland. It is not generally dominant, except in a few patches of up to an acre or so along higher levees. Yet historical sources from the Alabama Black Belt indicate that canebrakes formed a major vegetation type on some lowland soils (Urbo, Mantachie, Leeper, Houlka) and even calcareous upland soils (Vaiden, Oktibbeha, Sumter), before agricultural conversion (Barone et al. 2008). Groves of fruit and nut trees may also have been frequent in such transitions from low woods to grassland, as indicated by some early accounts—with presumed use by native people (Campbell and Seymour 2011b).

3. **Seasonally wet to damp grassland** (g2, g1): with local dominance of gamagrass (*Tripsacum dactyloides*), switchgrass (*Panicum virgatum*), and, at upper levels, just outside our two lowland sites, big bluestem (*Andropogon gerardii*). Only small patches and strips of such vegetation along roadsides were found within the two sites. But in transitions to uplands within a mile of both sites, there are more diverse remnants of native grassland along roadsides and in adjacent old pastures. Such grassland may well have occurred in the more disturbed old fields and roadsides at these localities, which now have abundant bushy broomsedge, *Andropogon tenuispatheus* (= *A. glomeratus* var. *pumilus*), and associated weedy species.

4. **Seasonally dry shores and shallow pond beds** (k1, k2, o3): with dominance of annual plants, including areas with "moist-soil management" for waterfowl. Although generally considered artificial and weedy, with little previous published botanical description, these areas are actually fascinating in terms of their botany and ecology. Several genera here contain easily confused native species, which need further examination across the region to sort out their relative occurrences and habitat preferences: especially *Ammannia*, *Cyperus*, *Echinochloa*, *Eleocharis*, *Eragrostis*, *Fimbristylis*, *Juncus*, *Ludwigia*, and *Persicaria*. It is possible that some of these species were associated with shallow beaver ponds, animal licks, and wallows in the region before artificial promotion for wildlife habitat.

Farther west, in the Yazoo Basin of Mississippi, Scharf (2010) has recently reviewed evidence that lowland woods dominated by oaks were partially converted to more open land by native people beginning 3,000 years ago. Sweetgum appears to have increased in subsequent regrowth, especially with the decline of native people after AD 1492. Similar changes may well have occurred in the Black Belt, to be investigated further as more archaeological and ecological information is gathered and synthesized.

Schauwecker et al. (2011) have recently analyzed historical patterns in forests of the flatwoods region adjacent to the Black Belt in the Noxubee National Wildlife Refuge, based on original surveys from the 1830s. Their work indicates that lowlands were dominated by woods similar to the drier oak woods outlined previously: with much willow oak (*Quercus phellos*), red oaks (*Q. pagoda*, *Q. falcata*) and sweetgum (*Liquidambar*), plus minor amounts of hickories (*Carya*), blackgum (*Nyssa*), ironwood (*Carpinus*), ashes (*Fraxinus*), and elms (*Ulmus*). But these areas were often described as having a thick undergrowth of "bushes," "briers," or "canes" in the original surveys. This type of transition to more open conditions may have been similar to the transition from drier oak woods on lowlands to grassland on uplands within the Black Belt itself.

During the past 100–200 years, there has probably been much reduction in the incidence of fire within these lowland oak woods and their drier transitions. In his 1910 notes on the Black Belt, C. E. Dunston (Lowe 1913) stated: "The bottom-lands of the large streams of the region were formerly forested with fine oaks, hickories, ash, yellow-poplar, cypress, red gum and less important species." But, he went on to generalize: "At present forest fires are frequent in the region during drought periods, and the damage from them is severe. This damage is preventable, and it would be for the benefit of the community and of the State to prevent it."

As well as fire, most of the original larger herbivores have become at least locally extinct, and their effects on the vegetation have generally not been simulated well by livestock. The influence of recovering deer and beaver populations today may be small compared to the original effects of bison and elk during the Holocene, plus more diverse megafauna during the Pleistocene—which probably included migrating mastodons (Kaye 1974; Kurtén and Kaye 1982; Hoppe et al. 1999). Nutritious forage such as cane (*Arundinaria*) could have been an important resource for larger herbivores on the relatively mesic fertile terraces of this region, especially during unusually dry or cold periods (S.

Platt and Brantley 1997). It is likely that effects of larger mammals were often concentrated in such productive—but relatively well-drained—transitions from lowlands to uplands, as well as in nearby licks, seeps, and wallows at the bases of chalky slopes. Livestock may have offered some ecological substitute when they ranged freely during early settlement, but they have become increasingly fenced out of lowland farms and woods.

So how low did the prairie go? In a few areas of the Black Belt, prairies or old fields do appear to have occurred on true alluvial lowlands, as indicated by original surveys (Barone 2005a) (fig. 7.1) and by some historical accounts (Campbell and Seymour 2011a; S. Cook 2012). Such low extensions were probably promoted by local disturbances from fires, herbivores (from mastodons to beavers), and humans, especially along regularly traveled traces. But unless disturbances were repeated at intervals of about 1–10 years, low openings would have rapidly become dominated by cane, brush, or deeper woods. During the past 500 years or more, shifting patterns of human occupation probably had a major influence on the pattern of openings (A. Johnson 2000; S. Cook 2012).

In the Pulliam (Campbell and Seymour 2011b) and DeMoville Tracts, the 1830s surveys (Barone 2005a) indicate that some patches of prairie extended from distinctly calcareous uplands down to the edge of largely alluvial lowlands (fig. 7.1). These mapped patches covered about 1–5 mi^2, with woodland prevalent on either side at the same low level on the chalk. Remnants of upland prairie do occur at the edge of alluvial lowlands at or near both sites. At the Barr Tract, original surveys do not indicate that prairie descended down to lowlands, but they do about 5–10 mi to the southeast (fig. 7.1). Moreover, the roadside adjacent to the Barr Tract has abundant gamagrass (*Tripsacum*) and the undescribed *Amsonia* sp. nov., which may well indicate some narrow remnants of original prairie. And on lowlands at all three sites, diverse native species are typical of marshy transitions to upland prairie, as detailed previously.

Floristic Diversity of Lowlands

About 513 species and varieties of vascular plants were found at the two lowland sites combined (table 7.3; see table 7.1 for complete list). Based largely on the two county floras that have covered the Black Belt of Mississippi, in part (MacDonald 1996; Leidolf et al. 2002), at least 40–50 other species are typical of lowlands in this region. Table 7.2 lists the most obvious 44 additions from those two studies, selecting only species that are typical of riparian zones or wetlands. Other species could be added with a broader review, especially for drier phases of the Grassland Complex and the Oak Woodland Complex that were outlined previously (in habitat class A of table 7.1).

Contrasts with adjacent regions. This whole flora of the Black Belt lowlands has a remarkable absence of many plants that are typical of more sandy, more acid, or less fertile soils in adjacent regions, as documented by Simmons (1968), MacDonald (1996), Hodges (1997), Leidolf et al. (2002), and Ervin et al. (2006). Largely absent woody species of this type include *Pinus* (all species), *Magnolia virginiana*, *Xanthorhiza simplicissima*, *Sassafras albidum*, *Itea virginica*, *Aronia arbutifolia*, *Crataegus pruinosa* (sensu lato), *Alnus serrulata*, *Betula nigra*, *Corylus americana*, *Quercus laurifolia*, *Nyssa biflora*, *Symplocos tinctoria*, *Viburnum dentatum* (sensu lato), and most Ericaceae (except the two species of *Vaccinium* that are uncommon here). Absent ferns include *Athyrium asplenioides*, *Polystichum acrostichoides*, *Woodwardia areolata*, and *Osmunda* spp. Absent forbs include *Brasenia schreberi*, *Ludwigia alternifolia*, *Viola primulifolia*, *Utricularia gibba*, *Polygonum sagittatum* [*Truellum sagittatum*], *Rhexia virginica*, *Chelone glabra*, *Solidago patula*, and *S. rugosa*. Absent monocots include *Melanthium virginicum*, *Iris virginica*, *Spiranthes cernua*, *Carex atlantica*, *C. crinita*, *C. joorii*, *C. lurida*, and several *Rhynchospora* spp. (except *corniculata*). Within the sites studied here, the only clear concentrations of acidophilous species were noted on relatively high, dry, fine-sandy levees and terraces in the Barr Tract. And even here, there were just small numbers of a few moderately indicative species: *Liriodendron tulipifera*, *Fagus grandifolia*, *Amelanchier arborea*, *Vaccinium arborescens*, *V. fuscatum*, *Ilex opaca*, *Mitchella repens*, *Carex striatula*, and *C. albolutescens*.

Aquatic or subaquatic habitats at the two studied sites are particularly depauperate in species. The few ponded areas with openings in the woodland are small (0.1–10 ac). They have probably not developed with sufficient depth, connectivity, or stability during recent millennia to support a diverse flora. Rapid erosion, especially after settlement, has probably led to filling of ponds with sediment. In the few permanent ponds, there is virtually no *Nelumbo*, Nymphaeaceae, or Alismatales (Araceae-Lemnaceae, Alismataceae,

Hydrocharitaceae-Najadaceae, Potamogetonaceae) except for a few Lemnaceae and *Potamogeton* cf. *pusillus*. Most of these aquatics are uncommon to absent elsewhere in the region—and only a few additional locally common species were listed by MacDonald (1996) (*Azolla*, *Callitriche*, and *Hydrocotyle* in table 7.2). The "lake-cress"—*Neobeckia aquatica*—provided the most interesting find in ponds at these sites. That species is widely scattered over eastern North America but is globally rare to imperiled. The few small ephemeral wooded ponds with cypress (*Taxodium distichum*), overcup oak (*Quercus lyrata*), or water elm (*Planera aquatica*) are also depauperate compared to other regions of Mississippi, without *Nyssa* spp., *Alnus serrulata*, *Forestiera acuminata*, *Itea virginica*, *Rosa palustris*, *Hypericum walteri*, or *H. tubulosum*.

Another notable group of generally "missing" species is typical of subhydric to submesic fertile soils on lowlands in more northern regions, but uncommon to absent in much of the Black Belt. This group includes *Acer saccharinum*, *Staphylea trifolia*, *Laportea canadensis*, *Amphicarpaea bracteata*, *Sicyos angulatus*, *Polygonum scandens* [*Fallopia scandens*], *Iresine rhizomata*, *Stachys tenuifolia*, *Scutellaria lateriflora*, *Scrophularia marilandica*, *Phryma leptostachya*, *Lobelia siphilitica*, *Symphoricarpos orbiculatus*, *Elymus macgregorii*, and *Muhlenbergia schreberi*. These species are locally common elsewhere on the Gulf Coastal Plain (Kartesz 2012), but they may be restricted in the Black Belt by periodic droughts or excessive disturbances—including browsing and burning in the past.

Contrasts with local uplands. At the Pulliam Prairie—our main upland reference—448 taxa were found, and at least 90 more occur elsewhere in prairies of the region (Campbell and Seymour 2012). All three sites combined have at least 750 taxa, and about 210 are in both lowlands and uplands. On uplands, there is a general lack of Aceraceae, Amaranthaceae, Betulaceae, Polygonaceae, and Urticaceae. Otherwise, the floras are similar at the familial level.

The lowlands have a much smaller proportion of regionally or globally rare species. At the Pulliam Prairie, 57 of the 385 (14.8%) undoubted native species are rare (up to S3S4 in the Natural Heritage system for Mississippi), including 11 that appear to be truly threatened (S1-S2S3). At both lowland sites combined, no more than 16 of the 430 (3.7%) undoubted native species appear to be rare (as summarized previously under "Notes on Rare Species"), while only *Hottonia* and *Neobeckia* appear to be threatened. This paucity of rare species reflects the general similarity in habitats and species composition here to those of lowlands throughout much of the Gulf Coastal Plain. In contrast, the uplands have unusual, distinctive types of grassland, shrubland, and open woodland, and several rare species are largely restricted to those habitats.

Yet several taxa of vascular plants in these lowlands remain poorly understood, and even undescribed, especially in transitional vegetation between lowlands and uplands. The two most obvious examples are as follows:

1. *Amsonia* sp. nov. of low grassland and thin woods (g2, h3, n1); this keys to *A. rigida* in existing manuals (Small 1933; Weakley 2012).
2. *Rudbeckia* sp. nov. of low grassland and thickets (g1, h2); this keys to *R. fulgida* in recent manuals, but it may be closer to *R. tenax* of Small (1933) and Fernald (1950).

Moreover, the *Rudbeckia triloba* in the DeMoville Tract has unusually small, early flowers and might be recognized as a new variety. Note also that the *Solidago* of mesic levees is a recently described taxon: *S. caesia* var. *zedia* (R. Cook and Semple 2004). And the correct names for several other plants in drier woodland or damper grassland on these lowlands have been overlooked or remain controversial (Campbell and Seymour 2012): *Amorpha croceolanata* (= *A. fruticosa* var. *croceolanata*), *Andropogon tenuispatheus* (= *A. glomeratus* var. *pumilus*), *Dichanthelium joorii* (= *D. commutatum* ssp. *joorii*), *Elymus glabriflorus* (= *E. virginicus* var. *glabriflorus*), *Fraxinus smallii* (= *F. americana* var. *subcoriacea*), *Heliopsis gracilis* (= *H. helianthoides* var. *gracilis*), and *Viola* cf. *langloisii* (or perhaps *V. missouriensis*). Varieties of *Acer rubrum* and *Fraxinus pennsylvanica* still need more attention in ecological studies. And several sedges have been poorly documented on lowlands of this region, including *Carex aureolensis* (described in 1855 but often confused with *C. frankii* until recent years), *C.* cf. *normalis* (here at the southern edge of its range and difficult to separate from *molesta*), *C. sangamonensis* (usually treated as a variety of *tribuloides*), *C. socialis* (described in 1969 but still often confused with related species), and *C. basiantha* (described in 1855 but still often confused with *C. willdenowii*).

Several other uncommon to rare species in the region were missed completely at the lowland sites studied here but have been well documented by MacDonald (1996), Leidolf et al. (2002), and others (McCook and Kartesz 2010; Kartesz 2012). General knowledge of the ecology of some of these species indicates that they are typical of varied transitions from wooded lowlands to drier grassy uplands, such as *Aesculus glabra*, *A. pavia*, *Astragalus canadensis**, *Apios priceana**, *Baptisia leucantha* [= *B. alba* var. *macrophylla*], *Camassia scilloides*, *Crataegus mollis* (or locally *C. meridionalis*), *Eryngium prostratum*, *Euonymus atropurpurea*, *Gentianella quinquefolia*, *Glandularia canadensis*, *Helianthus grosseserratus*, *Monarda citriodora*, *Osmorhiza longistylis*, *Perideridia americana**, *Quercus macrocarpa*, *Q. sinuata* (or locally *Q. austrina*)*, *Triosteum angustifolium*, *Veronicastrum virginicum*, and *Zanthoxylum clava-herculis*. Some of these species (as indicated by asterisks *) are rare across whole regions or globally, as are a few others found in this study (see table 7.3 and "Results"). A more thorough synthesis of available information, plus further fieldwork, would certainly improve estimates of the original floristic composition in these mysterious ecological transitions, and it would help guide efforts to recover these species.

Relevance for Conservation and Restoration

There is an intertwined nexus of issues in potential applications of data on flora and vegetation. Examples of these issues follow.

Further inventory and analysis. A more thorough inventory of native vegetation remnants on lowlands of the Black Belt would be useful for conservation and restoration planning. It would be relatively easy to locate better remnants of woodland, in terms of acreage and maturity, using aerial photos and remote sensing. However, a lot more driving and walking is needed to locate ecologically diverse remnants of grassland or transitions from grassland to woodland. Intensive fieldwork is needed to check the potential native composition of many low roadsides, old fields, and wildlife openings that could harbor significant remnants, albeit small in most cases. Inventory might be linked with a general search for the new species of *Amsonia*, which could well be globally threatened because of its apparently small range. The recent intensive inventory of upland prairie remnants in the Black Belt of Alabama (Schotz and Barbour 2009) is a model that might be applied to Mississippi, but prairie remnants in Mississippi are less restricted to xeric sites and often extend onto deeper soils with seasonal dampness.

Ideally, such effort would involve coordination of relevant agencies and other interested organizations. The whole "Upper East Gulf Coastal Plain" may be too large an ecoregion for regular coordination and meetings, but the Black Belt of Mississippi and Alabama plus adjacent land-type associations could become an excellent focus—including, more or less, the whole watershed of the Tombigbee River. An important initial step would be to improve the application of soil maps—from associations to series to local variants (USDA 2006, 2012a, 2012b)—in order to aid interpretation of vegetation data, and for planning plantings or management of existing vegetation. At sites of special interest, more detailed local soil surveys would be invaluable—for example, in order to enable a more detailed analysis of the two sites studied here and the Noxubee National Wildlife Refuge (Ervin et al. 2006; Schauwecker et al. 2011).

Ecological models for restoration. Better understanding of the original vegetation would help guide management of lowlands in the region for native wildlife, vegetation, and ecological processes. For restoration of native grassland, inventory of remnants can provide detailed lists of species to be propagated and tried. For example, this study has confirmed that gamagrass (*Tripsacum dactyloides*) and switchgrass (*Panicum virgatum*) are appropriate candidates for plantings on seasonally damp lowlands, but a more thorough floristic survey of the region would be useful to indicate the typical associates of these potentially dominant grasses. These two grasses have already received much attention in Mississippi for practical use as forages or biofuels. It is important to assess their local genetic diversity in native stands, as well as develop particular genotypes for varied applications (W. Anderson et al. 2008).

In reforestation, there is clear value from good understanding of relationships between original tree composition, soils, and hydrology. Despite a long history of forestry in bottomland hardwoods across the southeastern United States (Kellison and Young 1997; Guilfoyle 2001; Gardiner and Oliver 2004), much conservation practice along the Mississippi River arguably deserves improvement (Hodges 1997; Stanturf et al. 2001; Faulkner et al. 2011). Faulkner et al. (2011) summarized as follows: "All three of these [major] practices involve reforestation, primarily planting hard-mast species such as *Quercus* sp. and *Carya* sp. These plantings are likely to develop into even-aged stands of low tree diversity with little structural heterogeneity, which will impact future wildlife habitat. Since hydrology is a critical driver of wetland processes, the ability of a given conservation practice to restore wetland hydrology is a key determinant of how well it can restore ecosystem services. However, there is little to no follow-up monitoring of projects, so it is difficult to know how much variability exists for any given practice or the efficacy of specific practices."

Such concerns make it all the more important to advance simple, reliable ecological models that can be applied to specific sites. For example, the initial model presented here outlines truly riparian woods as having frequent flooding from active streams, but less pronounced seasonal drying than on more distant terraces. Such strictly riparian woods probably did not support general dominance of oaks or hickories in the original landscape (fig. 7.3).

Guidance for management practices. It is fortunate that traditional efforts to plant trees and shrubs outside their native ranges—often for wildlife—have dwindled during recent decades. But the new challenge in plantings is to select mixes of native species that are most appropriate for future management at each site. As well as considering hydrology and soil type, managers should plan ahead for cutting, browsing, or burning that would promote canebrakes (Gagnon 2009), native fruit-and-nut orchards, or more open grassland. Oaks and hickories would generally be more appropriate than more mesic or riparian trees, if such disturbances were planned.

Changes in drainage—from channelization to impoundment—have provided some of the most radical management on lowlands in the Black Belt. Yet, if a more "natural" condition is desired, there is often little clear basis for decisions at particular sites. In the pre-Columbian period, there was probably a dynamic pattern of temporary impoundments because of beaver dams and human influences, as indicated by several historical sources (Campbell and Seymour 2011a). Nairne (1708, 47–50) noted "multitudes of Beavor dams" among the Chickasaw villages near Tupelo. Black willow may have been a typical tree around more open wetlands within the grasslands (e.g., Romans 1775, 15–16). And several "swamps" were mapped in 1832 by surveyors among the old fields (S. Cook 2012).

Wildlife managers and restorationists today have increasing practical experience with changes in drainage, from artificial enhancement of moist-soil annuals for waterfowl, to much larger projects (Guilfoyle 2001). However, their work has rarely been linked with botanical surveys and ecological modeling (e.g., Ervin et al. 2006), and there are several important questions to address in research. For example, there would be considerable interest in an assessment of how the balance of native versus alien species responds to varied changes in drainage. More detailed analysis of environmental and biological relationships to hydrology would undoubtedly lead to more predictable outcomes from management.

Acknowledgments

This study would not have been possible without permission of the landowners, to whom we are deeply grateful: Dr. Joe Pulliam and his father, Sam D. Pulliam; Dexter Barr, and Margaret DeMoville. Also invaluable were the introduction to this site and continual guidance by Daniel Coggin of Wildlife Mississippi and John Gruchy of the Mississippi Department of Wildlife, Fisheries, and Parks, plus the fundamental support of Roundstone Native Seed, Inc. Other people have provided significant further assistance, including some anonymous reviewers. Curators at several herbaria helped greatly with access to their collections and floristic data: Dwayne Estes (APSU), Laura Lukas and Wendy Zomlefer (GA), Greg Wheeler and Sidney McDaniel (IBE), Lucille McCook (MISS), Jim Solomon and George Yatskievytch (MO), Heather Sullivan (MMNS), Lisa Wallace and Chris Doffitt (MISSA), Michael Vincent (MU), Carol-Ann McCormick and Alan Weakley (NCU), Mike Bevans, Lisa Fruscella, and Rob Naczi (NY), Mesfin Tadesse (OS), Harvey Ballard and Jason Lasson (BHO), John Nelson (USCH), Aaron Floden, Victor Ma, and Eugene Wofford (TENN), Donna Ford-Werntz (WVU), Deborah Bell, Vicki Funk, Harold Robinson, and Rusty Russell (US). Jeff Levy (Deptartment of Geography, University of Kentucky) assisted with mapping. And we are particularly grateful to Charles T. Bryson for identifying all our sedge (*Carex*) collections—we remain confused by only one or two taxa!

Table 7.1.

Annotated list of vascular plants.

Species are listed in alphabetical order, by family, genus, and species name.

[Common synonyms or other taxonomic notes are in brackets.]

*Adventive or uncertain alien/native status.

**Alien species in southeastern states.

SC: collection numbers; 1000 series for the Clay Co. site; 2000 series for the Monroe Co. site.

Numbers "1xxx" represent incomplete collections or sight records for Clay Co.

Numbers "2xxx" represent incomplete collections or sight records for Monroe Co.

For frequency: nea = nearby (along access roads but not seen within the main site blocks); pla = planted; rar = rare (only 1 or 2 observations); unc = uncommon; occ = occasional; sca = scattered; loc = local/locally; wid = widespread; fre = frequent; com = common; abu = abundant; dom = dominant.

A–F: broad habitat classes; these are simplified aggregates of the vegetation types described in figure 7.3, and the codes for those types are listed here in parentheses.

A = drier transitions to uplands, mostly thin woods, disturbed or open (g2, k1, n1)
B = typical seasonally swampy terraces, mostly wooded (m3, n2)
C = wetter swampy terraces, mostly wooded (n1, n3, p1, p2)
D = ponded areas, mostly open (k2, p4, p3)
E = riparian levees and banks, mostly wooded (m1, m2)
F = riparian zones along scoured banks or fresh alluvium, thin woods or open (o1, o2, o3)
u = upper/drier phases
w = wetter/lower phases
\+ indicates that the species is wide ranging, about as frequent in adjacent classes

Codes for the gradient from woodland to grassland are as follows:

1 = most frequent in woods with closed canopy
2 = most frequent in thin woods and on the inside of edges
3 = most frequent in transitions from grassland to woodland, thickets, or edges
4 = most frequent in grassland with little or no woody cover
5 = most frequent on partly bare ground with little or no woody cover
\+ indicates that the species has wide range, more than one unit either side of mode

---

Acanthaceae

Dicliptera brachiata (Pursh) Spreng., SC-2273: occ; Ew 1
Justicia lanceolata (Chapm.) Small [*J. ovata* (Walt.) Lindau var. *lanceolata* (Chapm.) R.W. Long], SC-1132, 2124: loc fre; D 3
Ruellia caroliniensis (J.F. Gmel.) Steud., SC-1266: occ; A+ 4
Ruellia strepens L., SC-1212, 2222, 2274: fre; Eu+ 3+

Alismataceae

Echinodorus cordifolius (L.) Griseb., SC-2305: occ; Dw 4

Alliaceae

Allium canadense L., SC-1067: occ; Eu+ 2
***Allium vineale* L., SC-2116: occ; A+ 4

Amaranthaceae

**Amaranthus hybridus* L., SC-2212, 2246: occ; A+ 5
***Amaranthus spinosus* L., SC-1548: rar; A+ 5
**Amaranthus tuberculatus* (Moq.) Sauer [*A. tamariscinus* auct.], SC-1407, 1567, 1568, 2255: loc fre; D+ 5
**Dysphania ambrosioides* (L.) Mosyakin & Clemants, SC-1584, 2275: occ; D+ 5

Amaryllidaceae

Hymenocallis occidentalis (LeConte.) Kunth, SC-1398, 2225: rar; B+ 2

Anacardiaceae

Toxicodendron radicans (L.) Kuntze var. *pubens* (Engelm. ex S. Wats.) Reveal, SC-1xxx: occ; A 2
Toxicodendron radicans (L.) Kuntze var. *radicans*, SC-1100: loc fre; B+ 2

Annonaceae

Asimina triloba (L.) Dunal, SC-1283: loc com; Eu 2

Apiaceae

Chaerophyllum tainturieri Hook., SC-1069: occ; A+ 5
Cryptotaenia canadensis (L.) DC., SC-1240, 2012: fre; Ew 2
Cynosciadium digitatum DC., SC-1xxx: rar; B+ 4
***Hydrocotyle sibthorpioides* Lam., SC-2167: rar; Du 5
Ptilimnium capillaceum (Michx.) Raf., SC-sight, 2008: loc fre; AD+ 4
Ptilimnium nuttallii (DC.) Britt., SC-1104, 2053, 2110: loc fre; C+ 4
Sanicula canadensis L., SC-sight, 2019, 2062, 2074: occ; A+ 3+
Sanicula odorata (Raf.) K.M. Pryer & L.R. Phillippe [*S. gregaria* Bickn.], SC-1237, 1415, 2071: loc com; Eu 2
Spermolepis echinata (Nutt. ex DC.) A. Heller, SC-1276: rar; A+ 4
Spermolepis inermis (Nutt. ex DC.) Mathias & Constance, SC-1103, 1134: occ; C+ 4
Thaspium trifoliatum (L.) Gray, SC-2144: rar; Eu 3
Trepocarpus aethusae Nutt. ex DC., SC-1174, 1222, 2061: loc fre; Eu+ 3
Zizia aurea (L.) W.D.J. Koch, SC-1217, 2017: com; Ew 2

Apocynaceae

Amsonia sp. nov. ["*ornata*" of S. McDaniel; see notes in text], SC-1020, 1200, 2126: occ; A+ 4

Table 7.1. *Continued*

Apocynum cannabinum L., SC-1268, 2211: com; A+ 4
Cynanchum laeve (Michx.) Pers., SC-1382, 2227: occ; B+ 3
Trachelospermum difforme (Walt.) Gray [*Thyrsanthella difforme* (Walter) Pichon], SC-1225, 2140: loc fre; C 3

Aquifoliaceae

Ilex decidua Walt., SC-1172, 2130: loc fre; C+ 2
Ilex opaca Ait., SC-1579: rar; Eu 1

Araceae

Arisaema dracontium (L.) Schott, SC-1139, 2040: occ; Ew 2

Arecaceae

Sabal minor (Jacq.) Pers., SC-1048, 1331: loc fre; A+ 3

Aristolochiaceae

Aristolochia serpentaria L. var. *hastata* (Nutt.) Duch [see text], SC-1145: occ; B 1
Aristolochia serpentaria L. var. *serpentaria* [*Endodeca serpentaria* (L.) Raf.], SC-1587, 2080: occ; Eu 1
Asclepias perennis Walt., SC-1131: occ; B 1
Asclepias viridis Walt., SC-1205, 1267: occ; A 4
Gonolobus suberosus (L.) R. Brown var. *granulatus* [*Matelea gonocarpos* (Walt.) Shinners, sensu lato], SC-1379, 2043, 2068: occ; Eu+ 3

Asparagaceae

Polygonatum biflorum (Walt.) Ell., SC-1247: occ; A+ 1

Aspleniaceae

Asplenium platyneuron (L.) B.S.P., SC-1042, 1093: occ; A+ 2

Asteraceae

Ageratina altissima (L.) King & H.E. Robins. [*Eupatorium rugosum* Houtt.], SC-1546, 2268: loc; Eu 2
Ambrosia artemisiifolia L., SC-1472, 2285: fre; A+ 5
Ambrosia trifida L., SC-1459, 2243: fre; F+ 3
**Baccharis halimifolia* L., SC-1300, 1403, 2190: com; A+ 3
Bidens aristosa (Michx.) Britt. [some suggesting transition to *B. mitis* (Michx.) Sherff], SC-1469, 1493, 1569, 2253: fre; C+ 5+
Bidens discoidea (Torr. & Gray) Britt., SC-1xxx: loc fre; Cw 3
Bidens frondosa L., SC-1364, 2254: fre; B+ 3
Boltonia diffusa Ell. var. *interior* Fern. & Grisc. [see text], SC-1454, 1475, 2236, 2278, 2298: fre; A+ 4
Cirsium altissimum (L.) Hill, SC-1573: occ; Eu 3
Cirsium horridulum Michx., SC-1038: occ; A+ 4
Conoclinium coelestinum (L.) DC. [*Eupatorium c.*], SC-1470, 2240: fre; C+ 4
Conyza canadensis (L.) Cronq. [*Erigeron c.*], SC-1363: fre; A+ 4
Eclipta prostrata (L.) L. [*E. alba* (L.) Hasskarl], SC-1554, 2177: fre; C+ 5
Elephantopus carolinianus Raeusch., SC-1467, 2291: fre; Eu+ 3
Elephantopus tomentosus L., SC-1367: occ; A 3
Erechtites hieraciifolia (L.) Raf. ex DC., SC-1476: loc com; A+ 3+
Erigeron annuus (L.) Pers., SC-1213: occ; A+ 5
Erigeron philadelphicus L., SC-1085: occ; Eu+ 3+
Erigeron strigosus Muhl. ex Willd., SC-2104: occ; A+ 5
Eupatorium capillifolium (Lam.) Small, SC-1477: loc com; A 5
Eupatorium perfoliatum L., SC-1242, 1293: occ; D 4
Eupatorium cf. *scabridum* Elliott [?*semiserratum* × *rotundifolium*; see text], SC-1356: rar; A 4
Eupatorium serotinum Michx., SC-1478, 2241: fre; B+ 4
Euthamia leptocephala (Torr. & Gray) Greene [*Solidago l.*], SC-1452, 2183, 2307: fre; B+ 4
Eutrochium fistulosum (Barratt) E.E. Lamont [*Eupatorium f.*], SC-1xxx: rar; C 4
Fleischmannia incarnatum (Walt.) King & H.L. Robins. [*Eupatorium i.*], SC-1239: rar; Eu 3
Gamochaeta argyrinea Nesom [segregate of *Gnaphalium purpureum* L.], SC-1007: occ; A+ 5
**Helenium amarum* (Raf.) H. Rock, SC-1521: occ; A+ 5
Helenium autumnale L., SC-1541, 2302: occ; F+ 4
Helianthus angustifolius L., SC-2306: occ; A+ 5
Helianthus cf. *resinosus* Small [or possible transition to *tuberosus*], SC-1540: rar; Ew 4
Helianthus tuberosus L., SC-2267: rar; Ew 4
Heliopsis gracilis Nutt. [*H. helianthoides* (L.) var. *gracilis* (Nutt.) Gandhi & R.D. Thomas], SC-1xxx, 2125: rar; Eu 3
Iva annua L., SC-1499, 2233: fre; C+ 5
Krigia cespitosa (Raf.) Chambers [*K. oppositifolia* Raf.], SC-1006, 1073: occ; A+ 5
Krigia dandelion (L.) Nutt., SC-1046, 1071: occ; A 2
Lactuca canadensis L., SC-1405: occ; B+ 5
Lactuca floridana (L.) Gaertn., SC-1466, 2xxx: occ; Eu 3
***Lactuca saligna* L., SC-1481: occ; A+ 5
Melanthera nivea (L.) Small [*M. hastata* Michx.], SC-1570: occ; Eu 2
Mikania scandens (L.) Willd., SC-1453, 2231, 2308: com; CD 3
Packera aurea (L.) Á. Löve & D. Löve, SC-2082: occ; Ew 2
Packera glabella (Poir.) C. Jeffrey, SC-1002, 1043, 1114, 2108: fre; C+ 5+
Pluchea camphorata (L.) DC., SC-1480, 2265, 2281: fre; CD 4
Pyrrhopappus carolinianus (Walt.) DC., SC-1377, 2135: loc; A+ 4
Rudbeckia hirta L. var. *pulcherrima* Farw. [*R. serotina* Nutt.], SC-1109: occ; A+ 4
Rudbeckia sp. nov. [aff. *R. tenax* C.L. Boynt. & Beadle; see notes in text], SC-2142, 2197: occ; AB 3
Rudbeckia triloba L. [unusual early, hairy, small-headed variant; see notes], SC-2072, 2111, 2220: loc; Eu 2
Silphium perfoliatum L., SC-1428: occ; Ew 3
Solidago altissima L. [*S. canadensis* L. var. *scabra* Torr. & Gray], SC-1471, 2242: com; A+ 4
Solidago caesia L. var. *zedia* R.E. Cook & Semple, SC-1571: loc; Eu 1

Continued on the next page

Table 7.1. *Continued*

Solidago gigantea Ait., SC-1327, 1483, 2277, 2295: com; B+ 4
****Sonchus asper* (L.) Hill, SC-1380: rar; A+ 5
Symphyotrichum dumosum (L.) G.L. Nesom [*Aster d.*; see notes], SC-1449, 1450ab, 1575, 1588, 2244: loc; A+ 4
Symphyotrichum lanceolatum (Willd.) Nesom [*Aster l.*], SC-1552: loc; Ew 2
Symphyotrichum cf. *lateriflorum* (L.) Á. Löve & D. Löve [*Aster l.*, probably × *dumosum*], SC-1455: occ; B+ 3
Symphyotrichum ontarionis (Wieg.) G.L. Nesom [*Aster o.*], SC-1456, 2293: occ; Eu+ 3
Symphyotrichum pilosum (Willd.) G.L. Nesom [*Aster pilosum*, sensu stricto], SC-2245, 2250, 2251: loc; A+ 4
Symphyotrichum racemosum (Ell.) Nesom [*Aster r.*], SC-1450c: rar; C+ 4
***Taraxacum officinale* L., SC-2xxx: occ; A+ 5
Verbesina alternifolia (L.) Britt. ex Kearney, SC-1544, 2269: com; Ew+ 4
Verbesina virginica L., SC-1545: occ; Eu+ 4
Vernonia gigantea (Walt.) Trel., SC-1424, 2184: fre; Ew+ 4
Vernonia missurica Raf., SC-1169, 1328: loc; A+ 4
Xanthium strumarium L. var. *canadense* (P. Mill.) Torr. & Gray, SC-1xxx: occ; Ew+ 5
Xanthium strumarium L. var. *glabratum* (DC.) Cronq., SC-1474, 2239: abu; Ew+ 4

Berberidaceae

Podophyllum peltatum L., SC-1090, 2134: loc; AB 1

Betulaceae

Carpinus caroliniana Walt. [not var. *virginiana* (Marsh.) Fern.], SC-1128: loc; Ew 1
Ostrya virginiana (P. Mill) K. Koch, SC-1369, 2083: loc; Eu 2

Bignoniaceae

Bignonia capreolata L., SC-1021: occ; Eu 3
Campsis radicans (L.) Seem. ex Bureau, SC-1339, 2120: loc; BC+ 3
Catalpa bignonioides Walt., SC-1422: rar [pla]; Ew 2

Boraginaceae (sensu lato)

**Heliotropium indicum* L., SC-1292, 2304: fre; D 4
Hydrolea uniflora Raf., SC-1309: abu; Dw 4
Myosotis macrosperma Engelm., SC-1023, 1208: loc; Eu+ 2

Brassicaceae

Cardamine bulbosa (Schreb. ex Muhl.) B.S.P., SC-1024: occ; Ew+ 2
***Cardamine hirsuta* L., SC-1058: occ; A+ 3
Neobeckia aquatica (Eat.) Greene [*Armoracia a.* (Eat.) Wieg.], SC-1164, 1352: rar; Dw 5
Rorippa sessiliflora (Nutt.) A.S. Hitchc., SC-1260: occ; Fw+ 5

Campanulaceae

Campanulastrum americanum (L.) Small, SC-2182: rar; Eu 3
Lobelia cardinalis L., SC-1446, 2292: fre; CF 3
Lobelia puberula Michx. var. *mineolana* F. Wimmer, SC-1581, 2283: rar; A+ 3
Triodanis biflora (Ruiz & Pavon) Greene, SC-1202: occ; A+ 5

Caprifoliaceae (sensu lato)

***Lonicera japonica* Thunb., SC-1323, 2218: abu; AB+ 3
Sambucus canadensis L., SC-1258, 2006: loc fre; Ew 3
Viburnum rufidulum Raf., SC-1254: occ; B+ 3

Caryophyllaceae

***Cerastium glomeratum* Thuill., SC-1004: occ; A+ 5
***Scleranthus annuus* L., SC-1310, 1511: com; DA+ 5

Celastraceae

Euonymus americanus L., SC-1371: rar; B+ 2

Ceratophyllaceae

Ceratophyllum echinatum Gray, SC-1426: com; Dw 5

Clusiaceae (sensu lato)

Hypericum hypericoides (L.) Crantz [not *H. stragalum* P. Adams & N. Robson], SC-1338: loc; A+ 3
Hypericum mutilum L., SC-1332: loc; AD 5

Colchicaceae

Uvularia sessilifolia L., SC-1246: loc; BC 2 [to be compared further with *U. floridana* Chapm.]

Commelinaceae

***Commelina diffusa* Burm. f., SC-1346, 1542: occ; C+ 4
Commelina virginica L., SC-1319, 2202, 2217: com; C+ 2
Tradescantia ohiensis Raf., SC-1157, 2054: occ; EF+ 3

Convolvulaceae

Ipomoea lacunosa L., SC-1538, 1539, 2247: com; CF+ 4
Ipomoea pandurata (L.) G.F.W. Mey., SC-1378, 2216: occ; A+ 4
***Ipomoea wrightii* Gray, SC-1586, 2280, 2297, 2308: occ; C+ 4

Cornaceae

Cornus drummondii C.A. Mey., SC-1218, 2024: loc; AB+ 3
Cornus foemina Mill. [*C. stricta* Lam.], SC-1243: fre; DC 3

Cucurbitaceae

Cayaponia quinqueloba (Raf.) Shinners, SC-2041, 2115: occ; Eu 3
Melothria pendula L., SC-1321: rar; BC+ 4

Cupressaceae

Juniperus virginiana L., SC-1294, 2025: occ; AB+ 2

Cuscutaceae

Cuscuta gronovii Willd., SC-1561: occ; DA+ 4

Cyperaceae

Carex abscondita Mackenzie, SC-1099: loc; B+ 1
Carex albolutescens Schwein., SC-1251: occ; B+ 3
Carex aureolensis Steud., SC-1107, 2014, 2188: fre; C 2

Table 7.1. *Continued*

Carex basiantha Steud., SC-2096: loc; Eu+ 1
Carex blanda Dewey [typical; see text], SC-1223, 2114: fre; BC+ 3
Carex caroliniana Schwein., SC-1034, 1115, 2xxx: fre; DA 4
Carex cherokeensis Schwein., SC-1014, 1137, 2022: fre; A+ 3
Carex corrugata Fern., SC-1053, 1138, 1215, 2098, 2099, 2015, 2100, 2158: fre; BC 2
Carex crebriflora Weig., SC-1081, 1124, 2159: fre; Eu+ 1
Carex crus-corvi Schuttlw. ex Willd., SC-1016, 2034: com; C 2
Carex debilis Michx., SC-1075: com; AC 1
Carex festucacea Schkuhr. ex Willd., SC-1035, 1111, 1171, 2055, 2103: abu; AD 4
Carex cf. *fissa* Mackenzie [or *C. annectans* (Bickn.) Bickn.; see text], SC-1030, 1065, 1070: loc; AD 4
Carex flaccosperma Dewey, SC-1026, 1049, 1080: loc; CA 3
Carex glaucodea (Tuckerman ex Olney) Kukenth., SC-1054, 1072: loc; A+ 4
Carex cf. *gracilescens* Steud. [or "minor" variant of *C. blanda*; see text], SC-1019, 1059, 2xxx: loc; Eu 2
Carex granularis Muhl. ex Willd., SC-2016: occ; A+ 4
Carex hyalinolepis Steud., SC-1036, 2077: com; C+ 3
Carex intumescens Rudge, SC-1031, 1151, 1187: loc; BC 2
Carex leavenworthii Dewey, SC-1045, 2050, 2106: loc; A 3
Carex louisianica Bailey, SC-1184: occ; BC 1
Carex lupuliformis Sartwell ex Dewey, SC-1360: loc; CF 4
Carex lupulina Muhl. ex Willd., SC-1130, 1150, 1204, 1219, 2164: abu; CD+ 4
Carex oxylepis Torr. & Hook., SC-1050, 1079, 1112, 1136: loc; B+ 3
Carex reniformis (Bailey) Small, SC-1032, 1098, 1170: loc; C+ 3
Carex sangamonensis (Clokey) Mohlenbr. [*C. tribuloides* Wahlenb. var. *sangamonensis* Clokey], SC-1110, 2036, 2109: abu; C 2
Carex socialis Mohlenbr. & Schwegman, SC-1025, 1044, 1060, 1077, 1097, 1158, 1252, 2065: com; C 1
Carex squarrosa L., SC-1160: loc; BC 2
Carex striatula Michx., SC-1088: occ; A 3
Carex triangularis Boeckl., SC-1113, 2056: abu; DA 4
Carex vulpinoidea Michx., SC-1089, 2039: com; CF+ 4
Cyperus acuminatus L., SC-2047: occ; DA+ 5
Cyperus compressus L., SC-1520: occ; DA+ 5
Cyperus croceus Vahl., SC-1441: loc; AD+ 5
Cyperus echinatus (L.) Wood, SC-1xxx: loc; AD 4
Cyperus erythrorhizos Muhl., SC-1307, 1506, 1506, 1536, 2238, 2257: dom; Du 4
Cyperus esculentus L. var. *leptostachyus* Boeckl., SC-2286: occ; A+ 5
***Cyperus iria* L., SC-1325, 1589, 2187: fre; A+ 5
Cyperus pseudovegetus Steud., SC-1177, 1361, 2048, 2195, 2230, 2299: com; Du 4
Cyperus retrorsus Chapm., SC-1273, 1306, 1484, 1508, 1537: loc; AD 4
Cyperus strigosus L., SC-1358, 1485, 1509, 1510, 2172: com; DA 4
Eleocharis compressa Sullivant, SC-1159: occ; DA 4
Eleocharis obtusa (Willd.) J.A. Schultes, SC-1153, 1179, 1317, 1333, 1529, 2032, 2097, 2105, 2173, 2199: com; D+ 5
Eleocharis quadrangulata (Michx.) Roemer & J.A. Schultes, SC-1269: abu; D+ 4
Eleocharis tenuis (Willd.) J.A. Schultes, SC-1086: occ; DA 5
Fimbristylis annua (All.) Roemer & J.A. Schultes, SC-1534: occ; DA 5
Fimbristylis autumnalis (L.) Roemer & J.A. Schultes, SC-1512, 1517: fre; DA 5
Kyllinga pumila Michx., SC-1xxx: occ; DA 5
Rhynchospora corniculata (Lam.) Gray, SC-1270, 2284: fre; D+ 4
Scirpus cyperinus (L.) Kunth, SC-1314, 2229: fre; Dw 4
Scirpus georgianus Harper, SC-1148, 2038: occ; Du 4
Scirpus pendulus Muhl., SC-2xxx: occ; AD 4

Ebenaceae

Diospyros virginiana L. var. *pubescens* (Pursh) Dippel, SC-1228: rar; B 3
Diospyros virginiana L. var. *virginiana*, SC-1227, 2156: fre; A+ 3

Ericaceae

Vaccinium arboreum Marsh, SC-1116: occ; A 3
Vaccinium fuscatum Ait., SC-1129: rar; AC 2

Euphorbiaceae

Acalypha rhomboidea Raf., SC-1416, 1516, 2271: fre; AB+ 3
Chamaesyce humistrata (Engelm.) Small [*Euphorbia h.*], SC-1431, 2213: fre; F+ 5
Chamaesyce maculata (L.) Small [*Euphorbia m.*], SC-1424: occ; AD+ 5
Chamaesyce nutans (Lag.) Small [*Euphorbia n.*], SC-1384: loc; A+ 5
Croton capitatus Michx. [perhaps var. *lindheimeri* (Engelm. & Gray) Müll. Arg.], SC-1348: com; A+ 4
***Euphorbia dentata* Michx., SC-1566: occ; A+ 5
Euphorbia spathulata Lam., SC-1198: rar; A+ 3

Fabaceae (sensu lato)

Amorpha croceolanata P.W. Wats. [*A. fruticosa* L. var. *croceolanata* (P.W. Wats.) P.W. Wats. ex Mouille], SC-1335, 2146: com; BC 3
Amorpha fruticosa L. [sensu stricto or possible *A. nitens* Boynt.], SC-1396: rar; F+ 3
Apios americana Medik., SC-1408, 2170: fre; EF+ 3
Cercis canadensis L., SC-1263, 2001: occ; Eu+ 3
Chamaecrista fasciculata (Michx.) Greene, SC-1437, 2203: loc; AB+ 4
Desmanthus illinoensis (Michx.) MacM. ex B.L. Robbins. & Fern., SC-1203, 2210: loc; A+ 4
Desmodium glabellum (Michx.) DC., SC-1447, 1543, 1547: loc; A+ 3
Desmodium paniculatum (L.) DC., SC-1465, 1505, 2290, 2300: loc; B+ 2
Desmodium perplexum Schub. [probably grading into *D. glabellum* (Michx.) DC.], SC-xxxx: occ; Eu+ 3
Dioclea multiflora (Torr. & Gray) C. Mohr, SC-1249, 2154: occ; BC 3
Gleditsia triacanthos L., SC-1101, 2157: occ; AB+ 3

Continued on the next page

*******Lespedeza cuneata* (Dum.-Cours.) G. Don, SC-1432: abu [pla]; A+ 4

Lespedeza violacea (L.) Pers. [sensu Fernald (1950) and most authors during 1950–2000; *L. frutescens* (L.) Ell.], SC-1448: rar; A 3

**Senna obtusifolia* (L.) Irwin & Barneby [*S. tora* auct.], SC-2002: occ; A+ 4

Sesbania herbacea (P. Mill.) McVaugh, SC-1501, 2171: fre; F+ 4

***Trifolium dubium* Sibthorp, SC-1214: occ; A+ 5

***Trifolium hybridum* L., SC-1233: occ; A+ 5

***Trifolium pratense* L., SC-1341: occ; A+ 4

***Trifolium resupinatum* L., SC-1008: occ; A+ 5

***Vicia sativa* L. ssp. *nigra* (L.) Ehrh. [*V. angustifolia* L.], SC-1068, 1206: occ; A+ 4

***Vicia villosa* Roth, SC-1207: occ; A+ 4

Wisteria frutescens (L.) Poir. var. *macrostachya* Torr. & Gray, SC-1376, 2163: loc; BC 3

Fagaceae

Fagus grandifolia Ehrhart, SC-1578: occ; Eu 1

Quercus alba L., SC-1253: occ; AB 1

Quercus lyrata Walt., SC-1255, 2042: occ; CD 1

Quercus michauxii Nutt., SC-1262, 2128: fre; BE 1

Quercus nigra L., SC-1397, 2091: abu; B+ 1

Quercus pagoda Raf., SC-1234, 2127: abu; BC 1

Quercus phellos [and possible hybrid with *nigra*], SC-1392, 2102 [2101]: dom; CB+ 1

Quercus shumardii Buckl., SC-1231, 2003: occ; Eu 1

Quercus stellata Wangenh., SC-1162: rar; A 2

Quercus texana Buckl., SC-1220 [pla], 2151: loc; C 1

Quercus velutina Lam., SC-1419, 2094: rar; A 1

Geraniaceae

Geranium carolinianum L., SC-1012: occ; A+ 4

***Geranium pusillum* L., SC-1194: occ; A 5

Haloragaceae

Myriophyllum cf. *heterophyllum* Michx., SC-1xxx: fre; Dw 5

***Myriophyllum* cf. *spicatum* L., SC-1xxx: fre; Dw 5

Proserpinaca palustris L., SC-1185: loc; DC 4

Hamamelidaceae (sensu lato)

Liquidambar styraciflua L., SC-1288, 2087, 2204: abu; CB 1

Iridaceae

Sisyrinchium angustifolium P. Mill., SC-1009: occ; Eu+ 4

Juglandaceae

Carya carolinae-septentrionalis (Ashe) Engler & Graebner, SC-2092: occ; A 1

Carya cordiformis (Wangenh.) K. Koch, SC-1285, 2020: fre; Eu 1

Carya glabra (P. Mill.) Sweet, SC-1126, 1144 [or transition to *car.-sep.*]: fre; B+ 1

Carya illinoinensis (Wangenh.) K. Koch, SC-1492 [pla], 2185: rar; C+ 2

Carya laciniosa (Michx. f.) G. Don, SC-1127, 2145: occ; C 1

Carya ovata (P. Mill.) K. Koch, SC-1256, 2095: occ; A+ 1

Juglans nigra L., SC-1429, 2090: occ; Eu 2

Juncaceae

Juncus acuminatus Michx., SC-1018, 1084, 1087, 1120, 2057: abu; DA 5

Juncus anthelatus (Wieg.) R.E. Brooks, SC-1173, 2076, 2129: fre; DA 5

Juncus biflorus Ell., SC-1149, 1296: loc; DA 4

Juncus brachycarpus Engelm., SC-1167, 1182: loc; A+ 4

Juncus coriaceus Mackenzie, SC-1123, 1146, 1278: abu; AD 4

Juncus diffusissimus Buckl., SC-2031, 2181: loc; DA 5

Juncus dudleyi Wieg., SC-2059: occ; D 4

Juncus effusus L. var. *solutus* Fern. & Wieg., SC-1013, 1056, 2037: fre; DC 4

Juncus interior Wieg., SC-1355: occ; DA 5

Juncus marginatus Rostk., SC-2052: occ; DC 4

Juncus secundus Beauv. ex Poir., SC-2049: occ; DA 5

Juncus tenuis Willd., SC-1062, 1064, 1122, 2136: fre; B+ 2

Juncus torreyi Coville, SC-2035: occ; AD 4

Juncus validus Coville, SC-1188, 1297: loc; AD 5

Lamiaceae

Blephilia ciliata (L.) Benth., SC-2067: occ; Eu+ 3

Callicarpa americana L., SC-1533: occ; A 3

***Glechoma hederacea* L., SC-1430: loc; Ew 2

Lycopus americanus Muhl. ex W. Bart, SC-2260: rar; AD 4

Lycopus rubellus Moench, SC-1564, 2169: loc; CwD 3

Lycopus virginicus L., SC-1482: loc; C+ 3

Physostegia angustifolia Fern., SC-1259: occ; A 4

Prunella vulgaris L. cf. var. *lanceolata* (W. Bart.) Fern., SC-1343: occ; BA+ 3

Pycnanthemum tenuifolium Schrad., SC-1117: occ; BA+ 4

Salvia lyrata L., SC-1005, 2143: occ; A+ 5

Scutellaria integrifolia L., SC-1101: loc; B+ 3

Stachys hispida Pursh [*S. tenuifolia* Willd. var. *hispida* (Pursh) Fern.], SC-2023: occ; B+ 2

Teucrium canadense L. var. *virginicum* (L.) Ait., SC-2221: occ; Eu+ 3

Lauraceae

Lindera benzoin var. *pubescens* (Palmer & Steyermark) Rehder, SC-1414: loc; Eu+ 1

Lemnaceae

***Landoltia punctata* (G. Meyer) D.H. Les & D.J. Crawford [*Spirodela oligorhiza* (Kurz) Hegelm.], SC-1287, 1427, 1444: dom; Dw 5

Lemna cf. *minuta* Kunth, SC-1xxx: loc; Dw 5

Wolfia brasiliensis Weddell, SC-1xxx: occ; Dw 5

Wolfia columbiana Karst., SC-1287, 1427, 1444: abu; Dw 5

Linaceae

Linum striatum Walt., SC-1133: occ; AC 4

Table 7.1. *Continued*

Loganiaceae

Mitreola petiolata (J.F. Gmelin) Torr. & Gray, SC-1381: occ; B+ 3
Spigelia marilandica (L.) L., SC-1245, 2028: loc; B+ 3

Lythraceae

Ammannia auriculata Willd., SC-1xxx: occ; DA 5
Ammannia coccinea Rottd., SC-1522, 1555: loc; DA 5
****Lagerstroemia indica* L., SC-1433: rar; Eu 3
Lythrum alatum Pursh var. *lanceolatum* (Ell.) Torr. & Gray ex Rothrock [*L. lanceolatum*], SC-1342, 2209: fre; AD 4
Rotala ramosior (L.) Koehne, SC-1553, 2201: loc; DA 5

Magnoliaceae

Liriodendron tulipifera L., SC-1286: occ; Eu 1

Malvaceae

Hibiscus laevis All., SC-1399, 2226: loc; DC 4
***Sida rhombifolia* L., SC-1440, 1445: occ; A+ 5
***Sida spinosa* L., SC-1494: rar; A+ 5

Melastomataceae

Rhexia mariana L. var. *interior* (Pennell) Kral & Bostick, SC-1345: rar; AC 4

Menispermaceae

Calycocarpum lyoni (Pursh) Gray, SC-1235: occ; Ew 3
Cocculus carolinus (L.) DC., SC-1351: occ; B+ 3

Moraceae

***Broussonetia papyrifera* (L.) L'Her. ex Vent., SC-1434: rar; Eu 3
**Maclura pomifera* (Raf.) Schneid., SC-1289, 1295, 2018: occ; B+ 3
Morus rubra L., SC-1365, 2118: rar; B+ 1

Oleaceae

Chionanthus virginicus L., SC-1125: rar; BA+ 2
Fraxinus pennsylvanica Marsh. var. *pennsylvanica*, SC-1290, 2033, 2186: fre; BEw 1
Fraxinus pennsylvanica Marsh. var. *subintegerrima* (Vahl) Fern., SC-1329, 1580: abu; CB 1
Fraxinus cf. *profunda* (Bush) Bush, SC-1359, 2131: occ; CD 1
***Ligustrum sinense* Lour., SC-1291, 1463, 2021: loc; B+ 2

Onagraceae

Ludwigia decurrens Walt., SC-1305, 1551, 2198: sca; loc fre; FD 4
Ludwigia glandulosa Walt., SC-1xxx: rar?; D 4
Ludwigia palustris (L.) Ell., SC-1180, 1304: com; Dw 4
**Ludwigia peploides* (Kunth) Raven var. *glabrescens* (Kuntze) Shinners, SC-1318: abu; Dw 4
Oenothera speciosa Nutt., SC-2027, 2208: occ; A+ 5

Onocleaceae

Onoclea sensibilis L., SC-1029: loc; C 2

Ophioglossaceae

Botrychium virginianum (L.) Sw., SC-1094: occ; B+ 1

Orobanchaceae

Agalinis cf. *purpurea* (L.) Pennell [suggesting var. *parviflora* (Benth.) B. Boivin], SC-1530: rar; B 4
Agalinis cf. *setacea* (J.F. Gmel.) Raf., SC-1451: rar; C 4
Agalinis tenuifolia (Vahl) Raf. var. *macrophylla* (Benth.) S.F. Blake, SC-2282: rar; A 4
Dasistoma macrophylla (Nutt.) Raf., SC-1261, 2123, 2215: occ; B+ 3
Mecardonia acuminata (Walt.) Small, SC-1347, 1515: occ; CA 4

Oxalidaceae

Oxalis dillenii Jacq., SC-1040: occ; B+ 3

Passifloraceae

Passiflora incarnata L., SC-1308: rar; A+ 4
Passiflora lutea var. *glabriflora* Fern., SC-1298: rar; A+ 2

Penthoraceae

Penthorum sedoides L., SC-1316, 2256: fre; CD 3

Phrymaceae

Mimulus alatus Ait., SC-1315, 2214, 2261: occ; C+ 4

Phytolaccaceae

Phytolacca americana L., SC-1274, 2155: occ; Eu+ 3

Plantaginaceae (sensu lato)

Gratiola neglecta Torr., SC-1140: occ; CD 5
Gratiola virginiana L., SC-1155: occ; C+ 3
Leucospora multifida (Michx.) Nutt., SC-2xxx: occ; A+ 5
Lindernia dubia (L.) Pennell var. *anagallidea* (Michx.) Cooperider, SC-1322, 1425: loc; Du 5
Penstemon alluviorum Pennell, SC-2058, 2059, 2069, 2070: loc; AB 3
Plantago aristata Michx., SC-1479: rar; A+ 5
***Plantago lanceolata* L., SC-1282: occ; A+ 5
Plantago rugelii Decne., SC-1372: rar; Eu 3

Platanaceae

Platanus occidentalis L., SC-1404, 2088: occ; Ew+ 2

Poaceae

Agrostis hyemalis (Walt.) B.S.P., SC-1066, 1457: loc; A+ 4
Agrostis perennans (Walt.) Tuck., SC-1582: occ; EB 2
Andropogon tenuispatheus (Nash) Nash [*A. glomeratus* (Walt.) B.S.P. var. *pumilus* (Vasey) L.H. Dewey], SC-1490, 2234: loc; AB+ 4
Andropogon virginicus L. [perhaps var. *virginicus*], SC-1486: occ; A+ 4
Aristida oligantha Michx., SC-1487: loc; A+ 5
Arundinaria gigantea (Walt.) Muhl., SC-1022, 2064: com; B+ 3
***Briza minor* L., SC-1041: occ; A+ 4
***Bromus japonicus* Thunb. ex Murr., SC-1147: occ; A+ 5
Chasmanthium latifolium (Michx.) Yates, SC-1248, 2192: com; Ew 3
Chasmanthium laxum (L.) Yates, SC-1334: com; BC 2
Cinna arundinacea L., SC-1461, 2193: fre; BEw 1
***Cynodon dactylon* (L.) Pers., SC-1387: abu; A+ 5

Continued on the next page

Danthonia spicata (L.) Beauv. ex Roemer & J.A. Schultes, SC-1165: rar; A 3
Dichanthelium acuminatum (Sw.) Gould & C.A. Clark var. *fasciculatum* (Tuck.) Freckmann, SC-1178: loc; A+ 4
Dichanthelium commutatum (Schultes) Gould [sensu stricto], SC-1063: occ; Eu 2
Dichanthelium joorii (Vasey) Mohlenbrock [*D. commutatum* (Schultes) Gould ssp. *joorii* (Vasey) Freckmann & Lelong], SC-1142, 1418, 1576, 2084: loc; Ew 1
Dichanthelium laxiflorum (Lam.) Gould, SC-1078: occ; A+ 4
Dichanthelium cf. *leucothrix* (Nash) Freckmann [perhaps grading into *D. acuminatum* (Sw.) Gould & C.A. Clark], SC-1xxx: occ; AD 4
Dichanthelium longiligulatum (Nash) Freckmann [perhaps grading into *D. acuminatum* var. *lindheimeri* (Nash) Gould & Clark], SC-1xxx: occ; CA+ 4
Dichanthelium microcarpon (Muhl. ex Ell.) Mohlenbrock [*D. dichotomum* (L.) Gould var. *ramulosum* (Torr.) LeBlond], SC-1105, 2060, 2194, 2219: loc; BC+ 3
Dichanthelium polyanthes (Schultes) Mohlenbr. [*D. sphaerocarpon* (Ell.) Gould var. *isophyllum* (Scribn.) Gould & C.A. Clark], SC-1106: loc; B+ 3
Dichanthelium scoparium (Lam.) Gould, SC-1272: loc; CA+ 4
**Digitaria ciliaris* (Retz.) Koel. [perhaps var. *chrysoblephara* (Fig. & De Not.) R.R. Stewart; *D. sanguinalis* (L.) Scop. var. *ciliaris* (Retz.) Parl.], SC-1385, 2266: com; AB+ 5
Digitaria filiformis (L.) Kohler, SC-1468: occ; A+ 5
***Echinochloa colonum* (L.) Link, SC-1401, 1500, 2196, 2258: loc; DFu 5
***Echinochloa crus-galli* (L.) Beauv., SC-1502, 1565, 2175: loc; A+ 5
***Echinochloa esculenta* (A. Braun) H. Scholtz [cultivar from *crus-galli*], SC-2252: occ; DA 5
Echinochloa muricata (Beauv.) Fern., SC-1175: occ; DA 5
***Eleusine indica* (L.) Gaertn., SC-1439: loc; Eu+ 5
Elymus virginicus L. [sensu stricto], SC-2011: dom; Ew 2
Eragrostis hirsuta (Michx.) Nees, SC-1458: occ; A 3
Eragrostis hypnoides (Lam.) B.S.P., SC-1513: com; DC 5
***Eragrostis japonica* (Thunb.) Trin., SC-1xxx: rar; Eu+ 5
Eragrostis pectinacea (Michx.) Nees ex Steud., SC-1531: occ; A+ 5
***Eragrostis pilosa* (L.) Beauv., SC-1438: loc; Fu 5
***Festuca arundinacea* Schreb. [*Schedonorus a.* (Schreb.) Dumort.], SC-1390: loc; A+ 4
Festuca subverticillata (Pers.) Alexeev, SC-1250, 2010: loc; Eu 1
Glyceria striata (Lam.) A.S. Hitchc., SC-1015, 2030: fre; EwF 3
**Hordeum pusillum* Nutt., SC-1110: occ; A+ 5
Leersia lenticularis Michx., SC-1464, 2294: occ; CD 3
Leersia oryzoides (L.) Sw., SC-1460, 2235: fre; FD 4
Leersia virginica Willd., SC-1370, 2270, 2296: fre; B+ 2
Leptochloa cf. *fascicularis* (Lam.) Gray [*L. fusca* (L.) Kunth ssp. *fascicularis* (Lam.) N. Snow], SC-1xxx: occ; Du 5
Leptochloa panicea (Retz.) Ohwi ssp. *brachiata* (Steudl.) N. Snow [*L. filiformis* (Lam.) Beauv.], SC-1324, 1557, 2207: occ; FDu 5
***Lolium perenne* L., SC-1119, 2051: loc [pla]; AB+ 4
Melica mutica Walt., SC-1017, 2122, 2138: loc; EuB 1
***Microstegium vimineum* (Tren.) A. Camus, SC-1556: loc; B+ 2
Panicum anceps Michx., SC-1320: loc; A+ 4
Panicum dichotomifolium Michx., SC-1503: loc; AD+ 5
Panicum longifolium Torr. [*P. rigidulum* Bosc ex Nees var. *pubescens* (Vasey) Lelong], SC-1xxx, 2288: loc; CA 4
Panicum rigidulum Bosc ex Nees, SC-1311, 2262: fre; DC 4
Panicum virgatum L., SC-1388, 2189: loc; A+ 4
***Paspalum dilatatum* Poir., SC-1160, 1560, 2228: abu; A+ 5
Paspalum dissectum (L.) L., SC-1559: occ; DFu 5
Paspalum floridanum Michx. var. *glabratum* Engelm. ex Vasey, SC-1386: loc; A+ 4
Paspalum fluitans (Ell.) Kuntz. [*P. repens* P.J. Bergius], SC-1525: loc; DF 4
Paspalum laeve Michx., SC-1389, 1489: loc; A+ 5
Paspalum pubiflorum Rupr. ex Scribn., SC-1558: occ; A+ 5
Paspalum setaceum Michx. [perhaps var. *longepedunculatum* (Le Conte) Wood], SC-1xxx: occ; AC 5
***Paspalum urvillei* Steud., SC-1353, 1462, 1473: loc; AC 4
Phalaris caroliniana Walt., SC-1201: occ; CA+ 5
Phanopyrum gymnocarpon [*Panicum g.*], SC-1400, 1527: loc; DC 4
Poa autumnalis Muhl. ex Ell., SC-1052: occ; B+ 1
Saccharum giganteum (Walt.) Pers. [*Erianthus g.* (Walt.) Beauv.], SC-1488, 2249: com; C+ 4
Setaria parviflora (Poir.) Kerguélen [*S. geniculata* auct.], SC-1391: loc; A+ 4
***Setaria pumila* (Poir.) Roemer & J.A. Schultes [*S. glauca* auct.], SC-1264, 1402, 2258: loc; A+ 5
***Sorghum halepense* (L.) Pers., SC-1xxx, 2205: loc; A+ 4
Sphenopholis intermedia (Rydb.) Rydb., SC-1047, 1074, 1156: loc; EuB 2
Steinchisma hians (Ell.) Nash [*Panicum h.* Ell.], SC-1265, 1303, 1526, 2200, 2289: fre; DC 3
Tridens flavus (L.) A.S. Hitchc., SC-1435: rar; Eu 4
Tridens strictus (Nutt.) Nash, SC-1232, 1349: occ; AC+ 4
Tripsacum dactyloides (L.) L., SC-1192, 2191: loc; B+ 4
**Urochloa platyphylla* (Munro ex C. Wright) R.D. Webster, SC-1368, 2174, 2287: abu; CD+ 5

Polemoniaceae

Phlox carolina L. [sensu lato; see text], SC-1091, 1102, 1244: loc; AB 3
Phlox divaricata L., SC-1241, 2073: loc; Eu 1
Phlox paniculata L., SC-1420, 1436: occ; Ew 3

Polygonaceae

Brunnichia ovata (Walt.) Shinners, SC-1299, 2206: abu; CD 3
Persicaria glabra (Willd.) M. Gomez [*P. densiflora* (Meisn.) Moldenke], SC-1496, 2279: loc; Du 4
Persicaria hydropiperoides (Michx.) Small, SC-1135, 1199, 2224: com; D 4
Persicaria lapathifolia (L.) S. F. Gray, SC-1497: occ; Du 5
***Persicaria longiseta* (de Bruijn) Kitagawa [*Polygonum cespitosum* Blume var. *longisetum* (de Bruijn) A.N. Steward], SC-1279, 1406, 2259: loc; B+ 3

****Persicaria maculosa* S.F. Gray, SC-1574: occ; A+ 5

Persicaria pensylvanica (L.) M. Gomez, SC-1495, 1549, 2264: loc; DC+ 5

Persicaria punctata (Elliott) Small, SC-1186, 2179, 2263: fre; BC+ 3

Persicaria setacea (Baldwin) Small, SC-1312: dom; Du 4

Persicaria virginiana (L.) Gaertn. [*Antenoron v.* (L.) Roberty & Vautier], SC-1375: loc; Eu+ 2

***Polygonum aviculare* L., SC-1336: occ; A+ 5

Rumex altissimus Wood, SC-1166: occ; Fu 4

***Rumex crispus* L., SC-1108, 2029: fre; A+ 4

***Rumex pulcher* L., SC-1xxx: loc; AC 4

Rumex verticillatus L., SC-1277, 2044: loc; C+ 4

Portulacaceae

Claytonia virginica L., SC-1011: occ; B+ 2

Potamogetonaceae

Potamogeton cf. *pusillus* L., SC-1xxx, 2141: loc; Dw 5

Primulaceae (sensu lato)

Hottonia inflata Ell., SC-1562: rar; DF 5

***Lysimachia nummularia* L., SC-1330: occ; CB 2

Samolus parviflorus Raf., SC-1163, 2162: occ; EwF 3

Ranunculaceae

Clematis crispa L., SC-1037, 1152, 1191, 1209, 1210: loc; B+ 3

Clematis virginiana L., SC-1423: occ; Eu 3

Ranunculus carolinianus DC. [*R. hispidus* Michx. var. *nitidus* (Chapm.) T. Duncan], SC-1083, 2137: occ; B+ 3

Ranunculus pusillus Poir., SC-1033: occ; CD 3

***Ranunculus sardous* Crantz, SC-1001, 1281, 2107: fre; A+ 5

Thalictrum cf. *pubescens* Pursh [or *T. dasycarpum* Fisch. & Avé-Lall.], SC-1257, 2132: occ; Ew 3

Rhamnaceae

Berchemia scandens (Hill) K. Koch, SC-1337, 1395, 2161: fre; AB+ 3

Rosaceae

Agrimonia parviflora Ait., SC-1498: occ; C+ 3

Amelanchier arborea (Michx. f.) Fern., SC-1143: rar; Eu+ 2

Crataegus crus-galli L., SC-2147: rar; AB+ 3

Crataegus engelmannii Sarg. [*C. berberifolia* Torr. & Gray var. *engelmannii* (Sarg.) Egglest.], SC-2139: rar; A+ 3

Crataegus marshallii Egglest., SC-1183: occ; AC+ 3

Crataegus mollis Scheele, SC-1583, 2133: occ; Eu+ 3

Crataegus reverchonii Sarg., sensu lato [perhaps var. *discolor* (Sarg.) Palmer], SC-2xxx: rar; B+ 3

Crataegus viridis L., SC-1491: rar; C+ 3

Geum canadense Jacq., SC-1211, 1238: occ; Eu 3

Potentilla simplex Michx., SC-1027: occ; A+ 3

Prunus americana var. *lanata* [?*P. mexicana* S. Wats.], SC-1176, 2093: occ; A+ 3

Prunus angustifolia Marsh., SC-1216: rar [pla in part]; AB+ 3

Prunus serotina Ehrh., SC-1413, 2xxx: rar; Eu+ 2

***Rosa multiflora* Thunb. ex Murr., SC-1061: occ; B+ 3

Rubus argutus Link, SC-1140, 2066: loc; AB+ 3

Rubus trivialis Michx., SC-1082, 1196, 2113: loc; AB+ 3

Rubiaceae

Cephalanthus occidentalis L., SC-1xxx, 2152, 2168, 2232: abu; CD 3

Diodia virginiana L., SC-1302, 2178: loc; DC+ 5

**Galium aparine* L., SC-1028: loc; Ew 3

Galium circaezans Michx., SC-1095: occ; BEu 1

Galium obtusum Bigelow var. *filifolium* (Wieg.) Fern., SC-1055, 1092: loc; CB 1

***Galium parisiense* L., SC-1275: rar; A 4

Galium tinctorium L., SC-1326, 1350, 2165: fre; CD 3

Galium uniflorum Michx., SC-1374: rar; BA 1

Mitchella repens L., SC-1373: occ; BA 1

Oldenlandia boscii (DC.) Chapm., SC-1518, 1528, 2xxx: loc; Du 5

**Richardia scabra* L., SC-1514, 1523: occ; Du 5

Rutaceae

Ptelea trifoliata L., SC-2121: rar; B 3

Salicaceae

Populus deltoides Bartr. ex Marsh., SC-1221, 2117: occ; Ew 2

Salix nigra Marsh., SC-1193, 2112: dom; Fu 2

Sapindaceae (sensu lato)

Acer floridanum (Chapm.) Pax [*A. barbatum* auct.], SC-1284: loc; Eu 1

Acer negundo L., SC-1392, 2013: com; Ew 1

Acer rubrum L. var. *drummondii* (Hook. & Arn. ex Nutt.) Sarg. [*A. drummondii*], SC-1154: occ; CD 2

Acer rubrum L. var. *trilobum* Torr. & Gray ex K. Koch, SC-1121, 1161: fre; CB+ 1

**Cardiospermum halicacabum* L., SC-1409, 2237: loc; CA+ 3

Sapotaceae

Sideroxylon lycioides L. [*Bumelia lycioides* (L.) Pers.], SC-2009: occ; BA 3

Saururaceae

Saururus cernuus L., SC-1118: dom; CF 2

Smilacaceae

Smilax bona-nox L. [including much var. *hederifolia* (Bey.) Fern.], SC-1383, 2026: abu; B+ 3

Smilax hispida Muhl. ex Torr. [*S. tamnoides* auctores; see Weakley (2012)], SC-1410, 2149: com; Eu 3

Smilax lasioneura Hook., SC-2081, 2160: occ; Eu 2

Smilax rotundifolia L., SC-1051, 2150: abu; B+ 3

Smilax smallii Morong, SC-1366: loc; A+ 2

Smilax cf. *walteri* Pursh, SC-1xxx: rar; C 2

Continued on the next page

Table 7.1. *Continued*

Solanaceae

Physalis angulata L., SC-1585, 2283: occ; Fu+ 5
Physalis heterophylla Nees, SC-1563: occ; A+ 4
Physalis longifolia Nutt. var. *subglabrata* (Mackenzie & Bush) Cronq. [*P. subglabrata*], SC-1394, 2005, 2046: loc; B+ 4
Solanum carolinianum L., SC-1195: occ; A+ 5
Solanum ptycanthum Dunal, SC-1572: occ; B+ 3

Styracaceae

Styrax americana Lam., SC-1421: loc; CD 3

Taxodiaceae

Taxodium distichum (L.) L.C. Rich., SC-1577 [pla?], 2301: rar; CF 2

Typhaceae (sensu lato)

Sparganium americanum Nutt., SC-2276: occ; Fw 4
Typha latifolia L., SC-1362: occ; D 4

Ulmaceae (sensu lato)

Celtis laevigata Willd., SC-1550, 2004: com; CB+ 1
Planera aquatica J.F. Gmel., SC-1532: occ; CD 2
Ulmus alata Michx., SC-1230, 2153: com; AB 1
Ulmus americana L. [tending to var. *floridana* (Chapm.) Little], SC-1301, 2119: fre; BEw 1
Ulmus rubra Muhl., SC-2085: occ; Eu 1

Urticaceae

Boehmeria cylindrica (L.) Sw., SC-1340: loc; Ew 1
Pilea pumila (L.) Gray, SC-1417, 2272: occ; Eu 1

Valerianaceae

Valerianella radiata (L.) Dufr., SC-1003: loc; Ew+ 3

Verbenaceae

Phyla lanceolata (Michx.) Greene, SC-1189, 2045: fre; CD 3
***Verbena brasiliensis* Vell., SC-1229, 2176, 2303: loc; A+ 4

Violaceae

Viola cf. *langloisii* Greene, SC-1076, 1096, 1236, 2079: fre; Ew+ 2

Viscaceae

Phoradendron leucarpum (Raf.) Reveal & M.C. Johnston, SC-1xxx: rar; B+ 2

Vitaceae

Ampelopsis arborea (L.) Koehne, SC-1057, 2086: com; AC+ 3
Ampelopsis cordata Michx., SC-1357, 2089: loc; Ew+ 3
Parthenocissus quinquefolia (L.) Planch., SC-1411, 2007: loc; Eu 2
Vitis cinerea (Engelm.) Millard, SC-1197, 1344, 1443, 2075: fre; BC+ 2
Vitis rotundifolia Michx., SC-1181, 1224: loc; AB+ 2
Vitis vulpina L., SC

TABLE 7.2.

Riparian or wetland species that were listed by MacDonald (1996) or Leidolf et al. (2002) for the Black Belt or nearby, but were not found during the fieldwork of this study.

Family and species name	Monroe County (MacDonald 1996)	Oktibbeha County (Leidolf et al. 2002)
Aceraceae: *Acer saccharinum*	occ/com; alluvial woods	—
*Aizoaceae: *Mollugo verticillata*	abu; open ground	— (occ; ditches/canals IF)
Alismataceae: *Sagittaria latifolia*	com; ditches to rivers	—
Apiaceae: *Cicuta maculata*	abu; ditches, etc.	— (loc abu IF)
Apiaceae: *Eryngium prostratum*	com; low wet open area	— (unc IF)
Apiaceae: *Hydrocotyle ranunculoides*	com; aquatic	—
Asteraceae: *Eutrochium fistulosum*	com; low wet drainages	—
Asteraceae: *Rudbeckia laciniata*	occ; low wet drainages	— (IF)
Azollaceae: *Azolla caroliniana*	loc abu; aquatic	occ; aquatic
Callitrichaceae: *Callitriche heterophylla*	com; aquatic	— (loc com IF)
Campanulaceae: *Lobelia appendiculata*	unc; low moist open BB	— (loc abu IF)
Caprifoliaceae: *Symphoricarpos orbiculatus*	com/abu; low woods/edges BB	occ/com; uplands
Cuscutacae: *Cuscuta compacta*	abu; low open, edge of water	+? (anonymous)
Cyperaceae: *Carex brevior*	occ; ditches BB	occ; uplands
Cyperaceae: *Carex grayi*	—	unc; low wet woods
Cyperaceae: *Carex longii*	abu; ditches, etc.	+ (Bryson); low woods
Cyperaceae: *Carex pigra*	—	+ (Bryson)
Cyperaceae: *Cyperus odoratus*	—	com; ditches, etc.
Cyperaceae: *Eleocharis elliptica*	—	+ (Bryson); BB
Cyperaceae: *Eleocharis montevidensis*	—	+ (Bryson); BB
Cyperaceae: *Eleocharis smallii*	—	+ (MacDonald); BB
Cyperaceae: *Fimbristylis decipiens*	—	+ (MacDonald); BB
Cyperaceae: *Fimbristylis vahlii*	occ/com; wetlands	—
Fagaceae: *Quercus sinuata* [*durandii*]	occ; roadsides, low woods	+ low woods BB
Gentianaceae: *Gentianella quinquefolia*	rar; creek banks, terraces BB	—
Haloragaceae: *Myriophyllum aquaticum*	abu; aquatic TR	+ (Jones)
Hypericaceae: *Hypericum tubulosum*	occ; low wet woods	— (IF)
Hypericaceae: *Hypericum walteri*	abu; shores, low wet areas	— (IF)
Lemnaceae: *Lemna perpusilla*	com; aquatic	—
Lemnaceae: *Spirodela polyrhiza*	com/abu; aquatic	— (loc abu PR)
Malvaceae: *Hibiscus moscheutos*	com; ditches, shores	unc; roadsides BB
Menispermaceae: *Menispermum canadense*	rar; mesic-wet woods BB	— (occ/rar PR)
Onagraceae: *Ludwigia alternifolia*	abu; moist roadsides, fields	— (IF)
Orchidaceae: *Triphora trianthophora*	rar; mesic/alluvial woods BB	— (loc com IF)
Plantaginaceae: *Penstemon tubiflorus*	—	rar; mesic w/riparian

Continued on the next page

Table 7.2. *Continued*

Family and species name	Monroe County (MacDonald 1996)	Oktibbeha County (Leidolf et al. 2002)
Polypodiaceae: *Pleiopeltis polypodioides*	com; epiphyte	occ/com; epiphyte
Potamogetonaceae: *Potamogeton nodosus*	abu; aquatic TR	— (loc com IF)
Rubiaceae: *Spermacoce glabra*	rar; ditches, etc. BB	occ; ditches, etc. BB
Salicaceae: *Salix exigua*	unc; alluvial woods edges	rar; ditches BB
Staphyleaceae: *Staphylea trifolia*	unc; alluvial woods, drains	— (loc abu IF)
Tetrachondraceae: *Polypremum procumbens*	—	loc com; disturbed sites
*Typhaceae: *Typha angustifolia*	—	loc abu; ditches, etc. BB
Verbenaceae: *Glandularia canadensis*	rar; low wet-mesic woods BB	+ (Morris et al.)
Vitaceae: *Vitis palmata*	rar; low wet woods	—

Note: Abbreviations for frequency are the same as in table 7.3; they apply to whole counties, not necessarily to the Black Belt. The following abbreviations indicate that the species was reported from only one region: BB = Black Belt; IF = Interior Flatwoods (mostly west of PR); PR = Pontotoc Ridge (west of BB); TR = Tombigbee River area (mostly east of BB). Notes in parentheses indicate records from a non-BB region; and sources cited by authors.

*Adventive species or uncertain status.

Table 7.3.

Approximate division of flora at the two lowland sites into typical habitat classes.

Typical habitat	Total species	Aliens	Rare natives
Grassland Complex [A/B/C 4/5] (roadsides, ditches, edges of seasonal ponds)	155	38–44	2 (1)
Thickets, edges, other transitions [A/B/C 3] (transitions from grassland to oak woodland)	91	4–7	5 (1)
Oak Woodland Complex [A/B/C 1/2] (oak woods with more or less shady interiors)	77	3–3	2
Ponded Complex [D 3/4/5] (more permanent ponds and edges)	75	6–11	4 (1)
Riparian Complex [E/F 1/2/3/4/5] (shady mesic levees to open scoured banks)	115	6–7	3 (1)
Total	513	57–72	16 (4)

Note: All taxa are assigned here to a "typical habitat" in one of the five broadly defined classes. Each of these classes contains a range of more precise habitats, with codes indicated in parentheses; see table 7.3 for explanation of these codes and details of each plant species. The few extra varieties are added to the numbers of species, but apparent hybrids are excluded. Numbers under "Aliens" are ranges—from species with certain alien status (lower) to totals that include species with uncertain or adventive status (upper). "Rare natives" are listed in the text; numbers within parentheses refer to the small subset of species that may be somewhat rare globally (up to G3G4 status).

8

An Alternative Natural Distribution for Osage Orange, *Maclura pomifera*, Including the Prairies of the Southeast

Jennifer L. Seltzer

Introduction

While preparing for their journey through the Louisiana Purchase, Lewis and Clark were introduced to a tree by Peter Choteau, who had spent time with the Osage nation. Choteau described it as an Osage plum or apple, and Lewis quickly described and shipped slips of the tree back to Thomas Jefferson, hoping that they would survive the journey back East. On March 26, 1804, Meriwether Lewis wrote to Jefferson from his camp outside Saint Louis, describing the "slips of Osages Plum and Apple" in a letter marking the first introduction of Osage into the lives of European Americans. Lewis noted at the time that the wood was used by the indigenous tribes and was highly sought after, and it was also used as a food source by the native prairie fauna. From this point forward, the Osage-orange tree has held a unique place in American history and lore.

Maclura pomifera (Raf.) Schneid. today is commonly known as Osage orange but also has many other descriptive common names: hedge or hedge-apple, bodark, bois d'ark, bowwood, Osage apple, horse-apple, prairie hedge-plant, mock orange, and yellow wood (J. Ball 2000; Peattie 1953; J. Smith and Perino 1981).

Osage orange (fig. 8.1) is a unique tree with many distinguishing characters. The leaves emerge after the male trees produce small pompom-like structures that disperse pollen in midspring. Female trees are easily identified when fruiting because of the unusual appearance of the fruits: grapefruit-sized, yellowish-green balls up to 15.24 cm (6 in) in diameter. The fruit produces milky latex when damaged, leading to the belief that the fruit is toxic; however, this is not true (Burton 1973; J. Smith and Perino 1981). Sprouts from the tree are also very distinct, with prominent thorns that have led to the tree's historical use as a natural fencerow.

Uses

Historically, the economic importance of Osage orange has been based on the strength, durability, and dye properties of its wood. Antifungal properties of the wood keep it from decomposing, making it one of the top three native woods for decay resistance and leading to its use for fence posts (Panshin and de Zeeuw 1980; J. Smith and Perino 1981). The wood is known to be very strong, with a crushing strength of 689 kg/cm^2 (9,800 lb/in^2) (T. Hamilton 1972). The wood and roots of Osage orange were traditionally used to create a variety of yellow to brown dyes, which were used until they were replaced with synthetic dyes (J. Ball 2000; J. Smith and Perino 1981).

Prehistoric uses for Osage orange included making bows from the wood. The strength and tension properties of the wood and the limbs of the Osage-orange tree make it ideal for use as bow wood (T. Hamilton 1972), presumably increasing the status of the Osage Indians, because of their access to the wood and their ability to expertly craft bows from it (Schambach 2003). This tree was also used by the

Figure 8.1. (A) *Maclura pomifera* tree located on the campus of Mississippi State University. (B) Close-up of fruit of *M. pomifera*. Photo by Jennifer Seltzer.

Natchez for ceremonial fires near the protohistoric Village of Valleur, prior to the French destruction of the village in February 1730 (Bushnell 1927).

Range

Current literature on this tree places its native range in the Red River drainage between Texas and Oklahoma (J. Smith and Perino 1981) (fig. 8.2), with some maps extending the range as far north as southern Missouri, as far east as the Mississippi River, and south into northwestern Louisiana (Betts 1929; Burton 1973; Collingwood 1939; Hough 1924). This limited distribution is in stark contrast to the

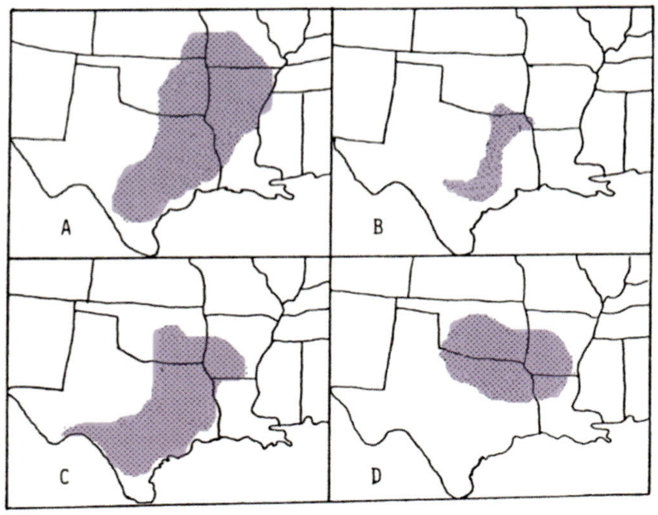

Figure 8.2. This map by J. Smith and Perino (1981) shows the hypothesized natural distributions of *M. pomifera*.

wide distribution that Osage-orange experienced prior to the Pleistocene (Barlow 2001a, 2001b; Blackwell et al. 1983; Schambach 2003). During this time, Osage orange flourished throughout North America, with fossil evidence placing it as far north and east as Ontario (Peattie 1953) and as far south as Mississippi (Blackwell et al. 1983). It has been suggested that the wide distribution was supported by megafauna, such as mastodons, horses, and mammoths, which consumed the fruits and spread the seeds as they traveled (Barlow 2001a, 2001b). Osage orange, along with the Kentucky coffee tree (*Gymnocladus dioicus)*, honey locust (*Gleditsia triacanthos*), pawpaw (*Asimina triloba*), and persimmon (*Diospyros virginiana*), is considered to be an "anachronistic" plant that no longer has its natural dispersal agents (Barlow 2001a, 2001b). Schambach (2003) proposed that the extinction of most of its megafaunal dispersers after the Ice Age caused the population of Osage orange to decline across the continent, and that its survival was a result of the unique relationship of streams and prairies in the Red River drainage, where the streams acted as dispersers and the openness of the prairies allowed the seeds to germinate.

This more restricted natural distribution (fig. 8.3) was suggested by Schambach (2003) to be related to the wealth of the prehistoric Spiro culture. He noted that the Spiro and Sanders Sites, located near blackland prairies in the Red River drainage, both have a high proportion of high-status trade items from the Lower Mississippi Valley (Schambach

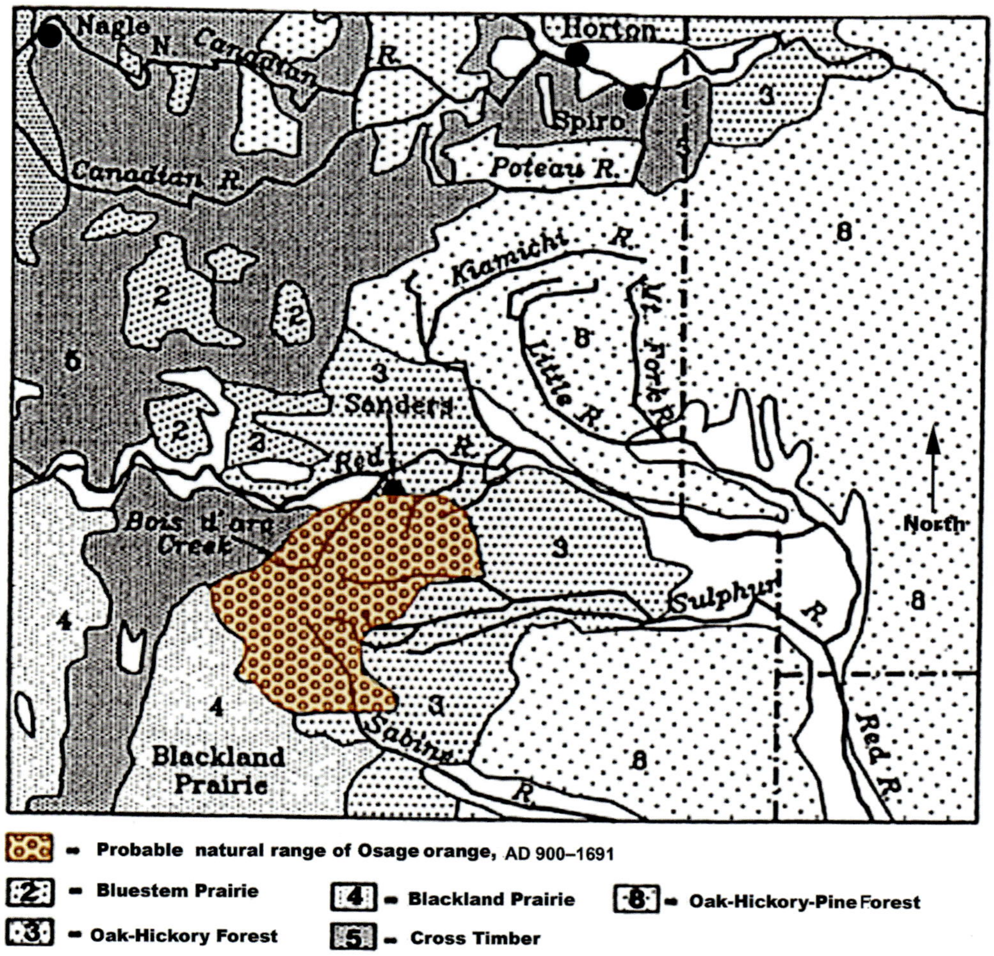

Figure 8.3. Map of *M. pomifera* distribution (Schambach 2003). The area in orange is in the Red River drainage in Oklahoma; highlight added by the author.

2003). Schambach proposed that the restricted distribution of Osage orange in a small area below the bluff where the Sanders Site is located gave the Spiroans unique access to the bottoms of Bois d'Arc Creek, where Osage orange flourished, and a monopoly on the wood for trading. By trading the wood or wood products with the Plains Indians for buffalo products, the Spiroans were then able to trade with the Indians of the Lower Mississippi Valley, creating a highly successful trade network (Schambach 2003).

Methods

Site History

Lyon's Bluff (22OK520) is a palisaded, single-mound Mississippian site (fig. 8.4) (Alvey et al. 2004; Lolley 2000; J. Rafferty et al. 2003), occupied between AD 1200 and 1650 (R. Marshall 1985c; Peacock and Hogue 2005), that has been a focus for research since M. Chambers's excavations in the 1930s (Galloway 2000). After Chambers's initial excavations at the site, R. Marshall supervised extensive excavations there during the 1967 and 1968 Mississippi State University Field Schools (R. Marshall 1985a, 1985b). More recent excavations have included the 2001 and 2003 Mississippi State University Field Schools and student research excavations in 2005, supervised by E. Peacock.

Lyon's Bluff is located within a unique region of the Southeast: the Black Belt Prairie physiographic province. The soils of this region support a flora and fauna similar to those of other blackland prairies and include disjunct species, such as the moth *Ceratomia hageni* (R. L. Brown 2003), that are otherwise found in the Great Plains states of Texas, Missouri, Kansas, and Arkansas. Blackland prairie habitat in Mississippi is made up of a mosaic of tallgrass prairies and various types of forests (Barone and Hill 2007;

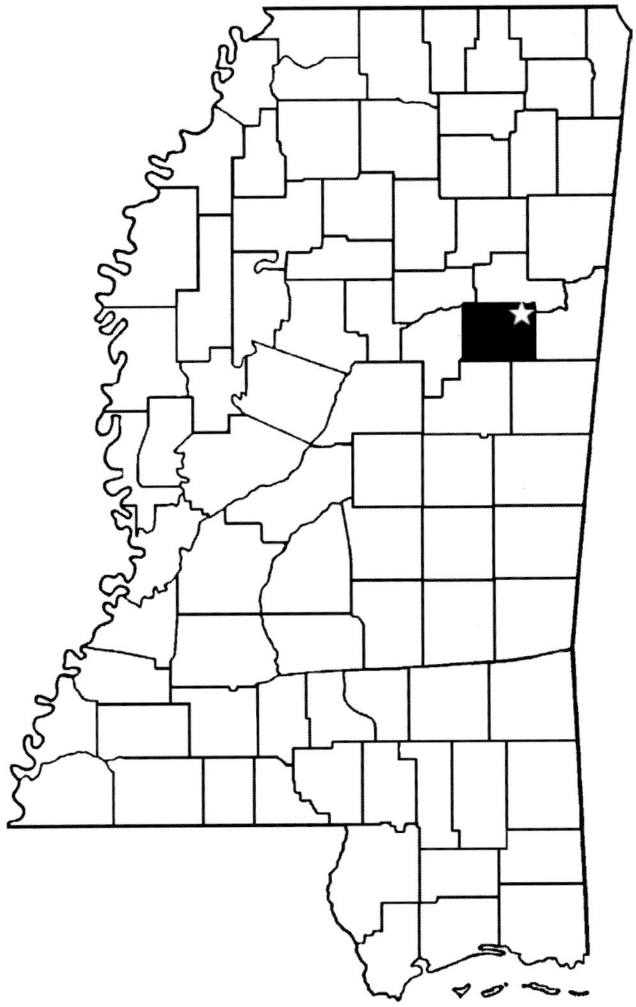

Figure 8.4. Mississippi County map with Oktibbeha County indicated. The star represents the approximate location of Lyon's Bluff (22OK520).

J. Hill et al. 2009; Leidolf et al. 2002), supporting the development of Mississippian mound and occupation sites such as Lyon's Bluff (J. Rafferty 2003).

Charcoal Identification

During the 2003 Mississippi State University Field School, excavations took place at Lyon's Bluff (22OK520), including the excavation of Feature 27, a wattle-and-daub structure (Seltzer 2007). Charcoal was recovered from excavations at Lyon's Bluff, including material from Feature 27, which was radiocarbon dated to AD 1420–1460 (calibrated, 1-sigma range) (Seltzer 2007). The charcoal was initially identified using standard charcoal identification techniques (Dimbleby 1967; Lopinot 2003; Rossen and Olson 1985), which included exposing three surfaces: the cross section (X), the radial surface (R), and the tangential surface. However, because of the fragile nature of the charcoal, identification was limited to exposed surfaces, typically the cross section of the wood. On exposed surfaces, key characteristics were found for identification: the arrangement of early and late wood pores, the number of rays, and the presence or absence of tyloses. Combinations of these characteristics are typically unique to each species, leaving few trees indistinguishable from each other (Hoadley 1990; Panshin and de Zeeuw 1980). Identifications were made using a comparative microscope to examine the features of the charcoal. Samples were then compared to prepared wood identification slides, the charcoal reference collection housed at the Cobb Institute of Archaeology, and wood identification guides. This technique was successfully used to identify the majority of charcoal samples recovered from excavations at Lyon's Bluff (Seltzer 2007); however, some samples from Feature 27 required additional methods to finalize their identification.

A few species of wood needed additional microscopic analysis for identification (Panshin and de Zeeuw 1980). Osage orange and black locust (*Robinia pseudoacacia* L.) appear identical in their internal structure when observed under a compound microscope. To separate these two species, a scanning electron microscope (SEM) is needed for the observation of vestured pits. Pits are internal features used for fluid exchange within the wood of trees, and in some species, these pits become vestured (fig. 8.5a). Vestures are small wartlike protrusions that form around the opening of the pits (Panshin and de Zeeuw 1980).

Vestured pits are present in black locust and not in Osage orange (fig. 8.5b) (Panshin and de Zeeuw 1980). Positive identification of the charcoal as either of these species would be of note, as it would place both species outside their known native ranges. SEM was used to verify the presence or absence of vestured pits in the charcoal samples. Because of the fragile nature of archaeological charcoal samples, exposure of the appropriate surfaces required the development of a new technique. A cryofracturing technique was created, which easily provided the necessary surface exposure (J. Seltzer, in prep.). The Feature 27 samples were then examined and compared with reference samples using SEM, and images were captured digitally.

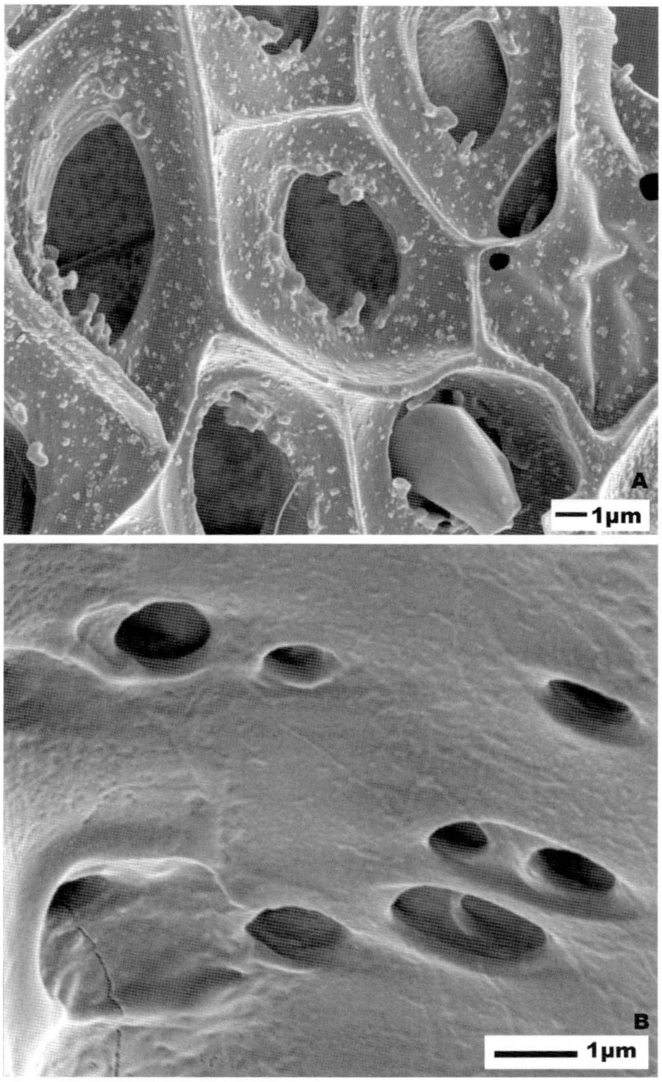

Figure 8.5. The vestured pits seen here (A) are from *Robinia pseudoacacia* reference charcoal. The nonvestured pits (B) are smooth, from *M. pomifera* reference charcoal.

Results and Discussion

The results of the charcoal examination confirmed that the samples from Feature 27 were Osage orange (fig. 8.6), and the samples were radiocarbon dated to AD 1420–1460 (calibrated, 1-sigma range). Additional charcoal recovered during excavation at several locations throughout the site from Lyon's Bluff was identified as Osage orange. Initial identifications were made from charcoal recovered from a wattle-and-daub feature located approximately 180 m north and 390 m west of the mound. Further identifications were made from units located 12, 20, and 40 m west of the mound. Many of these samples came from zones where

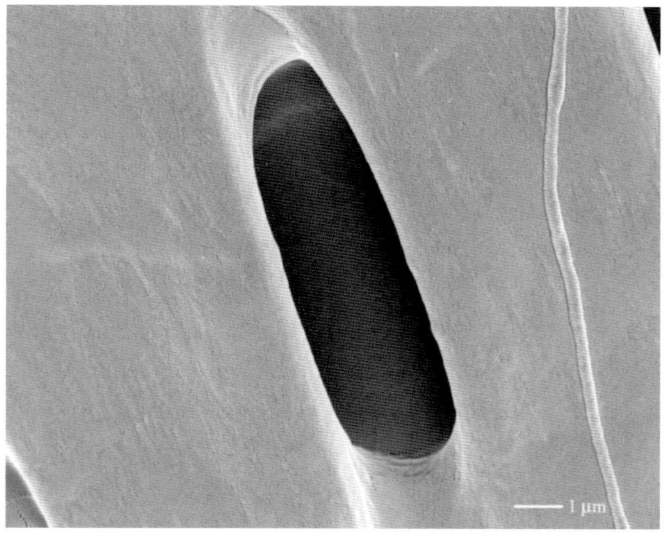

Figure 8.6. Charcoal sample from Lyon's Bluff, with nonvestured pit. This image shows that vestures have not been "knocked off," as the surface surrounding the pit is smooth and intact. This sample has been identified as *M. pomifera*.

charcoal was recovered from within a midden deposit. The presence of the charcoal across the site and within the midden layer indicates that the material was used for common, everyday purposes. This is less plausible if wood was brought in as a high-status trade item from the Red River drainage, and much more likely if the wood was found as part of the local forest. This indicates that Osage orange was present in the Black Belt Prairies of Mississippi prior to European contact.

Ann Early found Osage orange at the Winding Stair Site, Arkansas (Early 2000), identifying the species using SEM. Schambach has dismissed this charcoal, because of the small amount and its location within the excavation of a house structure, as being leftover material from bows, and therefore not informative regarding the distribution of Osage orange. However, when coupled with additional charcoal and pollen records from other sites, the data support a hypothesis that there was a broader, southeastern refugium for Osage orange. A search of current gray literature, published articles, and site reports generated examples of additional records of Osage orange in the Southeast. This includes charcoal from the Crosno Site in southeastern Missouri, dating between AD 1200 and 1400 (S. Williams 1954); charcoal from the McNight Site in northwestern Mississippi, dating between AD 200 and 1450 (Shea

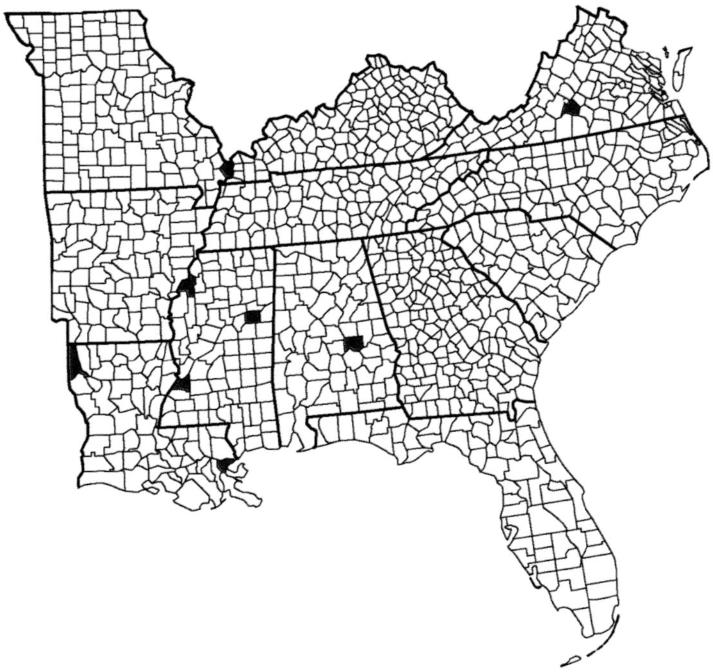

Figure 8.7. County map of the Southeast, indicating locations where *M. pomifera* artifacts have been found.

1998); fossil wood from Bayou Pierre, Mississippi (Blackwell et al. 1983); a bow from the Mounds Plantation Site, Caddo Parish, Louisiana, recovered from a burial dating between AD 1050 and 1070 (T. Hamilton 1972); journal entries and drawings by A. DeBatz from his journeys in Louisiana that include a drawing of a mature Osage-orange tree from 1732 (Bushnell 1927); and finally, pollen samples from the Fort Toulouse and Fort Jackson Sites in Alabama, dating to the mid-1700s (Waselkov et al. 1982) (fig. 8.7).

In addition to charcoal and pollen, other evidence suggests that Osage orange may have been present in the Southeast for a much longer period than implied by Schambach (2003). For example, very large and old individual Osage-orange trees are found in numerous locations, including champion trees at Red Hill, Virginia, the home of Patrick Henry, and at River Farm, a property owned by George Washington in Alexandria, Virginia. Champion trees are the largest in the country; a variety of measurements go into their determination and are taken by professional arborists. Originally, legend told that the Red Hill tree was brought as a cutting from the Lewis and Clark Expedition of 1805. However, when the tree was ring dated in 2003, it proved to be more than 330 years old, eliminating the possibility of any association with Lewis and Clark (Jon Kukla, pers. comm.). River Farm is the current home of the American Horticultural Society, and the Osage-orange tree on its property was the National Register of Big Trees Champion in 2011. Though this tree has not been ring dated, its size would suggest that it may be of equivalent age to the Red Hill tree. Another rival champion tree is just beyond the range of the Southeast, in Wilmington, Delaware, where a 350-year-old tree lives on the property of the Hagley Museum.

While Schambach's arguments that the blackland prairies of the Red River drainage were acting as a refugium for

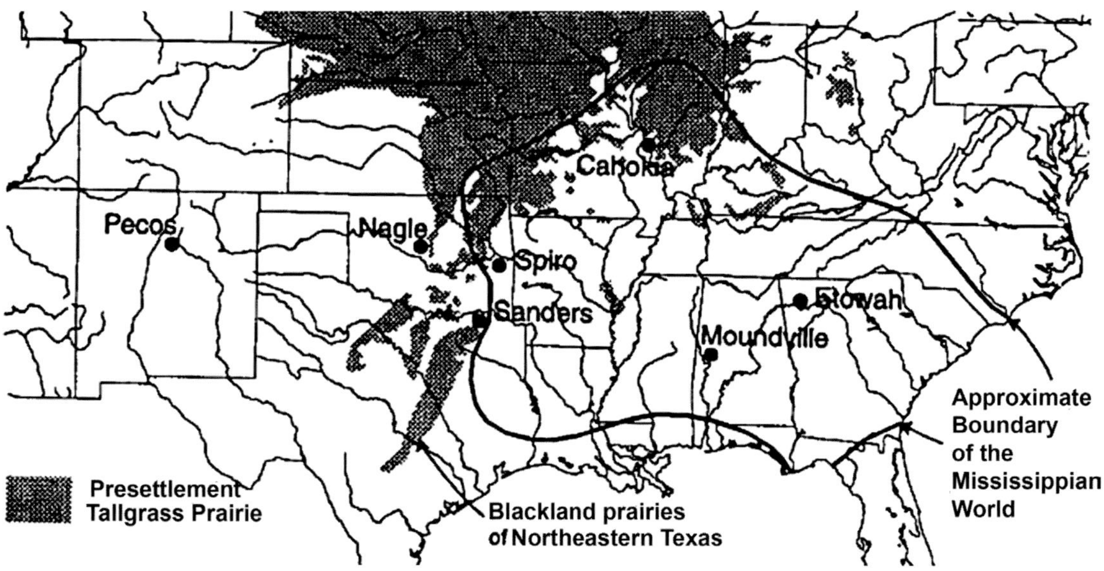

Figure 8.8. Map of blackland prairies that excludes those in Mississippi and Alabama. (Schambach 2003).

Osage orange may be accurate, he has excluded the blackland prairies of the Southeast (fig. 8.8) (Schambach 2003). Because the ecological conditions in these prairies were almost identical to those in the blackland prairies of Texas and Oklahoma, and because fossil evidence of Osage orange has been found in Mississippi (Blackwell et al. 1983), it seems likely they were an additional refugium for the species. Entomological data provide further support (R. L. Brown 2003). Studies in the blackland prairies of Mississippi on prairie endemic insects include data on *Ceratomia hageni*, a moth that feeds exclusively on Osage-orange leaves (fig. 8.9) (R. L. Brown 2003).

This moth is found in Oktibbeha County, Mississippi, and has population densities equivalent to those found in

Figure 8.9. The moth *Ceratomia hageni* feeds exclusively on the leaves of *M. pomifera*.

Figure 8.10. Map of natural resources (Swanton 1969). Swanton's proposed natural resources for the Lyon's Bluff area are highlighted in red, while his proposed area for "bow wood" is highlighted in orange.

western prairies, including the distribution area of Osage orange (R. L. Brown 2003). While the moth is found in other areas where Osage orange has been naturalized across the Southeast, the population densities do not compare to those of either of the blackland prairie populations discussed in R. L. Brown (2003). The combined archaeological and entomological evidence indicates that the blackland prairies around Lyon's Bluff likely had a native population of Osage orange (*M. pomifera*) prior to the European American use of Osage orange for the control of livestock, expanding the prehistoric range of Osage orange into the Southeast within the Black Belt Prairie.

To extend Schambach's refugium model to include the Osage-orange populations of Mississippi and Alabama, several paths of inquiry present themselves. If Moundville and its occupants were examined with the model that Schambach uses for Spiro, then one could ask whether Lyon's Bluff and other single-mound Mississippian sites in the Black Belt of Mississippi and Alabama acted in the role of entrepôts, or centers of trade and exchange. Placement of mound sites within the Black Belt may indeed have reflected areas where this resource was present and accessible for trade. When Swanton (1969) mapped out trade items for the Southeast (fig. 8.10), he did place Osage orange in the Red River drainage, but he placed pigeon roosts as the closest resources to Lyon's Bluff. With little archaeological evidence of this being traded from the Lyon's Bluff area, it would be interesting to test a model in which trade in Osage orange played a role in the development of mound centers within the Black Belt of Mississippi and Alabama.

9

Prehistoric Molluscan Faunas of the Mississippi Black Prairie

Evan Peacock

Blackland prairies (see Peacock and Schauwecker 2003a) have become an increasingly important target for habitat restoration and ecosystem management efforts over the last few decades. Coverage of the organisms involved has not been even, however; in particular, molluscan faunas have been largely neglected. Most work has focused on freshwater mussels in the Tombigbee River drainage, and this has been limited relative to what has taken place in other major watersheds in the Southeast. The aquatic and terrestrial gastropod faunas have been almost entirely ignored. Some work stemming from environmental archaeological research is beginning to fill these gaps, but much remains to be done. In this chapter, I briefly review what has been learned about paleoenvironmental conditions in the Mississippi Black Prairie as they existed prior to modern human impacts using archaeological and molluscan assemblages. Throughout, I use "Black Prairie" to refer specifically to the surface expression of soils and vegetation developed on Cretaceous chalk deposits in eastern Mississippi, while "Black Belt" refers to the arc-shaped physiographic province as a whole as it occurs in Mississippi and Alabama.

Land Snails

While "the conservation status of mainland North American land snails remains relatively unknown" (Minton and Perez 2010, 91; see also Minton and Perez 2005), it is particularly ironic that land snails receive little attention in conservation assessments of Black Belt habitats, given that calcium-rich blackland prairies can provide excellent habitat conditions for snails. General snail identification guides (e.g., Burch 1962) discuss spatial associations for species only at the state level. Pilsbry's classic monograph series (1939–1948) remains indispensable for species identification, but habitat data are essentially lacking in his work. Hubricht (1985) is better in that county-level occurrences are graphically displayed, but again, habitat information for species is very general, at best. The same is true for state-level monographs (A. Archer 1948; Walker 1928). Shelford (1974) described the changing gastropod taxa with successional changes in a cedar glade in Tennessee as well as species characteristic of deciduous forests in the Southeast, using mostly data from Meyer (1937). A recent, valuable compilation of museum records by Minton and Perez (2010) displays taxonomic richness across the state of Alabama; the Black Belt is shown in their work as being an area of medium richness. Systematic collections with improved data on habitat conditions have recently begun to appear for areas near the Black Belt (e.g., Waggoner et al. 2006).

From 2001 to 2003, I collected litter samples across different microhabitats in the Black Prairie (fig. 9.1) in order to obtain modern snail faunas to compare with archaeological ones. Sample locales were chosen subjectively to represent different general habitat types, including upland and lowland hardwoods, mixed hardwoods, cedar glades, prairies, canebrakes, silvicultural pine stands, isolated cedar hummocks in chalk blowouts (heavily eroded areas), the chalk blowouts themselves, and moist ground (drainage ditches

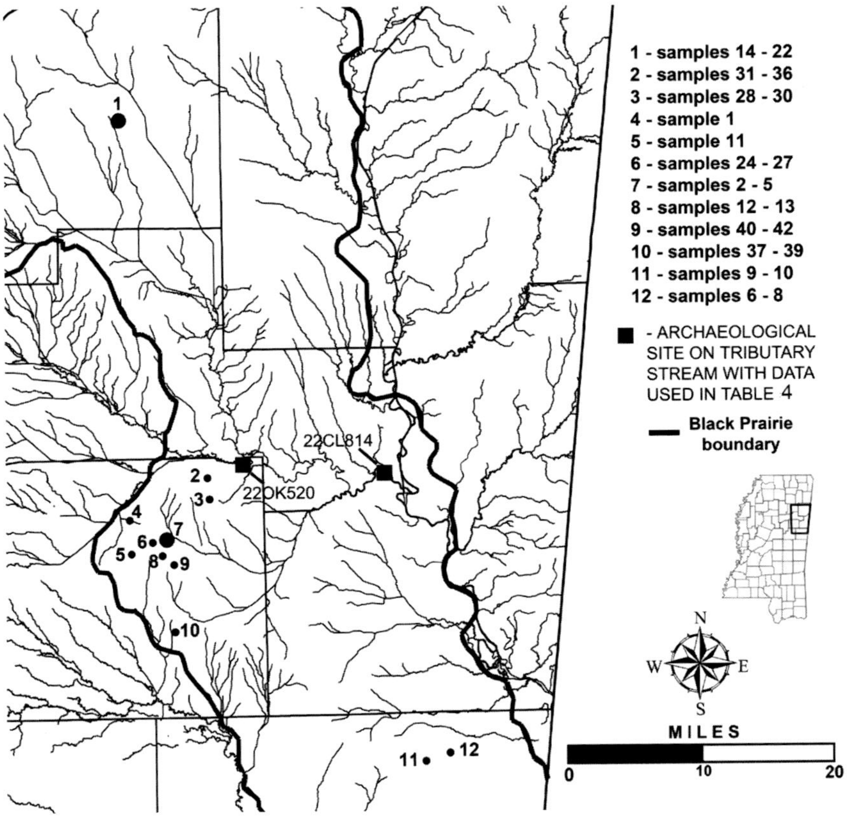

Figure 9.1. Location of modern snail samples and selected archaeological sites in the Mississippi Black Prairie.

of artificial ponds). Several samples taken at school playgrounds were meant to mimic conditions at prehistoric plazas (central, open areas at mound sites, thought to have been locales for public dancing, sports, etc.). While some of the snail and general habitat data have previously been reported, along with the results of matrix-based multivariate analysis of those data (Peacock and Gerber 2008; see also Peacock and Melsheimer 2003; Peacock et al. 2005), in this chapter I include previously unpublished taxonomic data as well as more detailed information on vegetation. The results of soil chemical analysis for the sample locales also are published here for the first time.

For vegetation, trees greater than 10 cm in diameter at breast height and within 10 m of each sample locale were recorded, along with general notes on each setting (table 9.1). To recover snails, vegetative litter and the top few centimeters of O/A horizon from an area of 1 m^2 were scraped with a trowel into plastic sample bags. Sample volume was measured in graduated cylinders in the lab, where samples were then washed thoroughly over nested screens down to 500 microns. Care was taken to wash individual leaves and pieces of woody debris during this process. The washed material was allowed to dry and snails were hand sorted from the detritus using 10X magnification on the 500-micron material. Identifications were made using standard guides, and a large number of specimens were verified via comparison with collections at the Field Museum of Natural History, Chicago. Results are given in table 9.2.

As can be seen in this table, the number of shells (including live snails) recovered per sample varied greatly, from 0 to 1,576. As there is no correlation between sample volume and number of specimens, abundance is assumed to be related to environmental conditions. Past studies have shown land snail abundance to be positively correlated with extractable calcium (Burch 1955; Gärdenfors 1992; Hotopp 2002; Skeldon et al. 2007), as land snails take up calcium from food and from absorption through the sole skin (Fournié and Chétail 1984). However, regression analyses evaluating abundance relative to extractable calcium showed no meaningful relationship, even when one outlier (Sample 12, which has over three times as many specimens as any other sample) is removed. Interestingly for omnivore-detritivores (no carnivorous snail species were recorded), there is likewise no correlation between abundance and the percentage of organic matter.

The only meaningful correlation noted is with pH

Table 9.1.

Modern snail litter sample volumes and sample locale characteristics.

Sample #	Volume (ml)	Setting	Soils	Trees within 10 m of sample*	Notes
1	4,000	ridgetop	3 cm dark gray silty clay A horizon; > 15 cm light brown silty clay E horizon	8 cedar, 1 persimmon, 1 dogwood, 1 winged elm	grass, a few hardwood saplings; a few dead cedars on ground
2	6,000	cane brake	> 15 cm very compact dark brown silty clay A horizon	none	medium-sized cane; some grass and clover
3	5,700	ridgetop	10 cm dark brown sandy loam A horizon; grades with 2 cm brown sandy clay loam E horizon; grades with red silty clay B horizon	4 Osage orange, 1 hackberry	very dense understory of greenbrier, mullein, beautyberry, sparse grass; lots of woody detritus
4	3,000	bottom	20 cm dark gray silty clay loam A horizon; sharp with > 10 cm gray silty clay E horizon	1 redbud, 5 sweetgum, 8 boxelder, 8 hackberry, 4 pecan, 5 Osage orange, 3 American elm, 1 black walnut, 1 honey locust, 5 winged elm	grass, greenbrier, scattered cedar saplings, privet
5	3,300	ridgetop	5 cm light brown sandy loam; irregular, abrupt with 5 cm compact, yellow sandy loam; smooth, abrupt with sandy loam B horizon	1 shagbark hickory, 7 other hickory, 1 black cherry	saplings of hickory, black oak, pecan, water oak, redbud, hackberry, cedar, winged elm; grass, beautyberry, privet
6	3,000	ridgetop	12 cm black silty clay; abrupt with chalk bedrock	4 Osage orange, 1 hackberry	thick grass up to 40 cm tall
7	3,600	ridgetop	> 30 cm dark gray silty clay A horizon	8 Osage orange	thick grass up to 40 cm tall; some woody detritus
8	2,000	ridgetop	> 30 cm mixed dark gray silty clay with chalk fragments	3 Osage orange	thick grass up to 40 cm tall; some woody detritus
9	3,700	upland flat	2 cm dark gray sandy loam A horizon; smooth, abrupt with light brown silty clay loam E horizon; abrupt with reddish-brown silty clay B horizon	4 shagbark hickory, 4 other hickory, 6 winged elm, 2 American elm, 2 post oak, 1 red mulberry, 1 huckleberry	dense cover of poison ivy, hickory, elm, and cedar saplings
10	3,200	upland flat	2 cm O horizon; 10 cm light brown silty clay loam E horizon; smooth, abrupt, with reddish-brown silty clay B horizon	4 shagbark hickory, 14 other hickory, 1 winged elm, 1 post oak, 1 other oak, 1 American elm	dense cover of poison ivy, hickory saplings
11	5,000	moderate western slope of broad ridge	4 cm grayish-brown silty clay E horizon; grades with > 15 cm reddish-brown silty clay B horizon	7 cedar, 1 pecan, 3 chinquapin oak	scattered grass, a few cedar and oak saplings
12	3,600	narrow ridgetop	12 cm mottled, brown silty clay loam with small chalk fragments; irregular, disturbed boundary with chalk bedrock	50 cedars	eroded; patchy grass and cedar saplings

Continued on the next page

Table 9.1. *Continued*

Sample #	Volume (ml)	Setting	Soils	Trees within 10 m of sample*	Notes
13	3,700	narrow ridgetop	12 cm mixed, mottled, silty clay loam with chalk fragments; irregular, disturbed boundary with chalk bedrock	similar to Sample 12	more shaded than Sample 12; hardwood saplings, clover, moss, ferns
14	5,600	upland flat	4 cm dark brown silty loam A horizon; smooth, even, with 4 cm brown silty clay loam E horizon; grades with reddish-brown silty clay B horizon	1 blackjack oak, 6 post oak, 16 Osage orange, 2 hackberry, 4 honey locust, 1 American elm, 1 black cherry	much open ground; some grass, poison ivy, small dead Osage-orange saplings; wooded clump in USFS prairie in Chickasaw County, MS
15	4,800	moderate eastern slope of broad ridge	> 30 cm yellowish-brown silty clay; clay content and yellowing increase with depth; platy at depth	1 cedar, 1 persimmon	grass, broomstraw, some bare ground; USFS prairie in Chickasaw County, MS
16	4,100	ridgetop	5 cm dark brown silty clay loam A horizon; smooth, abrupt, with > 20 cm yellow silty clay E horizon	none	grass, broomstraw, no bare ground; USFS prairie in Chickasaw County, MS
17	6,000	upland flat	5 cm dark brown silty clay loam A horizon; smooth, abrupt, with 4 cm yellowish-brown silty clay E horizon; grades quickly with reddish-brown silty clay B horizon	8 post oak, 21 Osage orange, 1 honey locust, 3 hackberry, 1 eastern hophornbeam, 1 cedar	some small cedar saplings; wooded clump in USFS prairie in Chickasaw County, MS
18	3,700	ridgetop	14 cm gray silty clay E horizon; sharp, smooth, with chalk bedrock	none	grass, broomstraw, no bare ground; USFS prairie in Chickasaw County, MS
19	5,200	upland flat	20 cm dark grayish-brown silty clay loam A horizon; grades with brown silty clay E horizon	15 loblolly pine	artificial pine stand in USFS prairie in Chickasaw County, MS
20	5,500	upland flat	2 cm brown silty loam A horizon; sharp, smooth, with 5 cm light brown silty clay loam E horizon; grades with reddish-brown silty clay B horizon	2 blackjack oak, 6 cedar, 1 black cherry, 22 Osage orange, 6 post oak, 1 honey locust, 1 hackberry	some cedar and hackberry saplings, small bushes, some grass; a lot of bare ground; wooded clump in USFS prairie in Chickasaw County, MS
21	3,200	upland flat	> 40 cm dark brown silty clay loam A horizon	none	grass, prairie flowers; USFS prairie in Chickasaw County, MS
22	4,800	upland flat	1 cm dark gray silty clay loam A horizon; sharp, abrupt, with 1 cm light brown silty clay E horizon; abrupt with mottled reddish-brown/light reddish-brown silty clay B horizon; reddens with depth	7 loblolly pine	artificial pine stand in USFS prairie in Chickasaw County, MS
23	3,900	upland flat	> 15 cm mixed dark gray silty clay loam and light brown silty clay; gravel and glass throughout	none	playground of Sudduth School, Starkville, MS; ca. 70% open ground, 30% short grass

TABLE 9.1. *Continued*

Sample #	Volume (ml)	Setting	Soils	Trees within 10 m of sample*	Notes
24	4,000	upland flat	same as 23	none	playground of Sudduth School; ca. 70% open ground, 30% short grass
25	4,100	upland flat	same as 23	1 pecan	playground of Sudduth School; ca. 70% open ground, 30% short grass
26	3,900	upland flat	same as 23	none	playground of Sudduth School; ca. 70% open ground, 30% short grass
27	3,400	upland flat	same as 23	none	playground of Sudduth school; ca. 70% open ground, 30% short grass
28	4,000	ridgetop	25 cm dark brown silty clay loam; sharp, smooth, with a red clay lens	none	playground of Alexander School, Oktibbeha County, MS; sparse grass, mostly bare
29	3,800	ridgetop	14 cm dark brown silty clay loam; grades with fragmented chalk C horizon	none	playground of Alexander School, Oktibbeha County, MS; sparse grass, mostly bare
30	4,850	ridgetop	10 cm black silty clay loam; sharp, smooth, with reddish clay and fragmented chalk C horizon	none	playground of Alexander School, Oktibbeha County, MS; ca. 50% grass, 50% bare
31	5,400	upland flat	chalk; no soil	none	eroded chalk area, Oktibbeha County, MS; no vegetation
32	6,000	upland flat	chalk; no soil	none	eroded chalk area; no vegetation
33	5,500	upland flat	chalk; no soil	none	eroded chalk area; no vegetation
34	4,250	upland flat	12 cm light brown silty clay loam E horizon; grades with fragmented chalk C horizon	14 small cedar	isolated cedar hummock, ca. 420 m2, in chalk blowout
35	3,900	upland flat	8 cm light brown silty clay loam; sharp, smooth, with light gray clay and fragmented chalk C horizon	5 small cedar, 1 small, unidentified hardwood	isolated cedar hummock, ca. 100 m2, in chalk blowout
36	4,600	upland flat	12 cm dark brown silty clay loam; grades with light gray clay and fragmented chalk C horizon	2 small cedar	isolated cedar hummock, < 100 m2, in chalk blowout; a little grass, ca. 80% bare
37	1,750	shallow ditch	2 cm saturated, brown silty clay; sharp, smooth, with wet, indurated, light gray clay	a few small willows	drainage ditch off small pond; rushes, grass
38	1,850	shallow ditch	2 cm saturated, brown silty clay; sharp, smooth, with wet, indurated, light gray clay	a few small willows	drainage ditch off small pond; rushes, grass

Continued on the next page

Table 9.1. *Continued*

Sample #	Volume (ml)	Setting	Soils	Trees within 10 m of sample*	Notes
39	2,000	shallow ditch	2 cm saturated, brown silty clay; sharp, smooth, with wet, indurated, light gray clay	a few small willows	drainage ditch off small pond; rushes, grass
40	2,350	upland flat	> 15 cm mucky, gray silty clay loam	2 cypress, 1 sweetgum sapling	south edge of Eckie's Pond, MSU campus; wet, seasonally inundated soil; lots of organic detritus
41	2,000	upland flat	> 15 cm mucky, gray silty clay loam	2 cypress, 1 sweetgum sapling	south edge of Eckie's Pond, MSU campus; wet, seasonally inundated soil; lots of organic detritus
42	2,250	upland flat	> 15 cm mucky, gray silty clay loam	2 cypress, 1 sweetgum sapling	south edge of Eckie's Pond, MSU campus; wet, seasonally inundated soil; lots of organic detritus

*Except for sample 1, for which trees within 30 m were recorded.

(fig. 9.2), with an exponential growth trend showing increased abundance with greater alkalinity (again with Sample 12 removed), something that may in part reflect better preservation of shells in less acid conditions. The failure to find clear correlations between soils and snail abundance is likely related to the nature of the samples: timed searches over larger surface areas, coupled with litter samples over multiple random plots, would probably provide more useful data in this regard. Along this line, one interesting absence is *Rabdotus dealbatus*, a species that can be locally abundant in eroded areas of the Black Prairie today (e.g., Peacock and Melshimer 2003), and which presumably would show up with larger or more spatially extensive samples.

Taxonomic richness also varies greatly, from 0 to 15 species per sample. When richness is viewed from the standpoint of general habitat type (using categories from Peacock and Gerber [2008] and assessments derived from the observations recorded in table 9.1), bottomland hardwoods and cedar glades are among the habitats with the most species represented; the one canebrake sampled is also relatively rich. Richness in cedar glades is higher than that reported by Meyer (1937) for cedar glades in the Nashville Basin, and further comparison of regional-scale samples would be of interest.

Prairies (grasslands with no tree cover) are intermediate to low in richness (fig. 9.3). Taxa characteristic of prairies, cedar glades, and isolated cedar hummocks include *Pupoides albilabris*, *Oligyra orbiculata*, Succinidae, *Triodopsis* spp., and *Lucilla* sp., as well as high numbers of species of *Gastrocopta* (see also Shelford 1974, table 2.4). *Daedalochila leporina* is also more common in prairies than in other habitat types. Hardwoods are characterized by *Hawaiia minuscula*, *Helicodiscus parallelus*, *Mesomphix* sp., *Strobilops* spp., and *Striatura meridionalis*; *Ventridens intertextus* and *V. ligera* appear to be bottomland hardwood species, while *Vertigo milium* is restricted to upland mixed or upland hardwood settings. Interestingly, no *Triodopsis* species were recovered from hardwood habitats, despite Shelford's (1974, 45) assessment of this genus being the "characteristic form" of land snail in deciduous forests. Osage-orange groves include high proportions of *Euconulus chersinus*, *Gastrocopta armifera*, *G. contracta*, and *G. tappaniana*. Pine stands are harder to characterize because of generally small samples, although the high proportion of *Striatura meridionalis* in one sample is striking.

Comparison of modern with archaeological snail faunas (recovered along with artifacts from excavations—see Peacock and Melsheimer [2003] and Peacock et al. [2005]

for description of contexts, recovery methods, and protocols for sorting modern from premodern specimens) clearly shows that Black Prairie locales favored for settlement by Indians were once hardwood (probably post oak) stands and that the current dominance of cedar in eroded areas of the Black Belt is a consequence of Historic-period land clearance and erosion. Another interesting find is the presence of an amphibious snail genus (*Fossaria*) at an archaeological site in Oktibbeha County, Mississippi. This genus was not found in any modern samples, but it may be taken as an indicator of wetter conditions when prehistoric agricultural settlement expanded into the Black Prairie around AD 1200, coincident with the Medieval Warm Period (Peacock and Gerber 2008; J. Rafferty and Peacock 2008). Although much work remains to be done in testing this hypothesis, a second example of this genus found recently at another archaeological site in the Black Prairie of Oktibbeha County (Peacock, n.d.) is encouraging.

Freshwater Mussels

Paleoenvironmental and Archaeological Considerations

The Black Belt archaeological record is rich but unevenly distributed in space and time. Systematic survey in the central part of the province (those areas farthest from major streams) has revealed little in the way of prehistoric occupations. As J. Rafferty (2003, 175) has noted, the longest-duration sites (those occupied over the longest spans of time) in the Black Prairie are located "primarily near the Tombigbee River and its major tributaries." She also noted, however, that a number of shorter-duration sites are located "along major tributary streams, along small streams, and in the upland areas well away from creeks." In Mississippi, there are two marked clusters of sites containing agricultural Mississippian (AD 1000–1540) to Protohistoric (AD 1540–1750) period artifacts—one in the uplands on the western edge of the province in the vicinity of Starkville (Atkinson 1979, 1987; J. Rafferty 2001, 2003), and one farther north in the vicinity of Tupelo (Atkinson 1987, 2004; J. Johnson 1997; J. Johnson et al. 1989; J. Rafferty 1995), the latter occupations of which are associated with the historically known Chickasaw tribe. A growing body of work demonstrates that Mississippian-period settlement in the Starkville area began to occur by about AD 1200, with subsequent accumulation of relatively small, short-term, sedentary occupations on ridgetops over the next several centuries (J. Rafferty and Peacock 2008). While a diverse economy is reflected in plant and animal remains, isotopic data from human and animal bones verify that these Black Prairie sites were part of a maize-based agricultural settlement system, and it therefore is appropriate that the hundreds thus far recorded are referred to as "farmsteads" (e.g., Hogue 2003; Hogue and Peacock 1995).

Historical descriptions of the Black Belt environment make the uplands of the province seem an unlikely place for a preindustrial agricultural settlement system to become emplaced. In particular, written accounts consistently mention a lack of water during the summer months (see table 9.3; see also Rankin 1974; Tower 1961, 481), to the extent that inhabitants were forced to carry water in jugs over long distances to fill large holding basins dug at settlements (a situation that persisted well into the Historic period). Creeks in the province certainly run dry today, and even higher-order streams may present a flat, white chalk bed with only isolated pools of water (if that) during the height of summer. How can a prehistoric maize-based economy have been selected for under such circumstances?

One hypothesis is that climatic conditions were wetter around AD 1200, when agricultural occupations began to appear in the province (J. Rafferty and Peacock 2008). Under this scenario, wetter conditions (broadly coincident with the Medieval Climatic Optimum) would have charged the solum with water; seepage at the contact between the solum and underlying clay beds on ridges would have fed small "seep springs," which in turn would have provided more-or-less perennial input into streams (e.g., Gavin 1843, 68; E. Smith 1883). Evidence supporting this hypothesis comes from spring-related toponyms on the western edge of the province (Schmitz et al. 2003) and, more directly, from freshwater mussel shells recovered from a number of archaeological sites (Peacock et al. 2011). Shell morphology and chemical testing indicate that these shellfish were recovered locally and were not imported (Peacock et al. 2010, 2012), implying that Black Belt streams that today are dry in the summers carried perennial flow when the sites were inhabited in prehistoric times, a suggestion bolstered by the *Fossaria* snails already discussed.

If optimum climatic and moisture conditions around AD 1200 enabled the spread of agricultural settlement into

Table 9.2.

Land snails recovered from different microhabitats in the Mississippi Black Prairie.

Sample #	Habitat	*Carychium exile*	*Daedalochila leporina*	*Euchemotrema leai*	*Euconulus chersinus*	*Gastrocopta armifera*	*Gastrocopta contracta*	*Gastrocopta pentodon*	*Gastrocopta procera*	*Gastrocopta tappaniana*	*Gastrocopta* sp.	*Glyphyalinia* cf. *indentata*	*Glyphyalinia wheatleyi*	*Hawaiia minuscula*	*Helicodiscus parallelus*	*Lucilla* sp.	*Mesodon clausus*	*Mesomphix* cf. *globosus*	*Oligyra orbiculata*	*Punctum minutissimum*
1	UM1			2	16	9		18	1		2	24		26				1	1	4
2	CANE			1	13	1	1		2			24	2	7	2					
3	UH1											9	3							
4	BH	2					9	4	1			4		11	2	1		1		
5	um											4						3		
6	OO1				6	3						1		2	2					
7	OO2				6	10		1				10		1						
8	OO3				16	17	14		3	9		12		7			1			
9	UH2				4			37		1	3	30			4			4		1
10	UH3								1	1		2			11					
11	cg				3	2	1	1				3		3					1	
12	CG1				29		33	24	293	4		78		16		6			96	1
13	CG2				3		3	36	28		90	5		12		3			27*	
14	UH4		1					2						21						
15	P1				4	2			12			7		7		3				
16	P2								7			3					1			
17	po		4						1			1		7						
18	P3		4						76		4	1								1+
19	pine		1	1					1			3								
20	UM2		1					9	3		1	3		23						
21	P4		4						1+		1									
22	PINE										1	2								
23	play								4											
24	play								2											
25	play																			
28	play								16					1						
29	play								18					2						
30	play								60					21						
34	ich				4	30		11	157					43			8		69	
35	ich								80											
36	ich																			
40	pdd																			

* Mistakenly tabulated under ***Punctum minutissimum*** in Peacock and Gerber (2008).

\+ Added since Peacock and Gerber (2008).

++ Removes 1 ***Oligyra orbiculata*** listed in Peacock and Gerber (2008).

Table 9.2. *Continued*

Pupoides albilabris	*Stenotrema hirsutum*	*Striatura meridionalis*	*Strobilops aeneus*	*Strobilops labyrinthicus*	*Strobilops texasianus*	Succineidae	*Triodopsis hopetonensis*	*Triodopsis* sp.	*Ventridens demissus*	*Ventridens intertextus*	*Ventridens ligera*	*Ventridens* sp.	*Vertigo milium*	*Vertigo rugolusa*	*Zonitoides arboreus*	Unidentified Oxychilidae	Unidentified Polygyridae	Unidentified Gastropoda	Total
1		30											10						145
1							2	7	8										71
									10									1	23
3			2		8	1				1	4	6						1	61
									3								1	2	13
					5										1		2		22
						11													39
2						8									2		13	1	105
		21			1	209			9				2				3	3	332
1		1							3								1		21
2																			16
680		2				182	41		44									47	1576
110						6	3		14					6			2	33	381
		7	3	1											1		14		50
86						47											1	4	173
58						16											7		92
						1													14
195		1				210													492
1						1													8
1	1	4															2	11	59
12						35										1	1		55
		20																2	25++
																			4
																			2
3																			3
77						6			1										101
47						2													69
43																			124
156						28													506
267						25													372
57						6													63
						2													2

Note: No snails were recovered in samples 26, 27, 31–33, 37–39, and 41–42.

For habitats, uppercase abbreviations represent numbered samples reported in Peacock and Gerber (2008); lowercase abbreviations are samples reported here for the first time. UM—Upland Mixed Forest; CANE—Canebrake; UH—Upland Hardwood; BH—Bottomland Hardwood; OO—Osage-Orange Grove; CG—Cedar Glade; P—Prairie; Pine—Silvicultural Pine Stand; PO—Post Oak Grove; PLAY—Playground; ICH—Isolated Cedar Hummock; PDD—Pond Drainage Ditch.

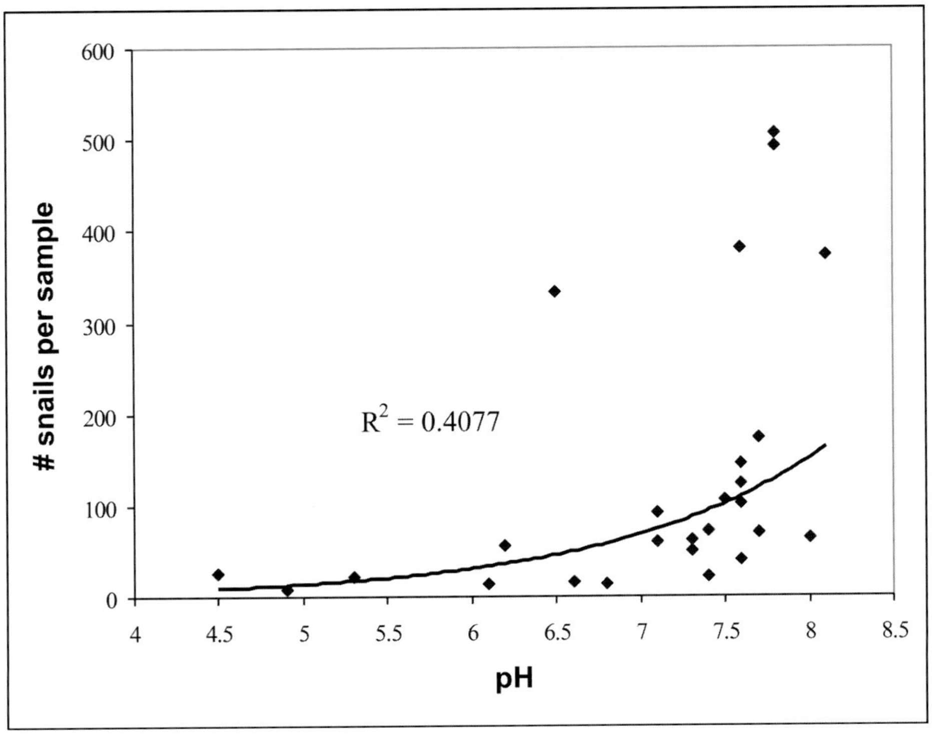

Figure 9.2. Snail species abundance versus pH in litter samples.

the Black Prairie, later drying (broadly coincident with the Little Ice Age—J. Rafferty and Peacock 2012), possibly exacerbated by the effects of soil loss resulting from Indian land clearance and agriculture (Peacock 1992; Peacock and Miller 1990), led to the rather dire conditions depicted in the historical records. If this hypothesis is correct, then the eighteenth- and nineteenth-century scenes recorded by European observers depict what were probably last-ditch efforts to maintain an economic system emplaced centuries before. Such drastic steps at settlement maintenance as those mentioned in table 9.3 suggest that even in relatively "good" climatic intervals, the Black Belt uplands likely were risky areas for long-term human settlement prior to modern times.

If it can be shown that the Black Prairie was a markedly fluctuating environment throughout the Holocene, then the area becomes an especially attractive one for evolutionary archaeologists. A variety of risk-buffering cultural practices that leave behind visible archaeological signatures may evolve in such "unstable" environments (e.g., Aranyosi 1999; Peacock and Rafferty 2013). Long-term palynological and dendroclimatological data are unavailable for the study area, but molluscan remains recovered from archaeological contexts do provide evidence related to long-term paleoenvironmental conditions in the Black Belt.

It has long been established that shell morphology (size, shape, surface sculpture) of freshwater mussels (Mollusca: Unionidae) is conditioned by local-scale environmental conditions (substrate composition, water depth, current velocity, etc.) (e.g., G. Ball 1922; Ortmann 1920; Tevesz and Carter 1980; but see also Hornbach et al. 2010), with differences between upstream and downstream phenotypes sometimes being signified by subspecific designations. The Black Belt is home to such morphologically distinct forms, some of a peculiar character. In a general discussion of phenotypic plasticity in mussels, Haag (2012, 19) noted: "In at least one case, sculpturing appears to be expressed locally in response to other habitat factors. In low-gradient, soft-bottomed streams of the Black Prairie region of east central Mississippi and west central Alabama, several species that do not elsewhere have shell sculpture (e.g., *Fusconaia cerina*, *Lampsilis straminea*, *Villosa* spp.) have a profusion of fine, concentric ridges. These ridges do not correspond to annual rings. . . . These are the only examples of concentric sculpture in North American mussels, and the reason for the occurrence here is unknown."

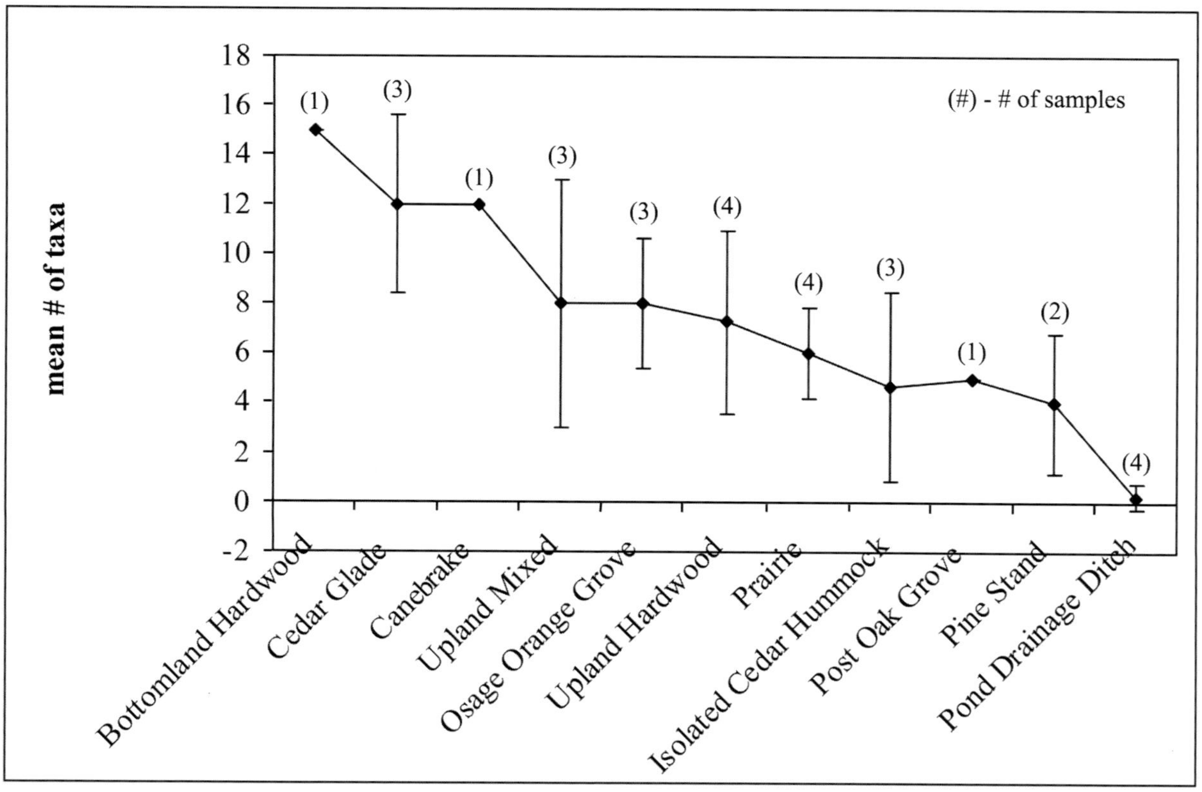

Figure 9.3. Snail species richness versus general habitat.

The occurrence of such a rare feature in more than one mussel species in tributary streams (but not in the Tombigbee River, the largest-order stream draining the Black Belt) clearly indicates that it arose because of selection rather than drift. To understand the function of these shell ridges, and to relate them to habitat conditions particular to the Black Belt, the function of similar structures in marine bivalves can be invoked. In such animals, ridging facilitates "rapid or deep burrowing. Concentric or divaricate ridges that are asymmetrical in cross section alternately abrade the sediment and serve as anchors during the rocking motion that accompanies burrowing in most species" (Haag 2012, 22). Haag further noted that "compressed and elongated shells allow faster burrowing than spherical or inflated shells by reducing drag in the sediment." The Black Belt unionid species mentioned above not only have distinct ridges but are also markedly compressed relative to their downstream counterparts (e.g., Peacock et al. 2010). Via analogy with marine bivalves, these features were selected for to facilitate burrowing. The question then becomes, what selective advantage did enhanced burrowing capability convey in Black Belt tributary streams?

Most mussel species spend most of their lives at least partly burrowed into the substrate (Haag 2012, 31). They do not burrow deeply, however, because they "lack true siphons," and stream sediments become anoxic within a few tens of centimeters (Haag 2012, 31). As with marine bivalves, burrowing is accomplished by rocking (see Haag [2012, 31] for a description of the biological actions by which this is accomplished). Although not well studied, burrowing seems to be a common response of some species to drought (Haag 2012, 108–9). In this regard, small streams are "more vulnerable to drought. Flow in large streams is sustained even during severe drought by the large size of the catchment, but small streams often dry completely or are reduced to isolated pools. Consequently, severe drought can nearly eliminate mussels from small streams, but assemblages in larger streams may be relatively unaffected" (Haag 2012, 107; see also Haag and Warren 2008).

It may also be that ridging helps keep mussels anchored in small streams when water is high. As Haag (2012, 107) noted, "Small streams are considered more unstable or variable habitats than larger streams. . . . In small streams, especially those with steep catchments, flow responds quickly

Table 9.3.

Historical descriptions of the lack of surface water in the Black Belt of Mississippi and Alabama.

Source	Description
Adair (1930, 384)	"They [the Chickasaw] have no running stream in their present settlement."
McGuire (1834, 95)	"In the summer and autumn, neither springs or wells are to be found, except below the rock."
de Richarville (1739)	"There are no rivers in any of the [Chickasaw] villages. They have only some springs where they dig wells that furnish them water."
E. Smith (1883, 266)	"In all the prairie country, the surface water is . . . often insufficient in quantity."
Roberts (1818; from Darby 1818, 133)	"The only objection to this part of the country seems to be the want of water."
Hawkins (1799; from Rostlund 1957, 401)	"There is notwithstanding these creeks a scarcity of water in the dry season, and all the creeks were dry in 1799 and not a spring of water was to be found."
Lyell (1849; from Rankin 1974, 13)	" . . . what is called prairie, or cane brake country, bare of natural wood, and where there is so great a want of water, that it was at first difficult for settlers to establish themselves upon it . . ."
Wailes (1854, 256)	"The prairie district is lamentably deficient in good water."
Gosse (1993, 80–81)	"There are prairies . . . of many miles in extent; the residents on which suffer greatly in dry seasons from the scarcity of water, a want that in a hot climate is peculiarly felt. There are no springs in the prairies, and the inhabitants depend on the rainwater, which, owing to the tenacity of the soil, does not soak into the ground, but accumulates in the hollows until evaporated by the sun."
Romans (1775, 124–25)	"This savannah at all times has but a barren look, the earth is very Nitrous, and the savages get their water out of holes or wells dug near the town . . . they live so far from waters, but they learn their children to swim in clay holes that are filled in the wet season by rain."
Nutt (1805; from Jennings 1947, 54)	"The country . . . for many miles affords good farming land, many prairies, no running water near the [Chickasaw] town, those who live in town use stagnant water and are unhealthy."

to rainfall, resulting in highly variable, flashy hydrographs, and subsequent brief floods are highly turbulent and violent because of the irregular nature of the streambed. . . . These events can radically reconfigure small streams and cause massive mussel mortality."

I hypothesize that the lower-order streams of the Black Belt, with their flat, impermeable chalk beds, may have been particularly prone to this sort of hydrological instability over geological time, and that enhanced burrowing and anchoring features were selected for under such conditions. Along these lines, it is interesting that *Corbicula fluminea*, a small introduced species with pronounced ridges, manages to persist today in very small tributary streams in the Black Belt (e.g., Catalpa Creek on Mississippi State University's South Farm, Oktibbeha County) where native mussel species do not occur (pers. obs.). I note, however, that Haag and Warren (2008) found that drought had a relatively more negative impact on *C. fluminea* than on native species in small streams on the Bankhead National Forest, Alabama.

Anthropogenic Alterations

The shells of freshwater mussels (Mollusca: Bivalvia: Unionidae) are common in archaeological sites throughout the

Black Belt, with abundance reflecting past availability (e.g., there is far more shell at sites on the main stem of the Tombigbee River than on lower-order streams) and preservation (generally reflecting soil pH, which is also affected by the amount of shell deposited per unit volume of soil). As of this writing, mussel shells have been analyzed from 19 sites in the Mississippi Black Prairie, and a minimum of 42 taxa have been recorded. (Peacock 2002, n.d.; Peacock and Seltzer 2008; Peacock et al. 2011). A number of archaeological sites in the Black Belt in Alabama have also produced mussel assemblages that have been analyzed (see Peacock 1998; Peacock et al. 2012). The data from all Tombigbee River drainage sites together provide an unparalleled look at mussel faunas at the drainage-basin scale before modern impacts (Peacock 2002; Peacock et al. 2012; Peacock et al. 2014). Quantified data (both original and those compiled from earlier sources) are presented in Peacock et al. (2012), and the discussions of relative abundance that follow derive from those data.

Prehistorically, the dominant species in the main river stem in Mississippi included *Pleurobema decisum*, a currently endangered species (see R. Jones et al. 2005), *Elliptio crassidens* and *E. arca*, and *Quadrula asperata*. *Epioblasma penita*, another currently endangered species, was also relatively common. Species that were rare prehistorically included *Plectomerus dombeyanus*, *Quadrula apiculata*, *Q. metanevra*, *Villosa lienosa*, *Strophitus subvexus*, *Pleurobema marshalli*, *Leptodea fragilis*, *Toxolasma parvum*, *Potamilus purpuratus*, *Megalonaias nervosa*, *Lampsilis straminea claibornensis*, *L. perovalis*, and *Ligumia recta*. The low numbers of *Leptodea fragilis* and *Toxolasma parvum* may represent preservation bias against small and thin-shelled species (but see Peacock et al. 2012). Although proportions changed considerably, a list of taxa recorded in the main stem of the Tombigbee prior to impoundment (see table 9.4) shows nearly complete overlap with archaeological records (see also Robison 1983). The main-river fauna has since been "nearly obliterated" by impoundment (Watters 2000, 263; see also J. Williams et al. 1992).

Tributary streams have also been severely negatively impacted in modern times, as comparisons of archaeological and modern faunas from Line Creek and Tibbee Creek (table 9.4) show. The large archaeological samples available from both main river and tributary locales (see Peacock et al. 2012) provide valuable reference data for stream rehabilitation efforts.

Other Molluscan Faunas

Even among the Mollusca, small, freshwater "fingernail," "pea," or "pill" clams (Sphaeriidae) have been little studied. These short-lived (three to five years) creatures are represented by three genera (*Musculium*, *Sphaerium*, and *Pisidium*). While they are occasionally reported from archaeological contexts (e.g., Rudolph 1983), it is assumed, because of their small-to-minute size (1–25 mm in length, average less than 6 mm), that they were not intentionally gathered for food or other uses by Indians but represent incidental transport and inclusion into the archaeological record. Alternatively, they may represent episodes of flooding at a particular site.

Early twenty-first-century excavations at the Lyon's Bluff site (state site number 22OK520) in the Black Prairie province in Oktibbeha County, Mississippi, provided what is, to my knowledge, a unique opportunity to document a Sphaeriidae community as it existed prior to Historic-period alterations of streams. Lyon's Bluff is a large, palisaded village site occupied from about AD 1200 to 1650 (Peacock and Hogue 2005). Geophysical (magnetic gradiometer) survey revealed the remains of several rectangular structures beneath the plow zone, and excavations showed that sand floors were a common feature within those structures. The sand, presumably obtained from the immediately adjacent Line Creek (see fig. 9.1), a tributary of Tibbee Creek, itself a major tributary of the Tombigbee River, was spread in a thin layer inside houses. Each such feature investigated contained large numbers of fingernail clams. Shovel tests placed on a grid over the main portion of the site produced many more specimens; geological screening of sediments showed that these were coincident with concentrations of sand and undoubtedly marked the location of house floors that have now been plowed through (Bierly 2008). Between excavation and shovel testing, hundreds of specimens have been recovered from the site. To date, no taxonomic work has been done with these specimens, which are available for study by interested scholars.

Aquatic snails are also routinely recovered from archaeological contexts (e.g., J. Clark 1969). These represent both food remains (especially the larger species of genera of Vivi-

TABLE 9.4.

Presence/absence of freshwater mussel species in the Tombigbee River drainage, Mississippi.

	Main Stem Tombigbee River			Line Creek		Tibbee Creek	
	Prehistoric	1935	1972–75	AD 1200–1650	AD 1994–2000	ca. AD 1480	AD 1997
Amblema plicata	X	X	X	X		X	X
Arcidens confragosus							X
Ellipsaria lineolata	X	X	X			X	
Elliptio arca	X	X	X	X		X	
Elliptio arctata	X	X				X	
Elliptio crassidens	X	X	X	X		X	X
Epioblasma penita	X	X	X			X	
Fusconaia cerina	X	X	X	X		X	
Fusconaia ebena	X	X	X			X	
Hamiota perovalis	X			X			
Lampsilis ornata	X	X	X	X		X	
Lampsilis straminea claibornensis	X		X	X		X	
Lampsilis straminea straminea				X	X		X
Lampsilis teres	X	X	X	X		X	X
Lampsilis sp.				X			
Lasmigona complanata alabamensis				X			
Leptodea fragilis	X	X	X			X	X
Ligumia recta	X	X	X	X		X	
Medionidus acutissimus				X			
Medionidus mcglameriae		X					
Megalonaias nervosa	X	X	X	X		X	X
Obliquaria reflexa	X	X	X	X		X	X
Obovaria jacksonia/unicolor	X	X	X	X			X
Obovaria sp.						X	
Plectomerus dombeyanus	X	X	X				X
Pleurobema decisum	X		X	X		X	
Pleurobema marshalli	X	X	X				
Pleurobema perovatum	X			X			
Pleurobema taitianum	X	X	X				
Pleurobema sp.						X	
Potamilus inflatus		X					
Potamilus purpuratus	X	X	X	X		X	X
Pyganodon grandis							X

Table 9.4. *Continued*

	Main Stem Tombigbee River			Line Creek		Tibbee Creek	
	Prehistoric	1935	1972–75	AD 1200–1650	AD 1994–2000	ca. AD 1480	AD 1997
Quadrula asperata	X	X	X	X		X	X
Quadrula metanevra	X	X	X			X	
Quadrula rumphiana/apiculata	X	X	X	X		X	X
Quadrula stapes	X		X			X	
Quadrula verrucosa	X	X	X	X		X	X
Strophitus subvexus				X		X	
Strophitus sp.	X	X	X	X			X
Toxolasma parvum	X	X		X			X
Truncilla donaciformis	X	X	X			X	
Uniomerus tetralasmus				X			
Uniomerus sp.	X						
Utterbackia imbecillis					X		
Villosa lienosa	X	X	X	X			
Villosa vibex			X	X			
Villosa sp.				X		X	

Source: Data from Hanley (1982); R. Jones et al. (1997); McGregor (2000); Peacock et al. (2011); van der Schalie (1939); J. Williams et al. (1992). Modern data from main river stem are from preimpoundment survey.

paridae and Pleuroceridae) and incidental inclusions in archaeological deposits (e.g., the *Fossaria* mentioned previously). Aquatic snails have been recovered from a number of sites in the Black Belt, but no detailed analysis of these remains has yet been undertaken. Because of the current poorly developed state of systematics, taxonomy, and conservation foci where freshwater gastropods are concerned (Lysne et al. 2008; Perez and Minton 2008), analysis of large archaeological collections by experts would potentially be useful.

Discussion and Conclusions

Environmental archaeology is, by its very nature, interdisciplinary, an approach that brings enormous promise for better understanding of past relationships between culture and nature and the application of zooarchaeological data to modern management efforts. As Peacock and Gerber (2008) noted, however, the challenges are considerable, and collaboration with specialists is needed because most archaeologists are not formally trained in the biological sciences beyond some basic course work. The data on archaeological and modern mollusks reported here and in earlier works represent an enormous investment of time and repeated checking of identifications with true experts in the field of malacology. Archaeological assemblages may be biased in various ways, but the problem this poses for the application of zooarchaeological data is routinely overstated; investigating and compensating for such biases is relatively straightforward (Peacock 2000, 2012) and is, in fact, a standard practice in zooarchaeology (Lyman 1994; Peacock et al. 2012). Because of the enormous numbers of molluscan remains that have been, and remain to be, recovered from archaeological contexts, there is no reason that collaboration between archaeologists and natural scientists should not provide clear targets for environmental reconstruction efforts in the Black Prairie and elsewhere.

Acknowledgments

The modern snail samples discussed in this chapter were obtained via a National Science Foundation grant (Grant No. 0003833). I gratefully acknowledge the guidance of many true experts, including the late Paul Parmalee, Jochen Gerber, Wendell Haag, Bob Jones, and Paul Hartfield: their patience and mentorship have meant a lot to me over the years. Jeffrey Alvey and Janet Rafferty provided assistance with preparing this chapter. I also thank JoVonn Hill for inviting me to participate in the excellent symposium that led to this volume.

10

The Grasshopper Fauna of Southeastern Grasslands

A Preliminary Investigation

JoVonn G. Hill

We came to a large prairie, that was about six miles across it, and in this I saw the trail which I knowed was made by bear, deer, and turkeys. We went on through it till we came to a large creek, and the low grounds were all set over with wild rye, looking as green as a wheat field.

—Quote from David Crockett describing a grassland landscape near the headwaters of the Choctawhatchee River in southern Alabama in 1815. Originally quoted in Crockett (1987) and annotated by Shackford and Folmsbee (1973).

When most people think of the historical landscape of the South, they likely think of vast forests and swamps stretching from the mountains to the coast. However, large areas of the Southeast were once covered by natural grasslands of various types, such as prairies, glades, savannas, marshes, sandhills, rock outcrops, and mountaintop balds (fig. 10.1).

For example, longleaf pine savannas covered approximately 89,600,000 ac (36,259,833 ha) from Virginia to eastern Texas (Longleaf Alliance 2013), and prairies were known to cover at a minimum 4,323,591 ac (1,749,695 ha) in Alabama, Arkansas, Florida, Louisiana, and Mississippi (Barone 2005a, 2005b, chapter 2, this volume; Barone unpublished data; K. Stephenson 2011). These grasslands contribute significantly to the biota of the region; for example, 927 plant taxa are endemic to the longleaf pine ecosystem alone (Sorrie 2011), and Black Belt Prairies support more than 1,160 species of moths (R. L. Brown 2003, unpublished data.). However, natural grasslands in the Southeast have decreased greatly since European settlement, and many are now considered highly imperiled as a result of anthropogenic disturbances, existing only as small remnants of what were once dominant landscape features (R. Anderson et al. 1999; Noss 2013). Because of their rich biodiversity and conservation importance, grasslands have been emphasized in an ongoing study of the grasshopper fauna of the southeastern United States, which is defined here to include Alabama, Arkansas, Florida, Georgia, Louisiana, Mississippi, North Carolina, South Carolina, and Tennessee. This definition is used largely for practical purposes, but also because the defined area has an identifiable fauna that can be distinguished from that of other regions.

In my opinion (and I admit that I am a bit biased), a book detailing the natural history of grasslands would be quite deficient if there were no mention of grasshoppers. I have been studying the grasshopper fauna of grasslands and other naturally open habitats in the Southeast for the last 13 years and have found that grasshoppers are often one of the most conspicuous members of the invertebrate fauna in these communities. Grasshoppers are relatively large (for an insect), are often very abundant in grasslands, and dazzle the eyes and ears with their colorful and crackling flight displays adding to the ambience of the habitat.

Figure 10.1. Landscape photographs of various southeastern grasslands: (A) Black Belt Prairie in Chickasaw County, Mississippi; (B) cedar glade in Davidson County, Tennessee; (C) longleaf pine savanna in Thomas County, Georgia; (D) sedge-dominated wetland, Bibb County, Alabama; (E) riverine sand dune in Emmanuel County, Georgia; (F) rock outcrop in Stone County, Arkansas.

Additionally, grasshoppers are important components of temperate grasslands, as they are often functionally dominant herbivores in grassland ecosystems (Joern 1979; Otte 1981; Branson et al. 2006), affect nutrient cycling (Belovsky and Slade 2000), are important food items for vertebrate and invertebrate predators (W. Hamilton and Pollack 1961; Cherrill and Begon 1989; Fowler et al 1991), and are vectors of some vertebrate diseases and parasites (J. Hill and Goddard 2012). In addition, some species are restricted to remnants of habitats that were once expansive, such as natural grasslands, and the reduction of these habitats raises concern about the continued existence of these species (Panzer et al. 2010). In writing on the importance of species and habitat conservation, Aldo Leopold (1953) stated that "to keep every cog and wheel is the first precaution of intelligent tinkering." Because grasshoppers play varied and important roles, it could be argued that they are a vital cog in the mechanism of grassland function. For this reason, I feel there is a strong basis for the inclusion of grasshoppers in biological inventories and studies of grassland ecosystem function.

Studies of the southeastern grasshopper fauna started in earnest with a paper by Samuel Scudder (1878) on the Orthoptera collected in Florida by J. H. Comstock. Thereafter, an impressive list of orthopterists contributed to knowledge of the regional fauna, including John Capinera (and students), Matt Dakin, J. J. Friauf, Morgan Hebard, Theodore Hubbell, J. McNeil, A. P. Morse, Dan Otte, and James Rehn. Though most of their work is descriptive in nature, enough conclusions can be drawn from it and some of my own work to (1) explore the uniqueness of the southeastern grasshopper fauna in relation to the North American fauna; (2) examine whether the grasshopper fauna of different types of grasslands in the region varies; and (3) explore and speculate on any biogeographic affinities of the fauna. However, the information presented here should be viewed as preliminary only, as a more thorough treatment awaits a firmer taxonomic framework (primarily of the Melanoplinae) and more detailed studies of the distribution, ecology, and phylogenetics of many of the southeastern species. Thus, again I stress that this chapter should be considered only a starting place, not the final word on this topic.

Here I use the definition of a grassland employed by Noss (2013) in his treatment of southeastern grasslands to include any natural community or ecosystem in which the herbaceous layer is dominated by grasses, other graminoid plants such as sedges, and associated forbs. This does not exclude the presence of trees, but trees should not shade out the herbaceous layer. Grasslands in the Southeast encompass many habitat types, such as prairies, glades, rock outcrops, sandhills, savannas, scrub, and several types of wetlands. The term "grasshopper" is commonly used to refer to various members of the order Orthoptera. Here I use the term to refer to species of "short-horned grasshoppers" and "lubber grasshoppers" belonging to the families Acrididae and Romaleidae, respectively. The Tettigoniidae, which are often called katydids, meadow grasshoppers, or long-horned grasshoppers, are likely an important group regionally, but they are not considered here.

The Southeastern Fauna

To answer the first question regarding the uniqueness of the grasshopper fauna of southeastern grasslands, one must examine the regional fauna as a whole and then examine which species inhabit grasslands. Data on grasshoppers associated with southeastern grasslands come largely from species accounts in numerous taxonomic works from the region, several faunistic studies, and my own personal collections. To be considered a grassland inhabitant, a species must be primarily associated with at least one of the habitats just described. Several taxa considered here as grassland inhabitants, such as *Syrbula admirabilis* (Uhler), are often associated with grasslands of anthropogenic origin, such as old fields or roadsides; however, these species are also found in naturally occurring grasslands and have been able to colonize and adapt to these new habitats. Thus, these species are still considered grassland inhabitants. Intuitively, it makes sense that grasshoppers would be associated with grasslands. However, several species groups, such as the *Melanoplus tribulus* and *M. nigrescens* groups, are near endemics or have their center of diversity in the Southeast, but they are denizens of forests, woodlands, and forest-grassland interfaces. Further, some species, such as *Romalea microptera*, are habitat generalists, meaning they occupy a wide variety of habitats, including forests, swamps, and various types of grasslands. Species that fit this category were not included in the grassland inhabitant list.

Grasshopper taxonomy is currently undergoing rapid changes as new morphological and molecular approaches

Table 10.1.

Proportion of North American and southeastern grasshoppers that occur in southeastern grasslands, are endemic to the Southeast, are endemic to natural grasslands in the Southeast, and are longleaf endemics.

	Number of species	% of North American fauna	% of southeastern fauna
North American fauna	751*	100	
Occur in the Southeast	211	28	100
Occur in open habitats in the Southeast (natural and disturbed)	173	23	82
Endemic to the Southeast	130	17	62
Endemic to natural grasslands of the Southeast	111	15	53
Endemic to longleaf pine system	71	10	34

*Acrididae and Romaleidae.

are being applied in revisions, but based on current taxonomy approximately 751 grasshopper taxa (Acrididae and Romaleidae) are documented from North America north of Mexico (Eades et al. 2013). Of these, 211 taxa (28% of the North American fauna) occur in the southeastern United States. Furthermore, more than half of the regional fauna, 130 taxa (62% of the southeastern fauna and 17% of the North American fauna), are endemic or near endemic to the Southeast, and a significant proportion (82%) of the endemic southeastern fauna is associated with grasslands of various types.

Thus, it appears that the natural grasslands of the southeastern United States are responsible for the majority of the grasshopper diversity in the region. Notably, 71 species (34% of the regional fauna and 56% of the regional endemics) are associated with longleaf pine savannas/sandhills, including the five southeastern endemic genera, *Aptenopedes*, *Floridacris*, *Floritettix*, *Eotettix*, and *Gymnoscirtetes* (see table 10.2).

Most of the endemism is found in the subfamily Melanoplinae, which includes many brachypterous (short-winged) or apterous (wingless) species that are flightless and are restricted to specific habitats, which often results in many of the regional endemics having small geographic distributions. For example, *Melanoplus dakini* Hilliard (fig. 10.2A) is known from only three parishes in Louisiana, where it inhabits open pine forests with a grassy understory; *Melanoplus ingrami* Hill (figs.10.2B and 10.5B) is endemic to the cedar glades in the Central Basin of Tennessee; *Melanoplus stegocercus* Rehn and Hebard (fig. 10.2C) is known only from dune or sandhill sites in east-central Georgia; and *Melanoplus pygmaeus* Davis (fig. 10.2D) is found only in sandhill/dune habitats in southern Alabama and northwestern Florida.

Two well-studied species groups containing short-winged, grassland-inhabiting species in the region are the *Melanoplus puer* species group and the *Melanoplus scudderi* species group (*sensu lato*). The *Melanoplus puer* species group consists of smaller flightless species that inhabit the sandhill/scrub habitats of sand ridges in peninsular Florida and longleaf pine ecosystems (Hubbell 1932; Folkerts et al. 1993; Squitier and Capinera 2002; Otte 2012). The sand ridges inhabited by these flightless grasshoppers are relatively elevated compared to most of Florida and were likely islands during periods of higher sea levels in the past. Although sea levels are much lower now, these grasshoppers are still isolated from each other by areas of habitats that are unsuitable to them, such as oak hammocks, swamps, and agricultural areas. As a result of its interesting biogeographic relationships, this group of grasshoppers has been used as a model of evolutionary divergence, serving as one of the strongest pieces of evidence of the relative isolation of various sand ridge systems in peninsular Florida (Deyrup 1996). The *Melanoplus scudderi* species group (*sensu lato*) contains several species with interesting distributions in the region, including five species that are endemic to sandhills along the Fall Line in Georgia and the Carolinas (fig. 10.3). Of these five species, two are largely sympatric and

Table 10.2.

List of grasshopper species that occur in the southeastern United States.

Species	SE endemic	Grassland specialist	Endemic and grassland specialist	Longleaf endemic
Romaleidae				
Romalea microptera (Beauvois)	X			
Melanoplinae				
Aptenopedes apalachee Hebard	X	X	X	X
Aptenopedes chefixico Otte	X	X	X	X
Aptenopedes chiaha Otte	X	X	X	X
Aptenopedes clara Rehn	X	X	X	
Aptenopedes menawai Otte	X	X	X	X
Aptenopedes rufovittata Scudder	X	X	X	X
Aptenopedes sphenarioides Scudder	X	X	X	X
Aptenopedes sphenax Otte	X	X	X	X
Aptenopedes yoholoi Otte	X	X	X	X
Booneacris variegata (Scudder)		X		
Campylacantha olivacea (Scudder)		X		
Dendrotettix australis (Morse)	X			
Dendrotettix quercus Riley				
Dendrotettix zimmermanni (Saussure)	X			
Eotettix hebardi Rehn	X	X	X	X
Eotettix palustris Morse	X	X	X	X
Eotettix pusillus Morse	X	X	X	X
Eotettix signatus Scudder	X	X	X	X
Floridacris etami Otte	X	X	X	X
Floridacris mariona Otte	X	X	X	X
Floridacris orangina Otte	X	X	X	X
Floridacris robusta (Hebard)	X	X	X	
Floritettix aptera (Scudder)	X	X	X	X
Floritettix borealis (Hebard)	X	X	X	X
Floritettix calusa Otte	X	X	X	X
Floritettix coquinae (Hebard)	X	X	X	
Floritettix floridana Otte	X	X	X	X
Floritettix hadjoi Otte	X	X	X	X
Floritettix holatamico Otte	X	X	X	X
Floritettix hubbelli Otte	X	X	X	X
Floritettix ocilla Otte	X	X	X	X
Floritettix osceola Otte	X	X	X	X

Continued on the next page

Table 10.2. *Continued*

Species	SE endemic	Grassland specialist	Endemic and grassland specialist	Longleaf endemic
Floritettix nigropicta (Hebard)	X	X	X	X
Floritettix saturiba (Hebard)	X	X	X	X
Floritettix simplex (Hebard)	X	X	X	X
Gymnoscirtetes morsei Hebard	X	X	X	X
Gymnoscirtetes pusillus Scudder	X	X	X	X
Hesperotettix floridensis Morse	X	X	X	X
Hesperotettix gemmicula Hebard	X	X	X	X
Hesperotettix osceola Hebard	X	X	X	X
Hesperotettix viridis brevipennis (Thomas)		X		
Hesperotettix viridis pratensis Scudder		X		
Melanoplus acidocercus Hebard	X	X	X	X
Melanoplus acrophilus Hebard	X	X	X	
Melanoplus adelogyrus Hubbell	X	X	X	
Melanoplus alabamae Hebard	X	X	X	
Melanoplus angustipennis (Dodge)		X		
Melanoplus apalachicolae Hubbell	X	X	X	
Melanoplus arkansas Hill	X	X	X	
Melanoplus attenuatus Scudder	X	X	X	
Melanoplus australis Morse	X	X	X	X
Melanoplus baronei Hill	X	X	X	
Melanoplus bispinosus Scudder		X		
Melanoplus bivittatus (Say)		X		
Melanoplus bonita Otte	X	X	X	X
Melanoplus cantralli Dakin	X	X	X	X
Melanoplus carabellae Otte	X	X	X	
Melanoplus carnegiei Morse	X	X	X	
Melanoplus carolinensis Hill	X	X	X	X
Melanoplus celatus Morse	X	X	X	
Melanoplus chattahoochee Hill	X	X	X	X
Melanoplus cherokee Hebard	X			
Melanoplus childsi Otte	X	X	X	X
Melanoplus clypeatus (Scudder)	X	X	X	
Melanoplus confusus Scudder	X	X	X	
Melanoplus coreyi Hill	X	X	X	X
Melanoplus dakini Hilliard	X	X	X	
Melanoplus davisi Hebard	X	X	X	X
Melanoplus deceptus Morse	X			X

Table 10.2. *Continued*

Species	SE endemic	Grassland specialist	Endemic and grassland specialist	Longleaf endemic
Melanoplus decoratus Morse	X			X
Melanoplus decorus Rehn and Hebard	X	X	X	
Melanoplus delaware Hebard				
Melanoplus devius Morse	X			
Melanoplus differentialis differentialis (Thomas)				
Melanoplus differentialis nigricans Cockerell				
Melanoplus divergens Morse	X	X	X	
Melanoplus eurycercus Hebard				
Melanoplus fasciatus (Walker)		X		
Melanoplus femurrubrum (DeGeer)		X		
Melanoplus foedus Scudder		X		
Melanoplus folkertsi Hill	X	X	X	
Melanoplus forcipatus Hubbell	X	X	X	X
Melanoplus foxi Hebard	X	X	X	X
Melanoplus furcatus Scudder	X	X	X	
Melanoplus gracilis (Bruner)				
Melanoplus gurneyi Strohecker	X	X	X	X
Melanoplus harrisi Hill	X			
Melanoplus hubbelli Hebard				
Melanoplus impudicus Scudder		X		
Melanoplus inconspicuus Caudell		X		
Melanoplus indicifer Hubbell	X	X	X	X
Melanoplus ingrami Hill	X	X	X	
Melanoplus keeleri keeleri (Thomas)		X		
Melanoplus keeleri luridus (Dodge)		X		
Melanoplus kissimmee Otte	X	X	X	X
Melanoplus longicornis Saussure	X	X	X	
Melanoplus mirus Rehn and Hebard	X	X	X	
Melanoplus mississippi Hill		X		
Melanoplus morsei Blatchley				
Melanoplus muscogee Hill	X	X	X	X
Melanoplus nanciae Deyrup	X	X	X	X
Melanoplus nigrescens (Scudder)	X			
Melanoplus nossi Hill	X			
Melanoplus nubilus Rehn and Hebard	X	X	X	X
Melanoplus ordwayae Deyrup	X	X	X	

Continued on the next page

Table 10.2. *Continued*

Species	SE endemic	Grassland specialist	Endemic and grassland specialist	Longleaf endemic
Melanoplus ouachita Hill	X	X	X	
Melanoplus ozarkensis Hill		X		
Melanoplus pachycercus Hebard	X	X	X	
Melanoplus packardii Scudder		X		
Melanoplus peatus Otte	X	X	X	
Melanoplus pegasus Hebard	X	X	X	
Melanoplus peninsularis Hubbell	X	X	X	
Melanoplus ponderosus (Scudder)	X	X	X	
Melanoplus primestivus Dakin	X			
Melanoplus propinquus Scudder	X	X	X	
Melanoplus puer (Scudder)	X	X	X	X
Melanoplus punctulatus arboreus Scudder				
Melanoplus punctulatus punctulatus (Scudder)				
Melanoplus pygmaeus Davis	X	X	X	X
Melanoplus quercicola Hebard	X	X	X	X
Melanoplus querneus Rehn and Hebard	X			
Melanoplus relictus Hill	X	X	X	X
Melanoplus rotundipennis (Scudder)	X	X	X	X
Melanoplus rusticus obovatipennis (Blatchley)				
Melanoplus sanguinipes vulturnus (Gurney and Brooks)		X		
Melanoplus savannah Hill	X	X	X	X
Melanoplus scapularis Rehn and Hebard	X	X	X	X
Melanoplus scudderi (Uhler)		X		
Melanoplus sebringi Otte	X	X	X	X
Melanoplus seltzeri Hill	X	X	X	
Melanoplus seminole Hubbell	X	X	X	X
Melanoplus serrulatus Hebard	X	X	X	
Melanoplus similis Morse	X			
Melanoplus stegocercus Rehn and Hebard	X	X	X	X
Melanoplus strumosus Morse	X	X	X	X
Melanoplus sylvaticus McNeil	X			
Melanoplus sylvestris Morse	X			
Melanoplus symmetricus Morse	X	X	X	
Melanoplus tepidus Morse	X			
Melanoplus tequestae Hubbell	X	X	X	X
Melanoplus texarkana Hill	X	X	X	X
Melanoplus tribuloides Morse	X			
Melanoplus tribulus Morse				

TABLE 10.2. *Continued*

Species	SE endemic	Grassland specialist	Endemic and grassland specialist	Longleaf endemic
Melanoplus tumidicercus Hubbell	X	X	X	X
Melanoplus tunicae Hebard	X			
Melanoplus viola Thomas		X		
Melanoplus vulnus Eades	X	X	X	
Melanoplus walshii Scudder				
Melanoplus withlacoocheensis Squitier and Deyrup	X	X	X	X
Melanoplus zento Otte	X	X	X	
Paratylotropidia beutenmuelleri Morse	X			
Paratylotropidia brunneri Scudder		X		
Paratylotropidia morsei Rehn and Rehn		X		
Paroxya atlantica Scudder		X		
Paroxya clavuliger (Serville)		X		
Paroxya hoosieri (Blatchley)		X		
Paroxya paroxyoides (Scudder)	X	X	X	
Paroxya recta Scudder	X	X	X	
Leptysminae				
Leptysma marginicollis (Serville)		X		
Stenacris vitreipennis (Marschall)		X		
Cyrtacanthacridnae				
Schistocerca alutacea (Harris)		X		
Schistocerca americana (Drury)		X		
Schistocera ceratiola Hubbell and Walker	X	X	X	X
Schistocerca damnifica (Saussure)		X		
Schistocerca lineata Scudder		X		
Schistocerca obscura (Fabricius)		X		
Schistocerca rubiginosa (Harris)	X	X	X	
Gomphocerinae				
Achurum carinatum (Walker)	X	X	X	
Ageneotettix deorum Scudder		X		
Amblytropidia mysteca (Saussure)		X		
Boopedon auriventris McNeil		X		
Chloealtis conspersa Harris				
Chorthippus curtipennis (Harris)		X		
Dichromorpha elegans (Morse)	X	X	X	
Dichromorpha viridis (Scudder)		X		
Eritettix obscurus (Scudder)	X	X	X	X

Continued on the next page

Table 10.2. *Continued*

Species	SE endemic	Grassland specialist	Endemic and grassland specialist	Longleaf endemic
Eritettix simplex (Scudder)		X		
Mermiria bivittata (Serville)		X		
Mermiria intertexta Scudder	X	X	X	
Mermiria picta (Walker)		X		
Orphulella pelidna (Burmeister)		X		
Orphulella speciosa (Scudder)		X		
Pseudopomala brachyptera (Scudder)		X		
Stethophyma celata Otte		X		X
Syrbula admirabilis (Uhler)		X		
Acridinae				
Metaleptea brevicornis (Johannson)		X		
Oedipodinae				
Arphia granulata (Saussure)	X	X	X	X
Arphia simplex Scudder		X		
Arphia sulphurea (Fabricius)		X		
Arphia xanthoptera (Burmeister)		X		
Chortophaga austrailor Rehn and Hebard	X	X	X	X
Chortophaga viridifasciata (DeGeer)		X		
Dissosteira carolina (Linnaeus)		X		
Encoptolophus soridus (Burmeister)		X		
Hippiscus ocelote (Saussure)		X		
Pardalophora apiculatus (Harris)		X		
Pardalophora phoenicoptera (Burmeister)		X		
Psinidia fenestralis (Serville)		X		
Spharagemon bolli Scudder				
Spharagemon crepitans (Saussure)	X	X	X	
Spharagemon cristatum (Scudder)		X		
Spharagemon equale (Say)		X		
Spharagemon marmorata picta (Scudder)		X		
Trachyrhachys kiowa (Thomas)		X		
Trimerotropis maritima (Harris)		X		
Trimerotropis saxatalis McNeil		X		
Totals: 211 species	130	173	111	71

Source: Compiled from J. Hill and Dakin (2011); Otte (2012); J. Hill (2014); and Otte (2014).

Note: Xs indicate that a species is endemic to the region, specializes in grasslands or open habitats, or is endemic to either grassland or longleaf pine savanna ecosystems. Taxonomy follows Eades et al. (2013).

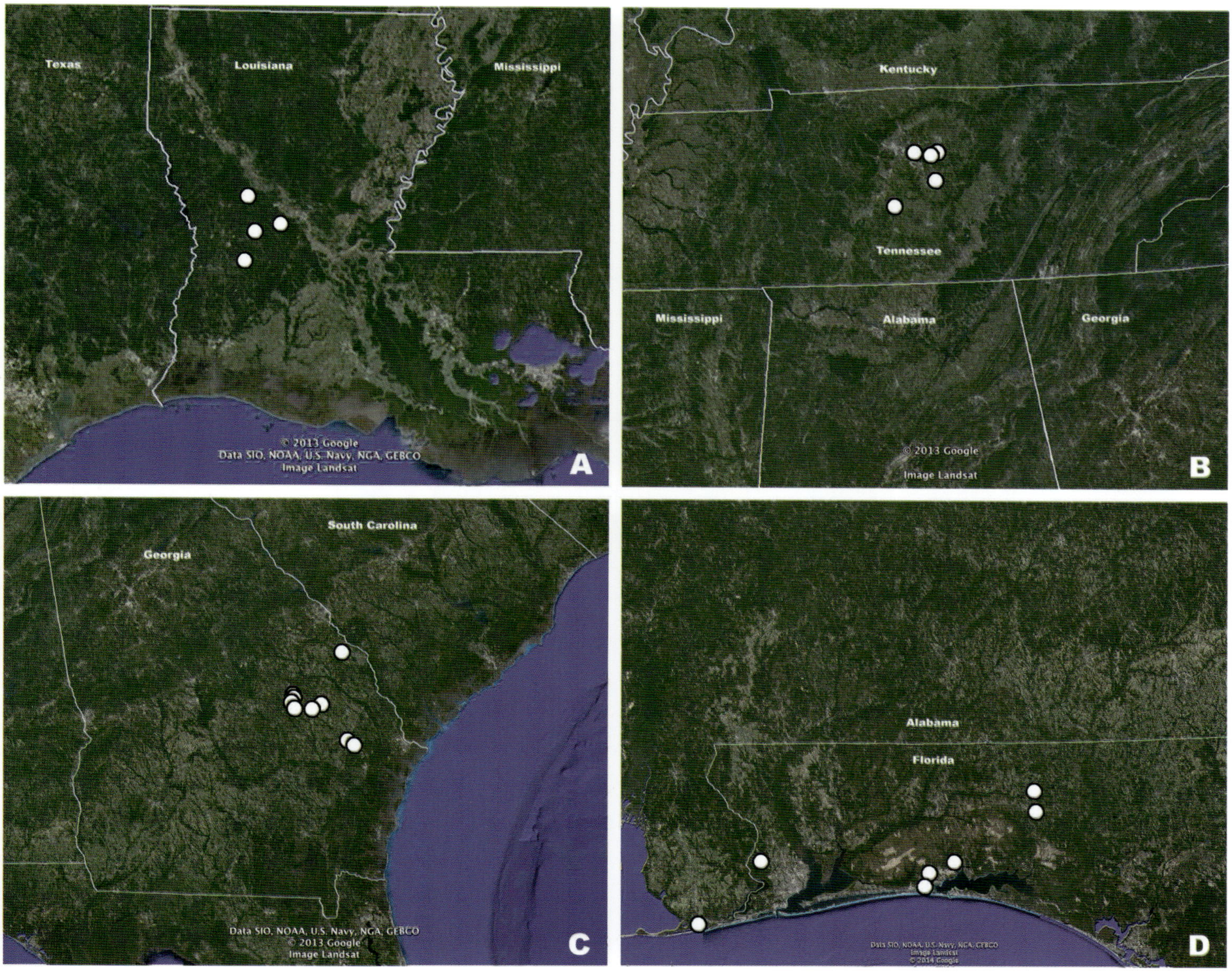

Figure 10.2. Restricted distributions of some flightless grasshoppers in the Southeast: (A) *Melanoplus dakini*; (B) *M. ingrami*; (C) *M. stegocercus*; (D) *M. pygmaeus*. Produced in Google Earth Pro.

the rest are allopatric, with their ranges being segregated by large river systems of the region (J. Hill 2015). Additionally, several other species in the *scudderi* group are associated with longleaf pine habitats on the Coastal Plain.

Affinities with Western Grasslands

Different grassland types within the southeastern region support distinct biotas that have biogeographic affinities with the grasslands of the western United States. Whereas the biota of the southeastern prairies is associated more with the tallgrass prairie, the biota of the sandhill habitats is associated more with the arid habitats of the southwestern United States. The sandhills of the Southeast often contain one or two species of a diverse western group of congeners, which suggests that the eastern species are derived from the western group (Folkerts et al. 1993). This pattern has been documented for numerous other taxa, both extant and extinct (see Noss 2013 for a summary). This relationship has been explained by a possible corridor of savanna habitat that developed during the Miocene and persisted into the early Pleistocene along the coast of the Gulf of Mexico. This corridor allowed an exchange of xeric-adapted animals be-

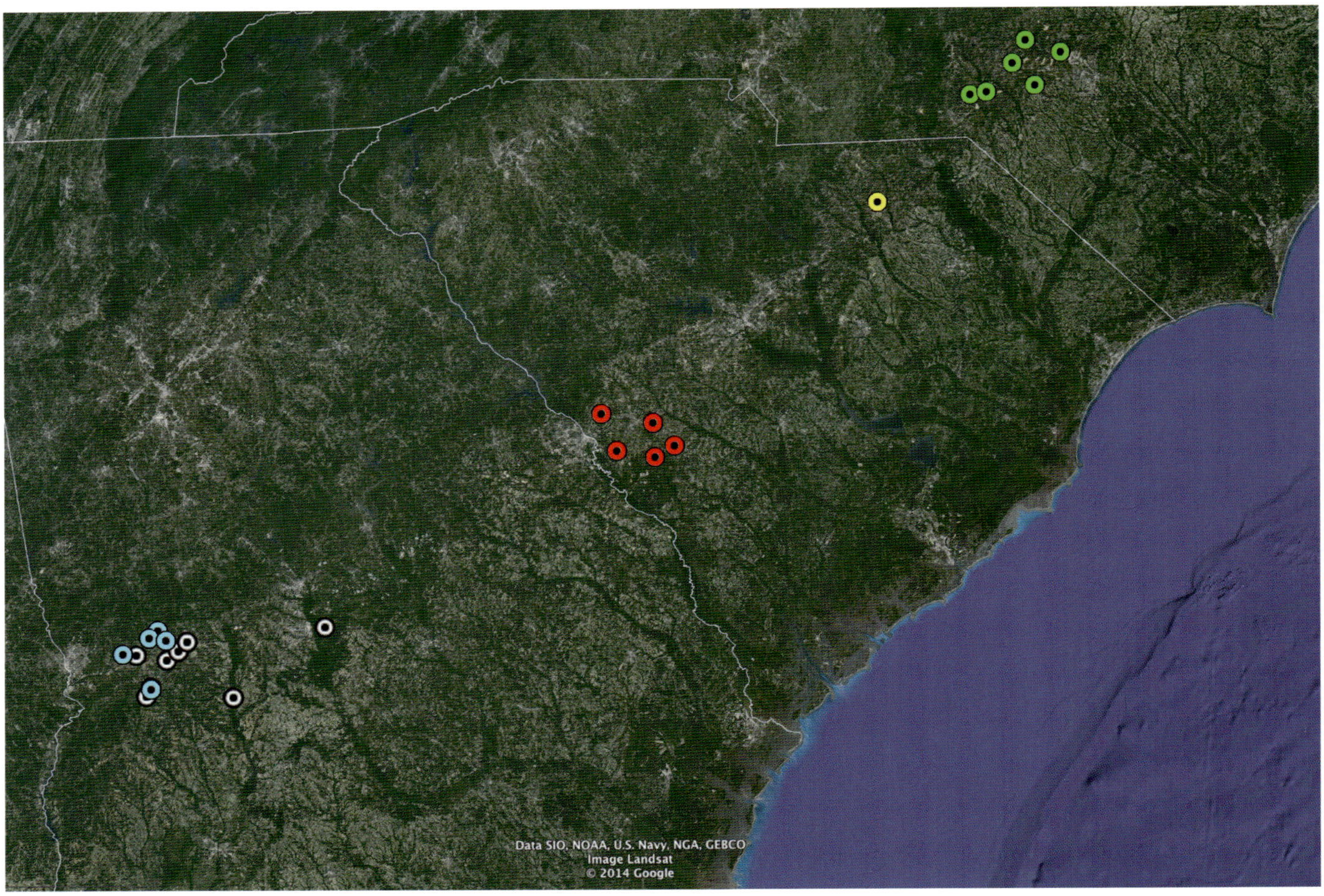

Figure 10.3. Members of the *Melanoplus scudderi* group (s.l.) endemic to sandhills along the Fall Line: white circles = *M. muscogee*, blue circles = *M. chattahoochee*, red circles = *M. savannah*, yellow circle = *M. carolinensis*, and green circles = *M. coreyi*. Produced in Google Earth Pro.

tween the semiarid West and the Southeast (S. Webb 1977; Emslie and Czaplewski 1999; Noss 2013).

Grasshoppers are no exception to this phenomenon of western species in the East. Several southeastern endemic grasshopper taxa are sole members of "western" or Neotropical genera or families, including *Achurum carinatum* (Walker) (fig. 10.6C), *Amblytropidia mysteca* (Saussure), and *Romalea microptera* (Beauvois). Additionally, at least 10 widely distributed western species have disjunct populations associated with grasslands in the Southeast. The primarily western genus *Hesperotettix* contains three sandhill-inhabiting species (*H. floridensis* Morse, *H. gemmicula* Hebard, and *H. osceola* Hebard) that are endemic to the Southeast. *Hespertotettix viridis* Thomas, a grassland species that is widely distributed in the West, has two subspecies, *H. v. brevipennis* (Thomas) and *H. v. pratensis* Scudder, that inhabit grasslands in the East. The genus *Eritettix* is most diverse in the western United States, but two species, *Eritettix simplex* (Scudder) and *E. obscurus* (Scudder) (fig. 10.6F), inhabit grasslands of the Southeast. Additionally, *Melanoplus angustipennis* (Dodge) and *M. bispinosus* Scudder are typically associated with sandy or arid habitats in the West and occur in sandhill habitats in the Southeast. Morse (1907) suggested that the genera *Arphia*, *Trimeroptropis*, and *Spharagemon* (fig. 10.7) were likely of western origin. Because of the habitat preferences of most of the species in these genera, I would agree with his assertion. See table 10.3 for a summary of taxa with disjunct distributions and their habitat associations in the Southeast.

The grasslands of the Southeast are well known for supporting a distinct flora and for harboring rare or endangered vertebrates such as the indigo snake (*Drymarchon couperi*), gopher tortoise (*Gopherus polyphemus*), and Bachman's sparrow (*Peucaea aestivalis*), but, as mentioned previ-

Figure 10.4. Some grasshopper species that inhabit southeastern grasslands: (A) *Floritettix borealis*; (B) *Gymnoscirtetes pusillus*; (C) *Aptenopedes apalachee*; (D) *Hesperotettix viridis brevipennis*; (E) *Campylacantha olivacea*; and (F) *Eotettix pusillus*.

ously, they also support unique invertebrate communities. One of the most common misconceptions about grasslands in the Southeast is that they are of anthropogenic origin, having arisen though burning and clearing by Native Americans. Some authors have pointed out that this is not the case because of the high level of endemism and biodiversity in these systems and the amount of time it would take for these systems to evolve (Noss 2013). The results from the analysis here underscore this conclusion.

Grasshopper Composition of Grasslands

The grasslands of the Southeast occur in a variety of environmental conditions (edaphic, hydrologic, elevational, etc.) and geographic regions, which would likely influence their faunistic compositions. For example, Barone (2005a) found that different prairie types in the eastern United States tended to have a floristic composition similar to that of their nearest neighbors. A database of grasshoppers was constructed with 117 species from 16 distinct grassland types to investigate the faunistic relationships of the various grasslands in the Southeast. For comparison, the database included a faunal list from the Konza Prairie in Kansas as an example of a western grassland, and a list from cedar-hardwood forests of the Central Basin of Tennessee as an open-forest type. The database was constructed using lists gathered from the literature (referenced in caption of fig. 10.8) as well as my own unpublished data.

A cluster analysis was performed on the database using PC-ORD software (McCune and Mefford 2011), with the Flexible Beta (β = -0.25) linkage method and the Sørensen (Bray-Curtis) distance measure, as recommended by McCune and Grace (2002).

Figure 10.5. Some grasshopper species that inhabit southeastern grasslands: (A) *Schistocerca rubiginosa*; (B) *Melanoplus ingrami*; (C) *Melanoplus pygmaeus*; (D) *Melanoplus stegocercus*; (E) *Melanoplus savannah*; and (F) *Paroxya clavuliger*.

All the grasslands, with the exception of the xeric limestone prairies of the Western Highland Rim of Tennessee and northwestern Alabama, branched independently, suggesting that each habitat type possessed a relatively distinct fauna for the most part (fig. 10.8). Geographic region appears to be important in structuring the dendrogram. The top cluster is made up of grasslands (prairies and glades) from Alabama, Mississippi, and Tennessee; the second cluster consists of the Arkansas glades and prairies and Konza Prairie; and the third major cluster comprises sandhill habitats from Florida and the southern half of Georgia. However, habitat type also appears to have an influence, as the sites represented in the first cluster are found on limestone or chalky clay soils, and the habitats in the third major cluster all have well-drained sandy soils. Balds, cedar-hardwood forests, and marshes are found on long branches of the dendrogram, which is an indicator of their very distinctive faunas. The fauna of ruderal grasslands of the Welka region in Florida includes habitat generalists from the region and appears to be a subset of the sandhill-scrub fauna. The cedar-hardwood forests and southern Appalachian balds are distantly clustered together, which is likely a result of both habitats having depauperate faunas composed partly of species that inhabit open forests or forest edges.

The cluster analysis illuminates some interesting patterns in the species composition of grasslands of the region, in which sandhills, prairies, balds, and forests all distinguish themselves. A more detailed look into the faunas of these grasslands is presented next.

Figure 10.6. Some grasshopper species that inhabit southeastern grasslands: (A) *Dichromorpha elegans*; (B) *Mermiria picta*; (C) *Achurum carinatum*; (D) *Pseudopomala brachyptera*; (E) *Metaleptea brevicornis*; and (F) *Eritettix obscurus*.

Calcareous Grasslands and Glades

The first large cluster of the dendrogram (fig. 10.8) is composed of calcareous grasslands, including blackland prairies (MS: Black Belt and Jackson Prairie); xeric limestone prairies (AL: Moulton Valley; TN: Central Basin and Western Highland Rim); glade habitats (AL: Bibb County; TN: Central Basin); Arkansas glades/prairies; and the Konza Prairie. There is some obvious geographic structuring of the cluster, with the Arkansas glades and the Konza Prairies lying at the easternmost limits of some "western" taxa, and the eastern grasslands containing widespread eastern species. The grasslands in this cluster share many grasshopper species, but one interesting aspect is the occurrence of native grassland specialists that are distributed predominantly in the Great Plains, namely *Paratylotropidia brunneri* Scudder, *Pseudopomala brachyptera* (Scudder) (fig. 10.6D), and *Trachyrhachys kiowa* (Thomas) (fig. 10.7B).

R. L. Brown (2003) examined the distributions of moths and butterflies (Lepidoptera) and a few other insects from Black Belt Prairies in Mississippi and observed numerous examples of species distributed primarily in the Great Plains. This observation led Brown to hypothesize that a grassland corridor existed between the Black Belt and the Great Plains prior to and following the Wisconsin Glaciation that provided dispersal routes for grassland species, and that the Black Belt may have served as a refugium for grassland species during this glacial episode. Further, his hypothesis proposed that the fauna of the tallgrass prairie in the postglacial Midwest was more likely derived from the

Figure 10.7. Some grasshopper species that inhabit southeastern grasslands: (A) *Psinidia fenestralis*; (B) *Trachyrhachys kiowa*; (C) *Trimerotropis saxatalis*; (D) *Spharagemon cristatum*; (E) *Arphia xanthoptera*; and (F) *Pardalophora phoenicoptera*.

refugium in the Black Belt than from the southwestern United States. The grasshopper data presented here appear to support Brown's hypothesis. Additionally, the flora of these southeastern prairies is similar to that of other eastern tallgrass prairies, with the dominant grasses being little bluestem (*Schizachyrium scoparium*), big bluestem (*Andropogon gerardii*), switchgrass (*Panicum virgatum*), Indiangrass (*Sorghastrum nutans*), and dropseed (*Sporobolus spp.*), and the common forbs being rosinweeds (*Silphium* spp.), prairie coneflower (*Ratibida pinnata*), prairie clovers (*Dalea* spp.), and blazing stars (*Liatris* spp.) (See Barone and Hill 2007; DeSelm and Murdock 1993; Lawless et al. 2006; and D. Webb et al. 1997.) This combination of these biotic associations indicates that the prairies of the Southeast should be included as part of the Eastern Tallgrass Prairie biome.

Longleaf Pine Ecosystems (Fall Line Sandhills, Ohoopee Dunes, Florida Sandhills/Scrub)

The second major cluster of the dendrogram is composed of sandhill/scrub habitats. Data for these habitats were taken from several studies, and it should be noted that these represent subsets of the longleaf pine ecosystem. The clustering of these habitats is likely due to the presence of widespread species that are restricted to xeric sandy soil, such as *Melanoplus angustipennis* (Dodge), *Melanoplus bispinosus* Scudder, *Psinidia fenestralis* (Serville) (fig. 10.7A), *Schistocerca rubiginosa* (fig. 10.5A), *Spharagemon cristatum* (Scudder) (fig. 10.7D), *Spharagemon marmorata picta* (Scudder), *Trimerotropis maritima* (Harris), and a high number of endemic species associated with sandhills on the Coastal

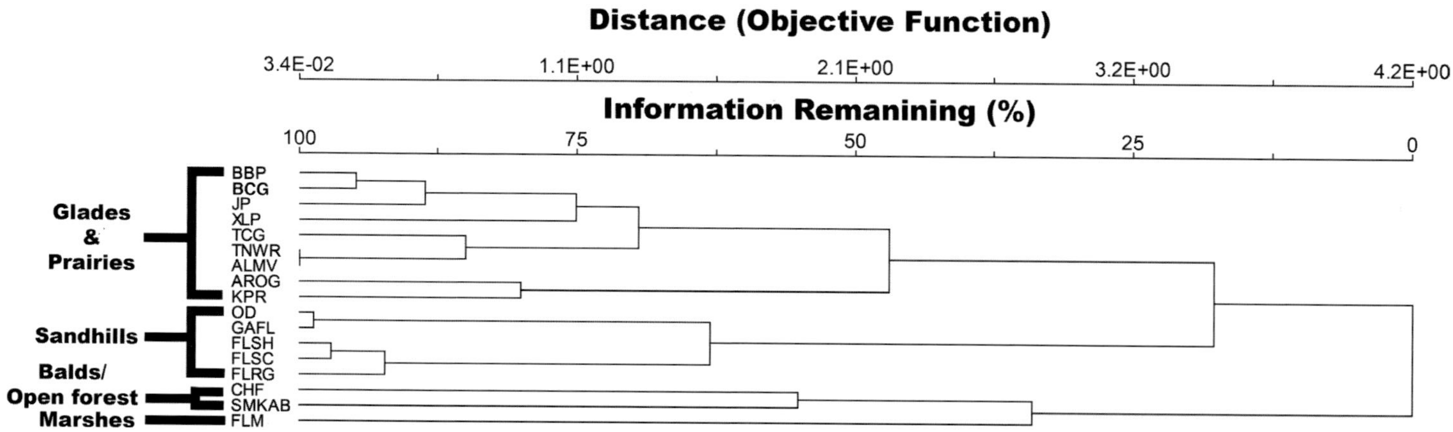

Figure 10.8. Cluster analysis results of grasshopper species composition from 15 different grassland types across the southeastern United States and Konza Prairie in Kansas. Data are for 117 species. Abbreviations: BBP = Black Belt prairie (J. Hill 2007); BCG = Bibb County glades (J. Hill 2007 and recent collections); JP = Jackson Prairie; XLP = xeric limestone prairies of the Central Basin of Tennessee (J. Hill 2012); TCG = Tennessee cedar glades in the Central Basin (J. Hill 2012); TNWR = xeric limestone prairies of the Western Highland Rim of Tennessee (J. Hill, in prep.); ALMV = xeric limestone prairies in the Moulton Valley of northwestern Alabama (J. Hill, in prep.); AROG = glades in the Ouachita Mountains of Arkansas (J. Hill, unpublished data; Morse 1907); KPR = Konza Prairie (Wright et al. 2005); OD= Ohoopee Dunes of Georgia (J. Hill 2009); GAFL = Fall Line sandhills in Georgia (J. Hill, in prep.); FLSH = sandhills in Florida (Friauf 1953; Squitier and Capinera 2002); FLSC = Florida scrub (Friauf 1953; Squitier and Capinera 2002); FLRG = ruderal grasslands in Florida (Friauf 1953; Squitier and Capinera 2002); CHF = cedar hardwood forests of the Central Basin of Tennessee; SMKAB = balds from Great Smoky Mountains National Park in Tennessee and North Carolina (J. Hill, unpublished data); and FLM = marshes in Florida along the St. Johns River (Friauf 1953).

Plain. Structure in this cluster is organized by geography, with the Georgia and Florida sites separated from each other. This is likely due to the high level of endemism associated with sandy habitats in Florida, as is seen in the *Melanoplus puer* group.

Species Shared between Prairies and Sandhills

The distant linkage of prairies and sandhills is likely a result of the sharing of many widespread grassland generalists, such as *Arphia xanthoptera* (Burmeister) (fig. 10.7E), *Dissosteira carolina* (Linnaeus), *Melanoplus sanguinipes vulturnus* Gurney and Brooks, and *Pardalophora phoenicoptera* (Burmeister) (fig. 10.7F), among others. Additionally, *Eritettix simplex* (Scudder), *Melanoplus impudicus* Scudder, *Mermiria bivittata* (Serville), and *Mermiria picta* (Walker) (fig. 10.6B) are patchily distributed grassland specialists shared between these broad habitat categories

Cedar-Hardwood Forests

The third branch is the cedar-hardwood forests of the Central Basin of Tennessee. This was the only forest type included in the analysis, and this was done because this habitat is relatively open and has a well-developed herbaceous layer. However, the grasshopper fauna of this habitat is composed of species associated with edges or closed canopies, such as *Melanoplus gracilis* (Bruner), *Melanoplus walshii* Scudder, and members of the *Melanoplus tribulus/obovatipennis* species group. This association is typical of most open forest and edge communities throughout the region.

Southern Applachian Balds

Clustering somewhat with the cedar-hardwood forests are the southern Appalachian balds. The data used in this survey were collected in Great Smoky Mountains National Park. These balds tend to be smaller and more disturbed than balds to the north (Noss 2013); thus these results may not be applicable to all balds in the eastern United States. Balds in Great Smoky Mountains National Park do not appear to have any grassland specialists, and the fauna is composed mostly of forest and edge species, such as *Melanoplus walshii* and members of the *Melanoplus viridipes* species group. Perhaps the cooler temperatures of these high elevations allow these sylvan species to escape the shade of the forest canopy and exploit the lush herbaceous vege-

Table 10.3.

Grassland-specific grasshopper taxa with notes on their distribution pattern and habitat preferences.

Taxa	Pattern	Grassland type
Achurum carinatum (Walker)	Sole member of a western genus in the SE	Various natural and anthropogenic grasslands on the Coastal Plain
Amblytropidia mysteca (Saussure)	Sole member of a western genus in the SE	Various natural and anthropogenic grasslands
Aptenopedes	Genus endemic to AL, FL, and GA that comprises 9 species	Open pine forests, prairies in Florida
Arphia	Genus likely of western origin represented by 4 species in the SE, one of which is endemic	Various natural and anthropogenic grasslands
Campylacantha olivacea (Scudder) (fig. 10.4E)	Sole member of a western genus in the SE; species widespread in the West, and restricted to natural grasslands in the Southeast	Prairies—associated with Ambrosia psilostachya
Eotettix	Genus that is endemic to the SE and comprises 3 species	Longleaf pine forests and sandhills
Eritettix obscurus (Scudder)	Member of a genus that is widely distributed in the West; endemic to Florida	Scrub and sandhills in peninsular Florida
Eritettix simplex (Scudder)	Widely distributed in the West, disjunct populations associated with grasslands in the SE	Prairies and sandhills
Floridacris	Genus of 4 species that are endemic to Florida	Sandhills and scrub
Floritettix	Genus endemic to Georgia and Florida that comprises 13 species	Longleaf pine savannas, sandhills, and scrub
Gymnoscirtetes	Genus comprising 2 endemic species	Longleaf pine savannas (often mesic areas) and sandhills
Hesperotettix	Genus likely of western origin represented by 3 species endemic to the SE, and 1 species that is widespread	Prairies, sandhills, and Florida glades and prairies
Melanoplus angustipennis (Dodge)	Widely distributed in the West, disjunct populations in the SE	Sandhills
Melanoplus bispinosus Scudder	Widely distributed in the West, disjunct populations in the SE	Sandhills
Melanoplus impudicus	Eastern United States	Various naturally open habitats
Melanoplus puer species group	Group that is endemic to the Coastal Plain	Sandhills and scrub
Melanoplus scudderi species group	Group that has apparently speciated in different grassland types in the SE	Various natural and anthropogenic grasslands
Mermiria bivittata (Serville)	Widely distributed and common in the West, disjunct populations associated with grasslands in the SE	Prairies and sandhills
Mermiria picta (Walker)	Widely distributed in the West, disjunct populations associated with grasslands in the SE	Prairies and sandhills
Paratylotropidia brunneri Scudder	Widely distributed in the West, disjunct populations associated with prairies in TN	Prairies
Pseudopomala brachyptera (Scudder)	Widely distributed in the West, disjunct populations associated with grasslands in the SE	Prairies

Table 10.3. *Continued*

Taxa	Pattern	Grassland type
Romalea microptera (Beauvois)	Sole member of a western family in the SE	Various habitat types, including some forested
Spharagemon	Genus likely of western origin represented by 5 species in the SE	Sandhills and open forests
Trachyrhachys kiowa (Thomas)	Widely distributed in the West, disjunct populations associated with grasslands in the SE	Prairies, glades, and rock outcrops
Trimeroptropis (Thomas)	Genus likely of western origin represented by 2 species in the SE	Sandhills and rock outcrops

tation of the balds. The grasshopper fauna of larger and less-disturbed balds should be studied to see whether this is the typical fauna or whether there are species that may be bald specialists.

Marshes

The final branch of the dendrogram represents *Spartina*/saw grass marshes along the St. Johns River in Florida. Though these data are from Florida, the species represented here are typical of those in many open wetlands across the Southeast, with species in the genera *Paroxya* (fig. 10.5F) and *Dichromorpha* (fig. 10.6A), and the species *Mermiria intertexta* Scudder being represented (pers. obs.). Further investigations of the grasshopper fauna of marshes and wetlands throughout the region will likely illuminate a greater diversity of faunal associations, as there are more species that inhabit wet grasslands throughout the region that are not present in these Florida marshes, including *Leptysma marginicollis* (Serville), *Metaleptea brevicornis* (Johannson) (fig. 10.6E), *Stenacris vitreipennis* (Marschall), and *Stethophyma celata* Otte.

Conservation/Management

The unique and diverse grasshopper fauna in the Southeast makes it apparent that the greatest threat to grasshopper biodiversity in the region is the loss and fragmentation of grassland systems. Indeed, many of the southeastern grasslands are considered endangered or imperiled as a result of human activities. Surveys conducted by the General Land Office in the 1830s suggest that prairies once covered at least 144,000 ha in the Black Belt of Mississippi and Alabama (Barone 2005a). Since that time, more than 99% of these prairies have been lost to agriculture and urban development (Noss et al. 1995). Prairie remnants are threatened by further development, erosion, and the encroachment of eastern redcedar, *Juniperus virginiana* L. (Cupressaceae), the latter probably as a result of fire suppression. Likewise, the cedar glades and xeric limestone prairies of the Central Basin of Tennessee are considered highly imperiled as a result of anthropogenic disturbances, including fire suppression, recreational driving, and the suburban sprawl of Nashville and Lebanon (Noss et al.1995; Niemiller et al. 2011). Even the once vast longleaf pine ecosystem that covered an area from southeastern Virginia to eastern Texas has been reduced to only 3% of its presettlement extent as a result of tar and pitch production, logging, clearing for agriculture, and modern forestry practices, namely site-preparation practices and stocking densities (Frost 1993; W. Platt 1999).

As mentioned, many of the flightless grassland specialists have small geographic distributions, and because of their narrow habitat breadth (such as sandhills), their populations are often isolated by inhospitable natural features such as forests, rivers, and swamps. Thus, these species often occupy relatively small areas, making them more vulnerable to extinction.

However, their reliance on a single habitat and small geographic area may also have positive conservation implications. Because grasshoppers are primary consumers, they are often abundant in many grassland systems, and a small habitat remnant can often be inhabited by a significant number of individuals, such that even establishing a small preserve, sometimes less than several hectares (or acres), could potentially protect an entire population. Moreover, a series of small preserves could conceivably provide adequate protection for the species, as well as save examples of its habitat and associated biota.

The effects of habitat management techniques on grassland species in the Southeast have not been well documented. Management practices aimed at maintaining grassland vegetation, such as burning and grazing, likely have a significant impact on the grasshopper fauna. For example, a species such as *Pseudopomala brachyptera* (a grassland endemic) that overwinters in the nymphal stage may be sensitive to the early-season burns that are currently favored by many land managers. There is little doubt that fire is important in maintaining the habitat for grassland species; however, for grasshoppers, the level, timing, and frequency of burning that will be beneficial or disruptive is unknown, and the answer likely depends on the species and grassland system in question. Even so, many grasslands are fire-maintained systems, and as members of that system, grasshoppers have several adaptations to cope with fire. For example, species that are capable of flight often disperse away from the fire. Large numbers of grasshoppers have been observed flying away from a prescribed fire in a longleaf pine savanna (Matt Elliott, pers. comm.). In addition, short-winged, flightless species may climb trees to avoid flames, particularly in longleaf savannas. During a visit to a sandhill at the Savannah River site in the fall of 2012, numerous individuals of *Melanoplus savannah* were observed perched about 4 m above the ground on longleaf pine trunks immediately after a burn (pers. obs.). Finally, though not a means to escape fire, grasshoppers may be adapting to frequently burned areas by developing darker coloration, which makes them more cryptic against the charred surface of their habitat. On several occasions, dark morphs of *Orphulella pelidna*, *Melanoplus coreyi*, and *Melanoplus strumosus* have been observed in areas that experience frequent burning.

Conclusion

Grasshoppers are ideal study organisms in temperate grasslands because of their biological importance, interesting distributions, relative abundance, and ease of sampling. The grasslands of the Southeast offer an astounding array of interesting questions for researchers from a variety of fields. Biogeographers and systematists could answer questions about origins of the fauna of these grasslands and the dispersal routes taken by disjunct species, and further research could investigate the faunal relationships between various grassland types. Ecologists could investigate habitat requirements and life histories of many grassland endemic species, particularly those that may be of conservation concern. Conservationists and management-oriented researchers could study the effects of various management issues, particularly those involving the effects of habitat fragmentation, the importance of preserve size and connectivity, the timing of prescribed burning, the effects of disturbance, and, in longleaf pine systems, the site preparation and stocking rate of new plantations.

This introductory synthesis of the grasshopper fauna of grasslands in the southeastern United States further demonstrates that these varied and distinct grasslands are rich in endemic species that contribute significantly to the biodiversity of the region and to that of North America, as well as providing further corroborating evidence that these systems are of natural origin. Thus the continued conservation of these systems along with the study of their biota is important, for they represent an important part of our regional natural heritage. It is my hope that the information presented here inspires further interest in researching and conserving southeastern grasslands.

Acknowledgments

I thank Matt Dakin, John Barone, and Richard Brown for their comments on this manuscript. Specimens in the Mississippi Entomological Museum were collected with financial support from the Georgia Department of Natural Resources, the William H. Cross Expedition Endowment in the Mississippi State University Development Foundation, and the Mississippi Agricultural and Forestry Experiment Station. This research was supported by the National Institute of Food and Agriculture, under Project No. MIS-012040.

11

Encroachment and Persistence of Trees in Southeastern Grasslands

J. Morgan Varner

We continued over these expansive illumined grassy plains, or native fields, above twenty miles in length, and in width eight or nine.

—William Bartram (1791), describing the treeless landscape near present-day Montgomery, Alabama

In a landscape typified by frequent, abundant rainfall and long growing seasons, the historical discontinuity of forest cover in the southeastern United States is striking. Trees were ephemeral or of minor importance in presettlement prairies, glades, grasslands, and savannas in the southeastern United States (A. Jones and Patton 1966; Rankin and Davis 1971; Barone 2005a; K. Stephenson 2011). Early descriptions (Bartram 1791) and public land survey records (Barone 2005a) attest to scattered trees on grassy expanses, in the Black Belt or blackland prairies (Peacock and Schauwecker 2003a) of Alabama and Mississippi (Bartram 1791; Hilgard 1860; R. Harper 1913b; A. Jones and Patton 1966; Rankin and Davis 1971), but also in other southeastern grasslands (e.g., Folkerts 1982; DeSelm and Murdock 1993; Barden 1997; K. Stephenson 2011).

The absence of trees in prairies, or scattered trees in savannas, contrasted sharply with the surrounding closed-canopy forests of the southeastern United States (R. Harper 1913b; A. Jones and Patton 1966; Rankin and Davis 1971; Barone 2005a). Existing remnants of these prairies and savannas differ from historical descriptions, as they frequently harbor trees, often at high densities. This widespread tree invasion or encroachment is a primary agent of degradation in these rare remnant landscapes (Wiygul et al. 2003; Weiher et al. 2004; J. Hill 2007a; J. Hill and Brown 2010). In this chapter, I will describe the ways that trees are able to invade grassland habitats, the characteristics of tree species that make them successful invaders, and different approaches to removing trees and restoring the open habitats in the southeastern United States.

How Trees Persisted in Southeastern Grasslands and Savannas

Several mechanisms work synchronously to prevent or slow tree canopy closure in southeastern grasslands. In spite of the plentiful rainfall (approximately 1,500 mm annually across the region) at the time of settlement, many landscapes of the southeastern United States were treeless prairies, had scattered trees in savannas, or were open-canopied woodlands (all three types are considered grasslands hereafter; as in W. Platt 1999). Many grasslands in the region are underlain by soils that lack fertility, as in the sands of much of the Lower Coastal Plain; or have extreme alkalinity that results in toxicity, as in several areas of the Black Belt (Hilgard 1860; Rankin 1974). The shrink-swell of the Black Belt soils may also limit tree establishment. Many

Figure 11.1. The pyrophytic trees *Quercus marilandica* and *Pinus echinata* persisting in a frequently burned prairie in Chickasaw County, Mississippi.

soils throughout the region are so coarse that they retain little water, or so fine that they retain too much water (Folkerts 1982; K. Stephenson 2011). These edaphic features filter the diverse tree flora of the region that might otherwise thrive in this warm, wet climate. When these edaphic limitations are combined with frequent fires, the number of tree species that can persist in grasslands, even at very low densities, is constrained even further (fig. 11.1).

Fires filter species composition via differential mortality as a result of fire-caused injuries (W. Bond and van Wilgen 1996; Beckage et al. 2009). Fires of sufficient intensity or duration fueled by the diverse herbaceous fuels from grasses and forbs can heat vital aboveground and belowground plant tissues (Vogl 1973; Gagnon et al. 2010). Where fires are frequent, these mechanisms prevent tree establishment. Where fires have been excluded, trees establish and cast leaf litter that slows the spread and diminishes the intensity of future fires.

Several mechanisms have been proposed that enable trees to survive fires: thickened bark, rapid pruning of lower branches, rapid height growth above the flaming zone, flammable foliage that promotes abbreviated duration of heating, and vegetative resprouting (e.g., W. Bond and van Wilgen 1996; J. Jackson et al. 1999, Veldman et al. 2013). Combinations of these fire-surviving adaptations characterize pyrophytic species, or "pyrophytes," that facilitate, survive, and prosper in fire-prone environments (table 11.1). Species that lack these adaptations, the so-called mesophytes (Nowacki and Abrams 2008), are therefore rare in these same fire-prone ecosystems.

One primary allocation strategy of pyrophytes is rapid accumulation of fire-protective bark, or so-called positive bark allometry (J. Jackson et al. 1999). Notable examples of southeastern trees with this strategy include most of the pines and several oaks (e.g., blackjack oak—*Quercus marilandica*; post oak—*Q. stellata*; and turkey oak—*Quercus laevis*; among others). Thick or thermally diffuse bark armors small trees against the heat release of surface fires (Hare 1965), reducing their rates of fire-caused mortality. Even without rapid accumulation, many mesophytes (e.g., white ash—*Fraxinus americana*) can accrue thick bark over time, enabling them to survive infrequent fires. Several southeastern species have exaggerated positive bark allometry and are common in fire-prone southeastern grasslands (J. Jackson et al. 1999; see table 11.1).

Another primary strategy of pyrophytes is vegetative resprouting from unaffected stem, crown, or root buds following heat damage to the crown that would otherwise result

Table 11.1.

Tree species that encroach southeastern United States grasslands and their fire-adaptive traits.

Species	Strategy*	Fire-adaptive traits†			Leaf habit
		Sprouting	Bark Allometry	Flammability	
Quercus alba	Pyrophyte	+	0	High	Deciduous
Quercus falcata	Pyrophyte	+	+	High	Deciduous
Quercus hemisphaerica	Mesophyte	+	–	Low	Semi-deciduous
Quercus marilandica	Pyrophyte	+	+	High	Deciduous
Quercus muehlenbergii	Pyrophyte	+	0	High	Deciduous
Quercus nigra	Mesophyte	+	–	Low	Deciduous
Quercus stellata	Pyrophyte	+	0	High	Deciduous
Pinus echinata	Pyrophyte	+	+	High	Evergreen
Pinus palustris	Pyrophyte	–	+	High	Evergreen
Pinus taeda	Pyrophyte	–	+	High	Evergreen
Pinus virginiana	Mesophyte	–	–	Low	Evergreen
Carya spp. (+ *Juglans*)	Pyrophyte	+	0	High	Deciduous
Juniperus virginiana	Mesophyte	–	–	Low	Evergreen
Celtis laevigata	Mesophyte	+	–	Low	Deciduous
Liquidambar styraciflua	Mesophyte	+	–	Low	Deciduous
Liriodendron tulipifera	Mesophyte	+	–	Low	Deciduous
Fraxinus americana	Mesophyte	+	0	Low	Deciduous
Cercis canadensis	Mesophyte	+	–	Low	Deciduous
Maclura pomifera	Mesophyte	+	0	Low	Deciduous
Acer rubrum	Mesophyte	+	–	Low	Deciduous

* Strategy refers to the suite of fire-adaptive traits, including sprouting, bark allometry, litter flammability, and others listed in the text.

† Fire-adaptive traits are based on published values in J. Jackson et al. (1999), Fonda (2001), Kane et al. (2008), Kreye et al. (2013), and the author's unpublished flammability data. In general, + corresponds to positive, 0 to intermediate, and – to negative values for each listed trait.

in mortality (W. Bond and van Wilgen 1996). Many of the southern hardwood trees, but few pines, are capable of such a response. Basal sprouting from the stem or sprouting from the root is common, resulting in rapid recovery of aboveground tissues. Bole epicormic and crown resprouting are less common but can refoliate injured stems or burned branches quickly, allowing them to attain their prefire stature over fire-free intervals. Most upland hardwoods have some capacity for bole and crown resprouting; among southeastern conifers, only shortleaf pine (*Pinus echinata*) and pond pine (*P. serotina*) have this capacity.

A less direct mechanism for pyrophytic tree species to survive frequent fires is by casting flammable litter into the surface fuel. Flammable litter is found in a diverse suite of species in fire-prone environments: many pines (Fonda 2001), other conifers (Fonda et al. 1998), oaks (Kane et al.

Figure 11.2A. Fire-adaptive trait of southeastern pyrophytic tree species: Vegetative resprouting following top kill of *Quercus marilandica*.

2008; Engber and Varner 2012), and other species worldwide (Scarff and Westoby 2006; Ormeño et al. 2009). While it may seem counterintuitive, flammable litter may actually "protect" (Gagnon et al. 2010) source trees by "killing thy neighbor" (sensu W. Bond and Midgely 1995). In southeastern grasslands and savannas, herbaceous fuels are highly flammable via their loads of fine (high surface area to volume) fuels and retention of senesced, dry thatch. Additionally, leaf litter cast by many pyrophytes is highly flammable, facilitating high-intensity fire in its surroundings (Fonda 2001; Kane et al. 2008) and spreading fire across adjacent grasslands and into forests. Mesophytic litter lacks these same fire-facilitating characteristics, retaining moisture longer than pyrophytic litter (Kreye et al. 2013) and impeding fire more when it does dry sufficiently to burn (Kane et al. 2008). Southeastern pyrophytic litters are notably among the most flammable litters observed. Among published work on pines, longleaf pine (*Pinus palustris*) reigns as most flammable (Fonda 2001); among oaks, turkey oak and southern red oak (*Quercus falcata*) are among the three most flammable (Kane et al. 2008). These flammability traits reflect a long evolutionary history involving fire, southeastern grasslands, and the few trees capable of tolerating chronic fires (W. Platt 1999).

A flammability-related, although less well-understood, mechanism for pyrophyte survival and persistence in fire-prone landscapes is canopy light transmittance and its effects on herbaceous fuels. Open crown architecture of isolated trees is a characteristic of savanna and woodland physiognomy worldwide (Scholes and Archer 1997; Veldman et al.

Figure 11.2B. Fire-adaptive trait of southeastern pyrophytic tree species: Epicormic sprouting by *Pinus echinata*.

2013). Trees and shade-intolerant herbs have well-known antagonistic relationships that affect the spatial patterns of herbaceous biomass and diversity in tree-grass communities (Scholes and Archer 1997). Many southeastern pyrophytes have open or translucent crowns that permit light transmittance (Canham et al. 1994; Battaglia et al. 2003; Veldman et al. 2013) to understory herbs that fuel southeastern grassland surface fire regimes. Open crown architecture is a characteristic of many shade-intolerant southeastern pyrophytic trees; many southern oaks (particularly turkey oak and bluejack oak—*Quercus incana*) and pines (particularly longleaf pine) are notable for the substantial amount of light they transmit (Battaglia et al. 2003). In contrast, most mesophytes are capable of retaining leaves and supporting branches in low light (beneath their own shade or the shade of other trees). Mesophytes that cast substantial shade effectively eliminate understory fuels (Weiher et al. 2004). In addition, the native invader eastern redcedar (*Juniperus virginiana*), while shade intolerant (Ormsbee et al. 1976), casts deep shade with its dense evergreen crown (Weiher et al. 2004). While this driver of differential survival is less well studied, the sharp contrast between fire-tolerant pyrophytes and fire-sensitive mesophytes in their ability to transmit light suggests that these differences may be adaptations.

While there may be physiological explanations for the architectural differences between trees of closed and open habitats, the seemingly correlated features of pyrophytes of the southeastern United States suggest that their persistence in open ecosystems is heavily reliant on their ability to pro-

Figure 11.2C. Fire-adaptive trait of southeastern pyrophytic tree species: Fire-resistant positive bark allometry of *Quercus laevis.*

mote the maintenance of fire-dependent surface vegetation while enhancing their own ability to survive frequent fires. Species that "self-prune" their lower branches, allocate carbon resources to thickened bark, and promote high surface flammability put themselves in a position not only to persist in prairies and savannas maintained by frequent fires, but to perpetuate the unique structure of these ecosystems.

Grassland and Savanna Degradation from Tree Encroachment

The widespread success of pyrophytes in southeastern grasslands is not apparent in the composition of many contemporary landscapes. Current estimates of loss or degradation of southeastern prairies are among the highest in North America; fewer than 1% of southeastern prairies remain (Noss et al. 1995; Barone 2005a), and many of those remnants are degraded (Barone and Hill 2007).

Managers of remnant southeastern prairies face several restoration and maintenance challenges. In intact and restored prairies, the invasion of mesophytic trees is a major driver of prairie degradation, altering the structure, composition, and processes of these remnants (Wiygul et al. 2003; Weiher et al. 2004). The mesophytes may generate positive feedbacks, impeding fire and altering the microclimate such that the establishment of other mesophytic invaders is facilitated and future fires are further inhibited (Nowacki and Abrams 2008; Maynard and Brewer 2013; Kreye et al. 2013).

The suite of mesophytic trees that encroach and modify southeastern grasslands is diverse (see table 11.1). In addition to fire-sensitive oaks (laurel oak—*Quercus hemisphaerica*; water oak—*Quercus nigra*; and others), several other hardwoods dominate, along with a few conifers; most notable among those is eastern redcedar. While most of these species resprout vegetatively (an exception is eastern redcedar), they all share a negative bark allometry (J. Jackson et al. 1999) and, save white ash and tuliptree (*Liriodendron tulipifera*), they rarely attain fire-resistant bark (Hare 1965). Another uniting characteristic of the mesophytes is the reduced flammability of their litter. The group either burns poorly (e.g., eastern redcedar and several semideciduous oaks) (Engle and Stritzke 1995; Kane et al. 2008) or forms such a dense litter mat that moisture loss is slowed (Engle and Stritzke 1995; Kreye et al. 2013), dampening their flammability (Vogl 1973). The majority of mesophytes have opaque crowns; eastern redcedar and to a lesser extent red maple (*Acer rubrum*), and even the naturalized Osage orange (*Maclura pomifera*), cast deep shade that not only intercepts understory light but also generates a compact litter on the forest floor that further impedes understory plant establishment (Vogl 1973; Weiher et al. 2004). The degree to which these species modify their local microclimate, retaining moisture and reducing solar heating, affects the spread and intensity of future fires, effectively accelerating the positive feedback (Nowacki and Abrams 2008; Beckage et al. 2009) toward fire-protected forest and farther away from pyrophytic grassland and savanna.

Mesophytic trees face impediments to their establishment and persistence in southeastern grasslands. For most trees and shrubs, germinating and growing in the often dense herbaceous cover of southeastern grasslands is a challenge. Competition from grasses and forbs, whether as shade or belowground competition for water and nutrients, can prevent establishment of tree seedlings. Where

Figure 11.2D. Fire-adaptive trait of southeastern pyrophytic tree species: Flammable litter under *Quercus marilandica* and *Q. stellata*.

disturbances (e.g., domestic livestock) have diminished the cover of herbs, trees and shrubs can establish. In savannas or woodlands where pyrophytic trees are present, many woody species are capable of establishing beneath the sparse shade cast by the scattered trees.

A few notable species have had great success in establishment in many southeastern grasslands. In blackland prairie, eastern redcedar, sugarberry (*Celtis laevigata*), and Osage orange are common mesophytic invaders. Eastern redcedar is notable in part because of its historical absence in presettlement prairies (R. Harper 1913b). There is much discussion over the apparent success of eastern redcedar in these settings, as it not only invades blackland prairie (Wiygul et al. 2003) but is a widespread invader throughout grasslands in eastern North America (e.g., Gehring and Bragg 1992; Briggs et al. 2005). Eastern redcedar has an animal-dispersed fruit (particularly by birds) and an affinity for the calcareous soils typical of many prairies (Lawson 1990). Its success is increased by its tolerance to high light and drought (Ormsbee et al. 1976; Lawson 1990), the latter a seasonal condition in blackland and many other types of prairies.

Reversing Mesophytic Tree Encroachment

Prescribed fire offers a promising tool for ecological restoration of many southeastern grasslands and savannas invaded by mesophytic trees. Fire is used in many forest and woodland restoration treatments, whether to kill small trees or to selectively kill large trees (Engle and Stritzke 1995; Engber and Varner 2012b; Ryan et al. 2013). Prescribed fires can kill small and large mesophytic trees, and subsequent fires can consume woody debris generated from initial mortality. In Black Belt Prairies, fire can reduce woody species substantially while simultaneously increasing forb richness (Schauwecker and MacDonald 2003). Impediments to using fire include the legal, social, and political issues (e.g., burner liability, public support along the wildland-urban interface) that face prescribed burning in general (Elsen and Wieland 2003; Ryan et al. 2013), plus the operational conundrum of raising intensity sufficiently to kill encroachers without simultaneously subjecting other plants or soils to excessive heating. There are few examples of using high-intensity prescribed fires in restoration treatments because of the obvious resistance to planned crown fires. Altered surface fuels, with reduced herbaceous fuel mass (Weiher et al. 2004) in combination with low-quality mesophytic litter (Kane et al. 2008), further complicate the burning problem in southeastern grasslands and other fire-prone ecosystems in the eastern United States (Nowacki and Abrams 2008; Maynard and Brewer 2013). Thoughtful prescriptions that balance the operational impediments with treatment effectiveness deserve increased attention in southeastern grassland restoration.

Several potential mechanical tools and approaches can be used to remove or diminish encroaching trees. Tree harvesting seems a simple approach, but the low commercial value of open-grown encroaching trees makes this option costly. When feasible, tree cutting and removal are effective and can have muted ecological effects on remnant sites (depending on skidding technique). Harvesting and subsequent burning of residual debris in piles or windrows are effective, but attendant soil heating beneath piles and along windrows can be substantial (DeBano et al. 1998). Mechanical mastication, or mowing, is an option used in southeastern savanna and woodland restoration to shred and scatter shrubs and small trees (Glitzenstein et al. 2006; Brockway et al. 2009). Similarly, roller chopping or drum chopping of small trees and shrubs has been used in dry prairie restoration in southern Florida grasslands (Watts and Tanner 2003; Menges and Gordon 2010). Drawbacks for mastication and roller-chopping methods are that the woody residues remain at the surface, likely stalling herbaceous recovery (Kane et al. 2010) and potentially causing belowground heating and undesired mortality to large pyrophytic trees when these dense woody residues burn (Glitzenstein et al. 2006; E. Knapp et al. 2011). As mentioned earlier, the majority of mesophytes also resprout vigorously; without follow-up herbicides or fire (see the following discussion of each combination), mechanical treatments are ineffective over time. In the few examples of these mechanical methods being used in pine savanna and dry prairie restoration, there have been mixed results when fire was not subsequently used (Menges and Gordon 2010). Where fire has followed these treatments, fewer dissimilarities with pristine or frequently burned remnants are found (Watts and Tanner 2003), but appropriate caution has been suggested for these treatments in degraded ecosystems (Menges and Gordon 2010).

Restoration treatments using herbicides are common in grassland and savanna restoration. Herbicides have the benefit of targeting specific taxa with minimal effects on nontarget species (Packard and Mutel 1997). Their shortcomings are that killed residues inflate short-term surface fuels (Brose and Wade 2002) and that suites of related taxa are eliminated despite potential differences in their function. In southeastern grassland and savanna restoration, herbicides have been used to kill invading trees and shrubs with success (J. Freeman and Jose 2009). Where herbaceous species have been excluded, seeding and planting of lost natives has been implemented (with attendant costs and issues related to provenance).

Restoration treatment combinations of fire, mechanical, and chemical methods are widely used to counterbalance the shortcomings of single methods. In Florida dry-prairie restoration, combining mechanical treatments with fire-reduced woody invasion resulted in greater herbaceous diversity than fire-alone or mechanical-alone restoration treatments (Watts and Tanner 2003). In blackland prairies, Schauwecker and MacDonald (2003) found similar effects from combinations using mechanical removal and fire. In areas where trees are large, mechanical treatments can be effective in restoring structure and composition, accelerating the pace of restoration, and eliminating the mesophytic feedbacks to subsequent fire. Grassland and savanna restoration treatments combining fire with herbicides are common, resulting in shifts in composition and changes in structure that persist until fire can be used (Rhoades et al. 2002; J. Freeman and Jose 2009). Treatment combinations that lack fire (i.e., mechanical and chemical combinations) are successful at restoring structure and composition, but without fire these treatment effects are ephemeral. Taking advantage of the positive outcomes of mechanical treatments (killing mesophytes and removing them) and burning (reducing litter and reinitiating herbaceous growth) may be most advantageous in regional restoration efforts.

The challenge of reversing encroachment is a critical issue across many southeastern grasslands (Elsen and Wieland 2003; Wiygul et al. 2003). The approach here was to help clarify the suites of adaptive traits that permit some trees to survive fires while others function to eliminate grassland and savanna physiognomy. Unlike in other southeastern ecosystems, scientists and managers lack a clear understanding of the structure and composition of pristine southeastern grasslands (Barone and Hill 2007), and we know even less about how disturbances functioned in these ecosystems. As our understanding of the history and future of fire in southeastern grasslands evolves, so will our understanding of trees in these dynamic and important landscapes.

12

Experimental Removal of Eastern Redcedar to Restore Black Belt Prairie Remnants

Effects on Plant, Ant, and Grasshopper Communities

John A. Barone, JoVonn G. Hill, and Lisa McInnis

Introduction

Prior to European settlement, the southeastern United States had several regions with abundant grasslands or prairies (Peacock and Schauwecker 2003a; Noss 2013). One such region, called the "Black Belt," extended in an arc from McNairy County, Tennessee, southeast across Mississippi, to Russell County, Alabama (fig. 12.1; Peacock and Schauwecker 2003a; Barone 2005b). Instead of being a continuous swath of grasslands, the Black Belt was a heterogeneous landscape of prairies and several types of forest. Land surveys from the 1840s show that prairies once covered about 144,000 ha (356,000 ac) in the region. Noss et al. (1995) estimated that more than 99% of these prairies have been lost, and the Mississippi Department of Wildlife, Fisheries, and Parks (2005) stated that Black Belt Prairie remnants are "critically imperiled" within the state because of their rarity as a result of agricultural development and land management practices.

A significant threat to remaining prairies in this region is the encroachment of woody species, which can trigger their succession into open woodlands and, later, forests (S. Archer et al. 1995; Briggs et al. 2005). This pattern of succession has been reported from grasslands worldwide (e.g., S. Archer et al. 1995; J. Brown and Carter 1998; Briggs et al. 2005; Moleele et al. 2002). For example, Briggs et al. (2005) reported that woody cover at the Konza Prairie Biological Station in Kansas had increased 72% in 63 years. The drivers leading to increased woody stem density in grasslands are not clear but may include a reduction in the intensity or frequency of natural fires, habitat fragmentation, grazing management, shifts in climate, or changes in the distribution of soil resources (Schlesinger et al. 1990; S. Archer et al. 1995; J. Brown and Carter 1998; Moleele et al. 2002; Briggs et al. 2005).

In North America, woody encroachment into prairies is often initiated by junipers, including *Juniperus virginiana* L., or eastern redcedar (Briggs et al. 2002a; Wiygul et al. 2003; Alford et al. 2012). Though native to the Southeast and long present in the Black Belt (Schauwecker and MacDonald 2003; Seltzer and Peacock, chapter 18, this volume), eastern redcedar has become dominant on many remnant prairie sites, perhaps because of a decrease in natural fires (Briggs et al. 2002b; Wiygul et al. 2003). Even otherwise undisturbed, high-quality prairie remnants in the Black Belt are slowly being transformed into woodlands by the spread of eastern redcedar (J. A. Barone and J. G. Hill, pers. obs.).

Here, we report on a four-year study that examined the efficacy of removing redcedars through clearing and burning with the goal of preventing succession and restoring open prairies. The experiment had four treatments: (1) a "burned" treatment, in which remnant prairie sites were burned to decrease the density and growth of eastern redcedar and other woody plants; (2) a "cleared" or mechanically thinned treatment, in which redcedar trees were cut

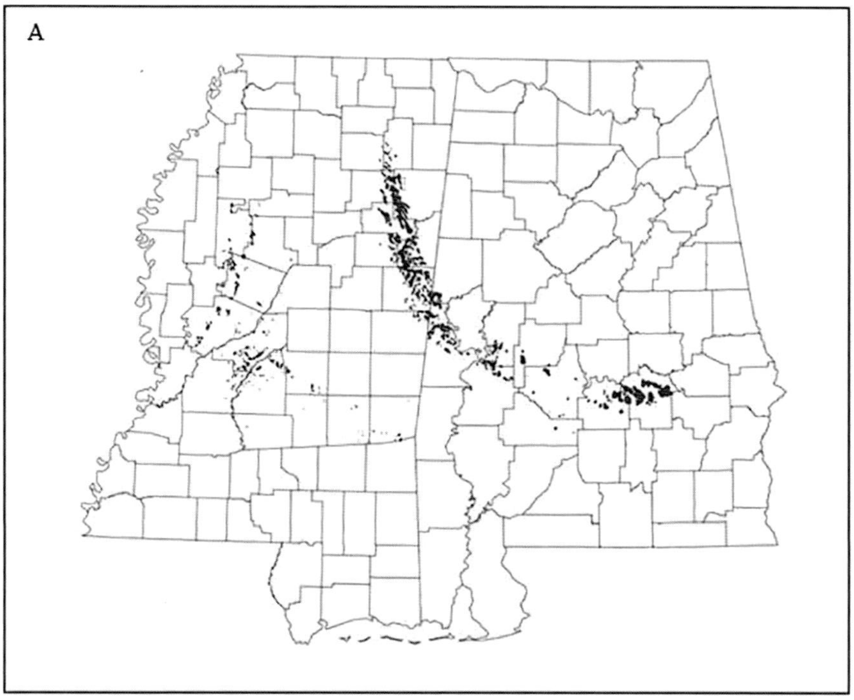

Figure 12.1. Historical distribution of prairies in Mississippi and Alabama according to federal land survey maps from the 1840s. The Black Belt extends from east-central Alabama through northeastern Mississippi to southwestern Tennessee. Modified from Barone (2005a, 2005b).

down and removed from the sites; (3) a "cut/burned" treatment, in which the redcedar trees were cut, left on-site, and burned; and (4) a control treatment, in which there was no active restoration.

The effects of these differing management strategies were evaluated using changes to the plant, ant, and grasshopper communities. We chose these taxa for several reasons. First, all three groups are locally diverse and include species that are largely or exclusively found on remnant prairies in this region (Leidolf and McDaniel 1998; J. Hill 2007a; Barone and Hill 2007; J. Hill and Brown 2010; Campbell and Seymour 2011a). Second, the composition of these three communities is quite well known for Black Belt Prairies. Third, the three groups represent different trophic levels and have different ecological roles. In grasslands, grasshoppers are the dominant invertebrate herbivores, and ants can be keystone species and are often valuable in detecting changes to habitats (King et al. 1998; Joern 2005; Jonas and Joern 2007; Alba-Lynn and Detling 2008). Examining these three groups simultaneously enabled us to assess the success of the different restoration strategies rather broadly.

Additionally, we evaluated the success of the treatments by determining which ones improved the remnant prairies in the most cost-effective manner. If two treatments had similar effects but one was cheaper to implement, obviously the lower-cost treatment would be more desirable.

Methods

Study Area

The study sites were located along the Natchez Trace Parkway, a two-lane roadway administered by the National Park Service that runs 740 km (460 mi) from Natchez, Mississippi, to Nashville, Tennessee. The parkway passes through the Black Belt in Chickasaw, Pontotoc, and Lee Counties in northeastern Mississippi (fig. 12.1). In July 2008, the land owned by the Park Service that is adjacent to the parkway was surveyed for prairie sites. A total of 38 remnants were identified, ranging from 18 to 56,260 m^2 (Barone and Hill

2008). Eleven other sites with small populations of prairie plant species were also located. It was noted that most of these remnants had been invaded and were increasingly dominated by redcedar.

Precipitation data for the years of the study came from the Verona Experimental Station, the nearest weather station to the sites (National Climatic Data Center 2013).

Plots and Treatments

Twenty-one 10 × 20 m plots were established in prairie remnants along a 22.5 km (14 mi) stretch of the parkway south of Tupelo, Mississippi, in the late winter and early spring of 2009 (fig. 12.2). Plots were assigned to one of the four treatments. There were five plots per treatment, except for the cut/burned treatment, which had six. Because of the logistical requirements of burning, parkway regulations, the accessibility of the plots, and the proximity of an urban area at the northern end of the study area, the plots receiving each treatment were generally grouped together geographically (fig. 12.2). Thus, plots were not assigned at random to treatments, meaning that initial conditions in the plots within a treatment were more similar than they otherwise would have been.

At the beginning of the study, all the woody plants on the plots with stems greater than 1 cm in diameter at 130 cm above the ground were identified and measured. Censuses of herbaceous plants (including grasses), ants, and grasshoppers began during the growing season of 2009, prior to the initiation of the treatments, in order to provide baseline data for comparisons.

In the initial experimental design, the treatments were to have been carried out during the winter of 2009–2010. However, wet weather prevented burns during that season, though the cutting and clearing of trees was accomplished. Therefore, by the beginning of spring 2010, the cleared treatment had been completed. However, the cut/burned plots were only half finished, as redcedars and other woody plants had been cut down but left on the plots unburned. Burned plots remained untouched as well. The burns were attempted in April 2011, but the fires did not carry very well, so the burns were repeated in March 2012.

Sampling

Censuses of forbs and grasses on the plots were performed monthly during the growing season, from April through October. The identities of all the plants presently or recently in flower were recorded during each visit. Voucher specimens were collected and placed in the herbarium at Columbus State University in Columbus, Georgia.

The ant fauna of each plot was sampled using three approaches. First, visual searching for ants took place for 15 minutes in each plot once a month. Second, a bait transect consisting of 10 small piles of cookie crumbs (Pecan Sandies by Keebler) placed 1.5 m apart was put on the ground and checked 15 minutes later, when the presence of any species was recorded. Third, leaf litter, grass duff, and the top layer of soil were collected and later placed in Berlese funnels for five days to extract ants. Baiting and litter sampling took place during October of each sampling year.

The grasshoppers on the plots were sampled by capturing specimens with an insect net 30 cm in diameter as they were flushed from the area within the plot. The identity of each specimen was recorded. To document any change in grasshopper abundance within a plot, a transect of 40 sweeps was performed once every two months in each plot, and the number and species of captured specimens were recorded. Sweep-net sampling is sensitive to technique (Evans et al. 1983), so the same individual (JGH) performed all grasshopper sampling.

To reduce the impact of our sampling on the fauna of the plots, a minimal number of voucher specimens were collected. Voucher specimens of ants and grasshoppers were deposited in the Mississippi Entomological Museum at Mississippi State University in Starkville.

The costs to implement the treatments were tallied as they were carried out. Two cost estimates were calculated. Our base costs included wages. The estimate without base included overtime, lodging per diem, fuel, and vehicle and equipment costs.

Analysis

Because the plots were not randomly assigned to experimental conditions and were grouped geographically, starting conditions in the plots in a particular treatment were likely to be fairly similar. Therefore, the species richness data were analyzed using a repeated-measures ANOVA, in which years were the within-subject factor and treatments were the between-subject factor. A significant year-by-treatment interaction indicated that the treatments differed across years in their effects on species richness. Because

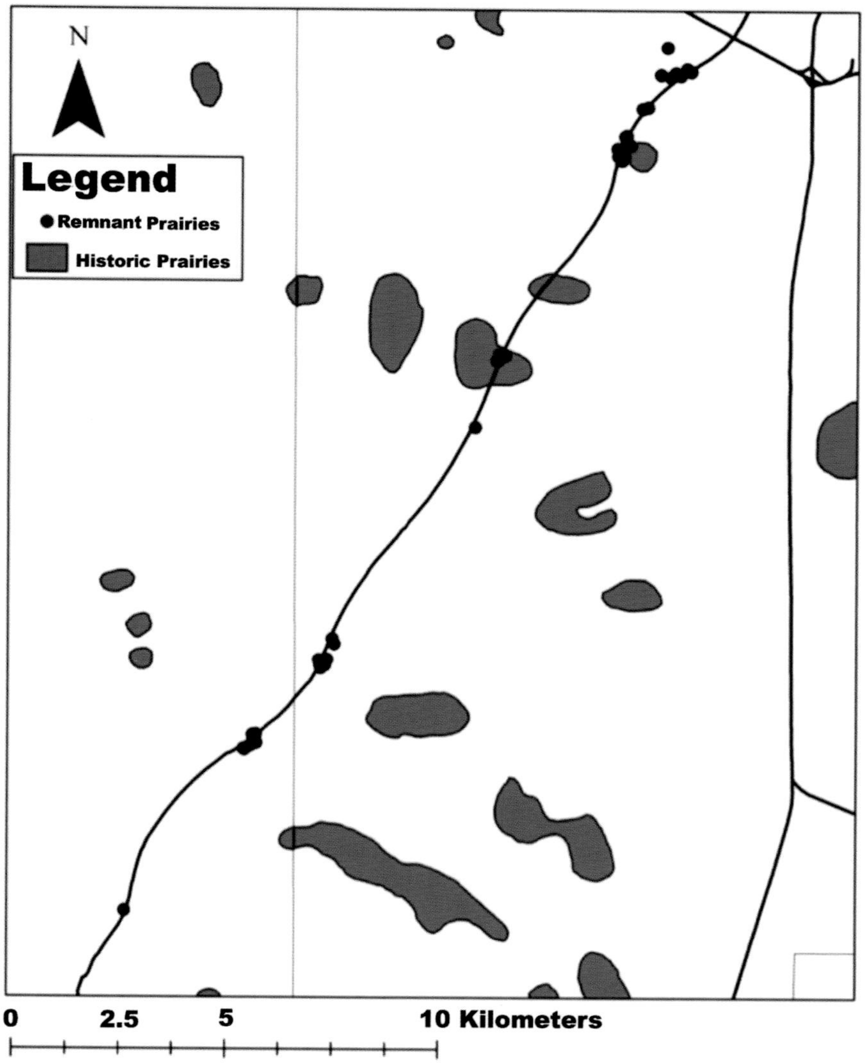

Figure 12.2. Location of test sites along Natchez Trace Parkway, where the study was conducted. Black points represent prairie sites found in Barone and Hill (2008).

three subsets of each data set were often analyzed separately, a Bonferroni correction was used to alter the critical value to P = 0.017 (0.05/3). A post hoc Bonferroni test was used to distinguish between treatments when the repeated-measures ANOVA was significant. When significant treatment effects were seen, paired t-tests were used to compare richness between the first year's pretreatment data and the final year of the study for a given treatment.

Changes to community composition in response to the treatments and years were analyzed using permutation-based multivariate analysis of variance (PERMANOVA), using the Sorensen index as the distance measure on presence/absence data for all three taxa, and with an additional analysis using the abundance of grasshoppers (M. Anderson 2001; Peck 2010). Since the technique requires a balanced design and nonzero values in samples, the sixth plot in the cut/burned treatment was excluded from the analyses, as were plots without grasshoppers. Nonmetric multidimensional scaling was used to visualize the changes to the communities, again using the Sorensen index as the distance measure (Peck 2010). These analyses were done with PC-ORD 6.08.

Results

Initial Conditions

Prior to the initiation of the treatments, there were 855 trees across all of the plots, about 90% (768) of which were redcedars. The treatments differed significantly in their initial density of trees (1-way ANOVA, $F_{3,17} = 9.27$, P = 0.001). There were 60.2 trees per plot in the control treatment, while the densities on the cleared and the cut/burned plots were 23% and 35% lower, respectively. The burned plots had only about a third of the stems (17.2 trees per plot) of the controls. Basal area also differed across the treatments initially, with the control and cleared plots having significantly lower basal area than the burned and cut/burned plots, which did not differ ($F_{3,17}$ = 3.74, P = 0.031).

We examined whether there were correlations between both the initial stem density and basal area on the plots and the species richness of the three taxa. No significant correlations were found for the plants. However, though overall ant richness was not affected, the richness of forest ant species was positively correlated with basal area (table 12.1). This is not surprising, since plots with greater basal area of woody stems were more forested. Grasshopper richness was not related to either stem density or basal area, but the grasshopper abundance on the plots was negatively correlated with stem density.

Precipitation Data

Rainfall varied considerably during the four years of study (fig. 12.3). The first year's rainfall was about 40% above normal, with a particularly wet late summer and autumn. In contrast, precipitation during the second year of the study (2010) was about 10% below the long-term average, but the late summer was very dry. Total rainfall during September 2010 was only about 18 mm (0.7 inch), 80% below normal. The April–June period of 2012 was also unusually dry, with only about 54% of average rainfall.

Plant Richness

In total, 158 plant species were collected on the prairie plots, including 140 native species. A repeated-measures ANOVA found that there was a significant year-by-treatment interaction, suggesting that plant species richness varied across the treatments as the experiment proceeded (fig. 12.4A, table 12.2). In the control plots, there was a consistent year-to-year decline in the number of species, leading to a significant difference between the first and fourth years of the study (paired t-test, t = 3.43, df = 4, P = 0.027). The cause of the drop was not obvious, but it may have resulted from greater shading with the continued increases in the size and density of woody species, especially redcedar, on these plots. In contrast, the cut/burned and cleared plots did not show significant changes in species richness between the first and final years of the study (P > 0.05). An increase in the number of species in the cleared plots in 2010 may have reflected a temporary release of species from the seedbank as well as an increase in the number of non-native species, but this change did not persist. The burned plots did show a significant decline in species richness between the beginning and end of the study (t = 3.60, df = 4, P = 0.038), after two years of modest increases. But since the burns did not carry very well on the plots, it is not clear that the change was associated with the treatment. Instead, the continued growth of the trees on the plot may have been the cause.

Because of concern about how disturbances associated with the treatments could alter the number of non-native species on the plots, separate analyses were performed on native and non-native plant species (table 12.2). For the native species, the results were identical to those of the previous analysis, with a significant treatment-by-year interaction. This was not surprising since the vast majority of the species are native. For the non-native species, only the treatment effect was significant. A post hoc Bonferroni test revealed that the burned plots had a persistently lower num-

Table 12.1.

Pearson product-moment correlations between initial stem density and basal area on the plots and species richness for plants, grasshoppers, and ants.

Stem density	r	P
Plant richness	−0.198	0.389
Non-native plant richness	0.382	0.087
Ant richness	−0.207	0.369
Forest ant species	−0.165	0.476
Grasshopper richness	−0.068	0.770
Grasshopper abundance	−0.587	0.005
Basal area		
Plant richness	−0.226	0.324
Non-native plant richness	−0.359	0.110
Ant richness	0.303	0.182
Forest ant species	0.527	0.014
Grasshopper richness	−0.208	0.364
Grasshopper abundance	−0.303	0.181

Note: N = 21.

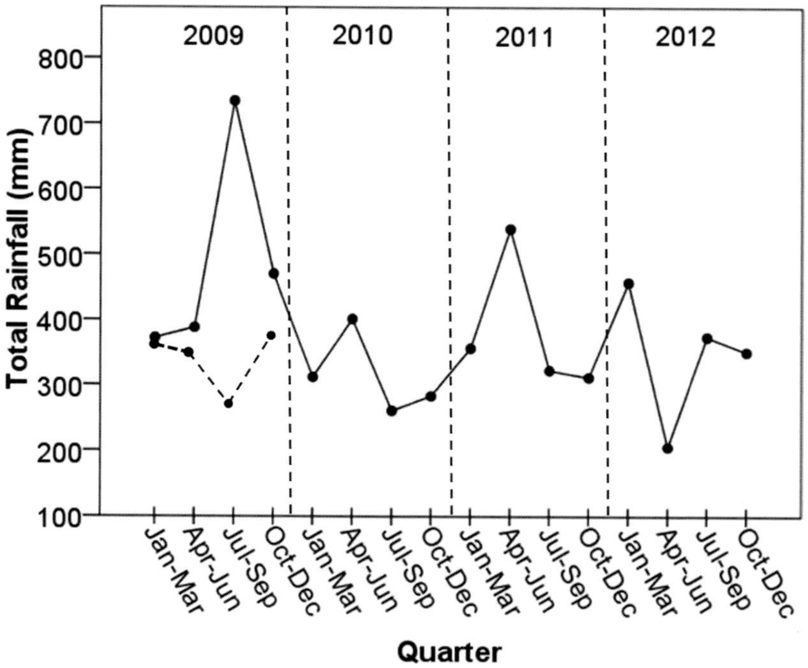

Figure 12.3. Quarterly rainfall totals for the Verona Experimental Station, the nearest weather station to the research sites (National Climatic Data Center 2013). The dashed line in 2009 shows the 30-year average. Annual totals were 1,965 mm for 2009, 1,258 mm for 2010, 1,531 mm for 2011, and 1,389 mm for 2012.

Table 12.2.

Results of repeated-measures ANOVA examining the effects of year and treatment on the species richness of plants, grasshoppers, and ants, as well as the abundance of grasshoppers.

Taxa/group	Treatment		Year		Interaction	
	F	P	F	P	F	P
Plant richness						
All species	3.03	0.058	5.92	0.002	3.71	0.001
Native species only	2.39	0.104	6.10	0.001	3.12	0.005
Non-native species	4.60	0.016	0.339	0.797	1.73	0.106
Grasshopper richness						
All species	5.45	0.008	3.39	0.025	1.32	0.250
Open/prairie species	11.9	< 0.001	9.32	< 0.001	4.89	< 0.001
Forest species	3.15	0.052	1.54	0.216	1.24	0.293
Grasshopper abundance						
Total abundance	11.93	< 0.001	9.32	< 0.001	4.89	< 0.001
Melanopus scudderi	8.22	0.001	4.41	0.008	4.56	< 0.001
Without *M. scudderi*	4.76	0.014	8.21	< 0.001	3.15	0.004
Ant richness						
All species	2.14	0.133	23.52	< 0.001	2.09	0.048
Open/prairie species	1.10	0.376	4.04	0.012	1.49	0.179
Forest species	1.81	0.184	21.06	< 0.001	2.61	0.015

Note: Degrees of freedom (df) for Treatment were 3,17; for Year 3,51; and for Interaction 9,51.

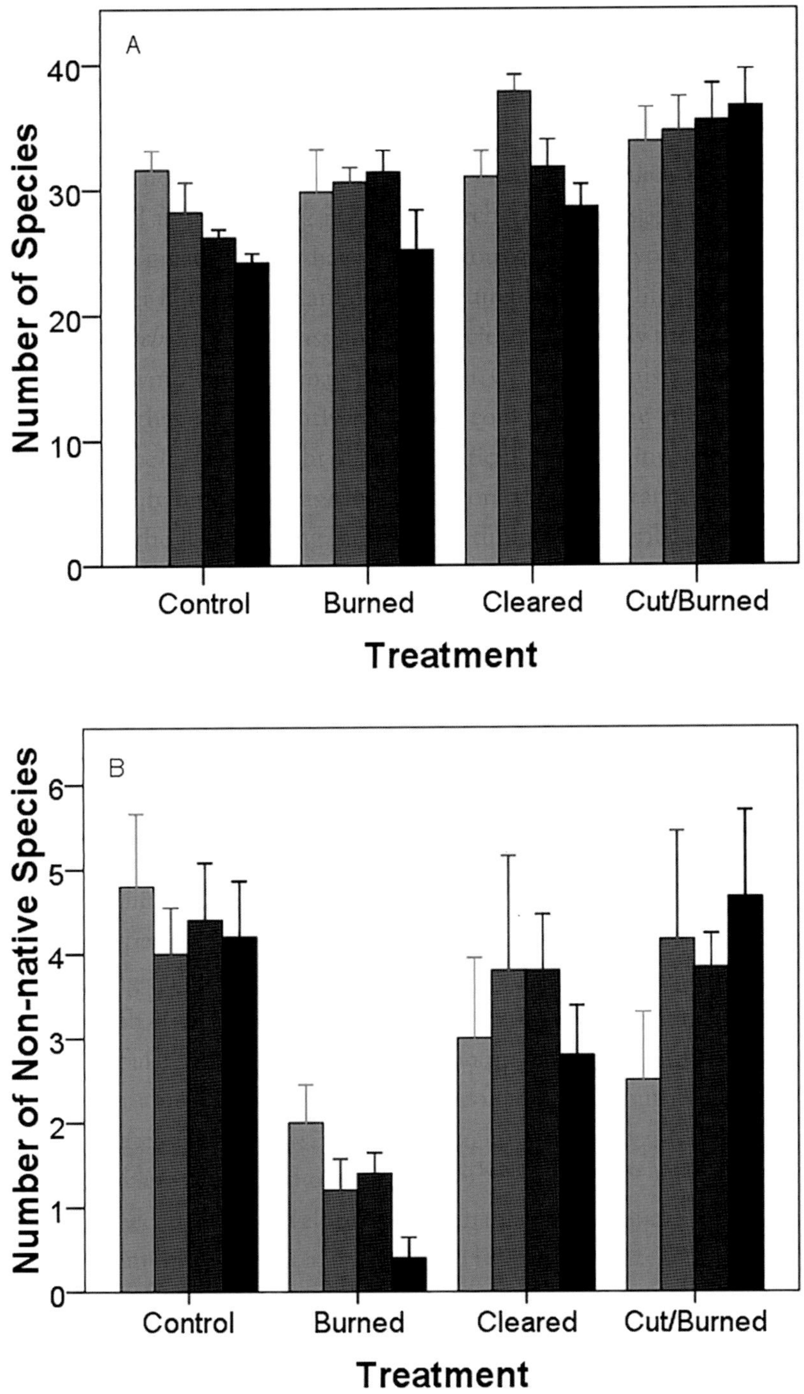

Figures 12.4A and B. Average (± 1 SE) number of plants species on prairie plots along the Natchez Trace Parkway. (A) Total number of species and (B) number of non-native species. Bars are in order from 2009 to 2012.

ber of non-native species than the other treatments, regardless of year (fig. 12.4B), perhaps because of their location.

In summary, plant richness declined in the control plots but was maintained in the cleared and cut/burned plots, suggesting that these treatments did help to maintain the diversity of the local plant communities.

Grasshopper Richness and Abundance

Nineteen species of grasshoppers, all native, were found on the plots. Three are considered prairie specialists in Mississippi, 4 are typically found in forests, and the remaining 12 species are generally found in open habitats (J. G. Hill, unpublished data). A repeated-measures ANOVA found

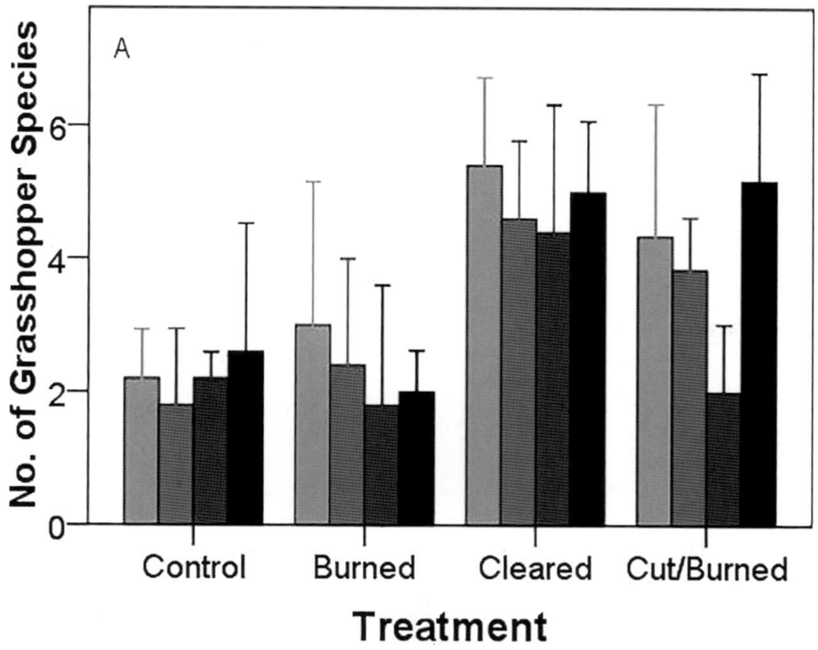

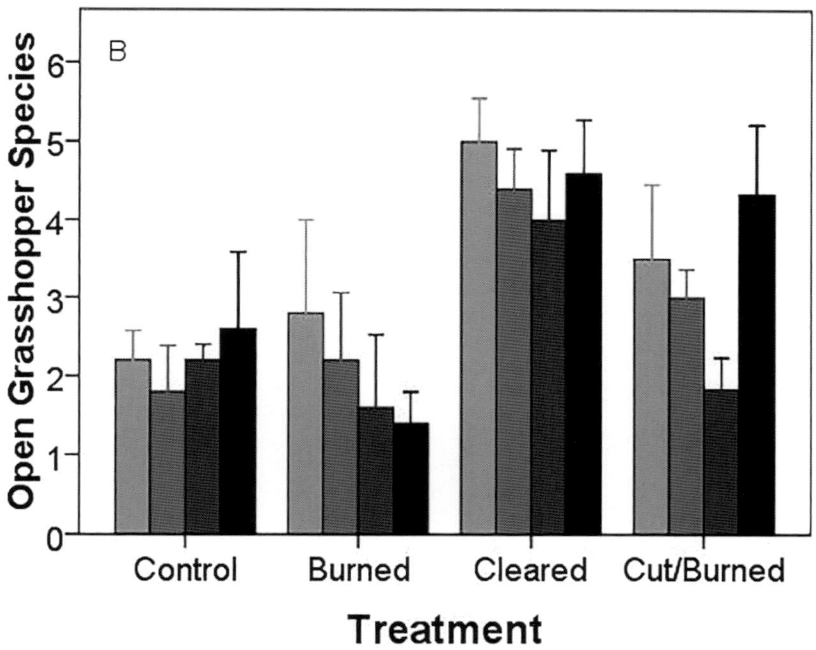

Figures 12.5A and B. Effects of treatments on average (± 1 SE) grasshopper richness for (A) all species and (B) open and prairie specialists only. Bars are in order from 2009 to 2012.

that the treatments had a significant effect on grasshopper species richness, but that time and the interaction were not significant (table 12.2, fig. 12.5A). According to a post hoc Bonferroni test, this reflects an initial and persistently lower number of species in the control and burned plots than in the cleared and cut/burned plots.

Since the treatments should make the sites more attractive to grasshoppers preferring open habitats, the analysis was repeated after removing forest species. In this case, the interaction was significant, along with year and treatment (table 12.2, fig. 12.5B). Again, initial conditions in the plots were likely important in the treatment differences. The post hoc Bonferroni test showed that the cleared plots had higher levels of richness than the control and burned

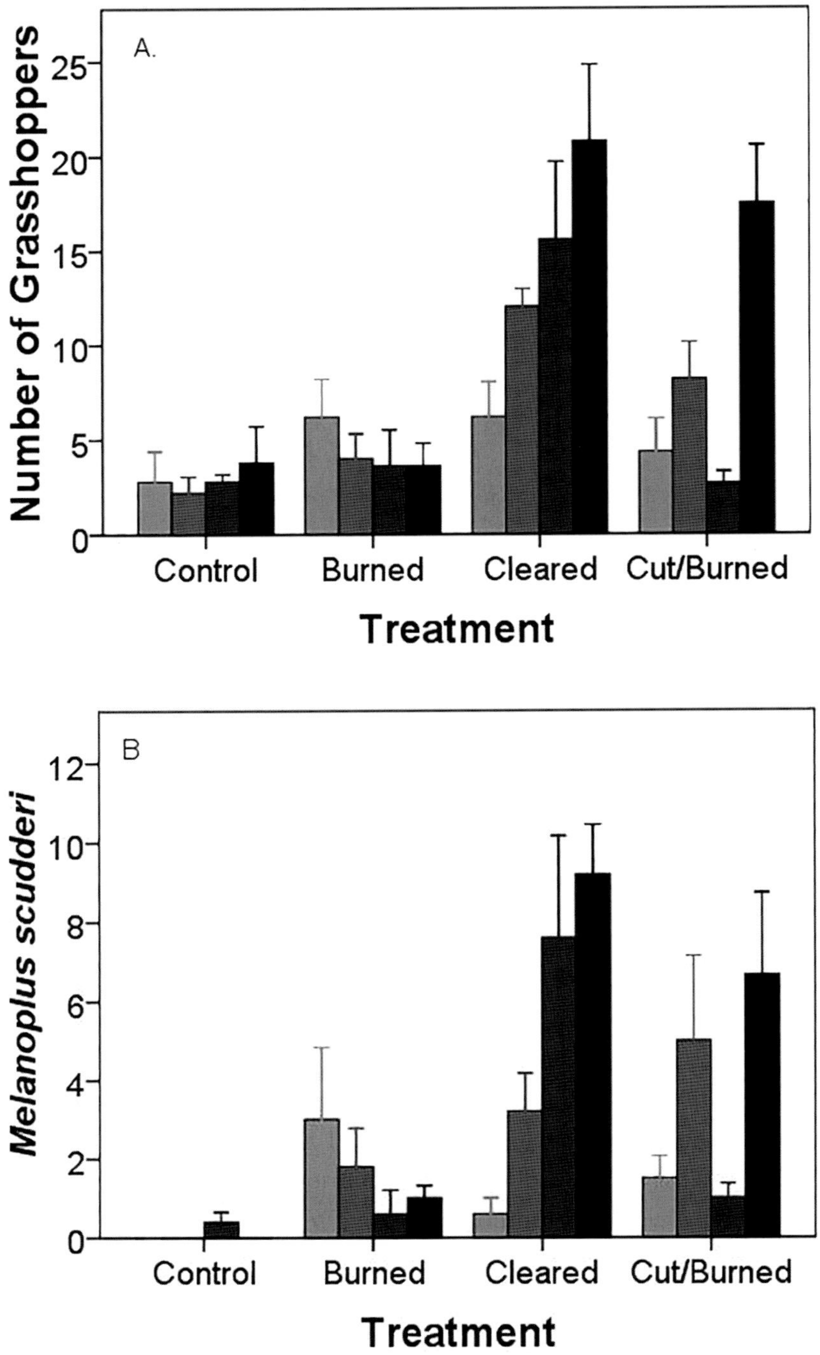

Figures 12.6A and B. Effects of treatments on average (± 1 SE) grasshopper abundance for (A) all species and (B) *Melanoplus scudderi* alone.

plots, though they did not differ from the cut/burned plots. Though both the burned plots and the cut/burned plots showed changes in the richness of open and prairie specialist species, t-tests showed that initial and final year samples were not significantly different. In general, there was little evidence that the treatments altered grasshopper species richness in a consistent, meaningful way.

In contrast, grasshopper abundance was strongly affected by the treatments, with a significant year-by-treatment interaction (fig. 12.6A, table 12.2). There was little change in abundance on the control and burned plots through time, but both the cleared and the cut/burned plots had three to four times more grasshoppers in 2012 than in 2009, differences that were significant (cleared: t-test, $t = -3.19$, $df = 4$, $P = 0.033$; cut/burned: $t = -3.82$, $df = 5$, $P = 0.012$). Across the treatments, these increases were due, in part, to a surge in the numbers of one species, *Melanoplus scudderi*, from 27 total individuals in 2009 to 91 in 2012 (fig. 12.6B). For this reason, the analysis was repeated without *M. scudderi* and for *M. scudderi* alone. For the cleared plots, the increase in *M. scudderi* from 2009 to 2012 was significant ($t = -43.3$, $df = 4$, $P < 0.001$; table 12.2), but the change for other species was not ($t = -1.82$, $df = 4$, $P = 0.14$). In contrast, the increase in *M. scudderi* was not significant in the cut/burned plots ($t = -2.40$, $df = 5$, $P = 0.061$), but the increase in the abundance of other species was ($t = -3.76$, $df = 5$, $P = 0.013$). Overall, grasshopper abundances increased in response to both the cleared and the cut/burned treatments, led by increases in *M. scudderi*, an open-habitat species.

Ant Richness

Fifty species of ants were sampled on the prairie plots, including 20 species largely found in forests, 8 species preferring open or prairie habitats, and 22 generalist species (J. Hill et al. 2008; J. Hill and Brown 2010). Two exotic species, the hybrid imported fire ant (*Solenopsis invicta* × *richteri*) and *Strumigenys hexamera* (Brown), were detected in the plots during the study. There was a significant time effect on species richness, but no treatment effect (fig. 12.7A, table 12.2). Year-to-year changes in richness were similar across treatments and may reflect annual variations in the weather. Because the treatments involved the removal of trees, separate analyses were also done for forest ant species and open/prairie species, to see whether the richness of these groups responded to the treatments in different ways. For open/prairie species, there was again a significant effect of year, but not of treatment (table 12.2). In contrast, forest ant species richness showed a significant interaction (fig. 12.7B, table 12.2), with large declines in the cut/burned treatment (paired t-test for first and final years of the study: $t = 6.57$, $df = 5$, $P = 0.001$) and the burned treatment ($t = 9.00$, $df = 4$, $P = 0.001$). Changes in the cleared and control plots were not significant. In summary, total ant richness was not affected by the treatments, but there was a significant decline in forest ant richness in the cut/burned and burned treatments.

Community Composition

The PERMANOVA for plants found highly significant differences in species composition across years and treatments, though the interaction was not significant (table 12.3). This pattern was also found when only native species were included in the analysis. Inspection of nonmetric multidimensional scaling (NMS) plots showed that initial differences in composition between the plots were likely responsible, at least in part, for the treatment differences (fig. 12.8A), but some changes did occur in apparent response to the treatments. Axis 1 appeared to be related to the level of openness at the sites. Factors affecting axis 2 were less clear, though level of disturbance could be involved. Both the control and burned plots (recall that the burns did not carry very well on these plots) saw more positive scores on both axes, likely reflecting changes related to the increasing dominance by redcedar and other woody plants. For both the cleared and the cut/burned plots, scores were more negative on axis 1, reflecting the greater openness after the treatments. In the final year of the study, plots in both treatments bounced back to having more positive axis 1 scores, likely reflecting a reinvasion of woody species into the plots. (See Discussion.) Overall, then, cut/burned and cleared treatments did see shifts in their plant communities, making them more "prairie-like," though these changes were not fully sustained.

The PERMANOVA analysis for the ant community showed highly significant effects of treatment and year but no interaction, a pattern that was robust even when exotic fire ants were excluded from the analysis (table 12.3). The ordination suggested that ant communities were responding to the treatments at least to a limited extent (fig. 12.8B). The distribution of ant species in the ordination space suggests that axis 1 is related to the openness of the habitat, with forest species having strongly negative scores and open-habitat species having less negative and positive scores. The control and burned plots saw community composition bounce about, maintaining similar ant communities from year to year. The cut/burned plots saw more dramatic shifts, as forest species disappeared from the plots and open-habitat species became more common from 2009 through 2011, though there was a small shift back in 2012, likely because of the return of woody plant species to the plots. The cleared plots saw smaller but similar changes on this forest–open habitat axis, but, perhaps because the community had fewer forest species initially, the shifts were not as dramatic. Changes on axis 2 were less obvious, and it is unclear what underlying ecological factors were related to this gradient. As was already noted, the cleared and the cut/burned treatments saw reductions in forest ant species

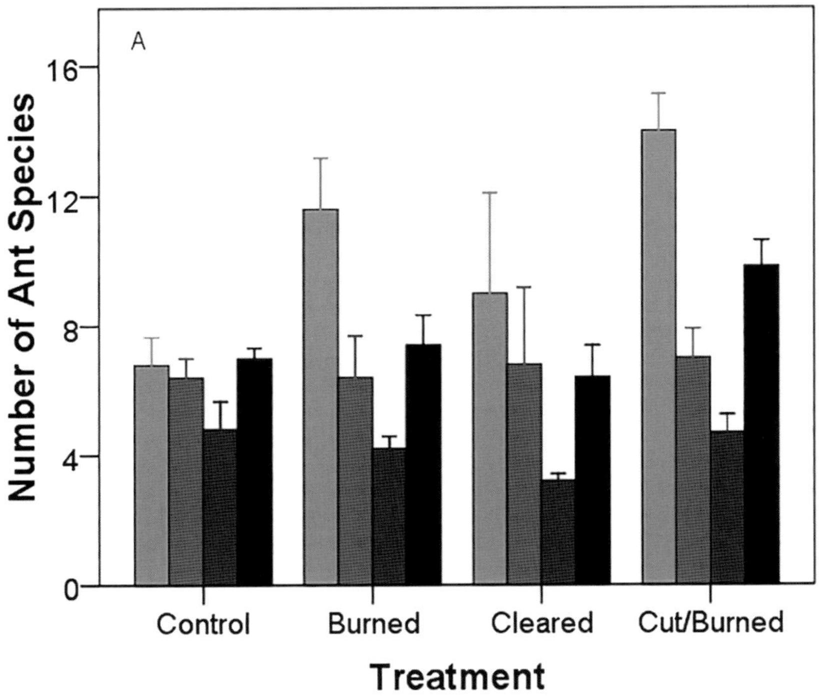

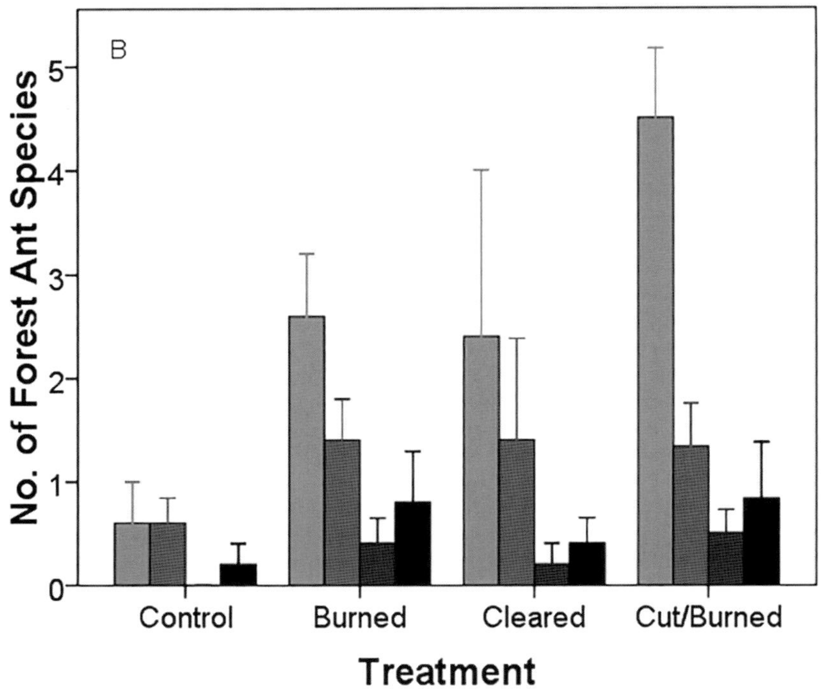

Figures 12.7A and B. Average (± 1 SE) ant species richness for (A) all species on prairie sites and (B) forest ant species only.

as the plots opened up, though in the final year some of these changes diminished.

For the grasshoppers on the plots, the PERMANOVA found a significant effect of treatment on community composition but not of year, nor was there an interaction (table 12.3). Nonmetric multidimensional scaling was not possible on the grasshopper data because of the relatively small sample size, so a principal components ordination was performed instead (fig. 12.8C). Initial community differences were clearly substantial. However, in the case of both the cleared and cut/burned plots, the communities did appear to respond to the treatments, though as was the case with

Table 12.3.

PERMANOVA analyses of responses in species composition to treatments and year.

Group/factor	df	SS	F	P
All plants				
Treatment	3	3.04	9.41	< 0.001
Year	3	1.99	6.16	< 0.001
Treatment × year	9	1.01	1.04	0.348
Native plants				
Treatment	3	2.99	9.57	< 0.001
Year	3	1.65	5.28	< 0.001
Treatment × year	9	0.95	1.01	0.44
All ants				
Treatment	3	4.33	11.5	< 0.001
Year	3	1.89	5.05	< 0.001
Treatment × year	9	1.14	1.01	0.461
Ants—no fire ants				
Treatment	3	4.21	10.4	< 0.001
Year	3	2.20	5.45	< 0.001
Treatment × year	9	1.34	1.11	0.299
Grasshoppers				
Treatment	3	4.61	6.62	< 0.001
Year	3	0.79	1.13	0.335
Treatment × year	9	1.17	0.56	0.986

plants, increases in woody stem densities in the final year seemed to undo some of these changes. It is not clear what variables are related to the axes.

Economic Costs

Removing redcedar trees from remnant prairies was expensive. Burning was clearly the cheapest and quickest option, and it resulted in the least mechanical disturbance to the prairie sites (table 12.4). The cost with the base was $61 per ha (or $150 per ac). The cut/burned treatment was the next-cheapest option, since it involved cutting the trees but burning them on-site. This treatment was obviously more labor intensive and led to more mechanical disturbance of the prairie sites. Clearing, which involved cutting down the trees, chipping them, and moving the chips off-site, was the most expensive of the three treatments tested here, with a base cost of $607 per ha (or $1,500 per ac).

Discussion

In spite of some delays in implementing the treatments, some clear trends were present in the experiment. First, on the control plots, the continued growth of trees appeared to reduce the diversity of prairie plants and kept the richness of ants and grasshoppers low. Second, in this setting, fire by itself had little noticeable effect on the plant or animal communities. To some extent, this result was unexpected, as numerous studies have shown that burns on prairies can increase the species diversity of forbs and grasses (e.g., S. Collins and Steinauer 1998; Brudvig et al. 2007; Bowles and Jones 2013), and of grasshoppers (Evans 1984; Joern 2005; Jonas and Joern 2007). In this study, fires did not carry particularly well on the plots despite two attempts, and they never burned hot enough to kill even moderately sized redcedar trees. It seems likely that this is because many of the sites were surrounded by forest that shaded the prairie grasses, reducing productivity. Additionally, many larger redcedars had a dense layer of litter around their base that resisted burning.

In contrast, the cut/burned treatments halted the decline of plant species richness that was observed in the control plots and shifted the composition toward more light-demanding species. When the cut trees were left on the plots, the fires burned much hotter, making them more effective. As already noted, fires have been shown to increase the species diversity of forbs and grasses (e.g., S. Collins and Steinauer 1998; Brudvig et al. 2007; Bowles and Jones 2013), and when it burns hot enough, fire is especially effective at removing redcedar, which cannot resprout (Briggs et al. 2002a).

Plots in this treatment also saw a dramatic rise in grasshopper abundance and a drop in the richness of forest ant species, again suggesting the effectiveness of this approach. Grasshopper communities have been shown to respond to fires, though the alteration is often mediated by changes in the plant communities (Evans 1984; Joern 2005; Jonas and Joern 2007). In this study, the cut/burned treatment removed redcedars and other woody plants (at least temporarily), opening up the sites. The plots saw a drop in grasshopper richness in the season immediately following the burns, though the community recovered in the final year. This likely resulted either from the direct effect of the fire or from the loss of vegetation. There was little evidence, however, of open-habitat or prairie-specialist species invading

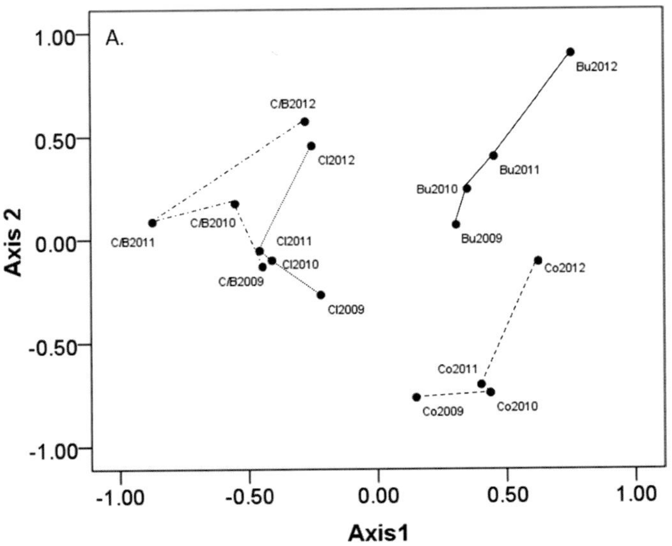

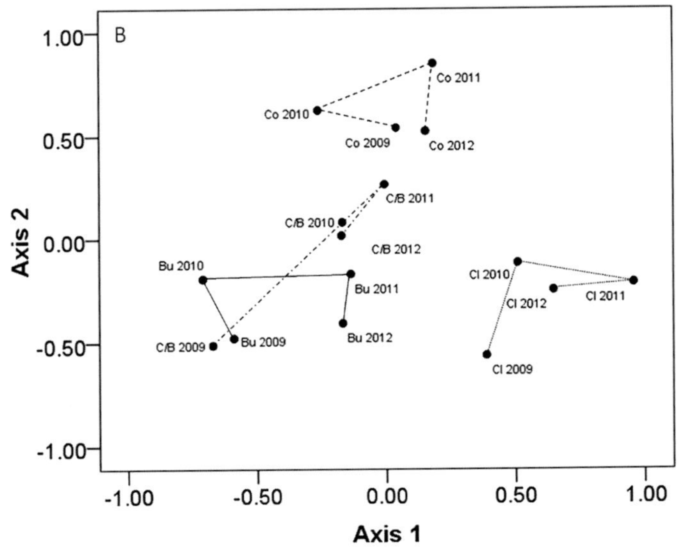

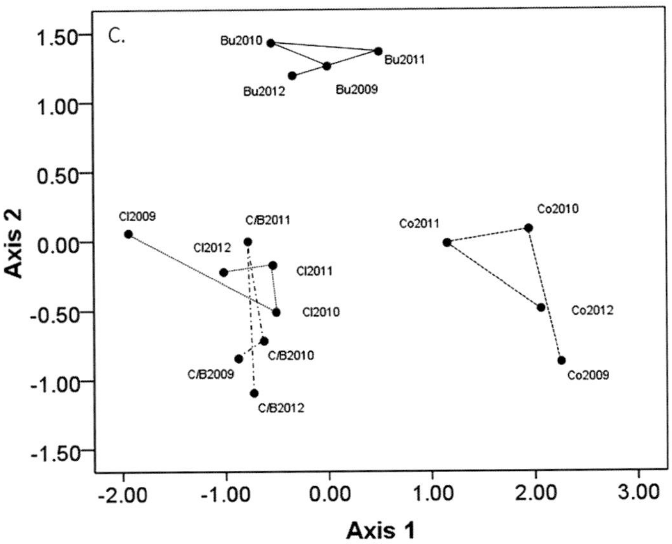

Figures 12.8A, B, and C. Ordination plots showing changes to (A) the plant, (B) ant, and (C) grasshopper communities in response to the treatment. (A) and (B) show the results of nonmetric multidimensional scaling; (C) shows the results of principal components analysis. To simplify the presentation, each point represents the average axis score for that treatment during a particular year. Co = control, C/B = cut/burned, Cl = cleared, and Bu = burned.

the newly opened sites, as composition changed relatively little. It seems likely that the relative isolation of many of the plots by the surrounding forest is slowing this movement. But if the sites remain free of woody vegetation, we expect an increase in grasshopper richness as more prairie and open-habitat species move in.

The cleared plots did not show a significant change in plant species richness between the initial and final years of the study, but there was a short-term increase during the first year after treatment. Alford et al. (2012) reported a similar uptick in herbaceous plant diversity on cleared prairies in Oklahoma and found that the change was more pronounced on sites with an initially higher cover of redcedar. The effect diminished in the second year after treatment, a change Alford et al. (2012) attributed to a loss of annuals as perennial forbs and grasses spread. In contrast, Bates et al. (2000) found that cutting juniper (*Juniperus occidentalis* ssp. *occidentalis* Hook) in Oregon woodlands resulted in persistent increases in forb and grass density, though the changes were mediated to some extent by variations in annual rainfall. Clearing, in general, does seem to result in a more abundant and diverse forb and grass community.

These changes to the plant community may also have led to the surge in grasshopper numbers on the cleared plots, similar to what was seen on the cut/burned plots. There was, however, little difference in the number of species or composition between the first and fourth years of the study. As with the cut/burned plots, we saw little evidence of immigration of open-habitat grasshopper species into the plots.

In this study, the cleared and cut/burned treatments led to a reduction in forest ant species, an unsurprising change, but open-habitat and prairie-specialist ants were not af-

Table 12.4.

Costs of the three treatments in chapter 12.

Treatment	Cost with base	Cost without base
Burned	$61/ha ($150/ac)	$20/ha ($50/ac)
Cut/burned	$101/ha ($250/ac)	$61/ha ($150/ac)
Cleared	$607/ha ($1,500/ac)	$567/ha ($1,400/ac)

fected by the treatments. Open-habitat ants, such as those found in grasslands and savannas, appear to be resilient to the effects of fire (Vasconcelos et al. 2008). For example, Parr et al. (2004) found that within different savanna habitats, fire frequency had no effect on ant richness or abundance, although there were compositional differences between burned and never-burned control plots.

One concern during this study was that the treatments would promote the invasion of imported fire ants (*Solenopsis* spp.), which are adapted to open or early-successional habitats (Tschinkel 2006; King and Tschinkel 2013a), into the plots, which in turn could affect the native ant communities enough to alter the results. In the Black Belt, imported fire ants are commonly found to varying degrees on prairie sites. J. Hill and Brown (2010) sampled the ant fauna at 23 Black Belt sites and found imported fire ants at all of them. The importance of imported fire ants versus habitat disturbance to native ant communities has long been a topic of debate (King and Tschinkel 2006, 2008; Stuble et al. 2013). However, even proponents who suggest that habitat disturbance is the most important factor in structuring native ant communities admit that fire ants may have at least some negative impact on native ant species (King and Tschinkel 2013b).

During the study, 14 of the 21 plots had the hybrid imported fire ant (*S. invicta* × *richteri*) during at least one year, but there was little evidence that treatments affected fire ant frequency. There was, however, considerable turnover in the presence of fire ants across sites during the study. Eleven sites were free of fire ants initially; only seven remained free of the ants throughout the following three years. In contrast, five sites lost fire ants during the study interval. There was no apparent pattern to these changes, and the high turnover suggests that imported fire ants may have trouble persisting in these prairie habitats. This also means that it is difficult to assess whether imported fire ants had any lasting effects on the native ant community.

Unexpected Consequences

Since the goal of this experiment was to assess the consequences of removing woody plants, we did not systematically record the presence of woody plants following the first year of the study. In addition, we did not anticipate that woody plants would, after the treatments, be an important part of the plant communities on the plots. We were wrong.

In both the cleared and the cut/burned treatments, woody species recruited onto the plots, though different species dominated. In three of the cleared plots, Chinese privet (*Ligustrum sinense* Lour.) grew in patches typically centered on sites where large redcedar trees had been. This species spread during 2011 until it occupied a substantial area in these plots. In 2012, its growth seemed to stop, perhaps because of dry conditions. Though there was no overall increase in non-native species on the plots, those that were present tended to be concentrated near the Chinese privet.

In the cut/burned plots, Chinese privet also moved in, but trees such as sumacs were more important. In addition, woody species on some of these plots regrew from cut stems. This situation was probably due, at least in part, to the one-year interval between the cutting of the trees on these plots and the burns. However, even after the burning in 2011, many of these woody species survived. We suspect that the presence of the cut trunks and large branches on these plots for a year increased the recruitment of woody species, perhaps by providing shade to seedlings and retaining moisture.

The presence of woody species on the plots was clearly the result of the treatments. Neither the control plots nor the burned plots had any Chinese privet, sumac, or other fast-growing, woody species move in during the study.

Management Considerations

While the implementation of the experiment did not go as expected, the results do provide some guidance to land managers working to restore small, degraded southeastern prairies. One positive outcome was that none of the treatments led to a substantial increase in invasive forbs or grasses.

Burning the prairies, in an effort to remove the redcedar

Figure 12.9. A burned plot after a burn attempt that was unsuccessful because of insufficient fuel and patches of bare soil.

Figure 12.10. A cut/burned plot in October 2012 showing the later reinvasion of woody species.

Figure 12.11. A cleared plot in October 2012. Note the stand of Chinese privet stretching across the rear of the plot.

trees, simply did not seem to work in this study. The sites were small and shaded and had relatively low productivity, making it difficult to burn them in a consistent fashion (fig. 12.9). Though inexpensive, this was also the least effective treatment. In the cut/burned treatment, the fires did carry across the plots. But treatments did not work as planned here. The year delay between cutting down the trees and burning them appears to have allowed the reinvasion of woody species (fig. 12.10). Moreover, the burns did not consume many of the larger trunks, creating shaded and moist refuges for the seeds from nearby forest plants. It is possible that if thinning had been quickly followed by burning, reinvasion would have been reduced, but the charred trunks would still have ended up on the sites, creating potential refuges.

In our view, the clearing treatment worked best (fig. 12.11). Much of this is practical: clearing can be done even when rainfall prevents burning, and it can be done at sites where burning is not an option because of proximity to urban areas. It is, however, the most expensive option. Additionally, clearing did seem to facilitate the invasion of Chinese privet onto these sites—a disturbing result.

For these reasons, we suggest that a range of approaches be used in removing redcedar trees from these prairie sites, depending on the size of the site, its productivity, and the size and density of the redcedar trees. When a site is relatively large, the density of redcedars is relatively low, and productivity is high, burning should be the first option. Such a site is more likely to sustain a burn, and the fire should be hot enough to kill or severely wound most of the redcedar trees present. Repeated burns may be necessary, however. When a site is small and surrounded by closed-canopy forest and has a high density of redcedars and/or low productivity, the best option would be to clear the site. Clearing would cost less on a small site and may be more efficient if there is a higher density of redcedars. Once a site is cleared and productivity rises, burns may be effective in keeping out redcedar trees.

Acknowledgments

We thank James Lewis, Charlotte Chesnut, and Justin Luke for assistance in the field and for insect specimen preparation. We also thank the Fire Management Division at the Natchez Trace Parkway for their efforts in implementing the treatments. This project was supported by the National Park Service, through the Cooperative Ecosystem Studies Units partnership, by the National Institute of Food and Agriculture, US Department of Agriculture, under Project No. MIS-012040, and by the USDA-ARS Areawide Management of Imported Fire Ant Project (Richard L. Brown, principal investigator).

13

Conservation and Management of Subtropical Grasslands in South-Central Florida

Elizabeth H. Boughton, Patrick J. Bohlen, Steve L. Orzell, Edwin L. Bridges, and Reed F. Noss

Objective

The objective of this chapter is to provide a general overview of the treeless part of the grassland/savanna region of south-central Florida (the Florida Dry Prairie); describe semi-native grasslands on a typical Florida cattle ranch in this region; and present the results of a 10-year experiment manipulating the season of fire, nutrient addition, and grazing to investigate drivers of plant community structure in semi-native grasslands. Insights from this long-term study have implications for management of semi-native grasslands in this region, and for other hydrologically altered subtropical grasslands. Although semi-native grasslands in Florida are biologically impoverished compared with many native grasslands, especially in terms of plant species richness, they serve as preferred habitat for some native animals and represent important opportunities for grassland restoration. Altogether, these vast landscapes, most of them private lands, are among the most important in the state for conservation of native species and natural processes. Plant species nomenclature follows Weakley (2015).

The Grassland Region of South-Central Florida

Florida is estimated to historically have had approximately 800,000 ha of subtropical grassland (most often referred to as Florida Dry Prairie) at the time of European settlement, with most of it occurring in the central peninsula on low, flat interdrainage areas straddling the Kissimmee River north of Lake Okeechobee and extending westward from the lake toward the Gulf of Mexico coast (Noss 2006; Orzell and Bridges 1999, 2006a, 2006b; Bridges 2006a, 2006b; K. Stephenson 2011). Despite the once vast coverage of peninsular Florida grassland/savannas and numerous publications that document their high biodiversity value and endemism (Abrahamson and Harnett 1990; Orzell and Bridges 2006a; Noss 2013; Noss et al. 2015; W. Platt et al. 2015), Florida's grassland/savannas have lagged behind in recognition compared to the attention given to the Florida Everglades. Today they remain largely an enigma at a global scale in comparison to other New World tropical and subtropical grassland/savannas, such as the Brazilian Cerrado, Venezuelan Llanos, or Argentine Pampas.

Bridges (2006a, 2006b) and K. Stephenson (2011) found that the essentially treeless grasslands in south-central Florida occupy flatland terrain with few drainage or topographic impediments to frequent fire and do not differ edaphically from pine savannas elsewhere in the region. Although the grasslands are flat and poorly drained, the varied soils, hydrology, fire regime, and underlying karstic geology have created a mosaic of seasonally wet grasslands and herbaceous depressional and wet-season sheet-flow wetlands (Orzell and Bridges 2006a, 2006b). Within this flat landscape, subtle changes in relative elevation can result in dramatic vegetation shifts (W. Platt 1999; S. Carr et al. 2009).

The pyrogenic, species-rich, seasonally wet, nutrient-poor, essentially treeless grasslands of the Florida Dry Prairie region are comparable to subtropical hyperseasonal, seasonal, and wet grasslands (see Amorim and Batalha 2007; Batalha et al. 2005; Cianciaruso et al. 2005) elsewhere globally, in a region that is otherwise capable of supporting pine savanna and subtropical forests, in the absence of a high fire-return interval (Bridges 2006a; W. Platt et al. 2006). The hyperseasonal, seasonal, and wet grasslands are partitioned almost imperceptibly along an edaphic-hydrologic gradient and have been classified using species composition and environmental characteristics (i.e., dry-mesic, mesic, wet-mesic spodic, wet-mesic alfic, acidic wet, and calcareous grasslands—see Orzell and Bridges [2006b] for a list of dominant, modal, and differential species by community type). The driest permeable sandy soils are considered to be somewhat poorly drained to poorly drained (based on soil taxonomic groupings) and are dominated by wiregrass (*Aristida beyrichiana*), runner oak (*Quercus minima*), and saw-palmetto (*Serenoa repens*) (Orzell and Bridges 2006a), with many long-lived, resprouting, often clonal shrubs (*Lyonia lucida, L. fruticosa, Gaylussacia dumosa, Licania michauxii, Vaccinium myrsinites, Hypericum tenuifolium*), bluestems (*Andropogon decipiens, A. brachystachyus, A. gyrans, A. cabanisii*), lopsided Indiangrass (*Sorghastrum secundum*), and a diversity of forbs, such as composites (*Trilisa paniculata, Pityopsis tracyi, Liatris gracilis, Carphephorus corymbosus*) and many monocotyledons in the Cyperaceae, Eriocaulaceae, and Xyridaceae (Abrahamson and Hartnett 1990; Orzell and Bridges 2006a, 2006b). The seasonally wet, very poorly drained communities occur on alfisols, alfic-spodosols, and spodosols with indurated horizons close to the soil surface, which create saturated or slightly inundated conditions during the wet season and are connected to lower-draining herbaceous wetlands that are themselves subject to periodic landscape wildfires during La Niña droughts (Slocum et al. 2010a; Slocum and Orzell 2013). The poorly drained wet-mesic to wet grasslands are heterogeneous in species composition because of differing edaphic-hydrologic characteristics, and they also differ in whether the substrate is acidic or calcareous. Typically, wet grasslands are characterized by less than half the shrub cover usually found in dry and mesic grasslands (Orzell and Bridges 2006c), and when shrubs are present, they are usually reseeder species such as St. John's-worts (*Hypericum fasciculatum, H. myrtifolium*), *Serenoa repens*, and *Quercus minima*. These are dominant in dry to mesic grasslands with wet seasonal conditions but are restricted to elevated microsites within the wet-mesic grasslands and are absent in the wet grasslands.

Acidic wet grasslands are rich in C4 grass cover (*Aristida beyrichiana, Ctenium aromaticum, Andropogon perangustatus, A. glaucopsis, A. dealbatus*), sedges (*Fuirena scirpoidea, Rhynchospora fascicularis, R. filifolia, R. nitens, R. latifolia, R. tracyi, Scleria georgiana, S. muehlenbergii*), pipeworts (*Eriocaulon compressum, E. decangulare, Syngonanthus flavidulus*), and various species of xyroids (*Xyris ambigua, X. floridana*, and especially Elliott's yellow-eyed grass—*Xyris elliottii*—which is dominant or codominant). Because of the prevalence of C4 grasses and the codominance of Xyridaceae, Cyperaceae, and Eriocaulaceae, acidic wet grasslands are more floristically similar to Neotropical savanna/grasslands than to other North American grasslands. The veredas, a wet grassland type in Brazil and Venezuela (Resende et al. 2013; Araujo et al. 2013), is comparable edaphically, hydrologically, and floristically to Florida wet grasslands. There are many similar plant genera and phylogenetic lineages (*Rhynchospora, Scleria, Xyris, Andropogon, Rhexia, Polygala*), and both are characterized by scattered palms (*Sabal palmetto* in Florida and *Mauritia flexuosa* in Brazil) over a graminoid- and cyperoid-rich ground cover.

Seasonally wet calcareous grasslands are dominated by purple muhly (*Muhlenbergia sericea*), Florida little bluestem (*Schizachyrium rhizomatum*), balsamscale (*Elionurus tripsacoides*), sedges (*Rhynchospora divergens, Scleria verticillata*), Ravenel's pipewort (*Eriocaulon ravenelii*), pineland rayless goldenrod (*Bigelowia nudata* var. *australis*), and Garber's blazing star (*Liatris garberi*). Wet calcareous grasslands in south-central Florida harbor a diverse assemblage of acidic, wetland, and calcareous taxa (from brackish marshes and sandhills) and have among the highest fine-scale vascular plant species richness values recorded in North America (Orzell and Bridges 2006c).

It is thought that the combination of (1) a pyrogenically evolved flora; (2) a very high fire-return interval (annual or biennial) of lightning wildfires that burn unimpeded by a lack of watercourses on a landscape level, particularly during La Niña conditions (Slocum et al. 2010a, 2010b; W. Platt et al. 2015); (3) dystrophic soils; and (4) a seasonal wet-dry climate that inhibits tree and woody establishment operate synergistically to maintain the essentially treeless aspect of Florida grasslands (W. Platt et al. 2006; Bridges 2006a, 2006b). The extent that grazing may have contributed to maintaining these grasslands in this region is less clear but is thought to be of lesser importance (Noss 2013).

Ranching in the Grassland Region

Today, cattle ranching is the dominant land use in the subtropical grassland region of south-central Florida (fig. 13.1). Grazing by large ungulates has long been a part of Florida grassland ecosystems, though historically it was by bison (*Bison bison*) and occurred at much lower densities and frequencies than it does in contemporary cattle production systems (Belue 1996). Prior to and overlapping bison, a rich assemblage of grazing and browsing megaherbivores occupied Florida grasslands and savannas for more than 10 million years before the extinctions at the end of the Pleistocene (S. Webb 1990; Noss 2013). Since the introduction of cattle into Florida in the 1500s, grazing has become a predominant land use in central Florida.

The vast majority of this prairie region is still rural and supports at least 88 federally listed and state-listed species as well as state species of greatest conservation need, many of which rely on large, open tracts of habitat, including grasslands. These include the crested caracara (*Caracara cheriway*), burrowing owl (*Athene cunicularia floridana*), Florida sandhill crane (*Grus canadensis pratensis*), Florida black bear (*Ursus americanus floridanus*), and Florida panther (*Puma concolor coryi*) (Maehr and Cox 1995; Maehr 2007; Everglades Headwaters National Wildlife Refuge proposal 2010; Noss 2013). Grassland bird species are the fastest declining group of birds in the United States, and privately owned Florida ranches are critical for their conservation, especially because only a small percentage of remaining grasslands are publicly owned and managed for conservation (Morrison and Humphrey 2001; Butcher and Niven 2007). Interestingly, some imperiled grassland birds, particularly the crested caracara, burrowing owl, and Florida sandhill crane, prefer improved and semi-native grasslands, and other short-grass habitats, to native Florida grasslands (e.g., Morrison and Humphrey 2001), which may reflect their evolution in an environment shaped by grazing megaherbivores, such as mammoths, which may have kept grasses short in stature in areas where they concentrated (Noss 2013).

At present, the Florida grassland region is a mosaic of native grasslands, pine and palm savannas, and working ranchlands that include both improved and semi-native pastures. Typical ranches in Florida have both improved pastures that

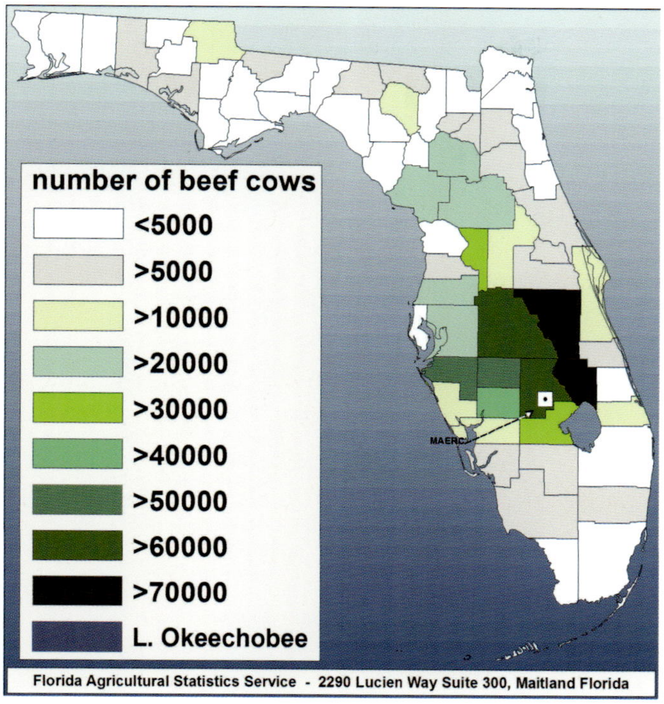

Figure 13.1. The number of beef cows is greatest in Polk, Highlands, and Okeechobee Counties, Florida, which overlap with the area that historically had the most prairie ecosystems.

are planted with non-native productive forage grasses, such as Bahia grass (*Paspalum notatum*), and semi-native grasslands that are used for grazing in the dormant season as a standing hay crop (Swain et al. 2013). Semi-native grasslands are defined as areas that have never been plowed or fertilized and contain variable levels of Bahia grass that were established by human-added seed. Semi-native grasslands, though hydrologically altered with some non-native forage grasses, contain many native species and heterogeneous habitats, supporting high biodiversity (Swain et al. 2007; Boughton et al. 2013). Depending on habitat type, soil, and hydrology, semi-native grasslands can range from open pinelands to dry and wet grasslands, and palm savannas with embedded depressional wetlands. Rare habitat types can often be found on Florida's ranchlands, including calcareous wet grasslands (Orzell and Bridges 2006b) and threatened cutthroat grass (*Coleataenia abscissa*).

Typical management of semi-native grasslands includes light to moderate grazing by livestock, usually in December through May, and dry-season prescribed fire (ignitions in January and February). Prescribed fires are used primarily to improve forage quality of native grasses during the dry season, when forage quality is at its lowest, and to reduce

Figure 13.2. A remnant of the Indian Prairie, a complex of vegetation that occupied the area between Lake Istokpoga and Lake Okeechobee, is found in the semi-native pastures at Buck Island Ranch. Here, purple muhly is in flower, surrounded by sand cordgrass and beggar ticks.

woody encroachment from competing with forage production (Hughes 1975; Kalmbacher 2002). In contrast, lightning wildfires historically occurred during the transition from the dry to the wet season (Abrahamson et al. 1984; Robbins and Myers 1992; Menges et al. 1993; Slocum et al 2010b, W. Platt et al. 2015). The transition season is characterized by drought, intense solar radiation, low humidity, and warm air temperatures (W. Platt et al. 2015). Some historical reports indicate that early Florida cattlemen burned within at least a portion of the lightning fire season (Eldredge 1911; Huffman and Platt 2014). The seasonal difference between natural and typical prescribed ignitions has consequences for biodiversity because fire seasonality can have a profound effect on species abundance, phenology, and reproduction (Robbins and Myers 1992; Howe 1994; Fill et al. 2012; K. Robertson and Hmielowski 2014; W. Platt et al. 2015). Despite the differences in fire season between ignitions set by today's ranchers and fires ignited by lightning, these fires have aided in maintaining grasslands on cattle ranches that would have otherwise fallen victim to the severe woody invasion that has occurred in fire-suppressed grasslands elsewhere on private lands in southern Florida. Therefore, Florida cattle ranches help sustain valuable grassland remnants, thereby presenting otherwise lost opportunities for restoration and conservation.

Very little is known about some historical vegetation types, and sometimes remnants on ranches are the only vestiges of natural vegetation in a region. For example, at the MacArthur Agro-Ecology Research Center (Buck Island Ranch; www.maerc.org), there are remnants of the Indian Prairie, a complex of wet grassland and wet palm savanna vegetation that covered about 250 mi^2 between Lake Istokpoga and Lake Okeechobee (Bridges and Orzell 2005; R. Harper 1927; see fig. 13.2). Within this remnant Indian Prairie, although impacted by grazing and hydrologic alterations on Buck Island Ranch, are some of the most-intact wet grasslands and wet palm savanna reminiscent of R. Harper's description (1921, 1927) of the Indian Prairie (Bridges and Orzell 2005). In Florida, rangelands and semi-native grasslands that contain remnant historical vegetation are often the most significant remnants in a region otherwise impoverished in biodiversity (Noss 2013).

Florida's semi-native grasslands are important for conserving the biodiversity of plants (Boughton et al. 2010) and birds (Morrison and Humphrey 2001; Willcox et al. 2010), and for maintaining water quality (Swain et al. 2007). Semi-native grasslands in Florida are under constant threat of conversion to more intensive land uses, abandonment (no fire), and disturbance by invasive feral swine (Swain et al. 2007; Boughton et al. 2013; Noss 2013). Over a 15-year period ending in 2003, there was a decrease in the amount of rangeland (from 74,000 ha down to 46,641 ha) and semi-native grassland (from 62,000 ha down to 33,453 ha) in the subtropical grassland region of south-central Florida (Hiscock et al. 2003). In the northern Everglades (the upper headwaters for the Everglades), conservation efforts now include private ranchlands to preserve biodiversity and ecosystem services (Bohlen et al. 2009; Everglades Headwaters National Wildlife Refuge proposal 2010). As more ranchland is included in conservation easements and as protected lands, there will be an increasing need for understanding how management of these lands affects conservation outcomes, particularly in semi-native grasslands that have the highest conservation value. There is also an interest on the part of ranchers in maintaining good stewardship practices on semi-native pastures. Plant composition in semi-native grasslands released from typical management (burning and light grazing) will likely shift to greater abundance of competitive species and increasing woody encroachment, with potentially negative effects on higher trophic levels. The rate and intensity of these shifts may differ depending on the degree of hydrological alteration and whether or not fertilizer has been added.

MacArthur Agro-Ecology Research Center at Buck Island Ranch

The MacArthur Agro-Ecology Research Center at Buck Island Ranch (MAERC) is a 4,170 ha commercial cattle ranch with approximately 3,000 cow-calf pairs, run by Archbold Biological Station (Swain 1998). The center is located about 30 km northwest of Lake Okeechobee. Its subtropical climate has distinct wet (May/June–October/November) and dry (October/November–April/May) seasons, and an average annual rainfall of approximately 1,300 mm, with an average temperature of 26°C in July (midsummer) and 13°C in January (midwinter). Most soils are poorly drained, acidic, sandy spodosols and alfisols.

The property contains approximately 50% improved pastures and 50% semi-native pastures, and over 600 seasonal wetlands. Two USDA Wetlands Reserve Program (WRP) sites are located on the property and encompass over 700 acres of restored wetlands. The west, east, and south marshes are semi-native pastures that are a mosaic of wet prairies and cabbage palm savannas composed of a mixture of native and planted forage grasses. Remnants of the Indian Prairie can be found in the east marsh WRP site. The ranch landscape is a mix of varied degrees of modified grasslands and seasonal wetlands embedded within the grassland matrix. There are also several large live oak (*Quercus virginiana*)/cabbage palm hammocks at several locations on the ranch. The heterogeneous mosaic of habitats supports many species, including the wide-ranging Florida black bear and panther, which commonly pass through the ranch.

Like most ranches in the region, MAERC contains two main pasture types (Boughton et al. 2010, 2013; Willcox et al. 2010): improved and semi-native pastures. Improved pastures are dominated by the non-native Bahia grass (*Paspalum notatum*), are extensively ditched, and are fertilized annually with nitrogen. Semi-native pastures are less extensively ditched, are never or very rarely fertilized, and are dominated by native perennial grasses, though they contain varying levels of Bahia grass. Native grasses include broomsedge (*Andropogon virginicus* var. *virginicus*, the old-field variant), redtop panicgrass (*Coleataenia longifolia* ssp. *longifolia*), and carpetgrass (*Axonopus fissifolius*). Common forbs of semi-native pastures include Carolina redroot (*Lachnanthes caroliniana*), thoroughwort (*Eupatorium recurvans*), meadow beauty (*Rhexia mariana*), and Virginia buttonweed (*Diodia virginiana*), and dominant woody herbs are dogfennel (*Eupatorium capillifolium*) and flat-top goldenrod (*Euthamia hirtipes*). MAERC pasture management includes prescribed burning of pastures in both types every two to three years, primarily during the dry season. On average, 20% of the property is burned each year.

Research on Semi-native Grassland Management at MAERC

Over the years, we (EHB and PJB) have accumulated data on semi-native grassland dynamics from several short-term and long-term studies. In 2002 we established one of our longest studies, known as FIREPLOTS, to examine interactive effects of fire and nutrients on grassland dynamics, with an additional emphasis on grazing in one of our experimental comparisons. The experiment was a split-plot design conducted in an area of relatively homogeneous mesic grassland that had been previously used as semi-native pasture. This particular site was dominated solely by native species, there was no Bahia grass present, and it had never been fertilized prior to the experiment. In October 2002, an area of 105 m × 80 m was fenced off to prevent cattle grazing (Boughton et al. 2013). Within the fenced area, 12 plots, each measuring 20 m × 20 m (four rows of three), were created and each was divided into four subplots, each measuring 10 m × 10 m. Outside the fenced area four additional plots, each measuring 20 m × 20 m, were established and exposed to cattle grazing, with each divided into four subplots, each measuring 10 m × 10 m. The main plot treatment in ungrazed plots was the burn treatment (wet season burn, dry season burn, and unburned control). Annually, from 2002 to 2007 and from 2009 to 2011, dry season burn plots were burned in January and wet season burn plots were burned in June. Grazed plots received only the dry season burn treatment and were open to grazing, with cattle densities (~ 2.1 ha per cow-calf pair) and timing determined by normal ranch practices in the surrounding large pasture, to simulate typical ranch burning and grazing management. This treatment was chosen because it was not practical to set up the entire replicated split-plot experiment with grazing as a factor, but we were interested in comparing results from the fenced plots to responses observed under typical burning and grazing regimes (i.e., dry season burn–grazed).

The subplot treatment was nutrient addition: nitrogen (N), phosphorus (P), nitrogen plus phosphorus (NP), and an unfertilized control plot. For N plots, N was added annually in March at 50 kg N ha^{-1} as NH_4NO_3 (35% N) from 2002 to 2004. Nitrogen application was cut back to 25 kg N ha^{-1} starting in March 2005–2007 and was switched to 25 kg N ha^{-1} as urea (46% N) in March 2008 because of difficulty in obtaining NH_4NO_3. Phosphorus was added annually in March at 14.3 kg P ha^{-1} as super triple phosphate (46% P_2O_5). Nutrient addition levels were chosen to simulate both typical historical agronomic fertilization rates for more intensively managed pastures (Swain et al. 2007) and variation in nutrient enrichment that might be expected in areas of semi-native grassland or abandoned pasture that differ in historical fertilizer use. For NP plots, combined N and P treatments were added annually in March of every year. Nutrient treatments were assigned randomly within each main plot. Prior to nutrient addition treatments, this area had never been fertilized.

The objective of this experiment was to understand how plant species composition responded to the season of fire under varying nutrient regimes when grazing was removed. In October 2002, one vegetation sampling point was randomly selected within each treatment subplot and permanently marked for annual sampling of species composition in November, at the end of the growing season. A 1 m buffer around each plot was applied to reduce edge effects. The random species composition point represented the center of a circle with an area of 10 m^2. At each sampling, the percent canopy cover of individual species was assigned to one of seven classes of a modified Daubenmire scale (Daubenmire 1959, 1968). Species were also classified into functional groups: graminoids (all species in Poaceae, as well as sedges and rushes: 24 total); forbs (monocotyledonous and dicotyledonous herbs: 14 total); and woody species (*Eupatorium capillifolium* and *Euthamia hirtipes*), and diversity was calculated for each of these groups. Detailed methods for monitoring species composition can be found in Boughton et al. (2013).

Response of Vegetation Structure to Season of Fire and Nutrient Addition

In Florida grasslands, several possible factors have been identified that maintain grassland structure and prevent encroachment by woody species. These include fire, seasonal flooding, and low-nutrient soils (Orzell and Bridges 2006b; K. Stephenson 2011; Noss 2013). Season of fire can also influence vegetation structure (Abrahamson 1984; W. Platt et al. 1988, 2015; Howe 1994; Main and Barry 2002; Towne and Kemp 2003; Slocum et al. 2003). At MAERC, we found that season of burn led to significant changes in species composition after 10 years of implementing treatments (Boughton et al. 2013).

An ordination of vegetation composition revealed that the season of burning (dry season vs. wet season) had significant consequences for plant composition, as did the exclusion of fire (fig. 13.3). When the study began, in 2002, there were no significant differences among the experimental plots in richness, diversity, or community composition, as indicated by the substantial overlap of all plots in a nonmetric multidimensional scaling ordination (fig. 13.3A). In 2010, after eight years of burning and nutrient addition treatments, the season of burn treatment resulted in strong separation of the plots (fig. 13.3B). The structural differences in the season of burn treatments were driven by shifts in species composition and dominance of species in relation to season of burn and exclusion of fire. Dry season burn plots contained flat-top goldenrod, while controls (unburned plots) were dominated by dogfennel or Carolina redroot. In 2010, wet season burn plots overlapped with dry season burn–grazed plots, suggesting similarity in species composition. However, in 2012, substantial change in species composition took place after an intense feral swine rooting event, resulting in a significant increase in Carolina redroot and a convergence of some wet season burn plots and unburned control plots (fig. 13.3C).

Feral swine broke through the fence in February 2012; prior to this rooting event there had been only very limited feral swine activity, restricted to one plot. Feral swine completely overturned the soil in 16 of the 64 species composition plots, which may have contributed to the increased heterogeneity of plots from 2010 to 2012. Feral swine disturbance occurred in all blocks, but activity was less in the recently burned dry season burn plots. In plots that were not rooted, vegetation shifted very little, while in plots that were rooted, Carolina redroot increased by 40% on average, and the dominant grasses significantly declined (Boughton and Boughton 2014).

Altering the fire and grazing regime caused shifts in grassland vegetation. Removing both grazing and burning resulted in yearly directional change in composition, suggesting succession in unburned control plots (fig. 13.4A). Wet season burn plots also exhibited directional change, probably as vegetation developed after removal of grazing and forbs responded favorably to the historical burn season (fig. 13.4B). The largest shift in these plots occurred between 2011 and 2012 after the feral swine disturbance. Dry season burn and dry season burn–grazed plots remained relatively stable through time, though dry season burn plots showed more change along one axis of the ordination (fig. 13.4C, D). Changes in these plots were most likely due to yearly precipitation differences and grazing pressure in grazed plots (fig. 13.4C, D). The main difference between grazed and ungrazed dry season burn plots was the greater abundance of sedges and mat-forming carpetgrass in the grazed plots and the greater abundance of flat-top goldenrod in the ungrazed plots.

The season of burn treatments also differed in total number of species: graminoid richness was lowest in plots with fire exclusion and highest in grazed plots (because of sedges), and the number of forbs was greatest in wet season burn plots. Additionally, graminoid richness was lowest in plots with both N and P addition. The percent cover of functional groups and dominant species also differed among treatments. Graminoid cover was highest in the dry season burn–grazed plots. Forb cover was highest in unburned control plots, mostly because of the monocot *Lachnanthes caroliniana*. Woody herb cover was highest in unburned control plots and dry season burn plots, although the species differed between these two treatments: dogfennel was most abundant in unburned plots and flat-top goldenrod was most abundant in dry season burn plots. Wet season burn plots had no cover of woody species.

At the beginning of the experiment, all plots had similar levels of broomsedge, a C4 perennial bunchgrass, and for the first five years of the experiment, broomsedge increased in unburned plots. After a drought year in 2007, most plots had low cover of broomsedge. In general, broomsedge cover was highest in dry season burn plots without nutrients added. At the start of the experiment, the percent cover of redtop panicum, another C4 perennial bunchgrass, was similar among plots, but after five years the cover of this species dropped dramatically in the unburned control plots while remaining similar in all other treatments (see fig. 13.5). Carpetgrass, a sprawling C4 grass, was most abundant in dry season burn–grazed plots with N. The most abundant forb, Carolina redroot, was most abundant in unburned, ungrazed (control) plots, and least abundant in dry season burn–grazed plots. This species also exhibited a lag response to the burn treatments, becoming abundant in unburned plots five years after the start of the experiment, and in the wet season burn plots after 10 years, possibly in response to rooting by feral swine.

In this grassland study system, forb richness was greatest in wet season burn plots, and the community overall was dominated by bunchgrasses. These results support those of other studies showing that many species rely on wet season burning (Main and Barry 2002; Orzell and Bridges 2006a; Fill et al. 2012; W. Platt et al. 2015). Our result of lowest richness in unburned plots is not consistent with results found in tallgrass prairie, where diversity and heterogeneity were highest in infrequently burned areas (S. Collins and Steinauer 1998). This difference may reflect the much higher lightning frequency in the South and the older and stronger evolutionary relationship between southern grassland plants and fire, compared to the Great Plains (Noss 2013). The lack of woody herbs in our wet season burn plots suggests that ignitions at or near the onset of the wet season are instrumental in mimicking a natural wildfire regime to successfully maintain biodiversity in southern grasslands.

What are the mechanisms behind these patterns? One obvious possibility for decreasing diversity in unburned plots is competitive exclusion. Unburned plots were dominated by either dogfennel or Carolina redroot.

Since we did not observe decreases in diversity in nutrient addition plots that were burned, this suggests that disturbance by burning may inhibit competitive exclusion by species that compete well for nutrients and that other factors, such as light or water, may be the limiting resource for many species. In all analyses, there were significant interactions between burning and year, implying that moisture availability may be crucial in mediating the response of grassland plants to fire, a result consistent with those of other studies on native grasslands (Towne and Kemp 2003; B. Foster et al. 2009).

Another potential mechanism for greater diversity in wet season burn plots is the direct positive influence of fire on germination and flowering by forbs. These forbs may be specifically adapted to germinate quickly after fire when light and temperature levels at the surface are increased. Water availability may also be higher after a burn because of decreased uptake by vegetation, and the increased rainfall associated with the onset of the wet season. That forb richness was highest in wet season burn plots in our study is in contrast to results found in North American tallgrass prairie where frequently burned grasslands resulted in increased competitive abilities in C4 grasses and consequentially reduced forb richness (Briggs and Knapp 2001). However, in Florida and elsewhere in the southeastern Coastal Plain, many species have adapted to transition-season burning to stimulate flowering and germination (Robbins and Myers 1992; Main and Barry 2002; Menges and Quintana-Ascencio 2004; Fill et al. 2012).

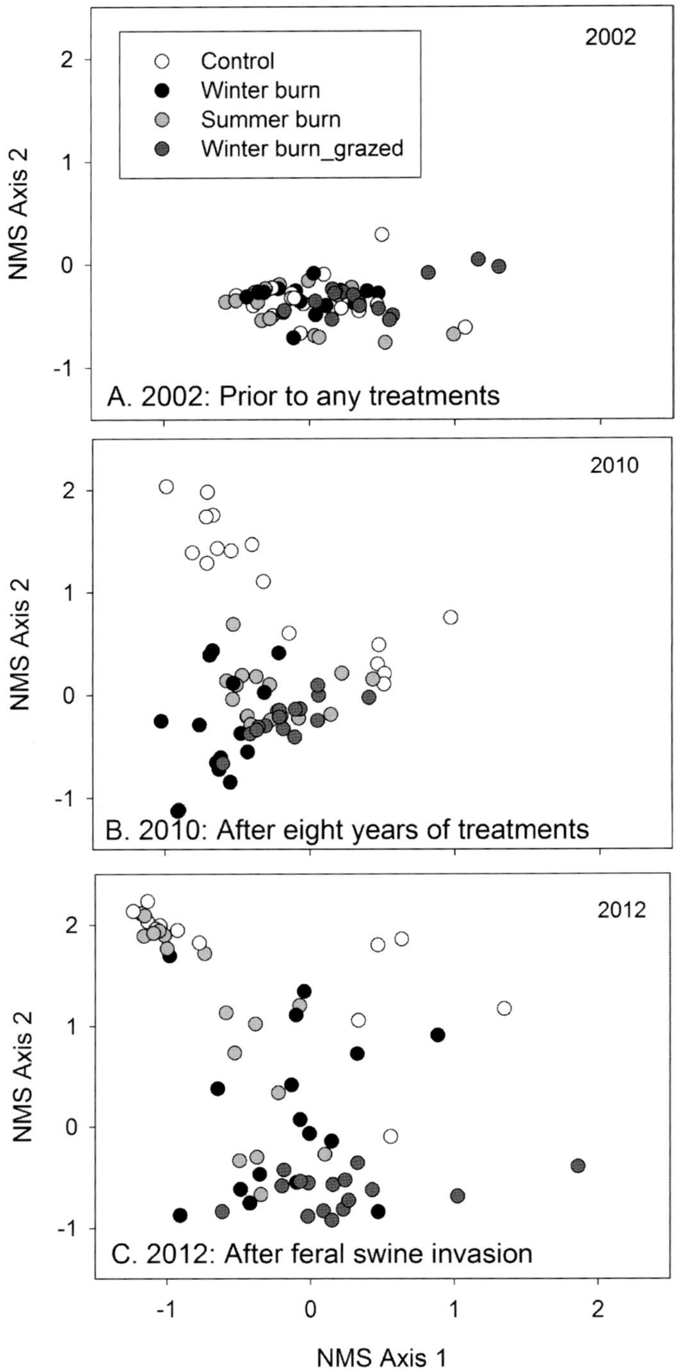

Figure 13.3. Nonmetric multidimensional scaling ordination showed that there was no difference in species composition among plots at the beginning of the experiment (A), but after eight years of experimental burning treatments, species composition diverged (B). Species composition also diverged after feral swine disturbance (C). The ordination had low final stress (13.56), and the two axes were significant (P = 0.019). The ordination represented 92% of the variance in the data (Axis 1, $R^2 = 0.59$; Axis 2, $R^2 = 0.33$).

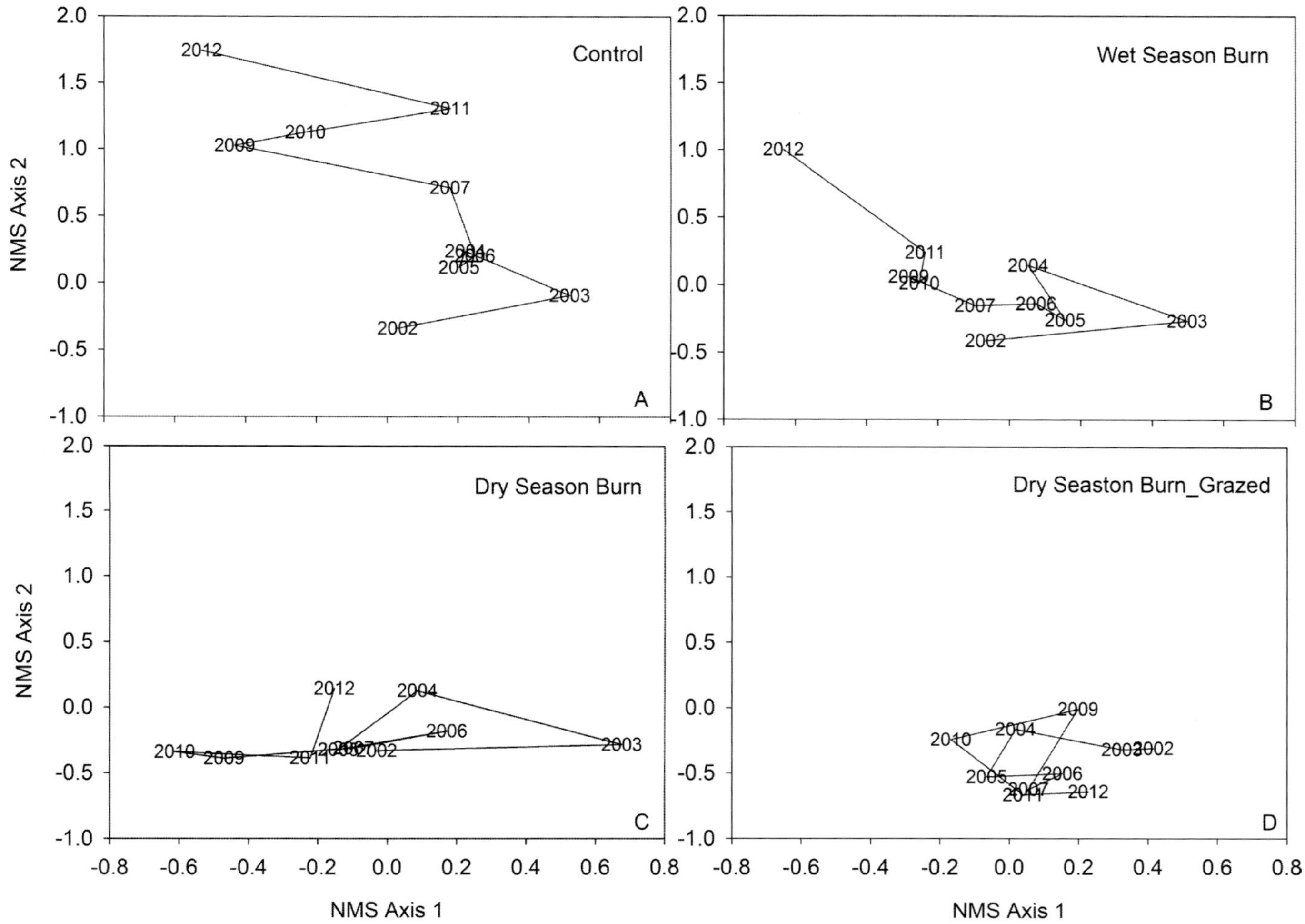

Figure 13.4. Nonmetric multidimensional scaling ordination of vegetation changes since 2002. Each year, four species composition plots were assessed in each of the four treatments: (A) control (unburned); (B) wet season burn; (C) dry season burn; (D) dry season burn–grazed. Aerial cover of plants is estimated in late summer. Directional change is occurring in control plots and wet season burn plots, but not in dry season burn or dry season burn–grazed plots.

Conservation and Management Recommendations

Five salient management points can be derived from the results of our 10-year study on season of burn and nutrient enrichment in subtropical semi-native grassland: (1) wet season burning promoted forb diversity; (2) wet season burning was the most effective regime in reducing woody herbs; (3) dry season burning can promote some woody herbs, especially flat-top goldenrod; (4) lack of burning can contribute to proliferation of a highly productive woody herb (i.e., dogfennel) during drought conditions; and (5) disturbance by feral swine can cause dramatic shifts in vegetation.

Our wet season burn treatment probably provided a good surrogate for historical transition-season fires because we attempted to ignite fires before wet season rainfall events began in earnest (Slocum et al. 2003, 2010b; W. Platt et al. 2015). Since the exact timing of the transition season varies from year to year and is strongly influenced by the El Niño–Southern Oscillation cycle, it should be determined using cumulative rainfall anomalies (see Slocum et al 2010b). Annual burning may also be useful in reducing woody herbs, which is consistent with results from nearby Myakka River State Park on dry prairie (Watts and Tanner 2006). Burning later during the wet season may produce different results; in fact, late wet season burning may be detrimental to native grass and forb species, for instance by top killing and then flooding before resprouting can occur

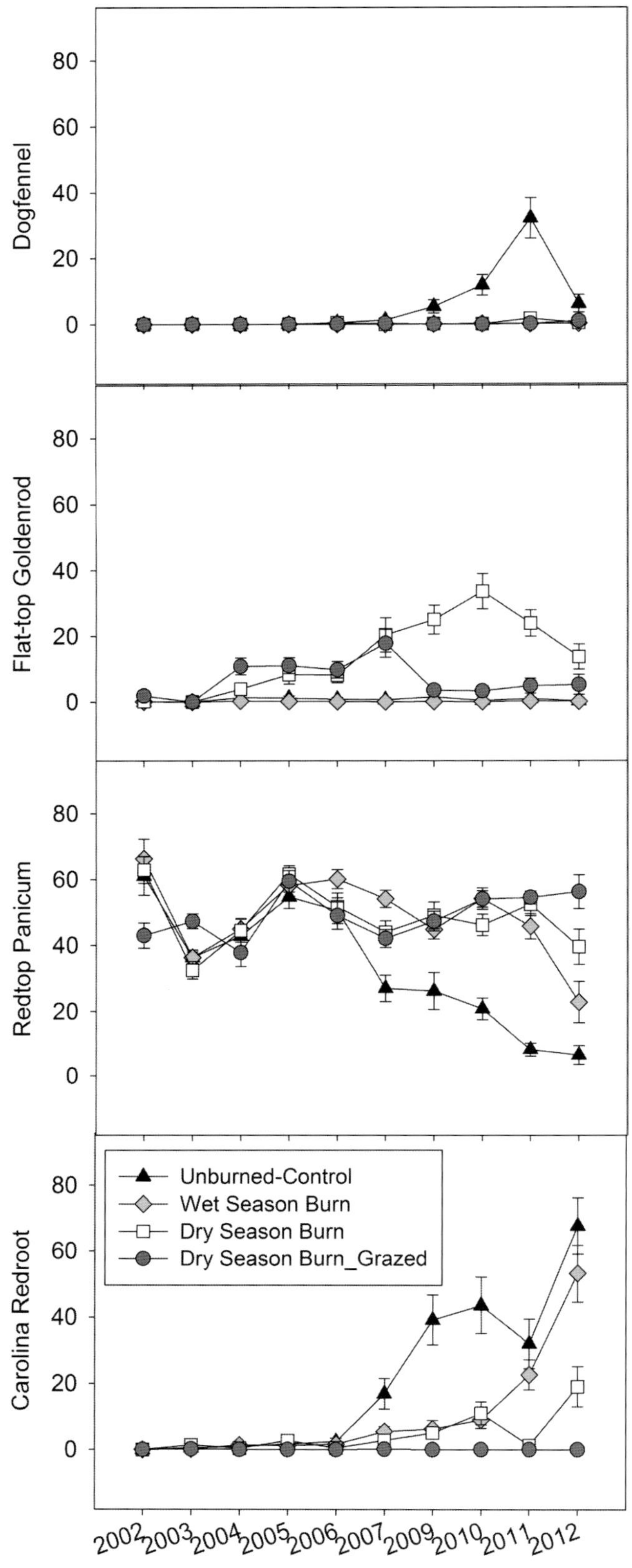

Figure 13.5. Treatment responses (mean percent cover and ± 95% CI) of four dominant species over the course of the study (2002–2012). Treatments are unburned-control, wet season burn, dry season burn; and dry season burn–grazed (n = 4 replicate plots for each treatment). Updated from Boughton et al. (2013).

(W. Platt et al. 2015). A short-term study of the effects of a late wet burn on broomsedge by MAERC research intern J. Patterson showed that a wet season burn conducted in August resulted in fewer live flowering stems and a decrease in rosette diameter (J. Patterson, unpublished data). Importantly, this fire was followed by a large rain event that flooded the area. This combination of fire and flooding is often detrimental to native plant diversity in the late wet season when many species are at their reproductive peak (S. Orzell, pers. obs.). However, this management regime may be appropriate if a goal is to reduce the cover of dominant species. For example, broomsedge can be considered weedy, and burning late in the wet season may reduce its abundance and potentially open up space for other species. More research should be done to understand how grassland communities with different initial species composition respond to transition-season burns and prescribed fires ignited at different times during the wet season.

Our results indicate that the typical ranchland management regime, which includes dry season burning and light grazing, can maintain grassland structure. This approach may be compatible with restoration efforts to maintain a mixture of grasses and forbs, although a shift to early wet season burning may be most beneficial. Wet season burn plots and dry season burn–grazed plots were similar in composition and richness, as shown by substantial overlap in NMS ordination.

Importantly, this result applies to grasslands that were previously grazed for many years, and we cannot tease apart the effects of removal of grazing from season of burn. Results showed that richness was more variable among years in grazed treatments, and this higher variability may be due to differences in grazing intensity. This variability in grazed plots is not surprising since cattle are selective grazers, and ranch manager decisions concerning the timing and density of grazing may differ each year. High dynamism in grazed vegetation compared to other management regimes has been found in other studies (Kahmen et al. 2002), but many studies have shown that moderate livestock grazing can result in increased forb cover and richness (McNaughton 1993; Willcox et al. 2010). Grazing at too high an intensity may result in decreased species numbers and reduced structural heterogeneity (Willcox et al. 2010; DeSelm and Murdock 1993). Nevertheless, grassland structure was maintained better in grazed plots than in dry-

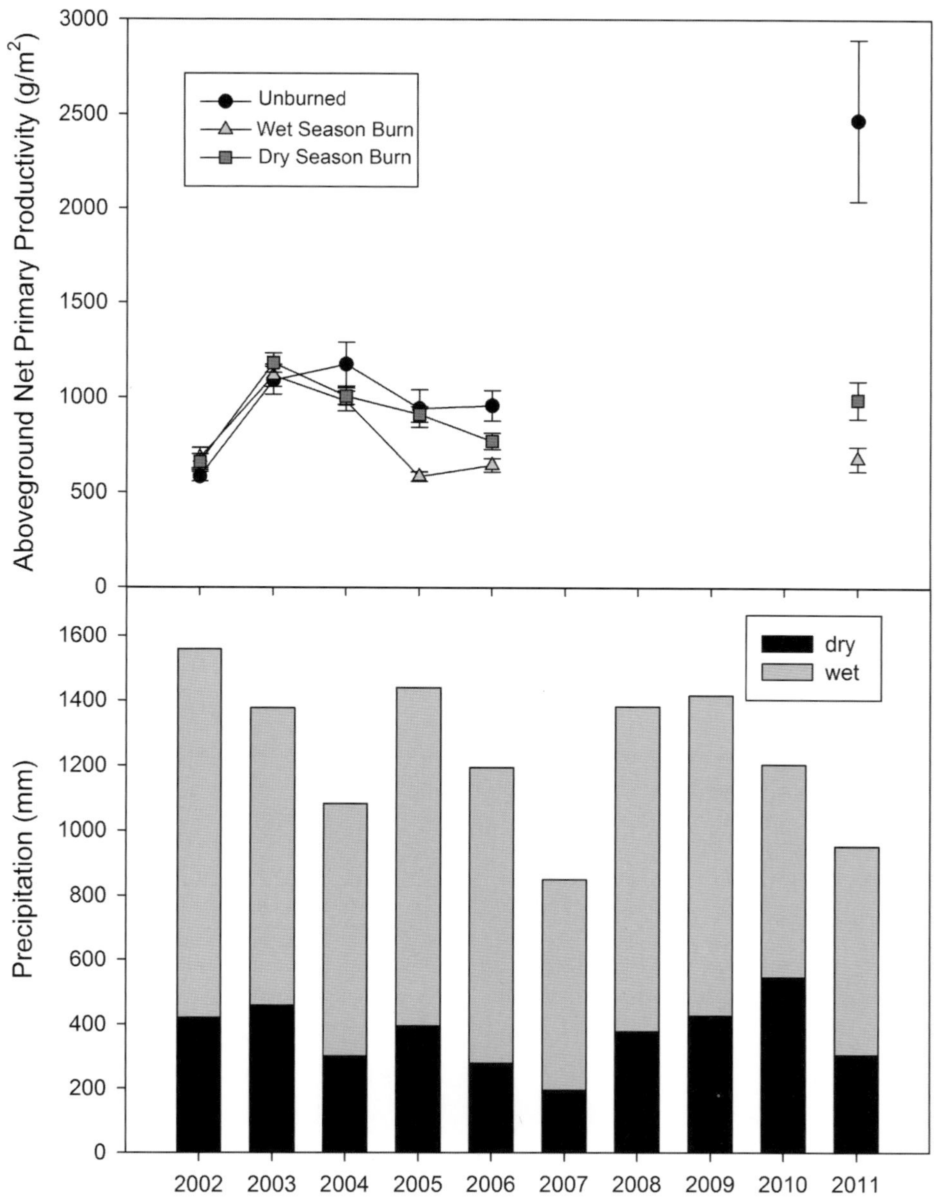

Figure 13.6. *Top*, aboveground net primary productivity (ANPP) in unburned, wet season burn, and dry season burn treatments over time. Data were not collected from 2007 to 2010. ANPP was similar between treatments until 2011, when we observed a large increase in unburned plots because of shrub encroachment. *Bottom*, annual rainfall during dry (black) and wet (gray) seasons over time.

season burned and unburned treatments because woody herb cover was low in grazed plots. Grazing appears to reduce woody herbs, especially dogfennel and flat-top goldenrod, although grazing may have allowed these species to establish in the first place by creating available microsites (Middleton 2002). Other work at MAERC has shown that grazing and roller chopping decrease woody herbs in large wetland tracts undergoing restoration, especially in wetlands that are heavily degraded prior to hydrological restoration (Bridges and Orzell 2013). Native plants in Florida's subtropical grasslands may be resilient to grazing at reasonably low stocking rates, given that they evolved with large herbivores (Noss 2013). If grazing is included in management plans to restore or maintain semi-native grassland structure, it should be low intensity, because high-intensity, persistent grazing is less likely to support the same level of species richness or conservation value (Willcox et al. 2010; Noss 2013).

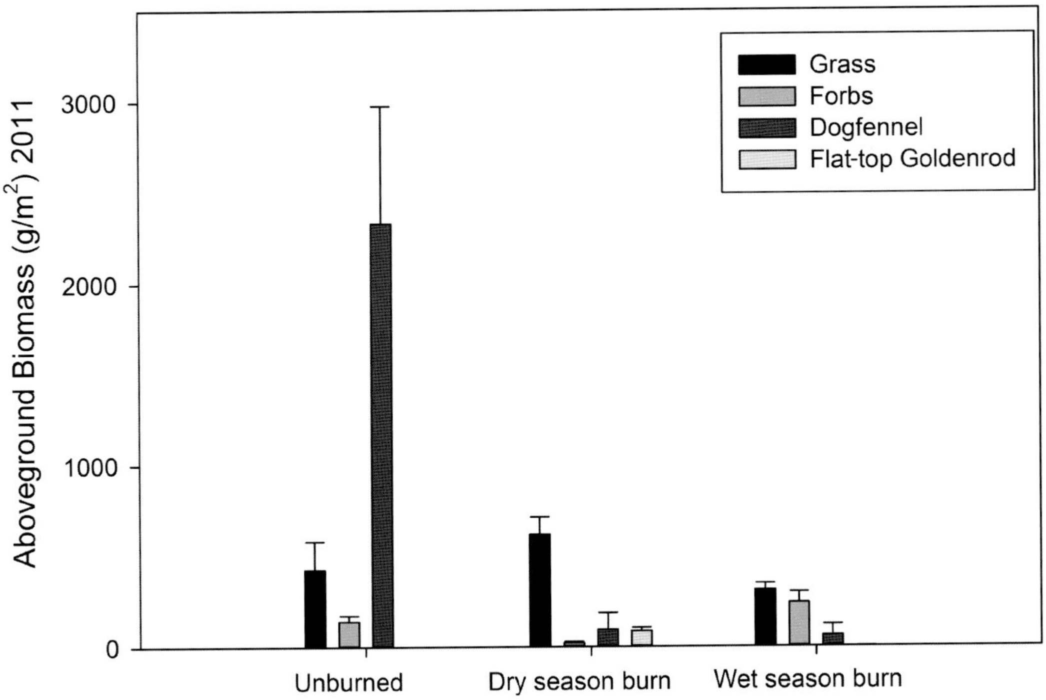

Figure 13.7. Aboveground biomass in 2011, split into four groups: grass, forbs, dogfennel, and flat-top goldenrod among the three treatments: control (unburned), dry season burn, and wet season burn.

Overall, our experiment suggests that grasslands released from typical ranchland management in Florida should be frequently burned to maintain grassland structure and reduce woody herb cover, and that the early wet season (May/June) is the most beneficial season for burning to reduce woody herb encroachment and encourage both graminoid and forb richness. The problem of woody encroachment in grasslands is a global issue (A. Knapp et al. 2008). Potential drivers of woody encroachment are several: increased grazing intensity, reduced fire frequency, alterations in management practices (i.e., altering season of burn), and rising atmospheric CO_2 concentrations. In southern Florida, another possible cause of woody encroachment is the landscape-scale hydrological alterations that have been implemented. Over 1,500 water-control structures and 2,600 miles of canals have been constructed in the greater Everglades to drain the landscape and provide flood control, water supply, and, in the past, navigation services. The main mechanism thought to be responsible for the treeless grassland region in southern Florida was the combination of frequent fires set by lightning in May and June, large undissected fire compartments, and the large-scale sheet-flow flooding that would occur during the wet season, which inhibited establishment of trees and other woody plants (Bridges 2006a; W. Platt et al. 2006; K. Stephenson 2011; Noss 2013). Now, flooding occurs only sporadically after tropical storm events. Even in large-scale wetland restoration projects, small-scale hydrological restoration may be ineffective in the larger context of a drained landscape with sandy soils, where water is easily lost when water tables are low. An additional problem is that burning is often prevented (e.g., by burn bans issued by the Florida Forest Service) when conditions are most conducive to intense fires that would be the most effective at removing woody species (Noss 2013).

In the grassland region of south-central Florida, it is crucial to maintain both pristine and semi-native grasslands and rangelands. Noss (2013) suggests that one of the most important conservation priorities for southeastern grasslands is to limit development on natural or semi-natural lands. Conserving and restoring these areas using a variety of conservation tools is necessary and could include easements through governmental programs (USDA Agricultural Conservation Easement Program, National Wildlife Refuge designation, and Florida Forever Program) and through other types of governmental programs such as the Conservation Stewardship Program offered by the Natural Resources Conservation Service, which works with ranchers to maintain biodiversity and environmental values on working landscapes. Providing programs and incentives for ranchers to maintain semi-native grasslands is necessary to prevent further intensification or conversion of these critically important ecosystems.

Acknowledgments

We would like to thank the staff and interns of the MacArthur Agro-Ecology Research Center for valuable logistical support and data collection: Gene Lollis, Gary Spencer, Gene Danford, Andy Yates, Raoul Boughton, Laura Calabrese, Chelsea Clifford, Bradley Christoferson, Stanley Gathumbi, Jan Kokemore, Julia Maki, Mike McMillian, Luz Morales-Castro, Rob Morrison, Lauren Peters, Adam Peterson, Wolfgang Prein, Robin Rossmanith, Jenna Smith, Christen Steele, Brooke Traynham, Rachel West, Annalisa Weiler, Chris Wilson, and Alissa Zimman. Hilary Swain and Julia Maki provided valuable comments on the manuscript.

14

Effects of Landscape History on Plant Communities in Semi-Natural Grassland Buffers

Jolie G. Dollar, Timothy J. Schauwecker, Samuel K. Riffell, and L. Wes Burger Jr.

Conservation and restoration of prairie remnants is critical for maintaining biodiversity (Packard and Mutel 1997). Many songbirds and game birds, butterflies, native bees, and small and large mammals use tallgrass prairie habitat to meet life history requirements (Brennan 1991; A. Knapp et al. 1998; J. Jones et al. 2007), including species that are endangered or threatened because of habitat loss (Panzer and Schwartz 1998; Peterjohn and Sauer 1999; Hunter et al. 2001; Brennen and Kuvlesky 2005; Sauer et al. 2008). Because relatively little prairie remains, grassland conservation and restoration are often accomplished by planting and managing "semi-natural" prairies on land that previously had other uses, such as agricultural fields, forests, airports, and parking lots (Doxon and Carroll 2007; Kutschbach-Brohl et al. 2010; Seigel and Lockwood 2010).

Native forbs are an important component of the grassland vegetation community, although less dominant than native perennial grasses (A. Knapp et al. 1998). The percent cover of forbs and the number of flowering stems can be related directly to the abundance and species richness of beneficial insects and grassland birds (e.g., Field et al. 2005; Lovell and Sullivan 2006; Davros et al. 2006; J. Davis et al. 2008). Forbs (and scattered brambles and shrubs) enhance nesting and brooding cover and produce a quality food source for many wildlife species (C. Harper 2007). Thus, forbs are both a component and a driver of diversity in grassland systems. To capitalize on these benefits, seed mixes for conservation plantings often recommend more forbs (Feber et al. 1996; Aviron et al. 2010). Additionally, more research on factors influencing abundance, richness, and persistence of forbs in semi-natural grasslands is needed to improve conservation and restoration of grassland systems

The historical Black Belt Prairie in Mississippi is thought to have comprised a mosaic of open grasslands, savannas, and wooded lowlands (Peacock and Schauwecker 2003a; Barone 2005b). Within the modern Black Belt landscape, grassland establishment and restoration occur on sites that are believed to have been historically grass dominated as well as on sites along the grassland-forest continuum. We investigated the influence of landscape history on forbs in recently established native grassland habitat. Restoration and establishment may be more successful in sites that were historically tallgrass prairie because of site-specific characteristics (e.g., soil conditions, microclimate, topographic factors) that are more favorable for prairie-associated forbs adapted to those conditions (Kline 1997). Sites that were not historical prairie may lack these conditions. In addition, prairie-associated forbs may persist in the seedbank.

Historical descriptions of land type (e.g., dominant vegetation) and disturbance patterns (e.g., wildfires) can greatly aid land managers in determining management options for present-day grassland restoration or semi-natural grassland establishment projects (P. White and Walker 1997; Antrop 2005; Johansson et al. 2008; Reitalu et al. 2010), but the effects of prairie history on present-day plant communities have not been widely investigated (but see Rankin and Davis [1971] and Barone [2005a, 2005b] for studies looking at historical descriptions of Black Belt Prairie). The vegetation, including forb communities, of the Black Belt Prairies in northeastern Mississippi have been documented (Foti 1989; DeSelm and Murdock 1993; Leidolf and McDaniel 1998; Forbs 1999; Barone and Hill 2007), but only a few floristic inventories on remnant Black Belt Prairies have been published (Schuster and McDaniel 1973; Leidolf and McDaniel 1998; Forbes 1999; Barone and Hill 2007; Campbell and Seymour 2012).

We compared forb communities on semi-natural grassland field buffers on historical Black Belt Prairie sites to those on sites with no previous prairie history. We predicted that sites with a history of prairie would have forb communities characterized by more prairie-associated species (e.g., *Aster* spp., *Liatris* spp.) (Leidolf and McDaniel 1998; Barone and Hill 2007; Packard and Mutel 1997). Conversely, we predicted that forb communities on sites lacking prairie history would have fewer prairie-associated forbs than sites with historical prairie.

Methods

Study Site

Grasslands were a scattered but ecologically distinctive part of the historical landscape of the southeastern United States (DeSelm and Murdock 1993; Barone and Hill 2007), and the southernmost extent of the Tallgrass Prairie region occurred in numerous prairie "islands" surrounded by forests in the nineteenth century in northeastern Mississippi (Rostlund 1957; A. Jones and Patton 1966; Barone 2005a, 2005b). These prairies of northeastern Mississippi are called the Black Belt Prairie and are an arcing band of black ultisols 60 km wide that extends from northeastern Mississippi to west-central Alabama. The alkaline soils overlie Cretaceous chalk, and historically the area comprised a mosaic of chalk grasslands and southeastern mixed forests (A. Jones and Patton 1966). Prairies likely existed in this region because of distinctive soil, fire (natural and anthropogenic from Native Americans), and grazing by native ungulates. Most of the Black Belt Prairie in Mississippi and Alabama has been lost to agricultural conversion and fire suppression (Peacock and Schauwecker 2003b). Current prairie remnants are dominated by *Schizachyrium scoparium* (little bluestem), which is commonly found with *Bouteloua curtipendula* (sideoats grama) or *Sorghastrum nutans* (Indiangrass). Common forbs include *Symphyotrichum dumosum* (rice button aster), *Liatris squarrosa* (scaly blazing star), and *Silphium laciniatum* (compassplant). Common legumes include *Neptunia lutea* (yellow puff) and *Dalea purpurea* (purple prairie clover) (Weiher et al. 2004).

We worked at B. Bryan Farms, a privately owned row-crop and grazing operation of 2,104 ha in Clay County, Mississippi. Portions of Clay County fall in the historical Blackland Prairie physiographic region. In spring 2004, 79 crop hectares were enrolled in a Conservation Reserve Program practice titled CP33, Habitat Buffers for Upland Birds, consisting of linear strips around 9–40 m wide of native warm-season grasses and forbs planted around perimeters of crop fields (USDA 2004). Although the primary goal of CP33 is to provide upland habitat for grassland birds, these native grass plantings supply resources for a variety of other taxa as well as ecosystem services (Davros et al. 2006; Lovell and Sullivan 2006; Olson and Wackers 2007; Adams 2011; Dollar 2011). The planting mix at Bryan Farms included native grasses—*Andropogon gerardii* (big bluestem), *Schizachyrium scoparium* (little bluestem), and *Sorghastrum nutans* (Indiangrass); and native forbs—*Chamaecrista fasciculata* (partridge pea), *Rudbeckia hirta* (black-eyed Susan), and *Helianthus maximiliani* (Maximilian sunflower). The buffer plant community also included many nonseeded species that were present in the seedbank or were introduced through wind and animal dispersion (see table 14.1). Late spring and summer (April–August) rainfall for the B. Bryan Farms area was 380 mm in 2007.

Plant Community Sampling

We used quadrats measuring 1 m^2 (Roberts-Pichette and Gillespie 1999) to collect plant data on the buffers (Driscoll et al. 2010). We sampled 30 quadrats per buffer along 200 m transects positioned in the center along the long axis of each buffer. We placed quadrats 10 m away from the transect line and alternated sides to avoid potential impacts of researcher disturbance on vegetation along the transects (these same transects were also used to monitor avian and insect communities).

Measurements were collected to the species level in August 2007 in the third growing season after planting. We recorded species and percent cover for all plants 1.0 cm or more in height and with two or more true leaves, along with the percent cover of bare ground and dead plant material (both upright and on the ground). Because of overlapping vegetation layers, cover sometimes exceeded 100% (Benson et al. 2007). We recorded the number of primary stems for all forb, vine, and shrub species. Species identification was based on field guides, dichotomous keys, online resources, and journal articles (e.g., Radford et al. 1968; P. Barbour 2006; Timme 2007; T. Murphy et al. 2007; USDA 2010–2013). These resources were also used to classify species as prairie associates (see table 14.1). Voucher specimens and/or in-field photographs were obtained for each species.

Determining Landscape History

To determine landscape history for B. Bryan Farms, we used 15 land survey maps and reports for Clay County, Mississippi, from the General Land Office records of the Bureau of Land Management (BLM). The documents show approximate locations of prairie habitat in Clay County in 1835 prior to increased European settlement and development (Rankin and Davis 1971). Portions of B. Bryan Farms were described as prairie in the 1835 BLM reports. We geo-referenced the 1835 map with aerial imagery from the 2007 National Agriculture Imagery Program (NAIP) and located our study sites in relation to the historical prairie (fig. 14.1). Two fields with 6 CP33 buffers were located in areas that had been wooded (outside the prairie). Two fields with 6 buffers were located on the edge of prairie and woodland. Four fields with 14 buffers were located within the historical prairie area.

Figure 14.1. A current aerial image overlaid with the outline of the 1835 prairie. Red shapes designate agricultural fields with native grass field buffers on B. Bryan Farms in Clay County, Mississippi.

Statistical Analysis

We used plant species cover estimates from August 2007 to determine the influence of prairie history on the community of recently established native grass buffers. Three indicator variables for buffers ("environmental" in multivariate terminology)—historically prairie, edge of prairie, or not prairie—were used in a redundancy analysis (RDA) in the CANOCO package (ter Braak 1990; ter Braak and Smilauer 2002) with the Monte Carlo permutation test. CANOCO analyzes repeated observations of species composition in a way equivalent to univariate repeated-measures ANOVA (Leps and Smilauer 2003). Using various combinations of explanatory variables and environmental variables in RDA with the permutation scheme in the Monte Carlo test (999 permutations), we constructed tests of significance.

We used RDA to test for the influence of the three environmental variables on the buffer plant community following Leps (1999) and Leps and Smilauer (2003). The first analysis (P1) tested the null hypothesis that there were no differences among landscape histories. This first analysis corresponded with the test of all within-subject effects in repeated-measures ANOVA (Leps and Smilauer 2003). The following three analyses (P2–P4) tested the effects of each environmental variable (i.e., prairie, not prairie, and edge) on the buffer plant community. These analyses cor-

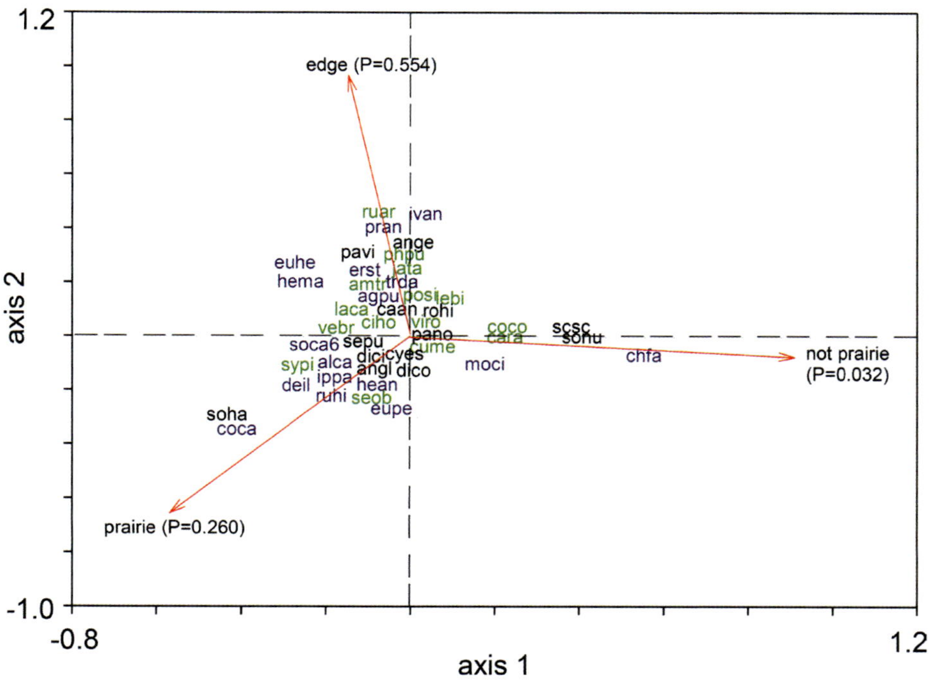

Figure 14.2. Biplot showing the results of RDA ordination to determine the influence of prairie history on buffer plant communities. Blue labels are prairie-associated forbs, vines, and shrubs. Green labels are other forbs, vines, and shrubs. Black labels are grasses, sedges, rushes, and trees.

responded to the tests of particular terms in a repeated-measures ANOVA, and the null hypotheses were that each environmental variable had no influence on the buffer plant community. To reduce the probability of type II errors, we set the a priori $\alpha = 0.10$ for all statistical tests.

Results

RDA rejected the P1 null hypothesis that there were no differences in species composition that were common for all environmental variables or specific for particular environmental variables (P = 0.001; table 14.2). Rejecting P1 required testing hypotheses P2–P4 about the individual environmental variables. We failed to reject the P2 null hypothesis that "historical prairie" had no effect on species composition (P = 0.260), rejected the P3 null hypothesis that "not prairie" had no effect on species composition (P = 0.032), and failed to reject the P4 null hypothesis that "historical prairie edge" had no effect on species composition (P = 0.554). Plant species composition was not the same among buffers in historical prairie, not historical prairie, and historical prairie edge. Buffers that were "not prairie" had the primary influence on those differences.

The RDA biplot shows that most plant species were associated with the "prairie" and "edge" variables (fig. 14.2). A mix of prairie-associated species (including forbs) and 20 nonprairie-associated species made up the groups of species associated with the "prairie" and "edge" variables. Eight of 10 forbs associated with the "prairie" variable were prairie-associated species. Seventeen forbs were associated with the "prairie edge" variable, and 7 were prairie associated. Only 6 species out of 82 were oriented toward the "not prairie" variable. Three of these species were part of the CP33 seed mix: *Sorghastrum nutans* (Indiangrass/sonu), *Schizachyrium scoparium* (little bluestem/scsc), and *Chamaecrista fasciculata* (partridge pea/chfa). *Monarda citriodora* (lemon beebalm/moci) was the only prairie-associated forb among the 6 species oriented toward the "not prairie" variable.

Discussion

In summer 2007, the semi-natural grass buffers at B. Bryan Farms were differentiated by their 1835 land cover (prairie, prairie edge, not prairie). Consistent with our predictions, sites that were not blackland prairie in 1835 had a plant community different from that in "prairie" or "edge" sites. We expected that the "prairie" sites would have greater species richness, with more prairie-associated forbs than the "edge" and "not prairie" sites. Most plant species present on buffers in summer 2007 were more abundant on "prairie" and "edge" sites, including multiple prairie-associated forbs and other forbs. The "not prairie" variable was significant primarily because many species were absent or had low abundance.

Three of the six species correlated with the "not prairie" variable were CP33 seeded species, possibly indicating that site conditions (e.g., soil fertility, soil moisture) allowed these seeded species to dominate species present in the seedbank. Two of the seeded species were C4 grasses that have wide ranges and broad ecological amplitudes (C. Freeman 1998). One nonseeded forb species, *Conoclinium coelestinum* (blue mistflower/coco), is a facultative wetland species (Tobe et al. 1998). Blue mistflower may have been associated with "not prairie" because some buffers included in the "not prairie" group were located near riparian areas. Another nonseeded forb species, *Campsis radicans* (trumpet creeper/cara), has invasive tendencies (Miller and Miller 2005), further indicating that the site conditions may have facilitated the dominance of a handful of species.

There are two ways that the historical conditions could influence current plant communities, the first of which is soils, because the physical and chemical properties of soil affect the kinds, amount, and spatial distribution of vegetation (Ransom et al. 1998). Soil usually contains and acts upon a more extensive part of the plant body than does the atmosphere, and often more than half of plant biomass is belowground (Weaver 1968). The two fields at B. Bryan Farms included in the "not prairie" variable were on Sessum silty clay soil. In contrast, the "prairie" and "prairie edge" sites were found mainly on Griffith silty clay and Kipling silt loam soils. All three of these soil types have a parent material of clayey marine or alluvial deposits, but the Sessum silty clay on the "not prairie" sites is in the "poorly drained" drainage class, whereas the Griffith silty clay and the Okolona silty clay on the "prairie" and "prairie edge" sites are moderately well drained to well drained. The poorly drained soils of the "not prairie" sites could have allowed higher moisture availability to plant roots, meaning that a handful of species (particularly, seeded species) could outcompete other species because moisture was not a limiting factor (as in upland, prairie sites).

The second way historical conditions could influence current communities is through the seedbank. Many nonseeded prairie-associated forbs (see blue labels in fig. 14.2) were found on buffers that had a history of prairie or were at the edge of historical prairie, indicating that germplasm from these species was present in the seedbank. The ability of many prairie-associated forb species to establish and compete with the dominant grasses on the "prairie" and "edge" sites adds support to the hypothesis that sites with historical prairie will have a plant community with more prairie-associated forbs compared to sites with no history of prairie presence (partly because of lower relative soil fertility and moisture).

Historical documents such as the Public Land Survey can be used to gain knowledge about historical land cover and land use, and this type of historical information can be very important for ecosystem restoration projects (Fritschle 2008; Rankin and Davis 1971). For example, grassland restoration and establishment projects might be more successful on sites with a history of prairie. The site conditions, including soil characteristics, might be better suited to prairie-associated plants (especially forbs) than conditions in sites with no prairie history. Therefore, landowners and land managers should consider historical information for a particular landscape when determining the most effective use and management for an area. Landowners should select primarily sites with a history of prairie (if available) to establish semi-natural grasslands. If the selected site has no history of prairie, then a diverse seed mixture including prairie-associated forbs in the same proportion as native warm-season grasses should be used.

Acknowledgments

We thank B. Bryan Farms and Prairie Wildlife, LLC, for allowing us to work on their property. Others who aided our efforts include B. Brzuszek (plant identification), J. Leps (multivariate analyses), and G. Ervin and two anonymous reviewers, who provided feedback on a previous version of this manuscript. Funding was provided by the USDA Natural Resources Conservation Service–Agricultural Wildlife Conservation Center (NRCS-AWCC), as well as the Mississippi Agricultural and Forestry Experiment Station, the Forest and Wildlife Research Center, and the College of Forest Resources at Mississippi State University.

Table 14.1.

Growth form and prairie association status of CP33 buffer plant species across Clay County, Mississippi.

Functional group/scientific name	Common name	Prairie associate
Native warm-season grasses		
Andropogon gerardii (ange)*	Big bluestem	Yes
Andropogon glomeratus (angl)	Bushy bluestem	Yes
Bouteloua curtipendula (bocu)	Sideoats grama	Yes
Panicum virgatum (pavi)	Switchgrass	Yes
Schizachyrium scoparium (scsc)*	Little bluestem	Yes
Sorghastrum nutans (sonu)*	Indiangrass	Yes
Tripsacum dactyloides (trda)	Eastern gamagrass	Yes
Native cool-season grasses		
Dichanthelium commutatum (dico)	Variable panicgrass	No
Digitaria ciliaris (dici)	Southern crabgrass	No
Native annual grasses		
Hordeum pusillum (hopu)	Little barley	No
Muhlenbergia capillaris (muca)	Hairawn muhly	No
Introduced grasses		
Echinochloa colona (ecco)	Jungle rice	No
Paspalum dilatatum (padi)	Dallisgrass	No
Paspalum notatum (pano)	Bahia grass	No
Paspalum urvillei (paur)	Vasey's grass	No
Schedonorus phoenix (scph)	Tall fescue	No
Setaria pumila (sepu)	Yellow foxtail	No
Sorghum bicolor (sobi)	Sorghum	No
Sorghum halepense (soha)	Johnsongrass	No
Urochloa mutica (urmu)	Para grass	No
Native perennial forbs		
Allium canadense (alca)	Meadow garlic	Yes
Asclepias tuberosa (astu)	Butterfly milkweed	Yes
Conoclinium coelestinum (coco)	Blue mistflower	No
Desmanthus illinoensis (deil)	Illinois bundleflower	Yes
Diodia teres (dite)	Poorjoe	Yes
Erigeron strigosus (erst)	Prairie fleabane	Yes
Eupatorium capillifolium (euca)	Dogfennel	No
Eupatorium perfoliatum (eupe)	Common boneset	Yes
Eurybia hemispherica (euhe)	Southern prairie aster	Yes
Helianthus angustifolius (hean)	Swamp sunflower	Yes
Helianthus maximiliani (hema)*	Maximilian sunflower	Yes
Hyptis alata (hyal)	Clustered bushmint	No
Lespedeza virginica (levi)	Slender lespedeza	Yes
Lobelia spicata (losp)	Palespike lobelia	Yes
Mimosa microphylla (mimi)	Littleleaf sensitive-briar	No
Mimosa strigillosa (mist)	Powderpuff	No

Continued on the next page

Table 14.1. *Continued*

Functional group/scientific name	Common name	Prairie associate
Oenothera speciosa (oesp)	Pinkladies	Yes
Oxalis stricta (oxst)	Common yellow oxalis	No
Phytolacca americana (pham)	American pokeweed	No
Plantago lanceolata (plla)	Narrowleaf plantain	No
Polygonum glabrum (pogl)	Denseflower knotweed	No
Potentilla simplex (posi)	Common cinquefoil	Yes
Pycnanthemum incanum (pyin)	Hoary mountainmint	No
Ranunculus fascicularis (rafa)	Early buttercup	Yes
Rudbeckia hirta (ruhi)	Black-eyed Susan	Yes
Rumex hastatulus (ruha)	Heartwing sorrel	No
Senna obtusifolia (seob)	Java-bean	No
Sesbania herbacea (sehe)	Bigpod sesbania	No
Solanum carolinense (soca3)	Carolina horsenettle	No
Solidago canadensis (soca6)	Canada goldenrod	Yes
Symphyotrichum pilosum (sypi)	Hairy white oldfield aster	No
Taraxacum officinale (taof)	Common dandelion	No
Verbena halei (veha)	Texas vervain	No
Vernonia gigantea (vegi)	Giant ironweed	Yes
Native annual forbs		
Agalinis purpurea (agpu)	Purple false foxglove	Yes
Amaranthus hybridus (amhy)	Slim amaranth	No
Ambrosia artemisiifolia (amar2)	Annual ragweed	No
Ambrosia trifida (amtr)	Great ragweed	No
Bidens bipinnata (bibi)	Spanish needles	No
Chamaecrista fasciculata (chfa)*	Partridge pea	Yes
Chamaecrista nictitans (chni)	Sensitive partridge pea	Yes
Chamaesyce cordifolia (chco)	Heartleaf sandmat	No
Cirsium horridulum (ciho)	Yellow thistle	No
Conyza canadensis (coca)	Canadian horseweed	Yes
Galium aparine (gaap)	Stickywilly	No
Gamochaeta purpurea (gapu)	Spoonleaf purple everlasting	Yes
Geranium carolinianum (geca)	Carolina geranium	Yes
Helenium amarum (heam)	Yellowdicks	No
Iva annua (ivan)	Annual marsh elder	Yes
Jacquemontia tamnifolia (jata)	Hairy clustervine	No
Krigia virginica (krvi)	Virginia dwarf dandelion	Yes
Lactuca canadensis (laca)	Canada lettuce	Yes
Lamium amplexicaule (laam)	Henbit deadnettle	No
Monarda citriodora (moci)	Lemon beebalm	Yes
Nuttallanthus canadensis (nuca)	Canada toadflax	Yes
Physalis pubescens (phpu)	Husk tomato	No
Plantago virginica (plvi)	Virginia plantain	Yes
Pseudognaphalium obtusifolium (psob)	Rabbit-tobacco	No
Pyrrhopappus carolinianus (pyca)	Carolina desert-chicory	No
Triodanis perfoliata (trpe)	Clasping Venus's looking-glass	Yes
Valerianella radiata (vara)	Beaked cornsalad	Yes

Table 14.1. *Continued*

Functional group/scientific name	Common name	Prairie associate
Introduced perennial forbs		
Rumex crispus (rucr)	Curly dock	No
Introduced annual forbs		
Cucumis melo (cume)	Cantaloupe	No
Daucus carota (daca)	Queen Anne's lace	No
Sonchus asper (soas)	Spiny sowthistle	No
Torilis arvensis (toar)	Spreading hedgeparsley	No
Verbascum thapsus (veth)	Common mullein	No
Verbena brasiliensis (vebr)	Brazilian vervain	No
Vicia sativa (visa)	Garden vetch	No
Native perennial vines		
Ampelopsis arborea (amar5)	Peppervine	Yes
Campsis radicans (cara)	Trumpet creeper	No
Ipomoea pandurata (ippa)	Man of the earth	Yes
Rubus trivialis (rutr)	Southern dewberry	No
Toxicodendron radicans (tora)	Eastern poison ivy	Yes
Vitis rotundifolia (viro)	Muscadine	No
Perennial sedges/rushes		
Carex annectens (caan)	Yellowfruit sedge	Yes
Carex striata (cast)	Walter's sedge	No
Cyperus echinatus (cyec)	Globe flatsedge	Yes
Cyperus esculentus (cyes)	Yellow nutsedge	Yes
Cyperus rotundus (cyro)	Nutgrass	No
Cyperus surinamensis (cysu)	Tropical flatsedge	No
Juncus tenuis (jute)	Poverty rush	Yes
Rhynchospora cephalantha (rhce)	Bunched beaksedge	No
Native shrubs/trees		
Callicarpa americana (caam)	American beautyberry	No
Pinus taeda (pita)	Loblolly pine	No
Prunus angustifolia (pran)	Chickasaw plum	Yes
Robinia hispida (rohi)	Bristly locust	No
Rubus argutus (ruar)	Sawtooth blackberry	No
Sambucus nigra ssp. *canadensis* (sani)	American black elderberry	No
Introduced shrubs		
Lespedeza bicolor (lebi)	Shrub lespedeza	No

Note: Names and groupings derived from USDA PLANTS Database (2010). Prairie association derived from Packard and Mutel (1997).

* Species included in CP33 planting mixture.

TABLE 14.2.

Results of RDA ordination to determine influence of prairie history on buffer plant communities (August 2007) in Clay County, Mississippi.

Analysis	Explanatory variables	% expl. 1st axis	P value	Null
P1	prairie, edge, not prairie	15.8	0.001	Reject
P2	prairie	7.5	0.260	Accept
P3	not prairie	13.2	0.032	Reject
P4	edge	2.8	0.554	Accept

15

Ground Cover Assessment of CRP Conservation Practice 36 in Georgia

James W. Tomberlin, Nicholas Brown, and Reggie E. Thackston

Under the Conservation Reserve Program of the USDA Farm Service Agency, the Longleaf Pine Initiative, known as Conservation Practice 36 (hereafter, CP36), began in 2006 in an effort to restore longleaf pine (*Pinus palustris*), native ground cover, and associated wildlife species. The longleaf pine ecosystem is estimated to have covered up to 36.4 million ha (90 million ac) from Virginia to Texas prior to European settlement (Van Lear et al. 2005). Through land-use changes it has been reduced to around 3% of its original range and is considered to be one of the most threatened ecosystems in North America (Frank and Platt 2006).

As of May 2012, 34,614 ha (85,530 ac) have been enrolled in CP36 within the longleaf pine conservation priority area in Georgia. Through CP36, private landowners receive financial incentives to retire cropland from row-crop production along with cost-share payments to help offset the cost of establishing longleaf pines and, if necessary, a native warm-season grass (NWSG)/forb mix (table 15.1) to expedite ground cover restoration and wildlife benefits.

A joint effort began in 2009 between the Georgia Department of Natural Resources, Georgia Forestry Commission, Georgia Farm Service Agency, and Georgia Natural Resources Conservation Service to conduct a coarse-level assessment of ground cover in fields enrolled in CP36. The study objectives were to (1) survey a representative sample of CP36 fields, after at least one growing season, and evaluate the effectiveness of the practice design and implementation in restoring native grasses and forbs; (2) assess the extent to which introduced pasture grasses (IPG) potentially dominate sites and negate programmatic intent; and (3) quantify and evaluate CP36 implementation results and provide feedback in order to facilitate optimization of costs and benefits.

Methods

Sampling was conducted between July 27, 2009, and October 15, 2009. The Georgia Farm Service Agency provided information on CP36 landowner contracts signed prior to September 30, 2007. We used the 2007 enrollment cutoff date to ensure at least one growing season for ground cover development prior to sampling. Contracts that met sampling criteria were grouped by county, and 20% were randomly selected for sampling. Counties with fewer than six total contracts signed during fiscal year 2007 were pooled, and 20% were randomly chosen for sampling. Field personnel randomly chose one field to sample if the contract included multiple fields.

Field personnel obtained background information on each of the contracts from county Farm Service Agency offices and by contacting the landowner and/or private contractor(s) who installed the NWSG/forb mix. Landowners were contacted prior to site visits in order to explain vegetation monitoring and gain permission for sampling.

Data collected included the percent cover of NWSG, other vegetation, legumes, vegetative litter, bare ground, woody stems (excluding planted longleaf pine seedlings), and introduced pasture grasses, or "IPG" (i.e., Bermuda-

TABLE 15.1.

Native warm-season grass/forb mix recommended by Georgia NRCS job sheet (645–647), Native Grasses for Wildlife Habitat, revised March 15, 2011.

Plant species	Seeding rate
Big bluestem (*Andropogon gerardii*)	1.7 kg PLS[a]/ha (1.5 lb PLS/ac)
Indiangrass (*Sorghastrum nutans*)	1.7 kg PLS/ha (1.5 lb PLS/ac)
Little bluestem (*Schizachyrium scoparium*)	1.7 kg PLS/ha (1.5 lb PLS/ac)
Forbs Partridge pea (*Chamaecrista fasciculata*)[b] Native lespedeza (*Lespedeza* spp.) Florida beggarweed (*Desmodium tortuosum*)[c]	1.1 kg/ha (1.0 lb/ac)
Total	6.2 kg/ha (5.5 lb/ac)

[a] PLS = Pure Live Seed.

[b] In 2011, NRCS changed variety from Lark to Comanche and revised seeding rate down to 0.28 kg/ha (0.25 lb/ac). This was in response to reports beginning in 2010 of the Lark variety reseeding prolifically and producing dense stands of partridge pea that overtopped longleaf seedlings and resulted in tree mortality on some sites because of development of rhizoctonia blight.

[c] Non-native.

grass, *Cynodon dactylon*; Bahia grass, *Paspalum notatum*; tall fescue, *Schedonorus arundinaceus*; and "other"). Introduced pasture grasses were identified to species. Additionally, data were collected on cropping history (i.e., type of crops grown in the past), last crop year, whether the NWSG/forb mix was planted, and chemical site preparation implemented to control existing IPG prior to beginning conservation practices.

Vegetation sampling began by selecting two rows (between planted longleaf pines) in the field; within each row, five equally spaced plots measuring 1 m^2 were systematically sampled by visually estimating the percent cover for each vegetation variable to the nearest 5% (Daubenmire 1959). If a vegetative species did not fit precisely into one of the seven categories, it was placed into the "Other Veg" category. Thus, some grasses, particularly Vasey's grass (*Paspalum urvillei*), panicgrasses (*Panicum* spp.), and crabgrasses (*Digitaria* spp.), along with some volunteer crops (e.g., peanuts, *Arachis* spp.; and grain sorghum, *Sorghum bicolor*), were placed into the "Other Veg" category. While grasses such as Vasey's and some crabgrasses may be introduced, they were not included in the "IPG" category simply because these grasses did not require chemical treatment prior to establishing longleaf pine and the NWSG/forb mix. Panicgrasses were not included in the "NWSG" category because they were not part of the NWSG/forb mix that received cost sharing to plant. Volunteer peanuts were not placed in the "Legume" category because they would not necessarily be available to the same suite of wildlife species as other legumes, particularly partridge pea (*Chamaecrista* spp.) or beggarweed (*Desmodium* spp.).

Relative to cropping history, we compared the percent cover of IPG between two biologically important groups regardless of chemical site preparation treatment: fields that were fallow one year or less prior to plantings (≤ 1 yr fallow) and fields that were fallow for more than one year (> 1 yr fallow). Regarding chemical site preparation, we compared the percent cover of IPG in fields treated with imazapyr to those treated with glyphosate or a glyphosate/imazapyr mixture. We omitted fields listed as "Other" (e.g., imazapic or imazapic/glyphosate) or "None" (received no chemical site preparation) from this particular analysis because it was assumed that the site preparation treatment for these fields was not prescribed to control IPG.

We used a Mann-Whitney U two-tailed t-test to determine statistical differences between treatments because the data were not normally distributed (Glantz 2002). Tests were considered significant at $\alpha = 0.05$.

Results

Field and Cropping History

Overall, 103 fields were sampled in 35 counties across Georgia's upper coastal plain (fig. 15.1). The average field size was 9.6 ha (23.9 ac), with a range between 0.7 and 66.3 ha (1.8 and 164 ac). Cropping history of sampled fields was most commonly cotton-peanut crop rotations (59%), followed by small grains (20%), "other" (16%), and pasture (5%). Most fields (91%) were planted with the recommended NWSG/forb mix and were typically planted between April and early June. "Other Veg" represented the largest proportion of ground cover in sampled fields regardless of chemical site preparation treatment (fig. 15.2).

Most fields (68%) were fallow up to only one year before establishing conservation practices. That is, most landowners elected to begin establishing practices in the fall through spring following the last crop rotation without allowing cropland to lie fallow for an entire growing season. We detected a significantly higher percent cover of IPG in fields that had been fallow one year or less than in fields that had been fallow for more than one year (≤ 1 yr fallow = 8.9%, n = 62; > 1 yr fallow = 2.1%, n = 28; P = 0.02).

Chemical Site Preparation

Of the 103 sampled fields, 18 (17.5%) were treated exclusively with glyphosate (usually more than one application; up to three were authorized for cost sharing); 12 (11.7%) were treated with a glyphosate/imazapyr mixture; 23 (22.3%) were treated with imazapyr; 11 (10.7%) were listed as "None"; 38 (36.9%) were listed as "Other" for the chemical treatment (i.e., imazapic or imazapic/glyphosate); and 1 (0.9%) provided incomplete data and was omitted from analyses. We detected no significant difference between fields treated with imazapyr and those treated with glyphosate or a glyphosate/imazapyr mixture (imazapyr = 6.7%, n = 23; glyphosate and glyphosate/imazapyr = 12.3%, n = 30; P = 0.20).

In total, 45 fields (43.7%) reported the presence of IPG across all site preparation treatments, with Bermudagrass being most commonly reported. Fourteen fields (13.6%) reported a significant IPG component (≥ 20% cover), and no significant difference was detected between chemical site preparation treatments (glyphosate and glyphosate/imazapyr, n = 7; imazapyr, n = 4; P = 0.73). Three of the 14 fields listed "Other" as the site preparation treatment and were omitted from this analysis.

Discussion

Chemical site preparation can greatly affect vegetation structure and composition. Introduced pasture grasses, especially Bermudagrass, are aggressive and make it difficult for competing native vegetation to establish and/or emerge through the sod layer. Consequently, IPG reduce diversity and ultimately result in low-quality wildlife habitat. Additionally, Bermudagrass has been shown to increase heat loads on northern bobwhite (*Colinus virginianus*) chicks to critical levels (Burkhart 2004). Even though IPG cover did not differ statistically between chemical site preparation treatments, it should be noted that its occurrence on glyphosate-treated fields was almost double that on imazapyr-treated fields (fig. 15.2), sample sizes were small, and treatments occurred during a time of drought, which can greatly reduce herbicide effectiveness. Research conducted by T. Bond et al. (2005) found that imazapyr was most effective for Bermudagrass control and that treatment by other chemicals (e.g., glyphosate and grass-selective herbicides) was ineffective in reducing Bermudagrass in the long term (i.e., one year). Likewise, the University of Georgia Extension Service recommends imazapyr for Bermudagrass control (D. Moorehead, University of Georgia, pers. comm.).

The Georgia Natural Resources Conservation Service recommends spraying fields with a preemergent herbicide (i.e., imazapic or imazapic/glyphosate) prior to NWSG/forb mix planting regardless of whether the field was treated for IPG. Imazapic herbicides are used for controlling competing vegetation when establishing the NWSG/forb mix or early-successional habitat (C. Harper et al. 2007), but they are not effective at controlling Bermudagrass (Butler and Muir 2006).

Chemical site preparation seemed to have little effect on other vegetative variables such as NWSG, nonlegume forbs, and legumes. Other variables, such as planting date

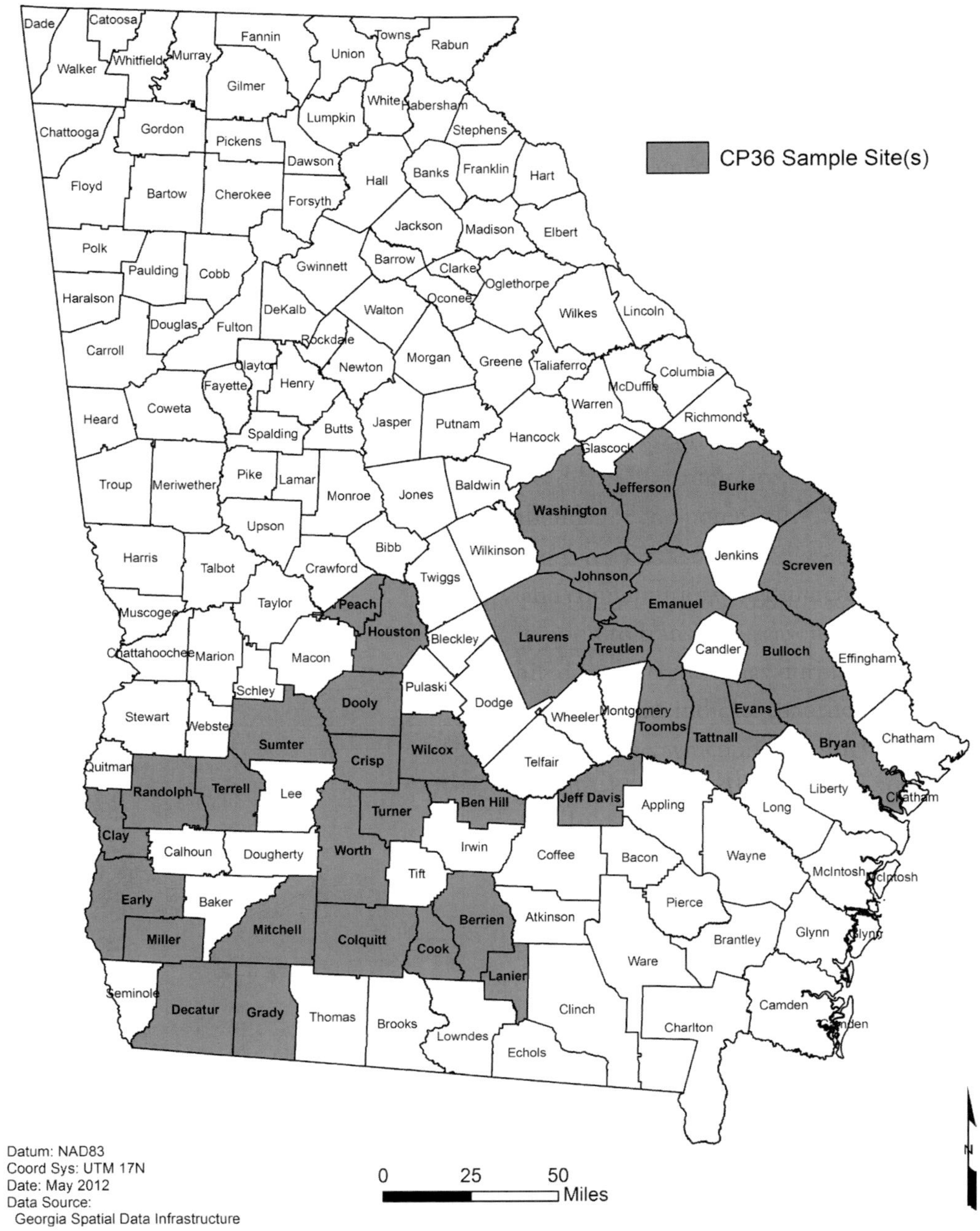

Figure 15.1. The 35 counties across Georgia's upper coastal plain where Conservation Reserve Program Conservation Practice 36 (Longleaf Pine) sites were sampled to assess the condition of ground cover restoration.

and method, weather, soil type, seed source/type, site history, and so on, most likely affected NWSG and legume cover more than chemical site preparation treatment. However, we were not able to quantify these variables or their potential impact on vegetative response.

Based on this study, empirical field observations, and previous research, we recommend the following steps to fulfill the intent of CP36 and optimize long-term natural resource benefits from CP36 implementation:

1. Provide CP36 participating landowners with a four-year window for longleaf pine and NWSG/forb establishment.

2. Require enrolled fields to lie fallow for one or more growing seasons prior to determining site preparation treatment. This will provide a more accurate assessment of species composition, and if IPG are present, sites can be treated more effectively prior to installing conservation practices.

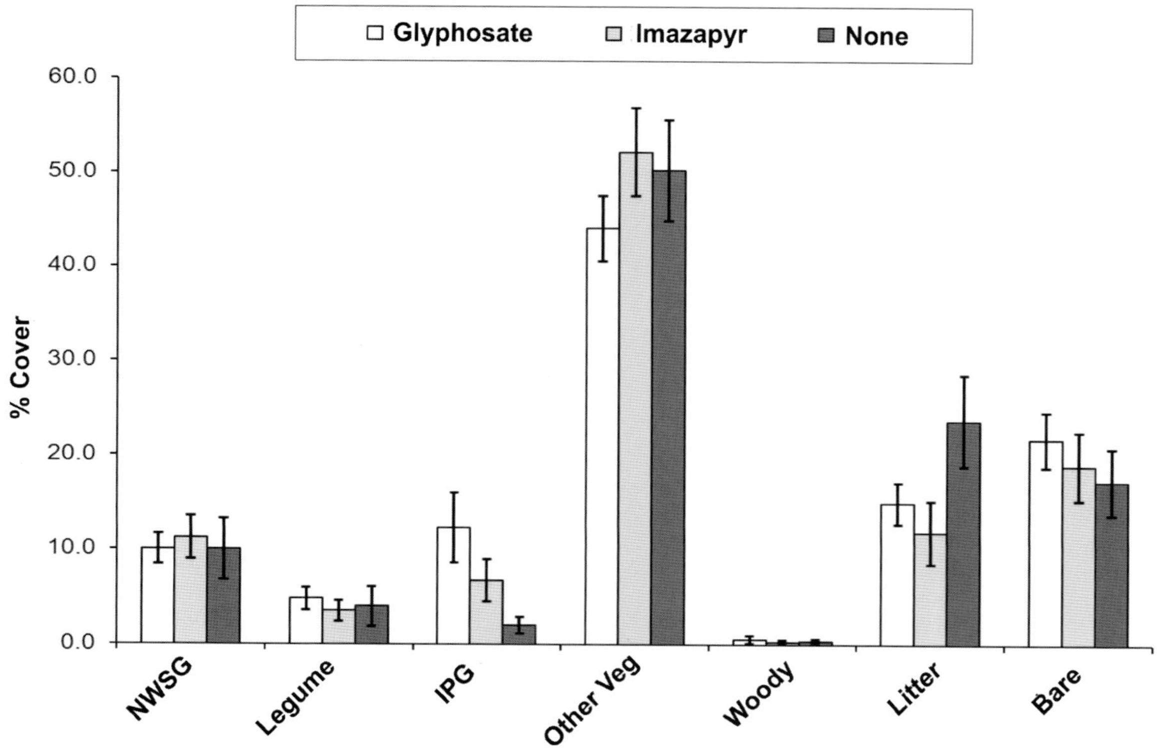

Figure 15.2. Mean percent cover (± SE) of vegetative variables in sampled CP36 fields by chemical site preparation treatment. "Glyphosate" represents fields treated with glyphosate or a glyphosate/imazapyr mixture. "None" represents fields that received no chemical site preparation treatment and were assumed to have no introduced grasses prior to planting. "NWSG" represents native warm-season grasses; "IPG" represents introduced pasture grasses (i.e., Bermudagrass, Bahia grass, and tall fescue), and "Other Veg" is a miscellaneous category that consists mainly of nonleguminous forbs.

3. If Bermudagrass is present, then as site feasible, treat with imazapyr at the specified label rates for Bermudagrass control:

- Year 1—Burn in the spring, followed by a summer application of imazapyr (0.84 kg active ingredient/ha [48 oz/ac]) while Bermudagrass is actively growing and not stressed (i.e., not during drought), and prior to seed head production (T. Bond et al. 2005).
- Year 2—Treat remaining Bermudagrass patches (i.e., "mop up") with imazapyr in spring. Prepare site and plant longleaf pine in the fall.
- Year 3—Plant NWSG/forb mix in the spring.

4. If fields are fallow for less than one complete growing season, then require treatment with imazapyr regardless of observed Bermudagrass. Bermudagrass rhizomes are commonly present and dormant under the soil surface, especially in fields where glyphosate-tolerant (e.g., Roundup Ready) crops have been grown, and the grass may not surface until the growing season following the last crop rotation. The soil activity of imazapyr herbicides should help control future emergence of Bermudagrass or other IPG.

If longleaf pine and NWSG are planted without first controlling the Bermudagrass, then control cannot be achieved without destroying the longleaf and NWSG, which undermines programmatic intent and increases the cost to the landowner and taxpayer. Additional studies are needed at later stages of CP36 contracts to quantitatively track longleaf pine and ground cover development and wildlife species response. These studies will help assess whether practice establishment and mid-contract management achieve stated resource goals.

Acknowledgments

We sincerely appreciate the assistance provided by Brett Martin and county office personnel at the Georgia Farm Service Agency; Steve Chapman and district foresters at the Georgia Forestry Commission; and Georgia Natural Resources Conservation Service state biologist Keith Wooster and county office personnel in conducting this study. We thank Georgia Department of Natural Resources biologist Drew Larson for review and comments. The mention of trade names is for ease of understanding and does not imply endorsement by the Georgia Department of Natural Resources.

16

Forb Community Response to Management of Grassland Buffers

Jolie G. Dollar, Timothy J. Schauwecker, Samuel K. Riffell, and L. Wes Burger Jr.

In north-central Mississippi during the nineteenth century, the landscape contained numerous prairie "islands" surrounded by forest (A. Jones and Patton 1966; Barone 2005b). Collectively called black prairie, these prairies were the southernmost extent of North America's tallgrass prairie ecosystem (Peacock and Schauwecker 2003a). Blackland habitats persisted in the landscape because of periodic disturbance such as fire (e.g., natural and anthropogenic from Native Americans) and grazing by native ungulates (Bachand 2001). These disturbances maintained early-successional vegetation and created habitat for numerous animal species adapted to grassland and edge habitats (Kline 1997; A. Knapp and Seastedt 1998).

The majority of Mississippi's black prairie ecosystem has been lost because of human development and suppression of natural disturbance regimes (Sampson and Knopf 1994; Bachand 2001). This has mirrored the loss of early-successional grassland habitat and biodiversity in North America (Peterjohn 2003; C. Harper 2007; Flynn et al. 2009) and Europe (Chamberlain et al. 2000; Robinson and Sutherland 2002). Many bird, mammal, and arthropod species need early-successional grassland habitat to meet life history requirements (Brennan 1991; A. Knapp et al. 1998; J. Jones et al. 2007). For these reasons, loss of grassland habitat has led to declines in abundance of some species, and in some cases, to listing as threatened or endangered (Panzer 1988; Hunter et al. 2001; Brennan and Kuvlesky 2005; Sauer et al. 2008).

Although less than 1% of natural Mississippi black prairie remains (Barone and Hill 2007), semi-natural grassland habitats created through the US Department of Agriculture Conservation Reserve Program (CRP) may offset the negative effects of grassland habitat loss (Heard et al. 2000; Reeder et al. 2005; Burger et al. 2006; Benson et al. 2007). The CRP is an agricultural set-aside program that provides financial incentives for producers to convert cropland to more natural habitats or to implement management practices designed to provide ecosystem services (similar to European agri-environment schemes; e.g., Ovenden et al. 2008). Certain CRP practices, such as upland habitat buffers, involve planting grasses, forbs, and shrubs around the perimeter of agricultural fields to benefit a variety of grassland-associated taxa (Le Coeur et al. 2002; E. Marshall and Moonen 2002; Lovell and Sullivan 2006; Olson and Wackers 2007). CRP practices involving establishment and maintenance of grassland habitat require landowners to perform mid-contract management activities. Two primary options for management in the southeastern United States are fall disking and spring burning (C. Harper 2007). Both disking and burning maintain early-successional plant communities by inhibiting growth of woody species. Consequently, such alterations can influence plant community composition (Greenfield et al. 2003; Spasojevic et al. 2010). For example, disking can decrease the competitive advantage of native perennial grasses and allow more forbs to germinate in subsequent years because disking breaks peren-

Figure 16.1. The distribution of historical blackland prairie in Mississippi and Alabama (adapted from Barone 2005b). The lighter-shaded area depicts the blackland prairie region. The darker-shaded "islands" depict historical blackland prairie sites. The arrow indicates the location of grassland buffers used in this study.

nial grass root systems and exposes bare ground (Benson et al. 2007; Olson and Wackers 2007; Spasojevic et al. 2010). Burning increases soil exposure because fire consumes dead plant material (Benson et al. 2007). It can also stimulate regrowth of native warm-season grasses by increasing available nutrients and soil temperature (Benson et al. 2007).

Knowledge of how individual forb species and forb communities respond to grassland buffer management is lacking. Such data are needed because forbs are a critical component of grassland habitat for many wildlife species. Abundant forbs can be directly related to the abundance and species richness of beneficial insects and grassland birds (Siemann et al. 1998; Symstad et al. 2000; Jamison et al. 2002; Field et al. 2005; Davros et al. 2006; Lovell and Sullivan 2006; J. Davis et al. 2008). The presence of forbs enhances nesting and brooding cover, and forb seeds are a food source for many wildlife species (C. Harper 2007). In addition, as forb density increases, so does plant diversity, thus maintaining ecosystem functioning and services (Isbell et al. 2011). Thus, forbs can be both a component of diversity and a driver of diversity in grassland systems.

Our objective was to investigate the effects of spring burning versus fall disking in forb communities of recently established grassland field buffers enrolled in the CRP. We predicted that forb communities in recently disked buffers would have more-abundant and species-rich annual components compared to recently burned and control buffers. Conversely, we predicted that recently burned buffers would have more native warm-season grass cover compared to recently disked and control buffers and would exhibit concomitant shifts away from annual components in favor of perennial forbs. Our results provide landowners, land managers, and conservation organizations with management recommendations for maintaining forb communities in early-successional grasslands in the southeastern black prairie region.

Methods

Study Site

We worked at B. Bryan Farms, a privately owned row-crop and cattle-grazing operation of 2,104 ha in Clay County, Mississippi. Portions of Clay County fall in the historical black prairie physiographic region (Barone 2005b; see fig.

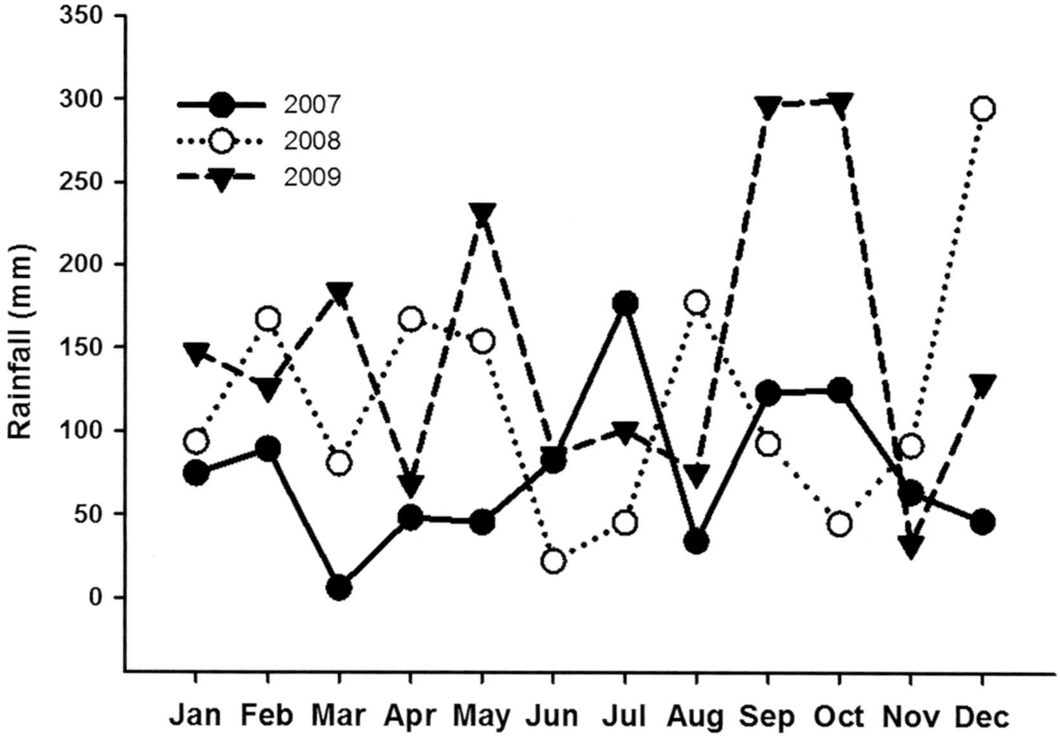

Figure 16.2. Monthly precipitation (mm) for B. Bryan Farms, Clay County, Mississippi. Data collected at Tibbee and Aberdeen stations. Yearly totals are as follows: 2007 = 918 mm, 2008 = 1,434 mm, 2009 = 1,782 mm.

16.1). Precipitation varied substantially from year to year (fig. 16.2). In spring 2004, 79 crop hectares were enrolled in a CRP practice titled CP33—Habitat Buffers for Upland Birds. CP33 buffers are linear strips 9–40 m wide of herbaceous vegetation (native warm-season grasses plus forbs) around the perimeter of agricultural fields. At that time, the following grasses were seeded: *Andropogon gerardii* (big bluestem), *Schizachyrium scoparium* (little bluestem), and *Sorghastrum nutans* (Indiangrass). The following forbs were seeded: *Chamaecrista fasciculata* (partridge pea), *Rudbeckia hirta* (black-eyed Susan), and *Helianthus maximiliani* (Maximilian sunflower). The buffer plant community also included many nonseeded species, which were present in the soil seedbank or introduced through wind and animal dispersal. At our study location, buffers were located primarily on Griffith, Okolona, and Sessums silty clays from parent material of clayey marine or alluvial deposits.

For CP33 buffers, mid-contract management is mandated to maintain early-successional plant communities (USDA 2004). Fifteen fields (with two, three, or four sides) at B. Bryan Farms containing 43 buffers were randomly assigned to one of three treatments (i.e., fall disk, spring burn, or no management). One randomly assigned buffer per field (i.e., one side of each field) received management each year. Thus, a rotational management regime was followed. In 2007 (pretreatment), we selected 8 fields with a total of 26 buffers (out of the total established 43 buffers) to collect vegetation and forb community data. These 8 fields were selected because they had relatively uniform establishment, had low invasive species cover, and were located on similar soil types. In 2007, 4 buffers (out of the 26 buffers selected in 2007) were lightly disked. In spring 2008, 4 buffers were burned. Eighteen buffers were unmanaged controls. In 2008, 4 additional buffers were disked in the fall, and 4 more were burned in spring 2009. Thus, 10 buffers served as controls in 2009. The 8 buffers disturbed in 2007–2008 represented two growing seasons postmanagement in 2009. We lightly disked (cut the existing vegetation, incorporating approximately one-half into the soil, exposing substantial soil; C. Harper et al. 2007) in the fall because disking at other times of year often facilitates undesirable species such as Johnsongrass (*Sorghum halepense*;

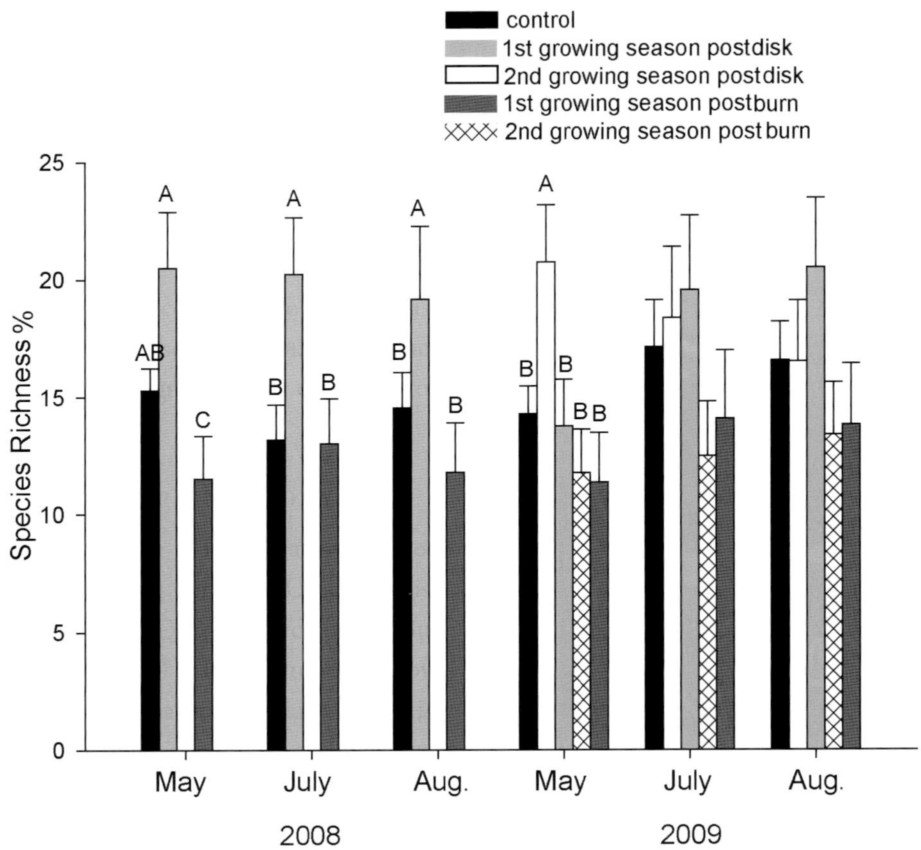

Figure 16.3. Least-squares means (SE) for total plant species richness by treatment and sampling period in native grass field buffers in Clay County, Mississippi. Letters designate significant differences at $\alpha = 0.10$.

C. Harper et al. 2007). We burned in early spring (i.e., March–April) to preserve winter cover for wildlife and because early spring burns produce less smoke than those later in the season.

Plant Community Sampling

We used standard quadrats measuring 1 m^2 for sampling buffer plant communities to the species level (Roberts-Pichette and Gillespie 1999; Driscoll et al. 2010). Buffers at B. Bryan Farms were an average of 36.5 m wide. We collected measurements from 30 quadrats per buffer, for a total sampling area per buffer of 30 m^2 for each sampling period, regardless of buffer size. We established 200 m transects on the long axis of each buffer. At 5 m intervals along each transect, we placed 1 m^2 frames 5–10 m away from transect lines and alternated sides to avoid any potential confounding disturbance along transect center lines (~ 22 m apart). In 2007 (pretreatment), we created a species accumulation curve (Dollar 2011), which indicated that 30 was a sufficient number of samples to estimate species richness in these assemblages (M. Barbour et al. 1980; Roberts-Pichette and Gillespie 1999).

Plants were identified to the species level in August 2007 (pretreatment); May, July, and August 2008; and May, July, and August 2009, capturing floristic diversity throughout the growing season and across years. We recorded the percent cover for each plant species that was at a physiologically mature stage to allow for identification (e.g., with two true leaves), along with the percentage of bare ground and the percent cover of dead plant material. Because of overlapping vegetation layers, percent cover sometimes exceeded 100% (Benson et al. 2007). For all forb and woody species, we recorded the number of primary stems. We used field guides, dichotomous keys, online resources, and journal articles to identify species (Radford et al. 1968; P. Barbour 2006; T. Murphy et al. 2007; Timme 2007; USDA 2010–

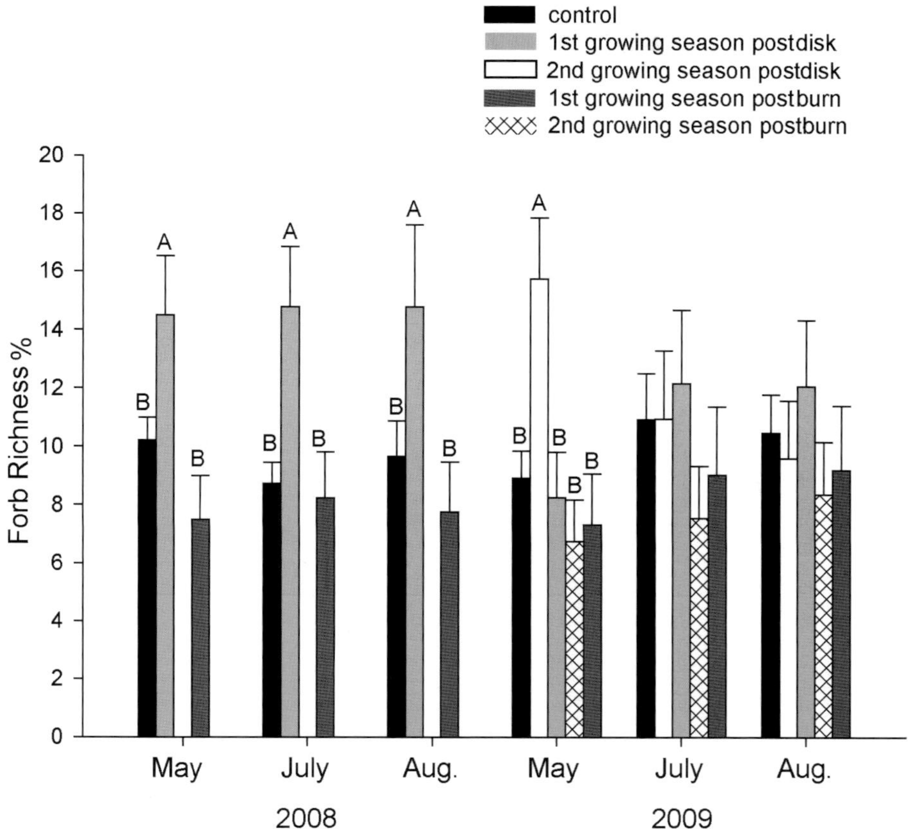

Figure 16.4. Least-squares means (SE) for forb species richness by treatment and sampling period in native grass field buffers in Clay County, Mississippi. Letters designate significant differences at $\alpha = 0.10$.

2013). Pressed specimens and/or in-field photographs were obtained for each species.

Statistical Analysis

We used general linear models in SAS Proc Mixed (SAS Institute 2007) to test for treatment effects on buffer plant variables (Littell et al. 2006). For richness variables that were counts, and hence Poisson distributed, we used Proc Glimmix (Littell et al. 2006). All plant variables were averaged over 30 samples for each buffer, except for species richness variables, which were summed over the 30 samples per buffer. Our explanatory variable was treatment type (i.e., disk 2007, disk 2008, burn 2008, burn 2009, control). Our response variables were percentage of bare ground, percent cover of litter, total species richness, percent cover of individual plant species, total forb stem counts, forb species richness, native perennial forb stem counts, native annual forb stem counts, and stem counts for individual forb species. Each collection period was analyzed separately, and we used field as a random effect.

In addition to general linear models, we used redundancy analysis (RDA) in the CANOCO package (ter Braak 1990) with the Monte Carlo permutation test to test for treatment effects on the buffer plant community following Leps (1999) and Leps and Smilauer (2003). Our first test (M1) corresponded to all within-subject effects in repeated-measures ANOVA. For M1, year and the interaction of year with each treatment (i.e., year*disk 2007, year*disk 2008, year*burn 2008, year*burn 2009, year*control) were used as explanatory variables, and buffer was a covariable. Covariables are explanatory variables that have an acknowledged (or hypothesized) influence on the response variables (Leps and Smilauer 2003). The null hypothesis for M1 was that there were no directional changes in species composition over time that were common to all treatments or specific to particular treatments. The second analysis (M2) tested

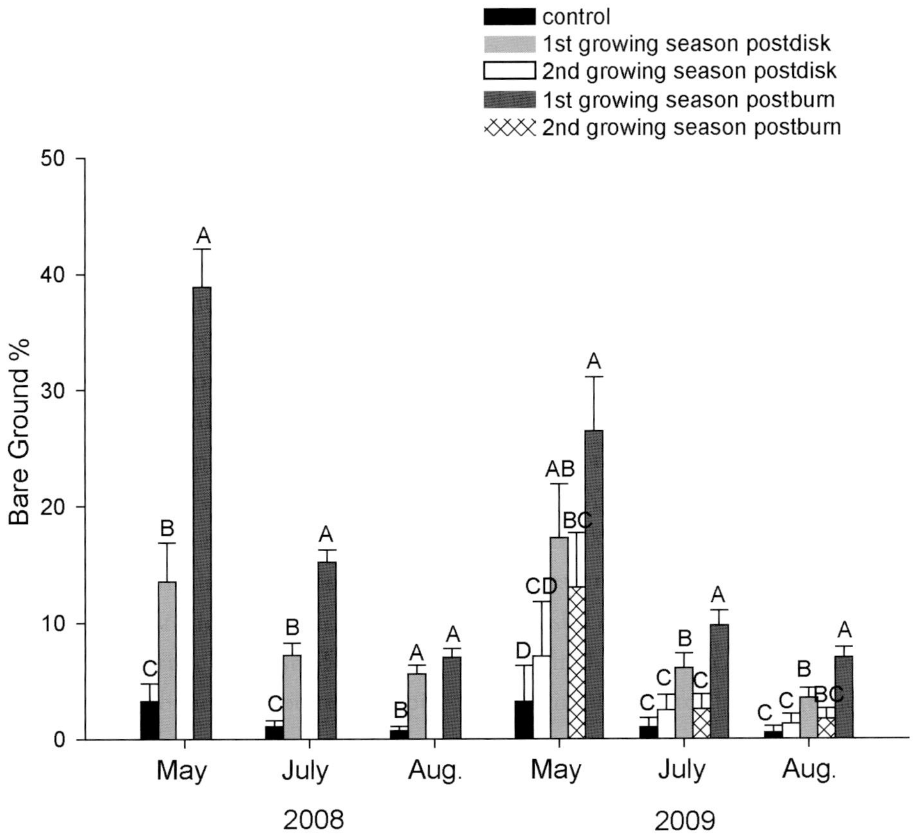

Figure 16.5. Least-squares means (SE) for percentage of bare ground by treatment and sampling period in native grass field buffers in Clay County, Mississippi. Letters designate significant differences at $\alpha = 0.10$.

the null hypothesis that temporal trends in species composition were independent of treatments. The M2 model was coded the same as the M1 model except year was included as a covariable instead of an explanatory variable. The subsequent four models (M3–M6) corresponded to tests of particular terms in repeated-measures ANOVA. These four models tested the effects of individual explanatory variables (i.e., disk 2007, disk 2008, burn 2008, and burn 2009) on the buffer plant community. All four models included year and buffer as covariables, along with the interacting terms of year and the three treatments not being tested.

Pretreatment Differences

Data collected during August 2007 allowed us to test for pretreatment differences among buffers that could be confounded with treatments. In SAS Proc Mixed (Littell et al. 2006), we used treatment levels from 2008 and 2009 with the 2007 response variables (pretreatment data) to identify response variables with significant pretreatment differences. In 2007, 9 out of 60 cover variables had significant treatment effects (Dollar 2011). These variables were percentage of bare ground; percent cover of dead plant material; and stem counts of *Chamaecrista fasciculata* (partridge pea); *Conyza canadensis* (Canadian horseweed); *Cyperus esculentus* (yellow nutsedge); *Desmanthus illinoensis* (Illinois bundleflower); *Erigeron strigosus* (prairie fleabane); *Helianthus angustifolius* (swamp sunflower); and *Ipomoea pandurata* (man of the earth). Nine out of 60 variables is only 1.5 times the number expected by chance alone ($60 \times 0.10 = 6$). For 2007, 7 out of 42 forb stem count variables had significant treatment effects. These variables were *Chamaecrista fasciculata*, *Desmanthus illinoensis*, *Erigeron strigosus*, *Helianthus angustifolius*, *Helianthus maximiliani* (Maximilian sunflower), *Ipomoea pandurata*, and *Senna obtusifolia* (java-bean). This is less than two times the number expected by chance alone. Nonetheless, we used 2007 data as covariates for analyses in 2008 and 2009 for those vari-

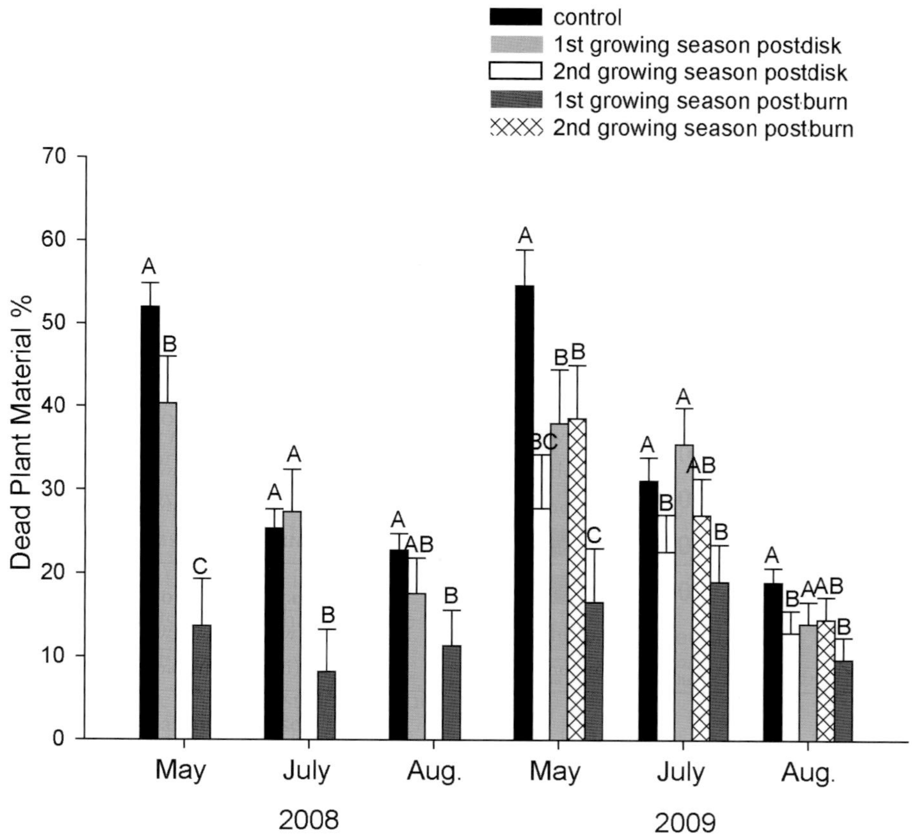

Figure 16.6. Least-squares means (SE) for dead plant material by treatment and sampling period in native grass field buffers in Clay County, Mississippi. Letters designate significant differences at $\alpha = 0.10$.

ables that had pretreatment differences. Because the conclusions were not changed (similar means and P values) for any variable when using pretreatment covariates, we report all analyses without 2007 covariates. In addition, principal components analysis (PCA) on 2007 species data did not indicate any grouping of samples by treatment (J. Dollar, unpublished data). Thus, it is unlikely that pretreatment differences influenced statistical tests for 2008 and 2009 data.

Results

We found 110 vascular plant species during 2007–2009 sampling periods (see the appendix). Total plant species richness was consistently greater on fall-disked buffers and lower on spring-burned buffers during the three sampling periods in summer 2008 ($P = 0.013–0.086$; fig. 16.3). In May 2009, buffers disked in fall 2007 (second growing season postmanagement) continued to have the greatest species richness ($P = 0.031$). However, there were no treatment differences in plant species richness for the July and August sampling periods in summer 2009. Forb richness was greatest on fall-disked buffers in all three summer 2008 sampling periods ($P = 0.009–0.039$; fig. 16.4). The trend of fall-disked buffers having the greatest species richness continued into 2009, but treatment differences were significant only in May 2009 ($P = 0.009$). The percentage of bare ground was greatest on spring-burned buffers and least on control buffers during all sampling periods in summer 2008 and 2009 ($P \leq 0.001–0.002$; fig. 16.5), but this effect was greatly diminished or nonexistent on burned buffers in the second year postmanagement. The percent cover of dead plant material was least on buffers burned the previous spring in the summers of 2008 and 2009 ($P \leq 0.001–0.068$; fig. 16.6).

Treatment Effects

Both fall disking and spring burning positively affected total forb stem counts in summer 2008. In summer 2009, buffers in the second growing season postdisk and the sec-

Table 16.1.

Least-squares means and standard errors for plant variables by treatment and sampling period at B. Bryan Farms, Clay County, Mississippi, 2008–2009.

	Control	Disking treatment		Burning treatment		P value
		Fall 2007	Fall 2008	Spring 2008	Spring 2009	
Total forb stem counts (#/m²)						
May 2008	16.12 (2.6) B	30.52 (5.5) A		18.33 (5.5) AB		0.075
July 2008	15.92 (3.0) B	28.64 (5.8) A		27.72 (5.8) A		0.040
Aug. 2008	17.40 (3.6)	28.72 (7.3)		24.01 (7.3)		0.275
May 2009	18.62 (3.7)	26.32 (5.9)	16.18 (5.97)	23.52 (5.89)	23.99 (5.9)	0.620
July 2009	15.50 (2.4) B	25.36 (3.8) A	11.74 (3.8) B	26.75 (3.8) A	19.61 (3.8) AB	0.017
Aug. 2009	14.15 (2.4) B	25.34 (3.8) A	11.71 (3.8) B	22.43 (3.8) A	14.13 (3.8) B	0.026
Native perennial forb stem counts (#/m²)						
May 2008	13.64 (2.6)	18.05 (5.2)		15.82 (5.2)		0.684
July 2008	11.89 (3.1)	14.64 (5.0)		14.87 (5.0)		0.704
Aug. 2008	14.63 (3.3)	18.69 (6.59)		17.52 (6.5)		0.779
May 2009	17.22 (3.8)	19.39 (6.0)	14.15 (5.0)	22.90 (5.0)	23.82 (5.0)	0.760
July 2009	13.69 (2.3) B	18.41 (3.5) AB	9.96 (3.5) B	23.37 (3.56) A	17.44 (3.5) AB	0.060
Aug. 2009	11.93 (2.2)	19.47 (3.5)	9.01 (3.5)	17.14 (3.5)	12.46 (3.5)	0.119
Native annual forb stem counts (#/m²)						
May 2008	2.05 (0.8) B	11.79 (1.4) A		2.27 (1.4) B		< 0.001
July 2008	2.90 (1.6) B	12.35 (3.3) A		12.51 (3.3) A		0.007
Aug. 2008	2.26 (1.0) B	9.58 (1.7) A		5.08 (1.7) B		0.001
May 2009	0.63 (0.7) B	5.96 (1.1) A	1.47 (1.1) B	0.73 (1.1) B	0.33 (1.1) B	0.004
July 2009	0.95 (1.0) B	6.22 (1.6) A	0.47 (1.6) B	3.19 (1.6) AB	1.81 (1.6) B	0.037
Aug. 2009	1.58 (1.3)	5.47 (2.1)	1.06 (2.1)	5.20 (2.1)	1.45 (2.1)	0.293

Note: Letters designate significant differences within rows based on least-squares means at $\alpha = 0.10$.

ond growing season postburn had the greatest forb stem counts (table 16.1). Native perennial forbs were not affected by management, except in July 2009, when buffers in the second growing season postburn had the greatest counts (table 16.1). In contrast, native annual forbs were positively affected by fall disking in all sampling periods except August 2009 (table 16.1). The effects of fall disking were greatest on buffers treated previous to summer 2008.

Eight native annual/biennial forbs had treatment effects (table 16.2). Except for one occurrence (i.e., *Ambrosia artemisiifolia*, common ragweed, in August 2008), these species responded positively to disking compared to burning and no management. Eleven native perennial forbs had treatment effects (table 16.3). As with annual forbs, these species generally responded positively to disking compared to burning and no management.

Redundancy Analysis (RDA)

May sampling period. For the early-summer postmanagement period (May), we rejected the first two null hypotheses (M1: P = 0.001 and M2: P = 0.001; table 16.4), which indicated that one or more environmental (treat-

Table 16.2.

Least-squares means and standard errors for annual/biennial plant variables (% cover of individual species) by treatment and sampling period at B. Bryan Farms, Clay County, Mississippi, 2008–2009.

	Control	Disking treatment		Burning treatment		P value
		Fall 2007	Fall 2008	Spring 2008	Spring 2009	
Ambrosia artemisiifolia (%/m²)						
May 2008	0.01 (0.0)	0.00 (0.0)		0.00 (0.0)		0.814
July 2008	0.06 (0.1)	0.06 (0.1)		0.22 (0.1)		0.376
Aug. 2008	0.04 (0.1) B	0.04 (0.1) B		0.37 (0.1) A		0.011
May 2009	0.03 (0.0)	0.07 (0.0)	0.00 (0.0)	0.00 (0.0)	0.00 (0.0)	0.736
July 2009	0.02 (0.1) B	0.21 (0.1) AB	0.00 (0.1) B	0.00 (0.1) B	0.43 (0.1) A	0.098
Aug. 2009	0.02 (0.1)	0.21 (0.1)	0.00 (0.1)	0.13 (0.1)	0.09 (0.1)	0.458
Ambrosia trifida (%/m²)						
May 2008	0.51 (0.7) B	4.32 (1.5) A		0.73 (1.5) B		0.068
July 2008	2.01 (1.6)	7.70 (2.8)		2.94 (2.8)		0.139
Aug. 2008	0.39 (1.3) B	7.18 (2.6) A		0.00 (2.6) B		0.074
May 2009	0.05 (0.7)	2.75 (1.1)	0.33 (1.1)	0.00 (1.1)	0.07 (1.1)	0.273
July 2009	0.72 (0.8)	3.34 (1.2)	0.45 (1.2)	1.18 (1.2)	1.60 (1.2)	0.231
Aug. 2009	0.62 (1.2)	4.62 (1.8)	1.83 (1.8)	0.36 (1.8)	0.19 (1.8)	0.348
Chamaecrista fasciculata (%/m²)						
May 2008	0.39 (0.4)	1.50 (0.7)		1.30 (0.7)		0.106
July 2008	0.47 (1.0) B	5.24 (2.1) A		2.95 (2.1) AB		0.085
Aug. 2008	0.75 (1.6) B	8.40 (3.4) A		5.25 (3.4) AB		0.091
May 2009	—	—	—	—	—	—
July 2009	0.00 (0.1) B	0.00 (0.1) B	0.30 (0.1) A	0.02 (0.1) B	0.00 (0.1) B	0.092
Aug. 2009	0.02 (0.1)	0.00 (0.2)	0.48 (0.2)	0.00 (0.2)	0.00 (0.2)	0.130
Conyza canadensis (%/m²)						
May 2008	0.27 (0.2) B	0.91 (0.3) A		0.17 (0.3) B		0.015
July 2008	0.52 (0.6) B	4.86 (1.2) A		0.69 (1.2) B		0.008
Aug. 2008	8.18 (1.6) A	4.43 (2.4) AB		3.98 (2.4) B		0.063
May 2009	0.00 (0.1)	0.42 (0.1)	0.00 (0.1)	0.00 (0.1)	0.00 (0.1)	0.108
July 2009	0.02 (0.2) B	1.02 (0.2) A	0.19 (0.2) B	0.01 (0.2) B	0.16 (0.2) B	0.024
Aug. 2009	0.07 (0.2)	1.00 (0.3)	0.45 (0.3)	0.05 (0.3)	0.51 (0.3)	0.132
Geranium carolinianum (%/m²)						
May 2008	1.29 (0.5) B	2.76 (0.7) A		0.54 (0.7) B		0.049
July 2008	0.56 (0. 7)	2.12 (1.4)		3.54 (1.4)		0.159

Continued on the next page

Table 16.2. *Continued*

	Control	Disking treatment		Burning treatment		P value
		Fall 2007	Fall 2008	Spring 2008	Spring 2009	
Aug. 2008	—	—		—		—
May 2009	0.33 (0.6) B	1.94 (0.9) AB	3.45 (0.9) A	0.93 (0.9) B	0.24 (0.9) B	0.079
July 2009	0.02 (0.0)	0.00 (0.0)	0.00 (0.0)	0.00 (0.0)	0.00 (0.0)	0.840
Aug. 2009	—	—-	—	—	—	—
Krigia virginica (%/m²)						
May 2008	0.00 (0.0) B	0.21 (0.1) A		0.00 (0.1) B		0.073
July 2008	—	—		—		—
Aug. 2008	—	—		—		—
May 2009	0.22 (0.0)	0.74 (0.2)	0.23 (0.2)	0.03 (0.2)	0.00 (0.2)	0.167
July 2009	—	—	—	—	—	—
Aug. 2009	—	—	—	—	—	—
Triodanis perfoliata (%/m²)						
May 2008	0.18 (0.3) B	2.13 (0.5) A		0.00 (0.5) B		0.005
July 2008	—	—		—		—
Aug. 2008	—	—		—		—
May 2009	0.00 (0.0) B	0.13 (0.0) A	0.06 (0.0) AB	0.00 (0.0) B	0.00 (0.0) B	0.049
July 2009	—	—	—	—	—	—
Aug. 2009	—	—	—	—	—	—
Valerianella radiata (%/m²)						
May 2008	0.01 (0.1) B	0.42 (0.2) A		0.00 (0.2) B		0.071
July 2008	0.02 (0.0)	0.00 (0.0)		0.00 (0.0)		0.814
Aug. 2008	—	—		—		—
May 2009	0.13 (0.2) B	1.32 (0.3) A	0.27 (0.3) B	0.06 (0.3) B	0.29 (0.3) B	0.022
July 2009	0.11 (0.1)	0.00 (0.1)	0.00 (0.1)	0.00 (0.1)	0.00 (0.1)	0.839
Aug. 2009	—	—	—	—	—	—

Note: Only species with significant differences for at least one sampling period are shown. Letters designate significant differences within rows based on least-squares means at $\alpha = 0.10$. Broken dashes indicate that the species was not detected during that sampling period. Blank cells indicate that no sampling was done that month.

ment) variables influenced the buffer plant community (M1) and that year was a factor in the influence (year effect, M2). We also rejected null hypotheses M3 (influence of first growing season postdisk, D1) and M4 (influence of second growing season postdisk, D2; P = 0.001 and P = 0.018, respectively), indicating that buffers in both the first growing season postdisk and the second growing season postdisk had different plant composition compared to other treatments. Null hypotheses M5 and M6 were not rejected (P = 0.167 and P = 0.404, respectively), indicating that buffers in the first growing season postburn (hypothesis M5, treatment B1) and second growing season postburn (hypothesis M6, treatment B2) did not have significantly different plant communities compared to disked and control treatments.

The ordination biplot visually represents the influence

Table 16.3.

Least-squares means and standard errors for perennial plant variables (% cover of individual species) by treatment and sampling period at B. Bryan Farms, Clay County, Mississippi, 2008–2009.

	Control	Disking treatment		Burning treatment		P value
		Fall 2007	Fall 2008	Spring 2008	Spring 2009	
Allium canadense (%/m²)						
May 2008	0.01 (0.0) B	0.50 (0.0) A		0.00 (0.0) B		< 0.001
July 2008	0.00 (0.0) B	0.16 (0.0) A		0.00 (0.0) B		0.001
Aug. 2008	—	—		—		—
May 2009	0.01 (0.0)	0.08 (0.0)	0.13 (0.0)	0.00 (0.0)	0.02 (0.0)	0.181
July 2009	0.00 (0.0)	0.02 (0.0)	0.08 (0.0)	0.00 (0.0)	0.00 (0.0)	0.293
Aug. 2009	—	—	—	—	—	—
Helianthus maximiliani (%/m²)						
May 2008	3.30 (2.5) B	12.50 (3.8) A		5.48 (3.8) B		0.040
July 2008	5.04 (3.3)	13.86 (5.3)		6.90 (5.3)		0.220
Aug. 2008	4.00 (2.3)	8.95 (3.5)		3.42 (3.5)		0.330
May 2009	0.92 (2.6)	9.83 (3.9)	4.27 (3.9)	5.38 (4.0)	9.10 (4.0)	0.145
July 2009	2.62 (2.6)	9.65 (3.7)	5.05 (3.7)	6.36 (3.7)	8.52 (3.7)	0.223
Aug. 2009	2.00 (1.5)	5.98 (2.3)	3.07 (2.3)	3.74 (2.3)	5.70 (2.3)	0.344
Ipomoea pandurata (%/m²)						
May 2008	0.03 (0.0)	0.13 (0.1)		0.13 (0.1)		0.221
July 2008	0.11 (0.1) B	1.11 (0.2) A		0.15 (0.2) B		< 0.001
Aug. 2008	0.14 (0.2) B	0.92 (0.4) A		0.83 (0.4) A		0.072
May 2009	0.00 (0.0)	0.00 (0.0)	0.00 (0.0)	0.00 (0.0)	0.11 (0.0)	0.287
July 2009	0.14 (0.2) B	0.00 (0.3) B	2.26 (0.3) A	0.01 (0.3) B	0.36 (0.3) B	< 0.001
Aug. 2009	0.41 (0.4) B	0.00 (0.6) B	6.08 (0.6) A	0.09 (0.6) B	0.55 (0.6) B	< 0.001
Mimosa microphylla (%/m²)						
May 2008	0.47 (0.2) B	1.47 (0.3) A		0.57 (0.3) B		0.020
July 2008	—	—		—		—
Aug. 2008	—	—		—		—
May 2009	—	—	—	—	—	—
July 2009	—	—	—	—	—	—
Aug. 2009	—	—	—	—	—	—
Oxalis stricta (%/m²)						
May 2008	0.15 (0.1) B	0.70 (0.2) A		0.04 (0.2) B		0.071
July 2008	0.01 (0.0) B	0.22 (0.1) A		0.00 (0.1) B		0.002
Aug. 2008	0.01 (0.0)	0.04 (0.0)		0.00 (0.0)		0.592

Continued on the next page

Table 16.3. *Continued*

	Control	Disking treatment		Burning treatment		P value
		Fall 2007	Fall 2008	Spring 2008	Spring 2009	
May 2009	0.12 (0.1) B	0.70 (0.2) A	0.02 (0.2) B	0.03 (0.2) B	0.00 (0.2) B	0.023
July 2009	0.00 (0.1)	0.00 (0.1)	0.26 (0.1)	0.00 (0.1)	0.00 (0.1)	0.280
Aug. 2009	0.00 (0.0)	0.00 (0.0)	0.06 (0.0)	0.00 (0.0)	0.00 (0.0)	0.450
Potentilla simplex (%/m^2)						
May 2008	0.00 (0.0) B	0.04 (0.0) A		0.00 (0.0) B		0.067
July 2008	—	—		—		—
Aug. 2008	—	—		—		—
May 2009	—	—	—	—	—	—
July 2009	—	—	—	—	—	—
Aug. 2009	—	—	—	—	—	—
Ranunculus fascicularis (%/m^2)						
May 2008	4.18 (1.4)	7.13 (2.7)		0.00 (2.7)		0.198
July 2008	—	—		—		—
Aug. 2008	—	—		—		—
May 2009	2.38 (2.2) B	7.77 (3.5) AB	14.30 (3.5) A	2.87 (3.5) B	5.73 (3.5) B	0.074
July 2009	0.05 (0.0)	0.00 (0.1)	0.00 (0.1)	0.01 (0.1)	0.01 (0.1)	0.794
Aug. 2009	—	—	—	—	—	—
Senna obtusifolia (%/m^2)						
May 2008	0.04 (0.1)	0.55 (0.2)		0.04 (0.2)		0.105
July 2008	0.20 (0.2)	0.75 (0.4)		0.00 (0.4)		0.343
Aug. 2008	0.55 (0.5) B	2.67 (1.0) A		0.01 (1.0) B		0.085
May 2009	0.00 (0.0)	0.01 (0.0)	0.00 (0.0)	0.00 (0.0)	0.00 (0.0)	0.267
July 2009	0.19 (0.1) B	0.19 (0.2) B	0.95 (0.2) A	0.04 (0.2) B	0.10 (0.2) B	0.003
Aug. 2009	0.56 (0.4)	0.90 (0.6)	2.08 (0.6)	0.00 (0.6)	0.01 (0.6)	0.151
Solanum carolinense (%/m^2)						
May 2008	0.01 (0.0) B	0.00 (0.1) B		0.17 (0.1) A		0.023
July 2008	0.21 (0.1) B	0.87 (0.2) A		0.07 (0.2) B		0.020
Aug. 2008	0.09 (0.1)	0.20 (0.1)		0.00 (0.1)		0.426
May 2009	0.05 (0.0)	0.02 (0.1)	0.00 (0.1)	0.00 (0.1)	0.09 (0.1)	0.582
July 2009	0.09 (0.1)	0.07 (0.1)	0.17 (0.1)	0.00 (0.1)	0.14 (0.1)	0.444
Aug. 2009	0.11 (0.1)	0.07 (0.1)	0.13 (0.1)	0.02 (0.1)	0.06 (0.1)	0.535
Symphyotrichum pilosum (%/m^2)						
May 2008	—	—		—		—
July 2008	—	—		—		—
Aug. 2008	0.26 (0.2)	0.08 (0.3)		0.79 (0.3)		0.233
May 2009	—	—	—	—	—	—

Table 16.3. *Continued*

	Control	Disking treatment		Burning treatment		P value
		Fall 2007	Fall 2008	Spring 2008	Spring 2009	
July 2009	0.67 (0.3)	0.16 (0.4)	0.12 (0.4)	1.19 (0.4)	0.02 (0.4)	0.162
Aug. 2009	0.47 (0.5) B	0.20 (0.8) B	0.28 (0.8) B	2.78 (0.8) A	0.00 (0.8) B	0.083
Verbena halei (%/m²)						
May 2008	—	—		—		—
July 2008	0.23 (0.2) B	0.07 (0.3) B		0.97 (0.3) A		0.080
Aug. 2008	—	—		—		—
May 2009	—	—	—	—	—	—
July 2009	0.33 (0.3)	0.91 (0.4)	0.00 (0.4)	0.53 (0.4)	0.08 (0.4)	0.495
Aug. 2009	0.12 (0.1)	0.34 (0.2)	0.00 (0.2)	0.24 (0.2)	0.03 (0.2)	0.539

Note: Only species with significant differences for at least one sampling period are shown. Letters designate significant differences within rows based on least squares means at $\alpha = 0.10$. Broken dashes indicate that the species was not detected during that sampling period. Blank cells indicate that no sampling was done that month.

of individual species on each treatment (fig. 16.7). The forbs *Allium canadense* (meadow garlic) and *Triodanis perfoliata* (clasping Venus's looking glass) were most abundant on buffers in the first growing season following fall disking. The forbs *Ampelopsis arborea* (peppervine), *Lespedeza virginica* (slender lespedeza), and *Ranunculus fascicularis* (early buttercup) were most abundant in the second growing season following fall disking.

July sampling period. For the midsummer sampling period (July), the first two null hypotheses (M1 and M2) were rejected (P = 0.001 and P = 0.001, respectively), meaning that at least one environmental variable influenced buffer plant communities and year was a factor (table 16.4). The only variable that statistically influenced the plant community was D1 for hypothesis M3 (the first growing season following disking; P = 0.001). *Ipomoea pandurata* was the most abundant forb in buffers disked the previous fall and drove the influence of the D1 treatment (fig. 16.8).

August sampling period. As with the above sampling periods, null hypotheses for M1 and M2 were rejected for the late-summer (August) sampling period (P = 0.001 and P = 0.001, respectively), meaning that at least one environmental variable influenced the buffer plant community (table 16.4). Additionally, the null hypothesis associated with M3 (testing the influence of the first growing season postdisk, D1) was rejected (P = 0.001). As with the July sampling period, *Ipomoea pandurata* drove the influence of the D1 treatment (fig. 16.9). We failed to reject the null hypotheses for analyses M4–M6 (P = 0.136–0.495).

Discussion

Effects of Disking

Disking influenced the buffer plant community, particularly during the first growing season following fall disking (figs. 16.7, 16.8, and 16.9). Positive effects of disking on plant species richness, forb species richness, total forb stem counts, and native annual forb stem counts are in accord with results from many studies (Buckner and Landers 1979; W. Webb and Guthery 1983; Madison et al. 1995; Paveglio and Kilbride 2000; Greenfield et al. 2002, 2003; Benson et al. 2007; Gruchy 2007). Native annual forb stem counts were more positively related to fall disking than were native perennial forb stem counts. This adds further support to the finding that disking has a stronger positive influence on annual forbs than on perennial forbs during the years following management (C. Harper 2007).

Effects of fall disking were highly variable between 2008 and 2009 for numerous plant variables. In 2008, for example, total plant species richness and *Ambrosia trifida* cover had strong positive responses from fall 2007 disking, but not in 2009 from fall 2008 disking (fig. 16.3; table 16.2).

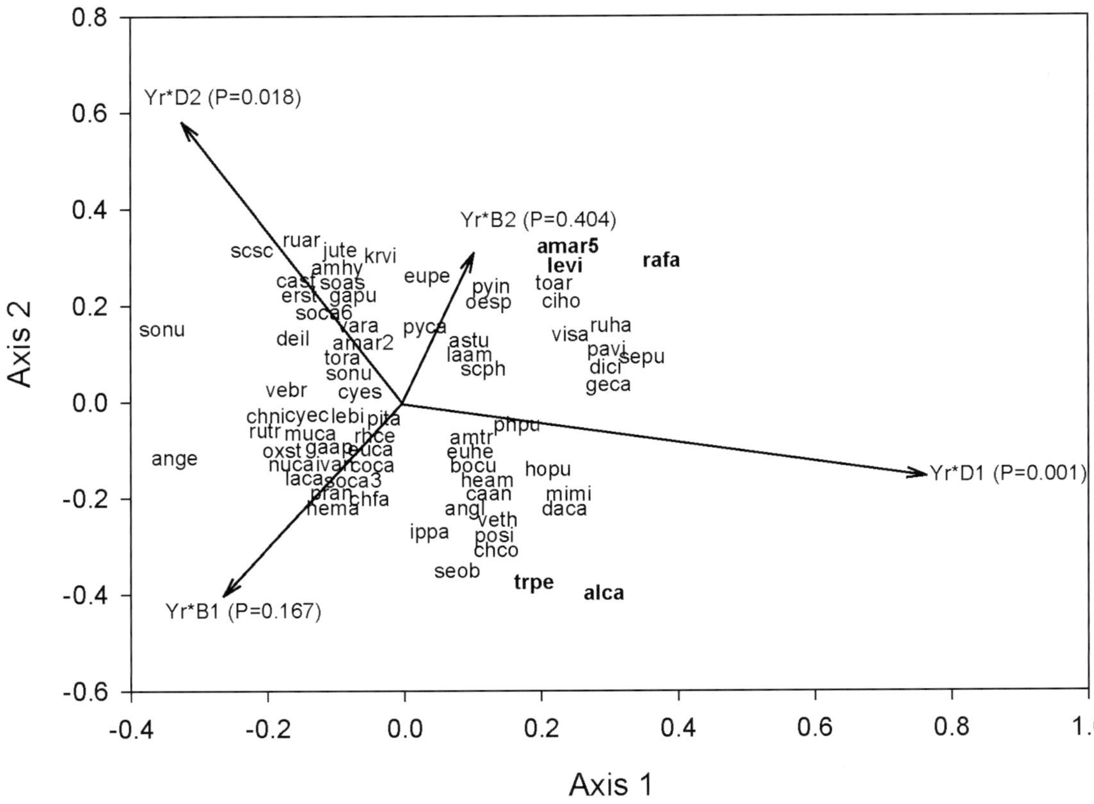

Figure 16.7. Biplot showing the results from RDA applied to species cover measurements during early-summer sampling periods (May 2008 and 2009) on grassland field buffers in Clay County, Mississippi. The bolded forb plant species most strongly influenced community response the year following disking (Yr*D1) and the second year following burning (Yr*B2). Species codes are in the appendix. Bolded species are alca = *Allium canadense*, amar5 = *Ampelopsis arborea*, levi = *Lespedeza virginica*, rafa = *Ranunculus fascicularis*, trpe = *Triodanis perfoliata*.

Additionally, many positive effects from disking in fall 2007 (e.g., *Conyza canadensis*, *Helianthus maximiliani*, *Triodanis perfoliata*, *Valerianella radiata*) persisted into summer 2009 (second growing season postdisk), but disking in fall 2008 had little or no effect on these species, even in the first growing season postdisk. However, other species (e.g., *Geranium carolinianum*, *Ipomoea pandurata*) increased in summer 2009 after the fall disking in 2008. Persistence of fall disking effects in the second growing season contrasts with the findings of Greenfield et al. (2002) that disking effects diminished by the second growing season.

Persistence of effects from fall 2007 disk treatments into summer 2009 (and the much smaller effects following fall 2008 disking) suggests that other factors may have influenced vegetation response. For example, rainfall variation between 2008 and 2009 (fig. 16.2) could have mediated vegetation responses. The higher level of rainfall in May 2009 compared to May 2008 could have dampened treatment effects on buffer vegetation in 2009. Also, the depth of disking may have varied between years (or interacted with rainfall differences), which could have led to interannual differences in vegetation response. Importantly, persistent effects of fall 2007 disking were observed in other taxa. A study of fire ants (*Solenopsis richteri* and *S. invicta*) on the CP33 buffers (Hale et al. 2011) and a study of butterflies (Dollar 2011) also found the persistence of 2007 disking effects into the second growing season postdisturbance (i.e., summer 2009). These invertebrate responses could have been related to the vegetation responses described here. Thus, our data confirm the importance of collecting data for multiple years on semi-natural grasslands to determine the influences of year-to-year variation in wildlife and vegetation responses to management.

For decades, disking has been used to control undesirable grass species and improve habitats for game birds (Kilbride and Paveglio 1999; Paveglio and Kilbride 2000; Madison

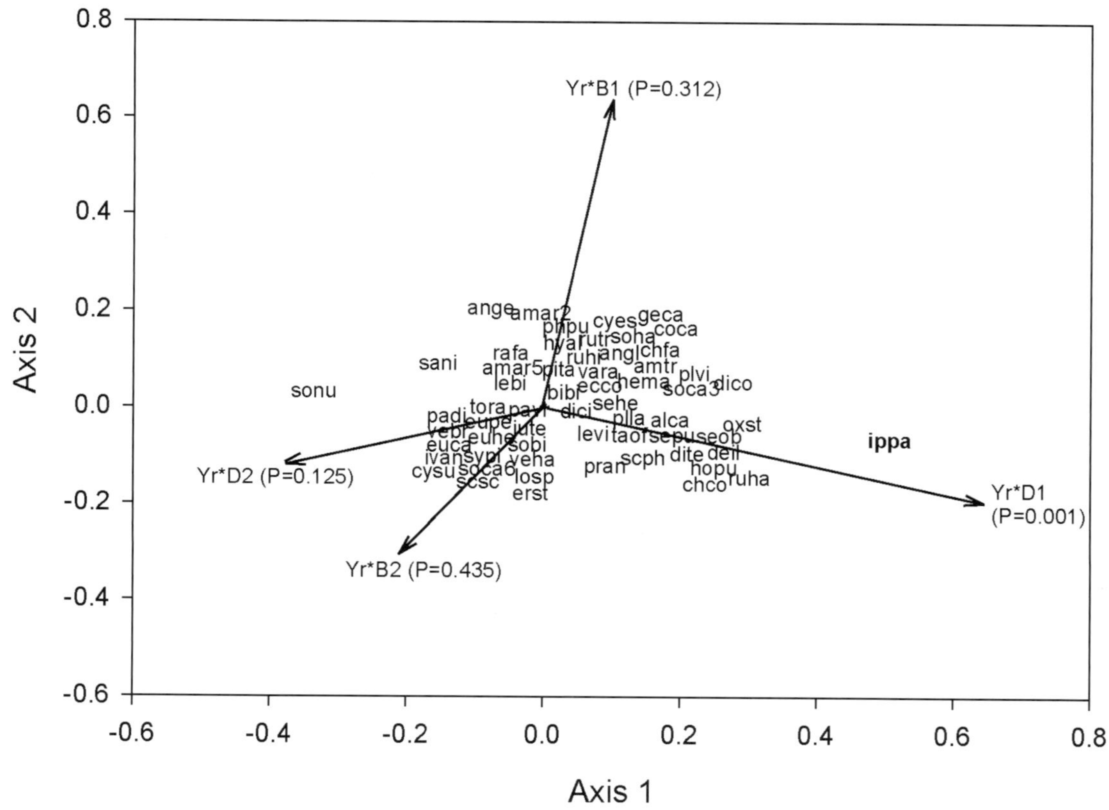

Figure 16.8. Biplot showing the results from RDA applied to species cover measurements during midsummer sampling periods (July 2008 and 2009) on grassland field buffers in Clay County, Mississippi. The bolded forb plant species most strongly influenced community response the first year following disking (Yr*D1). Species codes are in the appendix. The bolded species is ippa = *Ipomoea pandurata*.

et al. 2001). In our study of semi-natural grasslands, fall disking increased the abundance and richness of forb species, which in turn may benefit many wildlife species. This positive relationship between fall disking and forbs is likely due to the effects of disking on perennial grasses (the dominant species in the community). Disking is more disruptive to perennial grasses than burning because it turns the soil and damages the root mass of perennial grasses. Disking on perennial grasses allows forbs to establish and persist within the plant community. Disking also brings forb seeds to the soil surface, enabling germination. In this study, disking decreased native warm-season grass cover and increased forbs relative to burning (in 2008 but not 2009; see Dollar et al. 2013).

Effects of Burning

In accord with previous work (e.g., Kucera and Ehrenreich 1962; Hulbert 1988; Hartnett and Fay 1998; Benson et al. 2007; Vogel et al. 2010), dead plant material decreased and native warm-season grasses and bare ground increased with spring burning. Also, forb species responses to spring burning were variable, and only a few benefited from fire (Towne and Knapp 1996; Hartnett and Fay 1998). For example, *Ambrosia artemisiifolia* (annual ragweed) responded positively to burning in 2008 (table 16.2). The most recent burning did not impact this species in 2009.

A substantial amount of research has focused on plant responses to fire in tallgrass prairies (e.g., S. Collins and Wallace 1990; A. Knapp et al. 1998; Benson et al. 2007). Grasses and forbs respond differently to fire because of morphological and phenological variation (Hartnett and Fay 1998). In dominant C_4 grasses, fire stimulates rhizome development and tillering early in the season, resulting in greater tiller densities and thus greater biomass (Hulbert and Wilson 1983; Peterson 1983; Hulbert 1986; A. Knapp and Hulbert 1986). The magnitude of these responses varies considerably between years, with season of burning, and among species (D. Gibson 1988; Svejcar 1990; Towne 1995). Many annual and perennial forb buds (e.g., *Erigeron strigosus*, prairie fleabane) that initiate growth and flo-

TABLE 16.4.

Results of the analyses using RDA applied to species cover measurements during early-summer sampling periods (May 2008 and 2009), midsummer sampling periods (July 2008 and 2009), and late-summer sampling periods (August 2008 and 2009).

Analysis	Explanatory variables	Covariables	% expl. 1st axis	F ratio	P	Null
Early-summer sampling periods						
M1	Yr, Yr*D1, Yr*D2, Yr*B1, Yr*B2, Yr*Cont	PlotID	18.3	17.59	0.001	Reject
M2	Yr*D1, Yr*D2, Yr*B1, Yr*B2, Yr*Cont	Yr, PlotID	3.0	3.64	0.001	Reject
M3	Yr*D1	Yr, Yr*D2, Yr*B1, Yr*B2, PlotID	1.5	7.81	0.001	Reject
M4	Yr*D2	Yr, Yr*D1, Yr*B1, Yr*B2, PlotID	0.6	2.86	0.018	Reject
M5	Yr*B1	Yr, Yr*D1, Yr*D2, Yr*B2, PlotID	0.3	1.48	0.167	Fail to reject
M6	Yr*B2	Yr, Yr*D1, Yr*D2, Yr*B1, PlotID	0.2	0.97	0.404	Fail to reject
Midsummer sampling periods						
M1	Yr, Yr*D1, Yr*D2, Yr*B1, Yr*B2, Yr*Cont	PlotID	8.4	6.58	0.001	Reject
M2	Yr*D1, Yr*D2, Yr*B1, Yr*B2, Yr*Cont	Yr, PlotID	4.8	4.72	0.001	Reject
M3	Yr*D1	Yr, Yr*D2, Yr*B1, Yr*B2, PlotID	1.7	6.85	0.001	Reject
M4	Yr*D2	Yr, Yr*D1, Yr*B1, Yr*B2, PlotID	0.4	1.60	0.125	Fail to reject
M5	Yr*B1	Yr, Yr*D1, Yr*D2, Yr*B2, PlotID	0.3	1.14	0.312	Fail to reject
M6	Yr*B2	Yr, Yr*D1, Yr*D2, Yr*B1, PlotID	0.2	0.95	0.435	Fail to reject
Late-summer sampling periods						
M1	Yr, Yr*D1, Yr*D2, Yr*B1, Yr*B2, Yr*Cont	PlotID	18.3	17.59	0.001	Reject
M2	Yr*D1, Yr*D2, Yr*B1, Yr*B2, Yr*Cont	Yr, PlotID	3.0	3.64	0.001	Reject
M3	Yr*D1	Yr, Yr*D2, Yr*B1, Yr*B2, PlotID	1.7	8.08	0.001	Reject
M4	Yr*D2	Yr, Yr*D1, Yr*B1, Yr*B2, PlotID	0.3	1.45	0.495	Fail to reject
M5	Yr*B1	Yr, Yr*D1, Yr*D2, Yr*B2, PlotID	0.4	1.78	0.136	Fail to reject
M6	Yr*B2	Yr, Yr*D1, Yr*D2, Yr*B1, PlotID	0.4	1.91	0.348	Fail to reject

ral development early in the growing season are killed or harmed by fire and may be unable to set or produce seed and complete their life cycle (D. Gibson 1988). Forbs that are active late and flower in midsummer to late summer (e.g., *Solidago canadensis*, Canada goldenrod) show reductions in growth and vegetative reproduction in response to frequent burning (A. Knapp 1984; Hartnett 1990, 1991). Indirect effects of fire on most forb species are also likely (i.e., enhanced growth and competitive ability of the matrix grasses in the postfire environment; Hartnett 1991). Our study was not long enough to evaluate the long-term effects of burning on perennial forbs because these species can persist in the landscape for years without propagating. For example, we found no impact of either burning

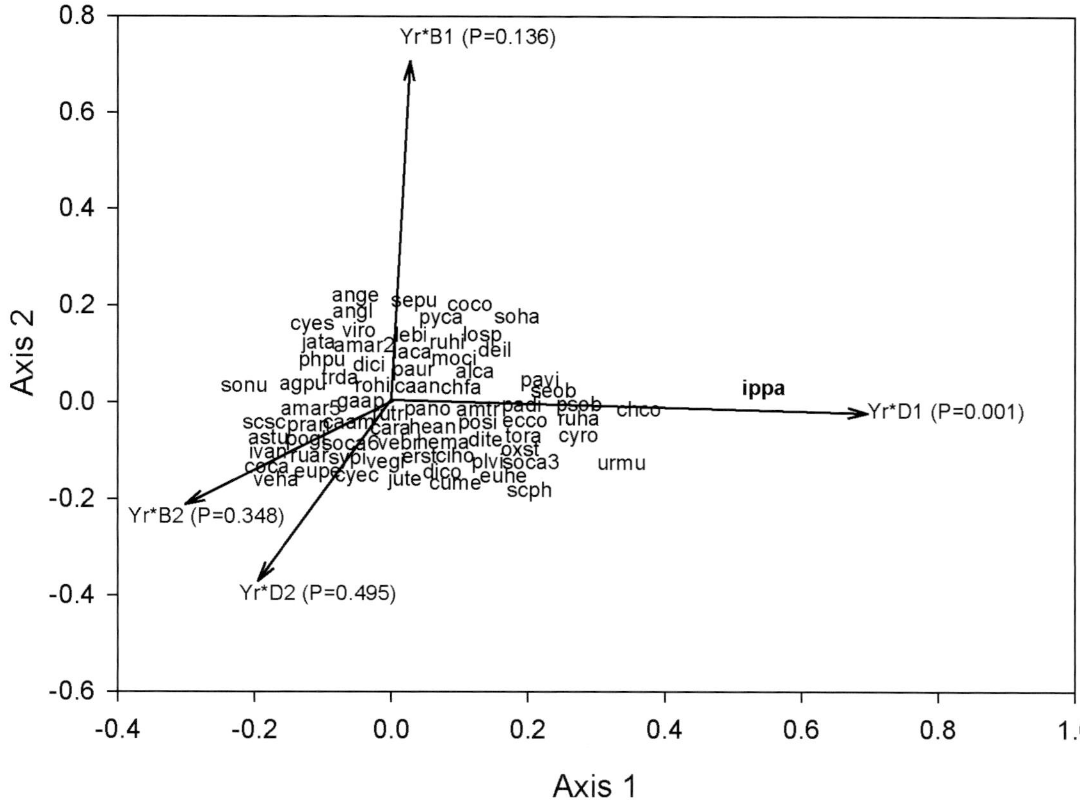

Figure 16.9. Biplot showing the results from RDA applied to species cover measurements during late-summer sampling periods (August 2008 and 2009) on grassland field buffers in Clay County, Mississippi. The bolded forb plant species most strongly influenced community response the first year following disking (Yr*D1). Species codes are in the appendix. The bolded species is ippa = *Ipomoea pandurata*.

or disking on the perennial *Solidago canadensis*. Furthermore, we recognize that when comparing fall disking and spring burning, we compare not only management techniques (i.e., disking vs. burning) but also seasonality of management (i.e., fall vs. spring). Variability in plant responses may therefore be due to management techniques and time of application.

Recovery of Buffers from Disturbance

There were no large or long-lasting effects on the buffer plant community, in terms of dramatic shifts in grassland-associated plant species, litter, or bare ground, with fall disking or with spring burning. The major influence of spring burning was to decrease dead plant material, increase bare ground cover, and stimulate growth of C_4 grasses and very few forbs. However, effects of spring burning on these variables greatly diminished by the second growing season post-disturbance, indicating that recovery time after spring burning is short in the Mississippi black prairie region. The major influence of fall disking was to increase plant species richness, mainly through the establishment and growth of forb species. The response of fall disking varied between years, with the fall 2007 disk treatment having a greater impact on the plant community than the fall 2008 disk treatment. Relatively longer growing seasons and greater precipitation in the southeastern United States (and hence productivity) may result in shorter recovery periods for plant communities following disturbance. This is in contrast to longer recovery times in more temperate or drier grasslands (e.g., Van der Valk 2007; Zedler 2007), where plant communities can take more than three years to recover after herbivore-related disturbance, partly because of shorter growing seasons (Speed et al. 2010). In dry environments, recovery from even moderate disturbance can take 60 years, mostly because of low available moisture (R. Webb et al. 1983). Results from the RDA and general linear models show that fall disking altered plant communities more than spring burning. Certain species (particularly annual forbs) responded positively to the more intense fall disking disturbance, thereby creating a more diverse plant community.

Conclusions and Recommendations

Although we would not advocate mechanical disturbance of remnant native prairie communities, fall disking of seminatural, established grasslands is warranted when increased forb richness and abundance are conservation goals. To enhance habitat heterogeneity, a rotational disking regime can be used (Benson et al. 2007). In our study, many forbs responded positively to fall disking, and an increase in forb species richness can greatly benefit other taxa, such as beneficial insects and grassland birds (C. Harper 2007). However, the effects of fall disking on vegetation may vary between years because of additional biotic and abiotic factors. Furthermore, fall disking may not be the preferred method of disturbance in habitats conducive to fire ants because fall disking may increase fire ant densities (Stiles and Jones 1998; Williamson et al. 2002; Hale et al. 2011). If fire ants are prevalent, management with fire may be a better alternative.

Prescribed spring burning may also have other advantages, in addition to potentially favoring some forb species during certain periods of the growing season. Plant communities, particularly the native warm-season grass component, tend to recover more quickly from burning than from disking (C. Harper 2007). If quick recovery in the first growing season (e.g., for grassland bird breeding habitat) is the primary conservation goal, then burning might be the best option. It is important to remember that the effects of burning (as with disking) may depend on abiotic and biotic processes, leading to temporal or spatial variability in species responses (Bradstock et al. 2005; Parr and Anderson 2006; Driscoll et al. 2010). Burning should be done at two- to three-year intervals and/or on a rotational schedule to minimize any negative effects on other taxa (e.g., grassland butterflies; Panzer 2003), promote habitat heterogeneity, and facilitate biodiversity conservation (Bradstock et al. 2005; Benson et al. 2007; C. Harper et al. 2007).

Acknowledgments

Our work would not have been possible without the cooperation of B. Bryan Farms and Prairie Wildlife, LLC. B. Brzuszek helped with plant identification. J. Leps helped with multivariate analyses. G. Ervin gave editorial feedback on a previous version of this chapter. We thank each of them for their contributions to this chapter. This research was funded by the USDA Natural Resources Conservation Service–Agricultural Wildlife Conservation Center (NRCS-AWCC). The Mississippi Agricultural and Forestry Experiment Station, the Forest and Wildlife Research Center, and the College of Forest Resources at Mississippi State University also provided financial and logistical support.

Appendix

Species Codes for RDA Analysis

Native warm-season grasses: *Andropogon gerardii* (ange); *Andropogon glomeratus* (angl); *Bouteloua curtipendula* (bocu); *Panicum virgatum* (pavi); *Schizachyrium scoparium* (scsc); *Sorghastrum nutans* (sonu); *Tripsacum dactyloides* (trda)

Native cool-season grasses: *Dichanthelium commutatum* (dico); *Digitaria ciliaris* (dici)

Native annual grasses: *Hordeum pusillum* (hopu); *Muhlenbergia capillaris* (muca)

Introduced grasses: *Echinochloa colona* (ecco); *Paspalum dilatatum* (padi); *Paspalum notatum* (pano); *Paspalum urvillei* (paur); *Schedonorus phoenix* (scph); *Setaria pumila* (sepu); *Sorghum bicolor* (sobi); *Sorghum halepense* (soha); *Urochloa mutic*a (urmu)

Native perennial forbs: *Allium canadense* (alca); *Asclepias tuberosa* (astu); *Conoclinium coelestinum* (coco); *Desmanthus illinoensis* (deil); *Diodia teres* (dite); *Erigeron strigosus* (erst); *Eupatorium capillifolium* (euca); *Eupatorium perfoliatum* (eupe); *Helianthus angustifolius* (hean); *Helianthus maximiliani* (hema); *Hyptis alata* (hyal); *Lespedeza virginica* (levi); *Lobelia spicata* (losp); *Mimosa microphylla* (mimi); *Mimosa strigillosa* (mist); *Oenothera speciosa* (oesp); *Oxalis stricta* (oxst); *Phytolacca americana* (pham); *Plantago lanceolata* (plla); *Polygonum glabrum* (pogl); *Potentilla simplex* (posi); *Pycnanthemum incanum* (pyin); *Ranunculus fascicularis* (rafa); *Rudbeckia hirta* (ruhi); *Rumex hastatulus* (ruha); *Senna obtusifolia* (seob); *Sesbania herbacea* (sehe); *Solanum carolinense* (soca3); *Solidago canadensis* (soca6); *Symphyotrichum pilosum* (sypi); *Taraxacum officinale* (taof); *Verbena halei* (veha); *Vernonia gigantean* (vegi)

Native annual forbs: *Agalinis purpurea* (agpu); *Amaranthus hybridus* (amhy); *Ambrosia artemisiifolia* (amar2); *Am-*

brosia trifida (amtr); *Bidens bipinnata* (bibi); *Chamaecrista fasciculata* (chfa); *Chamaecrista nictitans* (chni); *Chamaesyce cordifolia* (chco); *Cirsium horridulum* (ciho); *Conyza canadensis* (coca); *Galium aparine* (gaap); *Gamochaeta purpurea* (gapu); *Geranium carolinianum* (geca); *Helenium amarum* (heam); *Iva annua* (ivan); *Jacquemontia tamnifolia* (jata); *Krigia virginica* (krvi); *Lactuca canadensis* (laca); *Lamium amplexicaule* (laam); *Monarda citriodora* (moci); *Nuttallanthus canadensis* (nuca); *Physalis pubescens* (phpu); *Plantago virginica* (plvi); *Pseudognaphalium obtusifolium* (psob); *Pyrrhopappus carolinianus* (pyca); *Triodanis perfoliata* (trpe); *Valerianella radiata* (vara)

Introduced perennial forbs: *Rumex crispus* (rucr)

Introduced annual forbs: *Cucumis melo* (cume); *Daucus carota* (daca); *Sonchus asper* (soas); *Torilis arvensis* (toar); *Verbascum thapsus* (veth); *Verbena brasiliensis* (vebr); *Vicia sativa* (visa)

Native perennial vines: *Ampelopsis arborea* (amar5); *Campsis radicans* (cara); *Ipomoea pandurata* (ippa); *Rubus trivialis* (rutr); *Toxicodendron radicans* (tora); *Vitis rotundifolia* (viro)

Perennial sedges and rushes: *Carex annectens* (caan); *Carex striata* (cast); *Cyperus echinatus* (cyec); *Cyperus esculentus* (cyes); *Cyperus rotundus* (cyro); *Cyperus surinamensis* (cysu); *Juncus tenuis* (jute); *Rhynchospora cephalantha* (rhce)

Native shrubs and trees: *Callicarpa americana* (caam); *Pinus taeda* (pita); *Prunus angustifolia* (pran); *Robinia hispida* (rohi); *Rubus argutus* (ruar); *Sambucus nigra ssp. canadensis* (sani)

Introduced shrubs: *Lespedeza bicolor* (lebi)

17

Highway Right-of-Way Mowing Regimens in Northeastern Mississippi

Effects on Native Prairie Plant Species

Edward D. Entsminger, John W. Guyton, Raymond B. Iglay, and Jeanne C. Jones

Introduction

Prairies are one of the most imperiled ecosystems in Mississippi and are the least-protected biome in North America (Schauwecker and MacDonald 2003). Threatened and endangered plants, insects, and wildlife species reside in this relict habitat, and many populations of these organisms are rapidly declining as prairies are being destroyed by anthropogenic activities, such as agriculture, road developments, mowing, and fire suppression (Schuster and McDaniel 1973; Campbell and Seymour 2012).

Roadside rights-of-way (ROW) can provide prairie habitat, but current management practices have caused declines in native grasses and wildflowers (K. Hill and Horner 2005; Willard et al. 2010; Yager et al. 2011). Intensive herbicide application and mowing directly affect native plant communities by killing plants or preventing their successful reproduction (K. Hill and Horner 2005; Willard et al. 2010; Yager et al. 2011). General roadside management can also impact plants by altering soil fertility and structure, topography, and surface water movement and other hydrological factors (Greenfield et al. 2005). Over time, both direct and indirect impacts of management can cause ROW to become more susceptible to non-native plant invasion (Forman and Alexander 1998; Greenfield et al. 2005). Other negative impacts may include changes in adjacent ecosystems through modification of plant communities, changes in wetland hydrology, degradation of water quality, and dispersal of invasive non-native species (Yager et al. 2011). For wildlife, restriction of animal movement, habitat fragmentation and loss, population isolation, and increases in traffic-related wildlife mortality are additional concerns related to ROW management and roadways (Forman and Deblinger 2000; Harper-Lore and Wilson 2000; Ament et al. 2008). However, proactive management of ROW vegetation can enhance the establishment of native plant communities and improve their ecological function and structure for native wildlife (Harper-Lore and Wilson 2000; Arner and Jones 2009).

Enhancement of native plant communities can result in roadside beautification, and roadside wildflower programs have supported these efforts in many states over the past two decades (US Department of Transportation 2011). The Highway Beautification Act of 1965 emphasized the visual and aesthetic qualities of roadsides, resulting in the establishment of ROW as points of pride in communities throughout the United States (Aldrich 2002). Roadside wildflowers can also have positive effects on motorists, making them more relaxed and happier when driving (Cackowski and Nasar 2003; Transportation Research Board 2005). Furthermore, established native plant communities along roadsides can decrease the spread of invasive plants, reduce erosion, lower maintenance costs, protect water quality, and lessen visual obscurity (Forman and Alexander 1998; Green and Welker 2003; Yager et al. 2011; US Environmental Protection Agency 2012). Native plant cover within ROW also creates habitat for early-successional species, including small mammals, birds, herpetofauna, and insects (Bugg et al. 1997; Forman and Alexander 1998; Hopwood 2010). Although increasing native plant cover along highway ROW cannot completely mitigate the potential negative impacts to wildlife and native flora caused by roadways, modifications in the management of ROW can potentially lessen deleterious effects and benefit wildlife and humans.

However, there are some issues related to the enhancement of native plant communities as well, such as reductions in visibility along roadway ROW that can be an issue for motorists (Louisiana Department of Transportation and Development 2000). Also, concern exists over enhancement of ROW habitat for large mammals, which could increase wildlife-vehicle collisions (Michael and Kosten 1979; Dixon et al. 1984; McKee and Cochran 2012). There is a need for investigations of effective management options that benefit managing agencies, motorists, and native fauna and flora along roadway ROW.

Annual mowing in late fall is one low-maintenance approach that can potentially be used to improve native plant communities along ROW. Reduced mowing frequencies can also result in lower costs compared to multiple mowings during the growing season. An annual fall mowing can be conducted after seed maturation in native plants and following nesting seasons for ground-nesting wildlife. This approach also increases the availability of pollen and nectar-producing plants that provide food for birds, small mammals, and pollinating insects (L. Anderson 1996; Dickson and Wigley 2001). Reduced mowing also has the potential to create attractive prairie-like environments along Mississippi roadsides by increasing native grasses and wildflowers, especially on ROW that transect blackland and Jackson Prairie soil types.

Seeding of native wildflowers along roadside ROW is another practice for establishing native plant communities on ROW. This practice typically requires site preparation to remove existing non-native plant communities on ROW, selection of adapted plant species for seeding, and postplanting competition control methods (Young and Claassen 2007; Native American Seed 2011). Seeding of native wildflowers and warm-season grasses can be expensive because of site preparation, seed, and maintenance costs (Young and Claassen 2007). A more cost-effective approach to enhancement of native plant communities along ROW in Mississippi may be maintenance of a stable native plant community through a reduced mowing regimen (Bugg et al. 1997; Young and Claassen 2007).

We investigated the influence of modifications in mowing regimens along a state highway in central Mississippi on native plant community structure. We anticipate that this work will increase current understanding of potential methods for reclamation of roadside ROW in prairie regions as well as in other physiographic regions of Mississippi (A. Johnson 2000; Hopwood 2010). We compared plant percent cover, species richness, height of vegetation, and woody stem densities among three treatments (e.g., two mowing treatments and a reduced mowed-seeded treatment) in upland and riparian lowland ROW.

Study Area

This study was conducted in roadside ROW along a 48.3 km stretch of Highway 25, beginning at the intersection of Highways 12 and 25 in Oktibbeha County and continuing south 4.5 km into Winston County, in northeastern Mississippi. We had an average distance of 2.7 km between each site. The sites were regionally located on the outer edge of the Black Belt Prairie or within the Interior Flatwoods in northeastern Mississippi (33°12′N, 88°54′W; Township 15–18N, Range 13–14E) (Pettry 1977). Soil formation and vegetation communities were influenced by mild climatic conditions categorized as the humid subtropical climatic region of North America, with temperate winters (0°C–15°C) and hot summers (21°C–38°C) (Posner 2012). Annual mean temperature for the region is 16.7°C, and precipitation ranges from 127 cm to 165 cm across the state from north to south (Mississippi State University Department of Geosciences 2010).

Past highway ROW management for this area consisted of multiple mowings during the growing season (≥ four times/season) and vegetation control by herbicides, including imazapyr, triclopyr, and glyphosate herbicide mixtures for control of woody plants and exotic non-native invasive plant species (e.g., Johnsongrass, *Sorghum halepense*; kudzu, *Pueraria montana* var. *lobata*; and cogongrass, *Imperata cylindrica*). Prior to study initiation, the study area ROW plant communities were composed predominantly of non-native grasses such as Bermudagrass (*Cynodon dactylon*);

tall fescue (*Schedonorus phoenix*); Bahia grass (*Paspalum notatum*); and Vasey's grass (*Paspalum urvillei*). Surrounding vegetation types included agricultural fields, pastures, fallow fields, forests, and pine plantations over hilly and flatland topography. The study ROW was crossed by four third- to fourth-order streams differentiating upland and riparian lowland sites.

Methods

We randomly selected 10 (30.5 m × 30.5 m) research plots, including 5 upland and 5 riparian lowland plots. Following Li et al. (2008), we used a randomized complete block design by dividing each plot into three equal subplots (10.2 m × 30.5 m each) and randomly assigned one of three treatments to each subplot: (1) annual mowing during November; (2) annual mowing during November and seeding with wildflower seed mixture; and (3) mowing four or more times annually in May, July, September, and November. In treatment 2, we conducted one mowing during late November to reduce vegetation height prior to seeding. We broadcasted the seed mixture over existing mowed vegetation using a handheld seeder during March 2011. The wildflower seed mixture included black-eyed Susan (*Rudbeckia hirta*, 1.55 kg/ha), dense blazing star (*Liatris spicata*, 7.70 kg/ha), and lanceleaf tickseed (*Coreopsis lanceolata*, 7.70 kg/ha) (Native American Seed 2011).

We used a 30.5 m line intercept across each subplot to measure species richness and percent cover of woody and herbaceous plants during fall 2010 and 2011 and spring 2011 and 2012. We placed line intercepts 5 m or more from subplot edges to avoid potential edge effects and marked line initiation and end points using a Garmin E-Trex HCx Vista GPS unit. We also established one belt transect measuring 0.5 m × 30.5 m adjacent to the right side of each line transect to estimate stem densities of woody plants (stems/ha) (Hays et al. 1981). We measured percent cover and height (m) for all plants intercepting the line, categorized each as native or non-native, and grouped plant height data into three categories: less than or equal to 0.46 m (low), between 0.46 m and 0.91 m (mid), and more than or equal to 0.91 m (high) (Hays et al. 1981). We also summarized plant percent cover and species richness by height category and grouped vegetation by growth form and native category: native and non-native forbs, native and non-native grasses, native and non-native legumes, native and non-native sedges, native rushes, and native and non-native woody plants (i.e., trees, shrubs, and woody vines) (Hays et al. 1981). We grouped data by season (fall vs. spring) because of expected vegetation cover differences within treatment plots between seasons (e.g., taller plants in fall than in spring).

We tested the null hypothesis of no differences among treatments or between upland and lowland percent vegetation cover by growth form and native status with analysis of variance using distance matrices (ADONIS) in Program R's vegan package (M. Anderson 2001; Ott and Longnecker 2010). The analysis of variance using distance matrices partitioned sources of variation while fitting linear models to distance with permutation tests as pseudo-F ratios (M. Anderson 2001). We investigated the interactions of treatment, elevation, and year (fixed effects) and random effects of site and site by elevation with 999 permutations and Euclidean distances (Oksanen 2012; R Core Team 2012). When a significant interaction or simple effect occurred, we used the Wilcoxon signed-rank test for pairwise comparisons with sites as a blocking variable to determine how growth form cover differed with native or non-native status. We were then able to use Wilcoxon signed-rank tests for comparison of more than two groups because of significant effects within the two levels (McDonald 2009).

We tested the null hypothesis of no difference among treatments or between upland and lowland for species richness (total, native, non-native), height variation of vegetation (low, mid, high), and woody plant stem density (total, native, non-native) using mixed-model, univariate, repeated-measures analysis of variance (PROC MIXED) in SAS (SAS Institute 2007; Ott and Longnecker 2010). We investigated the interactions of treatment, elevation, and year (fixed effects) and random effects of site and site by elevation with year as the repeated measure. We selected models using the Akaike information criterion corrected (AIC*c*) to compare autoregressive (AR [1]), compound symmetry (CS), and unstructured (UN) covariance structures for each response variable under restricted maximum likelihood (REML) (Gutzwiller and Riffell 2007). We also investigated whether to include a random effect. We designated top model structures (i.e., best covariance structure and inclusion or exclusion of random effect) as models with $\Delta AICc \geq 2$ to the next best model (Burnham and Anderson 2002). We used Fisher's least significant difference (LSD) for pairwise comparisons of significant effects

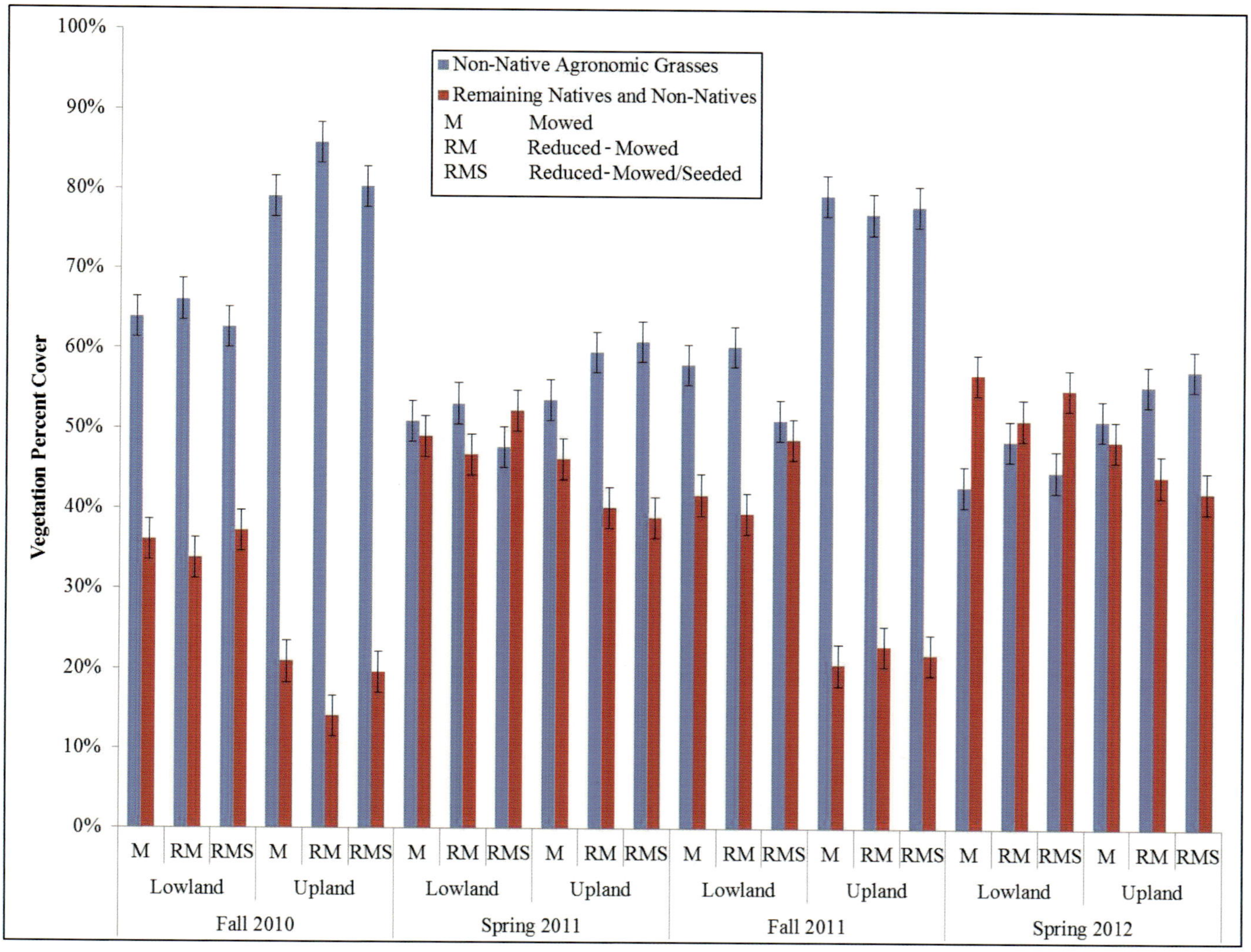

Figure 17.1. Mean percent cover of non-native agronomic grasses and the remaining native and non-native species cover among mowed (M), reduced-mowed (RM), and reduced-mowed/seeded (RMS) treatments in lowland and upland sites along Highway 25 rights-of-way in Oktibbeha and Winston Counties, Mississippi, measured using line intercepts from fall 2010 through spring 2012.

(Meier 2006). Our level of significance for all tests was $\alpha \leq 0.05$ (Ott and Longnecker 2010).

Results

Between April 2010 and June 2012, we recorded 277 plant species, 79% (219 species) native and 21% (58 species) non-native. Although non-native agronomic grasses made up less than 5% of the total species detected, they dominated vegetation cover in all treatments (mowed = 57.70%, SE ± 3.14; reduced-mowed = 60.87%, SE ± 2.87; reduced-mowed/seeded = 58.32%, SE ± 2.84) (fig. 17.1). We observed no treatment effects for any response variables in fall or spring. However, vegetation cover by growth form differed between upland and lowland sites in fall ($F_{1, 59}$ = 17.91, $P \leq 0.001$) and spring ($F_{1, 59}$ = 24.12, $P \leq 0.001$). In fall, cover of native forbs ($z = 4.81$, $P \leq 0.001$), native grasses ($z = 3.51$, $P \leq 0.001$), native legumes ($z = 3.37$, $P \leq 0.001$), and native rushes ($z = 4.54$, $P \leq 0.001$) was greatest in lowland sites (table 17.1). In spring, cover of native forbs ($z = 3.03$, $P \leq 0.01$), native legumes ($z = 2.62$, $P \leq 0.01$), non-native legumes ($z = 3.82$, $P \leq 0.001$), native rushes ($z = 5.00$, $P \leq 0.001$), and native sedges ($z = 4.15$, $P \leq 0.001$) was greatest in lowland sites (table 17.1).

Percent cover among vegetation height categories also differed between upland and lowland sites during fall and spring. Vegetation cover in the fall at mid heights and in the spring at high heights was greater in lowland sites, but vegetation cover in the spring at the low and mid heights

Table 17.1.

Percent cover of vegetation growth form categories with means and standard errors (SE) among upland and lowland elevations using pairwise comparisons along Highway 25 rights-of-way in Oktibbeha and Winston Counties, Mississippi, during fall (2010–2011) and spring (2011–2012).

	Fall 2010–2011				Spring 2011–2012			
	Upland		Lowland		Upland		Lowland	
Vegetation growth form	Mean	SE	Mean	SE	Mean	SE	Mean	SE
Native forb*	7.80	1.53	49.64	7.94	20.10	2.72	47.45	7.21
Non-native forb	2.54	0.54	2.22	0.46	3.83	0.96	8.07	1.52
Native grass*	1.84	0.47	4.93	0.94	2.08	0.75	5.87	1.27
Non-native grass	134.51	8.77	115.60	9.01	159.13	9.81	127.51	7.42
Native legume*	0.02	0.02	5.62	2.57	0.00	0.00	4.12	1.94
Non-native legume*	6.79	2.87	7.16	4.17	84.02	7.43	42.31	5.25
Native rush*	0.03	0.03	3.11	1.06	0.21	0.09	10.68	2.58
Native sedge*	0.33	0.11	0.89	0.25	1.51	0.47	7.17	1.30
Non-native sedge	0.21	0.17	0.91	0.36	0.00	0.00	0.00	0.00
Native shrub	0.52	0.40	0.34	0.11	0.19	0.13	0.66	0.29
Native tree	0.69	0.23	1.88	0.48	1.30	0.39	5.13	1.70
Native vine	12.02	4.09	2.49	0.76	5.92	1.48	5.02	1.42
Non-native vine	2.10	1.03	0.92	0.48	2.39	1.03	1.02	0.37

*Differed between upland and lowland elevations at $P \leq 0.05$.

was greater in upland sites (table 17.2). All species richness variables differed between upland and lowland elevations during fall and spring, with greatest species richness in lowland sites (table 17.2). However, woody plant stem density variables did not differ among treatments ($F_{2, 96} < 1.87$, $P > 0.17$) or between upland and lowland sites (= 1.60; $F_{1, 96} < 2.79$, $P > 0.11$).

Discussion

Low-frequency mowing with or without broadcast seeding of native plants did not affect the predominant vegetation composition of roadside ROW in Mississippi. Site elevation (upland versus lowland) was the only parameter with noticeable effects on vegetation cover and species richness. Greater percent vegetation cover and species richness within lowland sites may have been caused by greater moisture availability, quality of alluvial substrates, and deposition of vegetative propagules from flooding events (Bush and Van Auken 1989). Greater plant community diversity can be related to site productivity, and this effect may be related to nutrient loading from seasonal floods and increased water saturation during rainfalls (Greenfield et al. 2005; Huijser and Clevenger 2006).

Throughout our study, non-native grasses dominated sites, with over 70% ground cover on most plots. Non-native grasses exceeded cover of native forbs and grasses by more than three to five times in reduced-mowed treatments. With modifications of mowing regimes over a 10-year period in prairie sites along Wisconsin highways, Harrington (1995) observed an increased percent cover of native grasses, with more than 50% native grasses, along with more than 20% native forbs, and less than 40% non-native grass cover. Our findings were similar in terms of percent cover of native grasses, forbs, and woody plants, except we detected a continued dominance of non-native agronomic grasses over the two years. Barras et al. (2000) observed greater cover of non-native grasses (81% vs. 68%) in mowed sites compared to unmowed sites; they also reported an increase of native forbs (16% vs. 25%), native grasses (1% vs. 5%), and native woody plants (1% vs. 4%) in unmowed sites. We also found that non-native forb and grass cover remained consistent in the frequently mowed treatment across years, but the percent cover of non-native species decreased in the reduced-mowed treatments in lowland study sites over the study period. Because of our findings and those of previous studies, we suggest that reduction of mowing frequency can increase native plant species richness and enhance roadside aesthetics without loss of agronomic grass cover and subsequent compromise of erosion control from these grasses (Forman and Alexander 1998; Markwardt 2005; Transportation Research Board 2005).

Another concern with reduced frequency of vegetation management along road ROW is maintenance of good visibility on ROW for motorists. This visibility may be especially important along ROW that transect areas with dense populations of large mammals, such as white-tailed deer (*Odocoileus virginianus*) (Dixon et al. 1984; McKee and Cochran 2012). Mowing in the absence of herbicide application can lead to greater densities of woody plants that may grow to heights that impede visibility, in part because of the tendency of some tree and shrub species to resprout following mowing or prescribed burning (Arner 1979; Gruchy et al. 2006; R. N. Brown and Sawyer 2012). Greater abundance of woody plants and greater plant heights are both features that can impede visibility along highway and road ROW (Hamrick et al. 2007; Yager et al. 2011). In our study, we detected no difference in plant height or woody plant density between the frequent mowing and one-mowing-per-year treatment. We suggest that mowing once during fall is adequate to maintain good visibility along highway ROW, especially in areas where ROW are more than 100 m wide. Furthermore, most woody plants that we detected during our study were vine species, such as blackberry (*Rubus* spp.); Japanese honeysuckle (*Lonicera japonica*); greenbrier (*Smilax* spp.); purple passionflower vine (*Passiflora incarnata*); trumpet creeper (*Campsis radicans*); poison ivy (*Toxicodendron radicans*); and grapevines (*Vitis* spp.), and did not contribute much to the height of roadside vegetation. Our findings indicate that mowing could be reduced in frequency, cutting current expenditures for vegetation management on highway ROW by more than a third. Therefore, reduced mowing can result in budgetary savings to Mississippi's taxpayers without compromising visibility along ROW.

Reduced mowing frequencies could enhance conditions for the development of existing communities of native forbs, legumes, and grasses—a total of 219 species during our two-year study. Observed species richness in our study was comparable to that reported by Schuster and McDaniel (1973) and Leidolf and McDaniel (1998), who each observed more than 150 native plant species in prairie environments near ROW in Alabama and Mississippi. In our study, 15 species of native prairie indicator plants were observed exclusively in the two reduced-mowed treatments. These species included *Allium cernuum*, *Andropogon virginicus*, *Carex cherokeensis*, *Cocculus carolinus*, *Desmanthus illinoensis*, *Dracopis amplexicaulis*, *Hydrophyllum virginianum*, *Juniperus virginiana*, *Liatris aspera*, *Liatris mucronata*, *Panicum virgatum*, *Solidago stricta*, *Symphyotrichum novae-angliae*, *Tridens flavus*, and *Tridens strictus*. We submit that in addition to a reduced mowing regime, additional factors may have influenced our findings related to species richness of native plants. These factors include below-normal annual precipitation during 2010 and 2011, possible herbivory, and a lack of site preparation in seeded plots prior to planting and reseeding in the fall (Howard 1950; Chase et al. 2000; Young and Claassen 2007).

Modifications in mowing regimes can allow native plants to become established. A single mowing conducted during late fall can provide adequate time for plant maturation, root development, and seed production (Arner and Jones 2009). Production of flowers and seeds without loss from frequent mowings could enhance the seedbank and provide more plant foods for game and nongame wildlife, including insect pollinators (J. Jones et al. 2007). This approach could be more cost effective than seeding wildflowers along ROW because of the costs of site preparation, site management, and seeds and propagules. In our study plots, supplemental seeding of native plants combined with reduced mowing did not produce the density of plants that we anticipated; however, adequate site preparation to increase seed catchment and limit competition might have

Table 17.2.

Percent cover of three vegetation height categories and species richness of combined native and non-native, native, and non-native species in fall and spring, and in upland and lowland sites with summary metrics along Highway 25 rights-of-way in Oktibbeha and Winston Counties, Mississippi, during 2010–2012.

Season	Variable	Upland[a]		Lowland[b]			
		Mean[c]	SE[d]	Mean[c]	SE[d]	$F_{1, 96}$	P value
Fall	≤ 0.46 m height	126.24	11.49	125.22	11.49	0.99	0.326
	≥ 0.46 ≤ 0.91 m height	53.20	4.42	69.20	4.42	7.23	0.011
	≥ 0.91 m height	6.64	3.21	17.18	3.21	3.09	0.088
Spring	≤ 0.46 m height	226.96	9.74	216.79	9.74	29.04	≤ 0.001
	≥ 0.46 ≤ 0.91 m height	50.08	4.69	49.15	4.69	6.32	0.017
	≥ 0.91 m height	16.15	3.90	28.63	3.90	3.99	≤ 0.050
Fall	Combined native/ non-native species richness	10.87	1.13	22.13	1.13	67.61	≤ 0.001
	Native species richness	4.90	0.91	14.57	0.91	74.84	≤ 0.001
	Non-native species richness	5.97	0.40	7.57	0.40	8.29	≤ 0.010
Spring	Combined native/ non-native species richness	17.50	1.11	29.67	1.11	64.78	≤ 0.001
	Native species richness	7.60	1.05	18.37	1.05	48.60	≤ 0.001
	Non-native species richness	9.90	0.42	11.30	0.42	5.64	0.023

Note: Mean percent cover is > 100% because of overlap of species along each line transect.

[a] Upland elevation N = 5, hilly, slope, and dry soils.

[b] Lowland elevation N = 5, flat, periodic flooding, and moist soils.

[c] Mean percent cover of vegetation within specified height category.

[d] Standard error of the mean.

improved native plant cover in our seeded plots (Arner et al. 1976; Chase et al. 2000; Svedarsky et al. 2002; J. Jones et al. 2007). Others have reported low establishment rates (< 10% success rates) for native plants seeded into existing non-native grass cover (Burke and Grime 1996). Furthermore, most perennial grasses and forbs need cold-weather scarification and at least two years for maximum germination and establishment (Dana et al. 1997; J. Jones et al. 2007; Tallamy 2009). Other factors that can impede establishment rates of seeded native plants include depredation by rodents and birds (Howard 1950; L. Anderson 1996).

Management Implications

One mowing per year of roadside ROW retained agronomic plant cover for erosion control but did not significantly increase native plant species richness. Despite investments in personnel training and budgetary allocations associated with new management regimes (Daar 1994; Dana et al. 1996), reduced mowing offers budgetary savings with costs around $75/ha per year, a savings of more than 75% annually compared to the cost of current intensive mowing practices, which exceeds $100/ha with more than four mowings per year along Mississippi highway ROW (D. Thompson, Mississippi Department of Transportation). Reduced mowing combined with native plant seeding could enhance roadside beauty through establishment of native plant communities without decreasing roadside visibility (L. Anderson 1996; Barras et al. 2000; Kutschbach-Brohl et al. 2010). As suggested by Russell et al. (2005) and Hopwood (2010), ROW treatment regimens that include native plantings, selective herbicide use, manual removal of woody plants, and reduced mowing could lead to lower long-term maintenance costs. In prairie areas such as the Black Belt Prairie region of the southeastern United States, roadside ROW can provide areas in which native prairie plants can colonize and survive, serve as population sources for native plant species, and increase habitat quality along roads and highways for native fauna, including pollinating insects, small mammals, and birds (Svedarsky et al. 2002; J. Jones et al. 2007; Tallamy 2009). Future research efforts should consider and explore other appropriate management for ROW in prairies, such as additional maintenance methods for native plant establishment and retention, including chemical, prescribed fire, and seeding techniques.

Acknowledgments

We would like to thank Mississippi State University (MSU) and the Mississippi Department of Transportation (MDOT) for funding this project. We especially appreciated the visionary input and support of Dave Thompson and Cindy Smith of MDOT. We are also grateful to the College of Forest Resources, the Forest and Wildlife Research Center, and the Department of Biochemistry, Molecular Biology, Entomology and Plant Pathology at MSU, as well as county mowing personnel, MSU wildlife technicians, graduate and undergraduate students, and especially Drs. Raymond B. Iglay and Samuel C. Pierce for their help with long hours of statistical analysis. Special thanks go to JoVonn Hill in Entomology at MSU for his edits and comments and to the reviewers of this chapter. Finally, Edward D. Entsminger would like to thank his Savior Lord Jesus Christ, Christian family and friends in Starkville, Mississippi, and Robin L. Walden, Bronson K. Strickland, and the many others who provided professional and critical edits and comments during the review process.

18

Reconstructing Prehistoric Prairie Habitat Types Using Archaeological Data

Jennifer L. Seltzer and Evan Peacock

One of the biggest challenges facing biologists and ecologists is the lack of time depth in available data. This problem is especially important when determining the best course of action in prairie restorations, where it is often hard to determine which natural habitat type the restoration should be using as a goal. One solution is the use of archaeobotanical material, such as charcoal, to look at habitat and species types. This allows researchers to have a deeper time scale to build from. In this chapter, materials from Lyon's Bluff, a prehistoric to early Historic-period village site, will demonstate the ways in which archaeo-botanical data can be used to reconstruct past environments.

Site History

Lyon's Bluff (22OK520) is a Mississippian mound site in Oktibbeha County, Mississippi (fig. 18.1). Located within the Black Belt Prairie region of Mississippi, the site was first excavated by Moreau B. C. Chambers in 1934 and 1935 (Galloway 2000), and again by Richard Marshall in the 1960s and early 1970s (R. Marshall 1977, 1985, 1985a, 1985b, 1986, 1986a). More recently, Evan Peacock, with Mississippi State University, conducted excavations in 2001 and 2003. This site has produced AMS radiocarbon and thermoluminescence dates spanning most of the Mississippian and Protohistoric periods, from AD 1200 to 1650 (R. Marshall 1977, 1985b; Hogue 2000; Peacock and Hogue 2005). Overlapping radiocarbon dates and continuity in ceramic styles suggest that the site was occupied continuously throughout that time (Peacock and Hogue 2005). Geophysical survey and excavation at the site have shown the single mound to be rectangular and the flanks to contain layers of daub. The construction style indicates that the mound was built and used during the Mississippian period (J. Rafferty et al. 2003; Alvey et al. 2004; Peacock and Hogue 2005).

The first mention of the site comes from H. B. Cushman (Swanton 1911), who described Lyon's Bluff as the "most noted stronghold of the Chocchumas." In Smith's account of the tribes of Mississippi from 1540 to 1700, he discussed the possibility of de Soto's expedition having passed thought the area (M. Smith 1991). Lyon's Bluff loosely matches the description of the Chickasaw town of Chicaça, having a mound, a palisade, and numerous houses. However, this description could describe multiple Mississippian sites in the region. While a very few non-native artifacts have been found at the site (a piece of cobalt-blue glass and a coarse earthenware sherd) (R. Marshall 1977; M. Smith 1991; Peacock and Hogue 2005), there is at this time no clear link between Lyon's Bluff and de Soto (Peacock and Hogue 2008).

In this chapter, we focus on two areas, including the "main site," which contains the single mound and a surrounding deep midden deposit enclosed within a palisade trench (Peacock and Hogue 2005); and the Feature 27 (a single, short-term wattle-and-daub house) locale, well outside the palisaded area (Seltzer 2007) (fig. 18.2).

Figure 18.1. Map of Mississippi and Alabama Black Belt Prairie physiographic region, showing the location of Lyon's Bluff (22OK520).

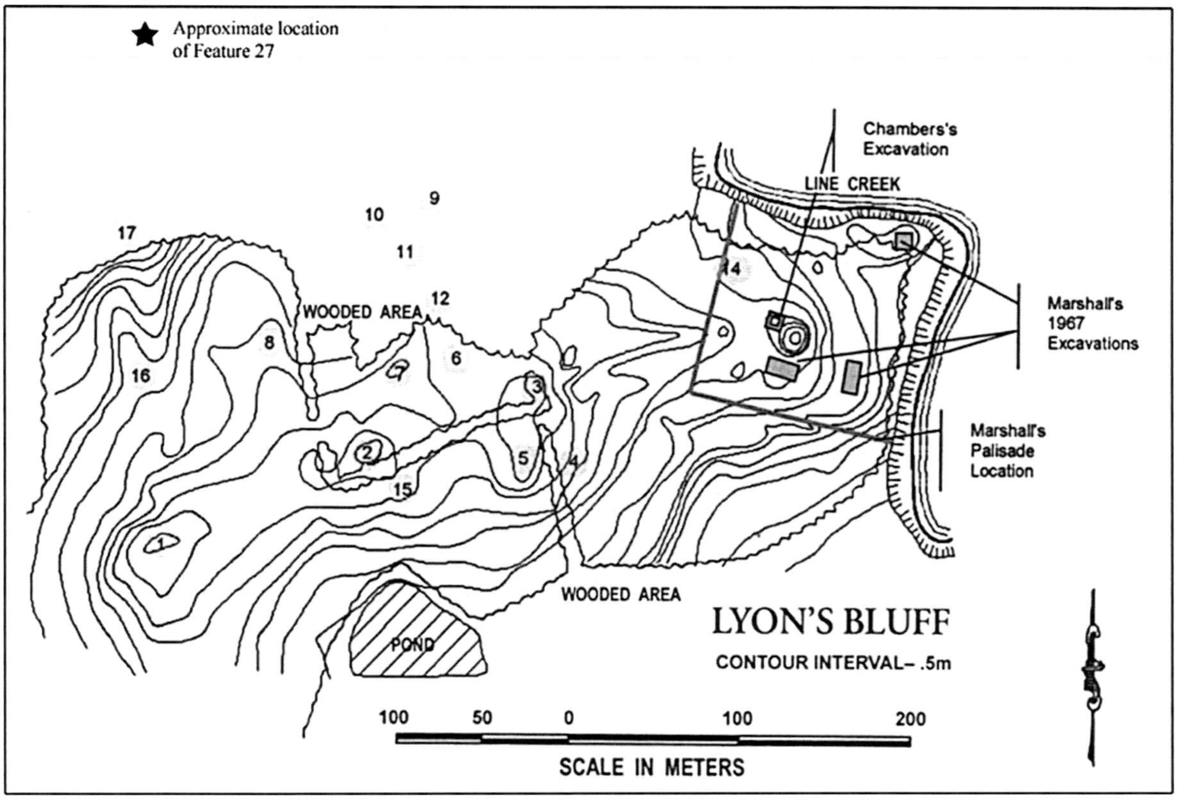

Figure 18.2. Topographic map (Lolley 2000) of the main site area of Lyon's Bluff showing Marshall's excavations and the estimated palisade location. The approximate location of Feature 27 is marked with a star to the north of "house" mound 17.

Figure 18.3. Leaf impression in daub from Lyon's Bluff.

Archaeo-Botanical Recovery

Recovery of botanical material from archaeological sites is often performed with the hope of obtaining a glimpse into the plant material used by site inhabitants. Such information can often provide insight into nutritional stress, population size and stability, cultural practices, and possibly the length of time over which sites were occupied. Botanical remains can be recovered from sites using various methods, including flotation, coarse and fine screening, and direct removal by hand during excavation. Flotation is typically better for recovering intact botanical materials (delicate seeds and charred materials) because of the gentle nature of the flotation process. Screening is more likely to cause fragmentation and loss of smaller, more delicate materials. Direct recovery during excavation is usually practical only when larger items are observed. Generally, botanical material from a site is recovered by a combination of these methods.

One other option for botanical recovery is available from Mississippian-period sites. Because of the long-term use of sites during the Mississippian period, settlement sites typically have evidence of structures, such as those constructed from wattle and daub. Daub is the fire-hardened clay that was used to create wattle-and-daub structures. During the building process, clay was mixed with grass and then worked into and onto the woven cane wattle, which formed the interior and/or exterior of the structure's walls. This process of mixing grass into the clay also led to the presumably unintentional (unbiased) addition of extraneous plant material that happened to be in the area during construction. When the structure was burned, either intentionally or by accident, the clay became fire hardened, preserving impressions of plant material within the daub (fig. 18.3). Exposure of these plant impressions is easily obtained by using hammers and small tools to crack and break the daub along the weakened areas created by the plant impressions (Peacock and Reese 2003; Seltzer 2007; Seltzer and Peacock 2011).

Seltzer (2007) compared these disparate data sets to determine the effects of both cultural and natural biases on each method. Sites where occupation was long term, with significant midden (anthrosol) development, are particularly appropriate for examining the interplay between human cultures and their natural settings over time, as the principle of least effort (PLE) had time to play out in the use of available materials that are represented in such settings. The PLE embodies the concept that effort and material choices work in balance to determine which sources are used first and which last. When a site is initially occupied, the materials used first are the prime materials with easiest access, followed by decreasingly desirable materials with easy access; access of all materials becomes more difficult over time until effort increases to a point where the occupants have to find a new location for occupation. The PLE is a concept that can be applied to many resources used by Native Americans; in this example, wood selection for fire use is analyzed (Seltzer 2007). When the PLE is used for analysis of firewood consumption, the expected progression is from choice woods (e.g., oak and hickory), which are easy to collect and use (when down and dry), to wood that is not ideal for burning (e.g., pine) but is still easy to collect, until the local area is drained of all usable wood. When sufficient time is represented, this approach provides a more complete record of trees occurring within gathering distance of the site and lessens the amount of bias from idiosyncratic firewood preferences.

The Environment Reconstructed

Environmental reconstruction in archaeology is often the stated intent of charcoal analysis at a given site. Because of the complex nature of the Black Belt and the limited amount of floral analysis previously conducted (Lowe 1913, 1919; Kaye 1955; A. Jones and Patton 1966; Leidolf 1995; MacDonald 1996; Leidolf and McDaniel 1998; Leidolf et al. 2002; J. Hill and Seltzer 2007; J. Hill et al. 2009), reconstruction based on charcoal alone has to be done with some discretion. Because fuelwood is thought to have been collected from the surrounding area, for sites within the Black Belt that means sampling a variety of habitats, including tallgrass prairies, xeric oak-hickory forests, red loam areas, red sand hills, stream bottoms, and lowland forests, all with specific characteristic floral and dendrological components. With this in mind, the analysis of charcoal at Lyon's Bluff, when used to reconstruct the possible environments during the time of occupation, should reflect multiple habitats neighboring the site.

The predominance of oak charcoal at the main site (table 18.1), along with high percentages of hickories and elms, would indicate that fuelwood was collected from a xeric oak-hickory forest (Lowe 1913, 1919; J. Hill et al. 2009). Lesser components of this habitat include several that are represented in the charcoal remains as well (e.g., pine, ash, cherry/plum, and sassafras) (table 18.1) (J. Hill et al. 2009). Charcoal from the main site also indicates that wood was obtained from the other habitats as well, and more species came from the bluffs and bottoms (e.g., maples, catalpa, yellowwood, honey locust, tuliptree, red mulberry, and willow) (Leidolf et al. 2002) than from the red sandy hills or red loam areas (e.g., persimmon, shortleaf pine, sweetgum, and sassafras) (table 18.1). It is likely that the limited amount of material from the red sandy hills and red loam areas may be a result of distance and location (Lowe 1913, 1915, 1919) if wood is typically collected following the PLE. Since the provenances of the materials recovered and analyzed from the main site were selected to provide a continuous picture of fuelwood use over time, the PLE does not limit or bias the reconstruction based on the total charcoal analysis. However, were the analysis to be based on the selection of only one provenance, such as the single hearth (Feature 8) or the submound midden (0N12W, Zone N), the analysis could be biased by choices made in fuelwood selection by the individual(s) gathering the wood. This would also be more likely to include wood from only a few habitat areas (or even a single one), thus limiting the general environmental reconstruction even further.

In the deepest excavation unit, the submound midden dating to the earliest occupation of the site in the main mound area, oak, American chestnut, and ash make up the greatest percentages of charcoal present. This indicates that the use of fuelwood at the site followed the PLE, and these hardwoods were selected first. These woods were not used again in the same percentages during the occupation of the site, indicating that they were less available during the site's occupation, and that other woods were increasingly used instead.

The charcoal from the Feature 27 area is indicative of the bottoms and stream bluffs of the Black Belt (table 18.1). Over half of the species represented in the charcoal analysis come from what Lowe (1913, 1915, 1919) and Leidolf et al. (2002) described as the stream bottoms and bottomland hardwood forests; the rest of the species come from a mix of other habitats (table 18.1). Because of the prox-

Table 18.1.

Comparison of tree species present at the site used to describe habitats within the Black Belt physiographic region.

Trees	Xeric oak-hickory forest	Red loam areas	Red sand hills	Stream bottoms	Stream bluffs and ledges	Daub Feature 27	Daub main site	Charcoal Feature 27	Charcoal main site
Acer sp. (maple)				X	X	X		X	X
Aesculus sp. (buckeye)									X
Betula nigra (river birch)				X					
Carpinus caroliniana (ironwood)								X	X
Carya sp. (true hickory)	X	X		X		X	X	X	X
Castanea dentata (American chestnut)					X			X	X
Catalpa bignonioides (southern catalpa)				X					X
Celtis sp. (hackberry)					X		X	X	
Cercis canadensis (redbud)	X			X	X				
Cladrastis kentukea (yellowwood)				X		X			X
Cornus sp. (dogwood)							X		
Crataegus sp. (hawthorn)							X		
Diospyros virginiana (persimmon)		X	X				X	X	X
Fagus grandifolia (beech)								X	X
Fraxinus sp. (ash)	X			X				X	X
Gleditsia triacanthos (honey locust)				X					X
Gymnocladus dioicus (coffeetree)									X
Ilex opaca (American holly)	X					X	X	X	X
Juniperus virginiana (eastern redcedar)	X				X	X	X	X	X
Liquidambar styraciflua (sweetgum)			X					X	X
Liriodendron tulipifera (tuliptree)				X					X
Maclura pomifera (Osage orange)						X		X	X
Morus rubra (red mulberry)				X	X			X	X
Nyssa sylvatica (sourgum)									X
Pinus echinata (shortleaf pine)	X	X	X				X	X	X
Platanus occidentalis (sycamore)				X					
Populus deltoides (eastern cottonwood)				X				X	
Prunus sp. (plum)	X	X				X		X	X
Quercus sp. (oak)	X	X	X	X		X	X	X	X

Table 18.1. *Continued*

Trees	Xeric oak-hickory forest	Red loam areas	Red sand hills	Stream bottoms	Stream bluffs and ledges	Daub Feature 27	Daub main site	Charcoal Feature 27	Charcoal main site
Rhamnus caroliniana (Carolina buckthorn)								X	X
Rhus sp. (sumac)			X						
Salix sp. (willow)				X		X		X	X
Sassafras albidum (sassafras)	X		X					X	X
Taxodium distichum (bald cypress)									
Ulmus sp. (elm)	X		X	X		X	X	X	X

imity of the Feature 27 area to Line Creek, this is the type of fuelwood that one would expect to find based on the PLE, given that the single structure investigated represents a short-term cultural deposit.

In addition to wood charcoal, daub impressions have been suggested as a possible source of data for environmental reconstruction. As nongrass impressions are believed to be the result of accidental inclusion in daub, they should be free from bias generated by cultural selection. Not only can these impressions be compared to charcoal to determine whether certain woods were excluded from use as fuel, but they also have the potential to provide an increased data set for a locale with limited time depth where the PLE will have conditioned the selection of wood to the most preferred (or remaining locally available) fuelwood choices. It is evident from the daub in the Feature 27 area that the incidental inclusions are similar to what is represented in the fuelwood selected for use at this locale. However, it is not what is similar, but what is missing in the daub impressions that provides a better impression of the habitat in which the structure was constructed. Seven of the 12 bottomland species found in the charcoal are absent from the daub impressions (table 18.1). This implies that the construction of the structure occurred in a bluff habitat and that the arbitrary leaf inclusions were from trees within this habitat (Lowe 1913, 1915, 1919).

When the daub impressions from the main site are examined, the first thing that is noticeable is that the number of nongrass impressions is significantly smaller than the number of impressions from the Feature 27 area: 53 versus 567. The few impressions that are present are associated with the xeric oak-hickory forest habitats and edges of prairies (table 18.1) (Lowe 1913, 1915, 1919; J. Hill and Seltzer 2007; J. Hill et al. 2009). The lack of leaf impressions from trees implies that the areas of construction were most likely open. Since the limited numbers of tree leaf impressions are from locations near the mound and within the palisaded area, it might be expected that this area was maintained in an open state because of occupation. However, daub that was examined from some of the deepest zones of the site (excavation unit 0N12W, Zone J, Levels 1–3; and unit 0N20W, Zones F–H) had no tree leaf impressions, which implies that the area had little tree growth, even early in the history of the site's use. Alternatively, the main site area may have been intentionally cleared of trees and other leafy vegetation prior to construction of housing.

Eastern Redcedar (*Juniperus virginiana*)

Eastern redcedar (*J. virginiana*) is often seen in modern prairie remnants throughout the Black Belt of Mississippi and Alabama. The basic soils of the prairies, especially the

Figure 18.4. Cedar impression in daub from Feature 27 area.

exposed chalk outcrops, are a perfect environment for these trees. Currently seen as an invasive problem species for the prairies of the Black Belt, cedars are often associated with areas of high soil erosion and disturbance. Cedars are much less common within intact prairie remnants but are often still a component.

It has been argued that Protohistoric sites (those occupied between initial European incursion and sustained European contact) in the Black Prairie were often situated in cedar glades, areas where eastern redcedar was dominant (J. Johnson and Sparks 1986), as opposed to oak-hickory stands interspersed with tallgrass prairies (Peacock and Miller 1990; Peacock 1992). The argument for the presence of cedar glades in the Protohistoric Black Prairie is based on Johnson's model for subsistence and settlement pattern change during this period (J. Johnson et al. 1984; Johnson and Sparks 1986; Johnson 1990). In this model, subsistence shifted from agriculture to reemphasize hunting of white-tailed deer (Johnson and Sparks 1986). Cedar glades enter into the model as a probable source of food (browse), based on research of food communities for white-tailed deer in the Arkansas Ozarks (Segelquist and Green 1968).

While cedar is currently abundant on the Black Prairie, its presence is generally attributed to Historic-period erosion and decreased fire maintenance of the prairies. Historical evidence shows more support for the oak-hickory model, as General Land Office (GLO) survey notes and historical descriptions of the area rarely mention the presence of cedar, with the exception of areas near waterways with constant disturbance (Peacock 1992, 1993). Plant impressions from Mississippian-period daub used in a pilot study by Peacock (1993) were compared to historical records and GLO notes for the Black Prairie. In this study, impressions from multiple sites within the Black Prairie appear to support the oak-hickory model for the Black Prairie during prehistoric times, with no evidence of cedar in the impressions. Peacock and Reese (2003) and Reese (2000) compared different sets of biotic data (charred plant material, pollen, and plant impressions in daub) from the Yarborough site (22CL814), a single Mississippian-period dwelling site in Clay County, Mississippi, to examine plant distributions in the Black Belt. These data also supported the oak-hickory model, as no evidence of cedar was present. It must be noted, however, that cedars are often more poorly preserved than oaks, meaning that charcoal from cedar may have been missed because of recovery techniques used at the time (Rossen and Olson 1985). While analysis of modern and archaeological land snails by Peacock and Gerber (2008) also supports the oak-hickory model, the data reported herein provide another opportunity to address this topic.

Daub impressions recovered from Lyon's Bluff indicate that eastern redcedar was present prehistorically within the prairie system (fig. 18.4) (Seltzer and Peacock 2011). Eastern redcedar made up 26.81% of the cedar impressions from one structure alone (Seltzer 2007). Around the mound, daub impressions of cedar were considerably less for eastern redcedar (0.98% of the impressions); however, the majority of the impressions from the mound area were from grass and weeds, and as noted above, very few tree leaf impressions were present at all (Seltzer 2007). This is what one would expect when a mound is constructed at a site where the habitat is predominantly prairie grasses and forbs. The few leaves that were recovered were likely included at the time of clay collection.

Charcoal from the Feature 27 structure area also included eastern redcedar, with a total of 37 pieces (0.2 g), making up 8.15% of the assemblage. The majority of this charcoal (75.6%) was recovered from the fine-screen samples; 23.6% was recovered from flotation samples; and no eastern redcedar charcoal was recovered from the water screening (0.25 in). This follows the findings of Rossen and Olson (1985) of poorer preservation in eastern redcedar relative to other charcoal types. Eastern redcedar remains were present within the area of the main site in both the "generalized" units selected from the area—represented by an ash dump, a daub-fall area, and midden levels—and a "specialized" unit, a single fire hearth. The proportion of eastern redcedar charcoal for the areas associated with general use was smaller than that found in the Feature 27 area. The total of eastern redcedar from general main-site contexts was 31 pieces. This represents 1.11% of the charcoal from this area. Feature 8, the fire hearth, had 280 pieces of eastern redcedar charcoal, representing 10.06% of the charcoal from the main site area.

These figures imply that eastern redcedar was present at Lyon's Bluff prehistorically. Most likely, this tree was present as it is today, in areas of high disturbance and erosion, and along the edges of waterways where the Cretaceous chalk is exposed. This is supported by the smaller representation of eastern redcedar from the area associated with the mound, as this area was most likely open tallgrass prairie originally, and by the greater representation of eastern redcedar at a structure near the edge of a waterway. Interestingly, the proportion of cedar seems to have stayed relatively level throughout the site's duration without any dramatic increases. It is unlikely that cedar glades were present during prehistoric times, as cedar makes up such a small percentage of the overall charcoal record. It is also unlikely that this tree was reserved only for ceremonial purposes, as it is found throughout the site both temporally and spatially, including in the midden layer.

Conclusion

Biological material recovered from archaeological excavations allows a more complete understanding of habitats and environmental conditions. This is just one example of how daub and charcoal can be used in conjunction to formulate a more accurate picture of the tree assemblage at Lyon's Bluff while it was occupied during the Mississippian period. Using empirical data to determine the amount of charcoal present in midden layers provides answers about the presence and use of eastern redcedar in the Black Belt Prairie prior to European settlement, as well as establishing that the central occupation and original mound construction occurred within a tallgrass prairie.

19

A Preliminary Study of Learning about Prairie Restoration Ecology

A Comparison between Biology Major and Nonmajor Students

Bruno Borsari and Malcolm F. Vidrine

Introduction

The tallgrass prairie that dominated much of the central North American landscape has been extirpated almost in its entirety (D. Smith 1992), leaving an void of irreplaceable knowledge about plant ecotypes that ecologists and prairie enthusiasts have been trying to fill from minuscule remnant patches for the last several decades (L. Jackson 1999; D. Smith 2012; Borsari et al. 2006; Vidrine et al. 2001). The general public remains mostly uneducated about the value of prairies and their conservation needs because school curricula seldom offer learning opportunities that emphasize these and similar topics. Now more than ever, there is an urgent need to engage students of all ages and academic interests in prairie restoration and reconstruction efforts. Borsari (2003) documented, for example, that by assessing students' knowledge about the coastal prairies of southeastern Texas and southwestern Louisiana, education could foster momentum in educating citizens about conservation and sustainability.

Introductory courses in biology, however, may not offer many opportunities to learn about the ecology, beauty, diversity, and resilience of prairies because the module on ecology, ecosystems, and the environment is usually presented in the last chapters of the textbook, and most professors teach it at the very end of the term. This may not leave sufficient time for learners to understand why prairies have vanished, or to reflect upon basic ecological principles that support conservation and restoration philosophies (Borsari et al. 2010). Students' aversion or apathy toward learning science in introductory biology courses often constitutes another challenge to professors who are committed to excellence in education. The severing of the links of "living off the land" for most people and an increasing depopulation of rural communities have contributed to an amplification of what Louv (2005, 2011) defined as "nature-deficit disorder," which has become a chronic condition in modern, industrialized society. On the other hand, place-based education can inspire teachers and their students to learn more about the landscape features of the specific bioregions (Orr 1992) in which they live, and within this context the study of prairies and grasslands could fulfill a very important curricular niche for US education (Borsari 2012; Noss 2013). Therefore, prairie habitats, whether they are reconstructed, restored, or preserved from preexisting remnant patches, could become valuable assets for a more ecologically based type of instruction in the upper Midwest, the Southeast, other regions of the United States, and elsewhere (Apfelbaum 2009; Apfelbaum and Haney 2010).

A potential ideal for the education process is to prepare a population of students (i.e., future landowners, landscapers, and developers) to appreciate the "ecological reserve paradigm" (Apfelbaum 2009), evidenced not only by the construction of individual prairie landscapes for homeowners but also by the creation of a conservation community. The latter is exemplified by Prairie Crossing—a community of 400 homes and over 400 adjacent acres restored to prairies, wetlands, and lakes, in Grayslake, Illinois, northwest of Chicago (for more information, see http://www.terrain.org/unsprawl/9/).

At the same time, emerging pedagogies aimed at attracting and retaining students in biology and life science curricula are gaining traction in colleges and universities. The case-study method proposed by Herreid (1997), for example, has gained support among educators who have been employing it with success to overcome students' boredom in large biology classes (Borsari et al. 2010). A case study is a story with a meaning, which is purposefully conceived to educate students. Case studies have been employed for a very long time by teachers on the assumption that storytelling captivates and holds the audience's attention (Herreid 1997). Herreid (2006) also suggested adoption of case studies that use a response system (clickers) in order to engage students a step further in the learning process, to make them understand thoroughly, and to learn critically and apply what is being taught. Duncan (2005) demonstrated the efficacy of clicker technology when it is employed in large introductory science courses. Drawing on his experience, we adopted the same technology, yet we decided to include a case study in addition to clickers, as suggested by Caldwell (2007). We compared the efficacy of this method of instruction with the lecture-clicker type of teaching in order to assess students' learning in introductory biology and ecology courses. A notable study that was conducted with a sample of more than 1,300 students across 11 institutions in the United States indicated that strong clicker cases can enhance students' understanding in large introductory biology classes more than the typical PowerPoint lectures (Lundeberg et al. 2011). Therefore, in an attempt to support case-study teaching and clicker technology, the authors used a "prairie clicker case" (Borsari 2011) in large sections of a nonmajor introductory biology course and with biology majors enrolled in ecology, through several semesters. The case study was constructed to challenge students' learning with concepts and issues about prairie restoration and ecology. The purpose of this work was twofold. The investigators wanted to assess the effect of using interactive technology (clickers) in combination with case-study teaching in large classes where case studies and lectures were employed in alternation as instructional methods. Second, the investigators wanted to verify possible differences in learning about prairie concepts between nonmajors and students majoring in biology.

Study Participants

In total, 456 undergraduate students were part of this study. Of these, 316 were nonmajor students in an introductory biology course, whereas 140 were biology majors enrolled in an ecology course. All participants in the introductory biology were either undeclared or nonscience majors and were enrolled in their first or second semester of studies. Participants in the ecology course were biology majors who had already earned eight or more academic credits in biology.

Materials

All students purchased an eInstruction Classroom Response System polling device, or "clicker," from the university bookstore. This small, handheld device allowed them to respond to multiple-choice, true/false, and yes/no questions posed by the instructor by pressing a corresponding button. Responses were transmitted to a receiver connected to the instructor's computer, and class results were displayed on a public screen. The instructor gave students a fixed amount of time to respond and could control when to reveal the results to students.

Research Design and Methodology

The case study "Prairie Garden of Troubles" was proposed by Borsari (2011), with the goal of achieving the following learning objectives:

1. Learn the basic ecological characteristics of a prairie ecosystem, including prairie ecological services (soil humus formation, water conservation, pollination, biodiversity).
2. Develop an appreciation for prairie habitat restoration and conservation.
3. Study plant biodiversity.
4. Be introduced to concepts of coevolution and symbiosis among plant species and pollinators.

The case was designed to engage students in a more active learning process about prairie restoration and ecology, and it was taught within a typical 50-minute class period over a course of four semesters, and in alternation with lectures about the same topic. This teaching method was employed by the same instructor for both courses and throughout the study period (table 19.1).

Table 19.1.

Schematic representation of the research design.

Semester/course	Fall 2009	Spring 2010	Fall 2010	Spring 2011
General Biology	Lecture	Case	Lecture	Case
Ecology	Case	Lecture	Case	Lecture

Students were exposed to five pretest questions prior to the presentation of the case or lecture, as proposed by Borsari and his collaborators (2010), regardless of students' knowledge about the topic. Students used their clickers to answer the five assessment questions. Every correct answer counted one point, whereas no answer was considered an incorrect response for which students earned no points.

1. What environmental factor do you think is most helpful to the reproductive success of prairie plant species that belong to the sunflower family (Asteraceae)?
   a. water
   b. light
   c. soil
   d. temperature
   e. an environment free from other Asteraceae

This question evaluates students' understanding of the relevance of abiotic factors in prairie habitats. It may be difficult to make a selection, as every factor could be important. However, we decided to ask this question because there is a clear hint about this topic in the case, and this would allow us to learn whether students had read the case with attention. **Correct answer: b.**

2. How do you think prairie plants keep under control the growth of weeds and exotic species? What is their best strategy to succeed?
   a. their massive root system occupies all soil niches
   b. their effective photosynthetic activities are superior to non-prairie plants
   c. their speed in germinating seed competes out non-prairie plants
   d. b and c
   e. none of the statements above is correct

The case emphasizes the difference between perennial and annual plants. Therefore, the rationale for this question was asking students to reflect on the growth of root systems of prairie species. Note that this is not always successful in disturbed sites, and it is often in these where alien species gain a hold and can then spread. **Correct answer: a.**

3. The monarch butterfly (*Danaus plexippus*) lays its eggs on different species of milkweed (*Asclepias* spp.) and the two species inhabit prairies. The plant offers a site to the insect where its offspring will find easy food (the plant itself), whereas the insect while visiting the several plants will pollinate its flowers and help milkweeds to reproduce. This is an example of:
   a. parasitism
   b. mutualism and herbivory
   c. competition
   d. commensalism
   e. synergism

The balance within a natural ecosystem is maintained by the synergies occurring among all its inhabitants. We thought this question was important to recall students' learning about the importance of mutualism in prairies. **Correct answer: b.**

4. Extracts of the prairie plant pale cone flower (*Echinacea pallida*) have provided medicine to enhance the capability of the immune system to resist infection. The ability to find this species of plant and discover these drugs is a strong argument for:
   a. genetic engineering
   b. overexploitation
   c. preserving biodiversity
   d. supporting pharmaceutical industries

e. developing a better technology in support of quality of life for humans

This question was designed to stress the need for conserving biodiversity. **Correct answer: c.**

5. All these ecological services BUT ONE can be performed by insects on a well-restored habitat (e.g., prairie).

a. pollination
b. sexual dimorphism
c. predation
d. parasitism
e. reproduction

The rationale for this question is to make students reflect on the importance of insects (in terms of ecological services) in natural ecosystems. An ecological service is a favorable performance of one species that is to the advantage of another one, or more. Think about the effect of beneficial insect species in agroecosystems so that the use of agrichemical products may be reduced or even avoided. **Correct answer: d.**

Then the case or the lecture was presented through the slides of a PowerPoint presentation. The process included (for the case only) more clicker questions to retain students' interest in the topic and to also allow small group discussion before students chose and submitted their answers through clickers. During the presentation of the case, students worked in groups to develop open-ended questions to be discussed with the whole class. Clickers were used again to select three to five main questions from all those that were proposed by the groups, and class discussion continued, leading to the solution of the case a few minutes before the end of class. The lecture mode of instruction was teacher centered instead and offered neither any discussion opportunities nor any kind of interaction among students. At the onset of the following class period, the same five clicker questions were proposed again (posttest) to all students in both case and lecture treatment conditions. The responses were later compared to the answers of the pretest questions, and an explanation of these questions was presented by the instructor to all students at this time.

T-tests for two independent samples and a two-way ANOVA were the statistical procedures employed to test the hypotheses under study.

Data Analysis and Discussion

For each treatment, lecture presentation and case presentation, 228 students were involved. Within each treatment group, 158 were students in introductory biology and 70 were students in ecology. The biology classes that were taught the prairie case achieved higher scores than the classes that were taught with the lecture method, and this difference was significant (t-test, $t_{314} = +5.84$, $P < 0.05$, two-tailed). Also, the ecology classes that were taught the prairie case achieved higher scores than the classes that were taught with the lecture method (fig. 19.1), and this difference was significant (t-test, $t_{138} = +7.35$, $P < 0.05$, two-tailed).

The two-factor ANOVA showed a significant main effect for both factors (teaching method and course type). More specifically, students who were taught through the case method achieved higher scores in both courses when they learned about prairie restoration ($F_{1,\,152} = 4.77$, $P < 0.05$). Also, biology majors enrolled in the ecology course did significantly better than the nonmajor students in introductory biology ($F_{1,\,152} = 5.89$, $P < 0.05$). However, the interaction between teaching method and whether or not students majored in biology was not significant ($F_{1,\,152} = 3.04$, $P > 0.05$), suggesting that the effectiveness of the teaching method (case vs. lecture) was independent from students majoring in biology or in other fields of study (see fig. 19.2).

It is necessary to point out that subjects in this evaluation were not assigned at random to the conditions prescribed by the study. Therefore, any interpretation of our data about students' learning in prairie restoration ecology should be qualified by this limitation because there could have been differences between groups that might not have been considered in this study.

Conclusion and Implications

This study explored whether the use of a case-based method for presenting prairie restoration and ecology concepts in an introductory biology course for nonmajors and in an ecology course for majors would result in improved learning outcomes when compared to a traditional lecture-based format, even when both classes were enhanced by clicker technology. Our results suggest that students who partici-

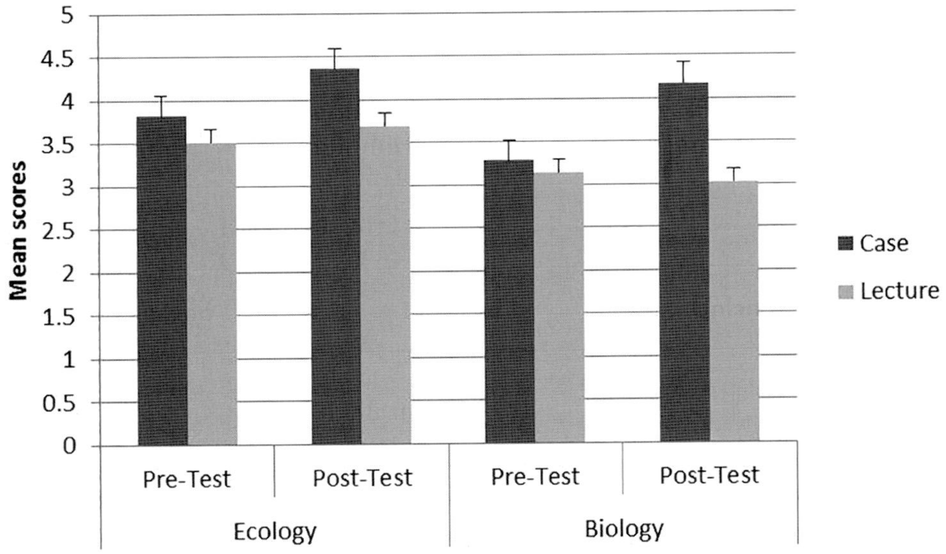

Figure 19.1. Students' average pretest and posttest scores (± 1 S.D.) using the prairie case versus lecture methods in the ecology and biology classes.

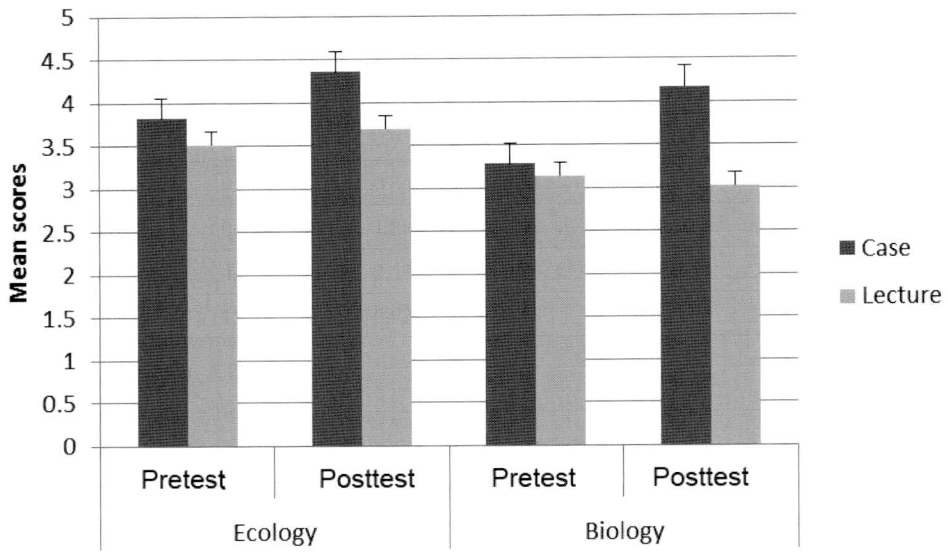

Figure 19.2. Lack of interaction effect between case and lecture teaching method in introductory biology and ecology courses.

pated in the case-based presentation performed better than students who were taught with the lecture method in both courses. Therefore, a combination of a case-based method with tools that allow all students to participate and engage in constructing knowledge is promising and beneficial to learning. In addition to this, the data did not produce a statistically significant interaction between the teaching method (case vs. lecture) and student type (biology major vs. nonmajor) in the ANOVA analysis, indicating that students' learning was independent of their academic interests and pursuits.

This initial study supports the idea that potentially all students can be engaged and achieve successfully when instruction becomes more student centered (Preszler et al. 2007) and curricula are more eco-driven (Borsari 2012). Technology can be valuable to engage large groups of students in learning about prairie habitat restoration and ecology. However, we remain cognizant of the limitations of technology in imparting knowledge of ecology and prairie restoration when educational encounters are limited solely to the classroom environment. Ultimately, the real experience of understanding the fascinating dynamics of prairie plant communities and other associated biota will always be gained in the outdoors. If technology serves as a vehicle to inspire students to learn more about prairie restoration while enhancing their desire to get involved with real, hands-on projects, we feel that there can be a viable niche for technology in the classroom. Various approaches can be employed in the restoration of prairies, depending on the bioregion (Vickery and Dunwiddie 1997), its climatic and abiotic attributes, and our knowledge of its plant species (Apfelbaum and Haney 2010; Van Bruggen 1992; Vidrine et al. 2001). Finally, future studies exploring the integration of clicker technology with alternative pedagogies about topics of conservation and restoration ecology might include control conditions where clickers are not used, and perhaps an outdoor activity like a prescribed burn or an effort to control invasive species.

In the end, the goal is to educate students to the point that a novel "conservation community" will be less novel and more "the way things are done." The students have a tremendous need to understand their local ecology in order to engage in professions that are potentially successful in this generation—a generation that is now teetering on the "cliff," not only economically but also ecologically (Louv 2011) and environmentally.

Acknowledgments

The authors would like to thank Clyde F. Herreid (University of Buffalo), Mary Lundeberg (Michigan State University), and Eric Ribbens (Western Illinois University) for their guidance and critical evaluation of the case. Ken Graetz and his staff gave valuable assistance for all issues concerned with technology use through this study.

20

Attempts at Converting a Southern Mississippi Bahia Grass Pasture to Diverse Prairie via Local-Provenance, Source-Certified Seed

Marc G. Pastorek, Malcolm F. Vidrine, Charles M. Allen, Bruno Borsari, and Gail Barton

Introduction

Grasslands, including prairies, once covered hundreds of thousands of acres in Louisiana and Mississippi. The development of industrial agriculture, the expansion of urban sprawl, and the absence of fire from the landscape have reduced these diverse habitats to a fraction of their original size (Allain et al. 2000). The Coastal Prairie of Louisiana, or "Cajun Prairie," the most southeastern extension of the tallgrass prairies of the central United States, has essentially been extirpated (Allain et al. 2003). A few strips along railroads and roadsides and patches on the edges of farmland remain, and these are becoming increasingly degraded (Vidrine and Hillman 2010; Vidrine et al. 2001a). It is estimated that less than 100 ac (40 ha) of this prairie vegetation still exists.

In 1997, 12.1 ac (4.9 ha) of land was purchased in the community of Henleyfield, Pearl River County, Mississippi, for the purpose of establishing a native seed field and preserving Louisiana tallgrass genetics. The effort was modeled after the Cajun Prairie Habitat Preservation Society's Cajun Prairie Restoration Project (CPRP) in Eunice, Louisiana (Vidrine et al. 2001a, 2001b). Since 1988, the society's work has focused on experiments to develop prairie habitat on ruderal lands using seed and sod from existing prairie remnants. This work proved feasible by the early 1990s, and scientific papers have reported the results of a variety of studies that examined prairie ecology. By the late 1990s, the CPRP site had developed into a rather diverse representation of high-quality prairie remnants. It has served as a design model and as a seed source for other restorations (Vidrine et al. 2001a).

The main goal of the project described here has been to produce seeds of native species from relict Cajun Prairies—to develop a museum of sorts with unique seed stocks (Schramm 1990; Allen and Grafe 2004). "Meadowmakers" is a name that was trademarked in 1999 with the intention of building a commercial market for ecological and grassland restoration services and for producing seed in the central Gulf Coastal area: Louisiana, Mississippi, and Alabama.

This chapter includes qualitative results and suggestions regarding the restoration of prairies, based on our efforts at Meadowmakers' Prairie Gardens (MPG) in southern Mississippi. These accounts provide a description of the operations that go into the making of prairie gardens, as well as the results. Such reports are essential and provide insights to those interested in re-creating a natural herbaceous landscape.

Methods

The initial concept for the MPG was based on the CPRP. The CPRP is the result of repeated efforts over many years to include as many prairie species as possible with high conservation value (i.e., those with the highest coefficients of conservation; see Allain et al. 2006; appendix 6 in Vidrine 2010). The hope was that the plants would find their niches and would spread and settle across the landscape as best they could, producing a facsimile of the few remaining remnant prairies, and placing plants into protected areas to ensure the survival of this threatened and vanishing genetic resource (Allen and Thames 2007). Similarly, the MPG project was intended to evolve naturally, aided by the manual removal of exotic invasive species and the use of controlled burns. The site had for the previous 20 years been managed as a Bahia grass (*Paspalum notatum*) pasture for grazing cattle. Prior to that, the land had been mechanically altered and sculpted with gentle terracing and planted as part of a tung tree orchard.

In the summer of 1998, seed collecting began for the project, with regular collections from some of the remaining Acadia Parish prairie remnants (Midland, Estherwood, Frey Prairie, etc.) and the CPRP. Each month or every six weeks, when seeds were ready to be collected, they were harvested by hand, put into paper bags to dry, and placed in an air-conditioned storage room. Plantings were done in the fall of the year, generally the first part of November. A machine-operated, handheld seed stripper was available to collect seed during 2001, which allowed larger amounts of seed to be collected. The seed stripper enabled the collection of a wide variety of samples from high-quality sites as well as the gathering of seeds from multiple species at a time. The machine is one of several mechanical products produced and sold by Prairie Habitats, Inc., of Argyle, Manitoba, Canada.

Approximately 8 ac (3.2 ha) were planted between 1998 and 2001, with 1 ac (0.4 ha) planted solely with Black Belt Prairie seed from prairie remnants in Sumter County, Alabama, and Oktibbeha and Noxubee Counties, Mississippi. Another 1.0 ac (0.4 ha) section was left untouched (i.e., unseeded and unplowed) as a control, in order to see what would regenerate naturally with only the introduction of fire.

The strategy for planting was very basic. The soil was plowed with an agricultural disk prior to planting. It was then leveled with a harrow. No herbicide was applied. Seeds that had been collected the previous growing season were distributed by hand broadcasting. At the lower ends of some slopes, plantings of running, colonizing species such as lowland mountain mint (*Pycnanthemum muticum*), narrowleaf mountain mint (*P. tenuifolium*), and ashy sunflower (*Helianthus mollis*) were planted to quickly establish vegetation. A light layer of field hay was used in these areas as an additional measure to help prevent soil erosion. Over 50 different plantings have been made since the project was initiated, with block sizes for different plantings ranging from 0.06 ac (0.025 ha) to 2 ac (0.8 ha) in size. A list of planted species can be found in table 20.1.

There have also been additional efforts to propagate and establish unique plants with specific traits and characteristics. For example, big bluestem (*Andropogon gerardii*) seems to produce very few viable seeds from Cajun Prairie wild stock, so there was a desire to establish a breeding block to create greater possibilities for cross-pollination and, potentially, better conditions for producing a viable seed crop than seeds from isolated populations that have been gathered from the wild. There has also been an ongoing effort toward propagating vegetative divisions and seedlings of multiple color variations or forms of downy phlox (*Phlox pilosa*), lanceleaf blanketflower (*Gaillardia aestivalis*), blue sage (*Salvia azurea*), big bluestem, Indiangrass (*Sorghastrum nutans*), yellow wild indigo (*Baptisia sphaerocarpa*), and whiteleaf mountain mint (*Pycnanthemum albescens*), to name a few.

Approximately 200 shovel-dug plugs of rescued prairie sod were divided and included in the first two plantings in 1998–1999. The plugs were approximately 8 in by 8 in (20.3 cm by 20.3 cm). Digging plants from prairies is *not* recommended. However, in the case of the remnants along the railroad track by Highway 90 in Acadia Parish and in the Frey Prairie, south of Eunice, the owners have continually insulted the prairie strips by spraying herbicide, plowing, excavating, and mowing, while invasive species (e.g., *Chloris* sp. and *Sapium sebiferum*) have moved in unimpeded. After permission was received from the property owners, rescued plants from these sites were replanted at the CPRP and MPG as a last recourse.

No further seeding activity was done from November 2001 until May 2006, when an effort was made to increase forb diversity using plants germinated in controlled nursery conditions. The plants were grown from seed in 4 in (10 cm) cups, 3,000 in all. In the field, strips 100 ft (30.5 m) long and 6 ft (1.8 m) wide were located, marked, and tilled using a tractor-driven tiller. The seedlings were transplanted and watered immediately, and then watered intermittently via soaker hose for about three weeks. The plants were left to their own devices after that, with no further watering or care. Many of these were large enough by fall to produce flowers and seed.

In April 2009, in an upland area approximately 80 ft by 30 ft (24.3 m by 9 m), where earlier plantings had failed, the vegetation was removed via tractor-driven mower (Bush Hog), and herbicide was first applied in May. The applications were sporadically repeated until the last application in October. No tilling or plowing was done prior to planting. In November, plugs for the project were planted. A heavy seeding rate (approximately 60 lb/ac or 67 kg/ha), including a diverse Cajun Prairie mix (100+ species), was applied. In 2010, another plot was treated the same and sown with a higher seeding rate (approximately 100 lb/ac or 111 kg/ha) and no nursery plugs.

In an area approximately 250 ft by 40 ft (76 m by 12.1 m) between slopes, where soil moisture was nearly constant and controlled burns could not to be carried out, Chinese privet (*Ligustrum sinense*) had become dominant and shaded the area, forming a thicket over the length of the swale. Some plants had become as much as 12 ft tall (3.6 m). The area was mowed with a tractor and Bush Hog in April 2011. The larger trees were cut by chain saw. Shortly after the mowing, applications of glyphosate were made. Foliar applications of glyphosate continued throughout the growing season and winter until April 2012. Switchgrass (*Panicum virgatum*) and eastern gamagrass (*Tripsacum dactyloides*) were dug from established clumps on the property and replanted on 2 ft (0.6 m) centers. No other fertilization or pesticides were used for any of the plantings. The intent was to introduce the wet-tolerant grasses so they would become dominant and fuel the fire through the bottom area, helping restrict growth of invasive woody species.

Prescribed fires have been used intermittently as a natural management tool with a rotational schedule of about one burn every three years. Fire lines of wide-mowed paths divide the site into several sections and are mowed three or four times per year. Prior to burning, the lanes are mowed particularly close to the ground, and in some cases, scratched with a disk harrow to further reduce fuel at the perimeter of the burn area.

Results

The overall look of the seeded fields is that of a natural landscape. If a closer inspection is made, clear lines between some of the individual plantings can be seen. Some areas have matured to the point where all species with high conservation value clearly dominate, producing the effect of a late-successional prairie worthy of Weaver and Fitzpatrick's description in their seminal 1934 work *The Prairie*: "kaleidoscopic in character." Throughout the year, the landscape changes, but summer is when much of the activity occurs.

Many visitors have come to the MPG over the years. Landscape designers commonly come to see firsthand the nature of the project. Field-day events have been held for native plant societies and other groups. The gardens have been a study site for the main author and have served as a harvest site for seed in the last several years. Over 10,000 plugs have been dug using a custom-made plugging tool. The MPG is built for educational purposes as well, reconnecting people in an urban setting to the land (Borsari et al. 2012). It is a shining example of Aldo Leopold's land ethic on a country-farm scale (Leopold 1949).

There were a few plantings that were only modestly successful, but for the most part the plantings succeeded, and each one tells its own story. For example, Bahia grass has persisted in sites with single-species or low-diversity species plantings. But in some instances, Bahia grass was pushed out completely by a single prairie species. In each area where multiple species were introduced, success in establishing targeted species was more quickly achieved. Some combinations of two or three species were able to sustain and dominate their planting area. Some species of medium to high conservation value are very aggressive early on and then decline as a result of pressure from late-successional species. Ashy sunflower, narrowleaf mountain mint, and lowland mountain mint act as a nurse crop, while less aggressive species of high conservation value produce sufficient roots to become more competitive. The term "pushy perennials" has been used to refer to perennial plants that are aggressive enough to outcompete weedy, early-successional plants so that they are naturally removed. This competitive ability has also been referred to as perennial "staying power" (Schramm 1990). Most species of high conservation value (7, 8, 9 or 10 on the coefficient of conservatism scale), such as big bluestem, compassplant (*Silphium laciniatum*), rosinweed (*Silphium gracile*), rattlesnake master (*Manfreda virginica*), slender bluestem (*Schizachyrium tenerum*), pineland dropseed (*Sporobolus junceus*), yellow Indian blanket (*Gaillardia pulchella*), and various orchids and lilies, can compete with other valuable species. The reduction of competition prior to planting and the introduction of fire seem to be two factors that positively influence successful establishment and management.

Some plantings failed to meet their intended objectives, but many target species have dominated their planting blocks, though some took years to do so. At least some level of representation by the targeted species has occurred in most every case. One plot was planted in 2001 with seed from the only known population of white wild indigo (*Baptisia alba*), in Pearl River County. Today, a few hundred plants thrive in this planting area, and the stand currently amounts to a population larger than the original roadside remnant population. The original population, which was about 150 plants when it was found 20 years ago, no longer exists.

A collection of local big bluestem has been established, along with over 80 ecotypes from Louisiana, Mississippi, and western Alabama. Some of these plants were shared by Charles Allen from his personal ecotype collection. Others are from seed collections made from plants found across the three-state region. There are many variations in form, color, and texture. Notable specimens include a yellow-and-green-stemmed variety, a red-and-green-stemmed variety, and a glaucous blue form selected by Malcolm Vidrine from the wild. There is also a form with gray-silver foliage from Allen's garden, and another that darkens to purple foliage, stems, and flower heads in October. This variety was discovered at the CPRP site. This collection of big bluestem ecotypes has been established to provide a breeding block for producing viable seed.

Three different plantings of compassplant were seeded. At first these efforts looked to be a failure, since Bahia grass and little bluestem (*Schizachyrium scoparium*) had become dominant in these areas. There was no sign of the compassplants for several years, until a single seedling was spotted, struggling in thick vegetation. A thorough search of the planting areas revealed a total of 70 seedlings. Several of these plants flowered in 2009. The seed for these blocks had been collected from the CPRP and from Black Belt Prairie remnants on roadsides in Oktibbeha County, Mississippi.

Two significant ecotypes of whiteleaf mountain mint were collected by Malcolm Vidrine from the CPRP and were planted early on as rooted cuttings. They have established as colonies in the landscape, having thrown off seeds and seedlings over the years. Both were chosen for their aromatic properties, as one has a stronger-than-typical camphor aroma and the other has a distinctly spearmint aroma. The latter is commonly referred to, by those who have known it, as "Malcolm mint."

Button snakeroot (*Eryngium yuccifolium*) took several years to become dominant from seed. There are several representations in mixed plantings, but one block was planted as a monoculture, and over the last few years it has become a substantial presence in the landscape.

Wild bergamot (*Monarda fistulosa*) and Lindheimer's beebalm (*M. lindheimeri*) are early and midsuccessional plants that are among the most common plants in the field besides the little bluestem. These species have covered at least 4 ac (1.6 ha) of the field and seem to spread more each year. In June, when they are in bloom, it is a very active and attractive scene.

An interesting combination of species is a planting done with only Lindheimer's beebalm and yellow false indigo. These two plants have dominated their 100 ft by 50 ft (30.5 m by 15.2 m) planting area. Few plants in the area have been able to compete with this two-plant combination.

One particularly pleasing seed-driven result is a planting block of smooth beard tongue (*Penstemon digitalis*), eastern smooth beard tongue (*P. laevigatus*), and nodding beard tongue (*P. laxiflorus*). This area has grown up beautifully, increasing in richness and target-species density each year. It has held its own in competition from little bluestem and Virginia bluestem (*Andropogon virginicus*) and with other plants that have happened in, along with some remnant Bahia grass. In spring, when in bloom, the area is a sight to behold.

Another block was planted with Indiangrass that had been propagated in 4 in (10 cm) pots. The plants were planted on 6 ft (1.8 m) centers. This species has dominated the entire area for many years, and the block has been a source of plugs for numerous projects. Indiangrass seems to benefit from the soil disturbance caused by the digging.

Mixed-Seed Plantings

Several planting blocks of mixed-species seed collections that were taken from the CPRP site were introduced in the third year of planting, 2001. These seed collections were made with a mechanized handheld seed stripper. The mixed-

Figure 20.1. "One-day" seed collection planting block, planted November 2001, M. Pastorek.

seed plantings are the most developed and mature-looking prairie representations in the field. After several years, little bluestem became dominant. These blocks are particularly encouraging. They resemble a high-quality natural prairie and seem to always have seasonal interest. All of the mixed-species planting blocks differ in composition.

The "One-Day" Seed Collecting Model

A most interesting and enlightening experiment was a mixed-species planting established from seed collected on one day in October 2001. The day was spent at Frey Prairie, a remnant strip just south of Eunice, collecting with a newly purchased seed stripper. About four hours were spent harvesting a diverse suite of fall-blooming forbs, composites, and grasses. The collection amounted to about 15 lb (6.8 kg) of seed and chaff. The mixture was planted on a 1.5 ac (0.6 ha) block in November 2001. This planting, for the first four or five years, seemed to have no target species. However, it eventually developed into the best representation of mature prairie on the property. A recent survey, in September 2012, counted 55 species in this block. The dominant species in the planting is little bluestem grass, followed by grass-leafed goldenaster (*Chrysopsis graminifolia*), with large patches of Indiangrass, muhly grass (*Muhlenbergia capillaris*), bluesage, rosinweed, and flowering spurge (*Euphorbia corollata*).

This "one-day" model of seed collecting is a very effective means by which to establish the herbaceous habitat. It can be done with relative ease using mechanical seed harvesters. Finding an exceptional example of prairie or pine savanna as a seed source to collect from is an important first step to restoration success. However, care should be taken, when "wild collecting" seed, to identify all species present before collecting with a mechanical harvester, in order to avoid including exotic-invasive or native-invasive species in the seed mix. The one-day model includes the use of machine-driven tools to sample from the best representations of flora possible, targeting all the herbaceous (mostly perennial) species that make up the habitat. The machine-driven seed-collecting equipment also gathers samples of bits and pieces of plants along with seed capsules, and an occasional stem. All of this excess material should be put into the planting field, since it likely contains spores of essential underground fungi (mycorrhizae) and other natural inoculants. With additional one-day sampling collections at different times during the growing season (July, September, October, etc.), one can dramatically increase the number of species produced for a prairie reconstruction project.

New Planting Blocks

Two new planting blocks (planted in 2009 and 2010), testing the effects of herbicides in controlling highly competitive, weedy species, have been very successful in demonstrating that this sort of preplanting effort is beneficial at least to a more successful transition from early- to late-successional species dominance (C. Harper et al. 2007). The repeated herbicide treatments have largely eliminated Bahia grass, Canadian goldenrod (*Solidago canadensis*), Japanese honeysuckle (*Lonicera japonica*), and Chinese privet. Little bluestem has become the dominant plant, providing fuel for the first controlled burn in February 2013. In one of these herbicide-treated blocks, nursery-grown plugs were planted. The plugs were propagated from seed and grown in 1 in by 1 in by 6 in (2.5 cm by 2.5 cm by 15 cm) trays with 50 per tray. Approximately 500 plugs were introduced in all. These plantings have produced a surprising display of diversity even though they are in the early stages of growth.

The 1 ac (0.4 ha) planting block that has been used as a control was never disturbed other than by introducing fire. There are many American bluehearts (*Buchnera americana*) and spiral orchids (*Spiranthes* spp.) scattered across this area. It has transformed into a beautiful wild meadow with many sedges and rushes (*Rhynchospora*, *Juncus*, *Carex*, *Cyperus*, *Fimbristylis*, *Scleria*) and grasses (*Andropogon glomeratus*, *Schizachyrium scoparium*, *Dicanthelium* spp.), punctuated by large, dense patches of Maryland meadowbeauty (*Rhexia mariana*), rough-leaf goldenrod (*Solidago rugosa*), and flat-topped goldenrod (*Euthamia leptocephala*).

Discussion

Each situation in the field presents a different set of circumstances and choices. Many factors are involved when attempting to develop a prairie meadow reconstruction. Prairie and herbaceous pine understory systems are good tools for landscaping in the natural fashion (i.e., prairie gardening, naturescaping, or wildscaping). Patience is needed, above all, to refrain from throwing in the towel when things do not immediately happen as planned. A period of a few years is the minimum time required for perennial plants to make sufficient roots to produce a flower. With heavy competition, this can take even longer, as the plant struggles for sunlight, water, nutrients, and root space. A lifetime of pleasant surprises will reward the person who sows the seed and lets time heal the earth's wounds.

Although some research in the Midwest has shown that some degree of management can be sustained with annual mowing, there seems to be no substitute for fire when transitioning land from agricultural weed species to conservative prairie species (Schramm 1990; Diboll 1984). Most often, invasive species, along with native shrubs and trees, need suppression regardless of the management method used, unless the existing soils have a natural capability to do so (as with clay, chalk, or rocky soils).

Diverse plantings are natural ground covers that dramatically change the ecology of the land and turn the botanical clock backward toward a more natural, historical landscape. A facsimile of the natural meadow can be established by using small amounts of locally collected (source-identified) seed. This type of gardening is particularly adaptable to poor soils and harsh environmental conditions. Seeding gardens like this provides a natural association of plants that support the native insect populations found only in high-quality prairie and pine herbaceous understory.

Meadowmakers' Prairie Gardens have afforded a unique insight into the process of reconstructing prairie using diverse local seed mixes. Further, they have provided a place of refuge where some of the pieces and parts of a habitat have been preserved (Bisset 2004). The plantings serve as a model for future reconstruction projects. The gardens have become a glimpse into the past and a glimmer of hope for the future, when humans and nature can become whole (Borsari 2005; Borsari and Vidrine 2005).

Table 20.1.
List of plants from Meadowmakers' Prairie Gardens.

Monocotyledons

Poaceae
Agrostis hyemalis (Walt.) B.S.P.
Andropogon gerardii Vitman
Andropogon glomeratus (Walt.) B.S.P.
Andropogon gyrans Ashe
Andropogon ternarius Michx.
Andropogon virginicus L.
Anthaenantia rufa (Nutt.) J.A. Schultes
Aristida purpurascens Poir.
Bouteloua curtipendula (Michx.) Torr.
Ctenium aromaticum (Walt.) Wood
Dichanthelium aciculare (Desv. ex Poir.) Gould & C.A. Clark
Dichanthelium commutatum (Schult.) Gould
Dichanthelium dichotomum (L.) Gould
Dichanthelium scabriusculum (Ell.) Gould & Clark
Dichanthelium scoparium (Lam.) Gould
Elymus virginicus L.
Eragrostis elliottii S. Wats.
Eragrostis refracta (Muhl.) Scribn.
Eragrostis spectabilis (Pursh) Steud.
Muhlenbergia capillaris (Lam.) Trin.
Panicum anceps Michx.
Panicum hemitomon J.A. Schultes
Panicum virgatum L.
Paspalum floridanum Michx.
Paspalum laeve Michx.
Paspalum plicatulum Michx.
Saccharum baldwinii (Walt.) Spreng.
Saccharum giganteum (Walt.) Pers.
Schizachyrium scoparium (Michx.) Nash
Schizachyrium tenerum Nees
Sorghastrum nutans (L.) Nash
Sporobolus junceus (Beauv.) Kunth
Steinchisma hians (Ell.) Nash
Tridens ambiguus (Ell.) J.A. Schultes
Tridens flavus (L.) Hitchc.
Tridens strictus (Nutt.) Nash
Tripsacum dactyloides (L.)

Dicotyledons

Acanthaceae
Ruellia humilis Nutt.

Apiaceae
Chaerophyllum tainturieri Hook.
Cicuta maculata L.
Eryngium yuccifolium Michx.
Polytaenia nuttallii DC.

Apocynaceae
Amsonia tabernaemontana Walt.

Asclepiadaceae
Asclepias tuberosa L.
Asclepias viridis Walt.

Asteraceae
Arnoglossum ovatum (Walt.) H. Rob.
Bidens aristosa (Michx.) Britt.
Boltonia asteroides (L.) L'Hér.
Chrysopsis mariana (L.) Ell.
Conoclinium coelestinum (L.) DC.
Coreopsis lanceolata L.
Coreopsis pubescens Ell.
Coreopsis tinctoria Nutt.
Coreopsis tripteris L.
Echinacea pallida (Nutt.) Nutt.
Erigeron philadelphicus L.
Erigeron strigosus Muhl. ex Willd.
Eupatorium hyssopifolium L.
Eupatorium perfoliatum L.
Eupatorium rotundifolium L.
Eurybia hemispherica (Alexander) Nesom
Euthamia leptocephala (Torr. & Gray) Greene
Euthamia tenuifolia (Pursh) Nutt.
Gaillardia aestivalis (Walt.) H. Rock
Helenium flexuosum Raf.
Helianthus angustifolius L.
Helianthus mollis Lam.
Liatris elegans (Walt.) Michx.
Liatris pycnostachya Michx.
Liatris spicata (L.) Willd.
Liatris squarrosa (L.) Michx.
Liatris squarrulosa Michx.
Mikania scandens (L.) Willd.
Oligoneuron nitidum (Torr. & Gray) Small
Pityopsis graminifolia (Michx.) Nutt.
Pluchea foetida (L.) DC.
Pseudognaphalium obtusifolium (L.) Hilliard & Burtt
Ratibida pinnata (Vent.) Barnhart
Rudbeckia grandiflora (D. Don) J.F. Gmel. ex DC.
Rudbeckia hirta L.
Rudbeckia texana (Perdue) P. Cox & Urbatsch
Silphium gracile Gray
Silphium integrifolium Michx.
Silphium laciniatum L.
Solidago odora Ait.
Solidago rugosa P. Mill.
Solidago sempervirens L.
Symphyotrichum dumosum (L.) Nesom
Symphyotrichum lateriflorum (L.) Á. Löve & D. Löve
Symphyotrichum patens (Ait.) Nesom
Symphyotrichum praealtum (Poir.) Nesom
Symphyotrichum pratense (Raf.) Nesom
Vernonia gigantea (Walt.) Trel.
Vernonia texana (Gray) Small

Campanulaceae
Lobelia puberula Michx.

Clusiaceae
Hypericum hypericoides (L.) Crantz ssp. *hypericoides*

Euphorbiaceae
Croton monanthogynus Michx.
Euphorbia corollata L.

Fabaceae
Baptisia alba (L.) Vent.
Baptisia bracteata Muhl. ex Ell. var. *leucophaea* (Nutt.) Kartesz & Gandhi
Baptisia nuttalliana Small
Baptisia sphaerocarpa Nutt.
Centrosema virginianum (L.) Benth.
Chamaecrista fasciculata (Michx.) Greene
Dalea candida Michx. ex Willd.
Desmanthus illinoensis (Michx.) MacMillan ex B.L. Robinson & Fernald
Erythrina herbacea L.
Galactia volubilis (Michx.) MacMill. ex B.L. Rob. & Fernald
Lespedeza capitata Michx
Lespedeza virginica (L.) Britt.
Mimosa hystricina (Small ex Britt. & Rose) B.L. Turner
Neptunia lutea (Leavenworth) Benth.
Orbexilum pedunculatum (P. Mill.) Rydb.
Orbexilum simplex (Nutt. ex Torr. & Gray) Rydb.
Tephrosia onobrychoides Nutt.

Hydrophyllaceae
Hydrolea ovata Nutt. ex Choisy

Lamiaceae
Hyptis alata (Raf.) Shinners
Monarda fistulosa L.
Monarda lindheimeri Engelm. & Gray ex Gray
Monarda punctata L.

Table 20.1. *Continued*

Physostegia digitalis Small
Physostegia intermedia (Nutt.) Engelm. & Gray
Prunella vulgaris L.
Pycnanthemum albescens Torr. & Gray
Pycnanthemum muticum (Michx.) Pers.
Pycnanthemum tenuifolium Schrad.
Salvia azurea Michx. ex Lam.
Salvia lyrata L.
Scutellaria integrifolia L.
Teucrium canadense L.

Malvaceae
Callirhoe papaver (Cav.) Gray
Hibiscus aculeatus Walt.
Hibiscus moscheutos L.

Melastomataceae
Rhexia mariana L.

Onagraceae
Gaura lindheimeri Engelm. & Gray
Gaura longiflora Spach
Ludwigia hirtella Raf.
Ludwigia linearis Walt.

Passifloraceae
Passiflora incarnata L.

Polemoniaceae
Phlox pilosa L.

Portulacaceae
Claytonia virginica L.

Scrophulariaceae
Agalinis fasciculata (Ell.) Raf.
Agalinis purpurea (L.) Pennell
Buchnera americana L.
Penstemon digitalis Nutt. ex Sims

Verbenaceae
Verbena halei Small

References Cited

Abrahamson, W. G. 1984. Species response to fire on the Florida Lake Wales Ridge. *American Journal of Botany* 71:35–43.

Abrahamson, W. G., and D. C. Harnett. 1990. Pine flatwoods and dry prairies. In *Ecosystems of Florida*, edited by R. L. Meyers and J. J. Ewel, 103–50. Orlando: University of Central Florida Press.

Abrahamson, W. G., A. F. Johnson, J. N. Layne, and P. A. Peroni. 1984. Vegetation of the Archbold Biological Station, Florida: An example of the southern Lake Wales Ridge. *Florida Scientist* 47:9–250.

Adair, J. 1930. *History of the American Indians*. New York: Promontory Press.

Adams, H. L. 2011. Agricultural conservation buffers for breeding grassland birds in eastern Mississippi. PhD diss., Mississippi State University.

Alba-Lynn, C., and J. K. Detling. 2008. Interactive disturbance effects of two disparate ecosystem engineers in North American shortgrass steppe. *Oecologia* 157:269–78.

Aldrich, J. 2002. Factors and benefits in the establishment of modest-sized wildflower plantings: A review. *Native Plants Journal* 3:67–86.

Alford, A. L., E. C. Hellgren, R. Limb, and D. M. Engle. 2012. Experimental tree removal in tallgrass prairie: Variable responses of flora and fauna along a woody cover gradient. *Ecological Applications* 22:947–58.

Allain, L., L. Smith, C. Allen, M. F. Vidrine, and J. B. Grace. 2006. A floristic quality assessment system for the coastal prairie of Louisiana. *Proceedings of the 19th North American Prairie Conference* 19:1–18.

Allain, L., M. Vidrine, J. Faust, C. C. Lockwood, G. Ross, J. Boundy, J. Thomas, and R. Greco. 2003. Our vanishing wild Cajun Prairies of Louisiana. Poster. Baton Rouge Earth Day, Inc.

Allain, L., M. Vidrine, V. Grafe, C. Allen, and S. Johnson. 2000. *Paradise Lost? The Coastal Prairie of Louisiana and Texas*. 2nd ed. US Fish and Wildlife Service and US Geological Service with Coastal Prairie Conservation Initiative, TX.

Allen, C. M. 1993. *Ecological Assessment of the Keiffer Prairie Research Natural Area in the Kisatchie National Forest*. Alexandria, LA: US Forest Service.

Allen, C. M., and V. Grafe. 2004. The Duralde Prairie restoration project, Cajun Prairie on a federal refuge. In *Proceedings of the Third Eastern Native Grass Symposium*, edited by J. Randall and J. Burns, 114–16. Chapel Hill, NC: Omnipress.

Allen, C. M., and S. Thames. 2007. Observation on vegetation changes in Cajun Prairie, a Coastal Plain flora in southwest Louisiana. *Journal of the Botanical Research Institute of Texas* 1:1141–47.

Allen, C., S. Thames, L. Chance, and C. Stagg. 2006. Flora of the Anacoco Prairie on Fort Polk, Louisiana. *Proceedings of the 19th North American Prairie Conference* 19:19–22.

Allen, C. M., and M. F. Vidrine. 1989. Wildflowers of the Cajun Prairie. *Louisiana Conservationist* 41:20–25.

Allen, C. M., M. F. Vidrine, B. Borsari, and L. Allain. 2001. Vascular flora of the Cajun Prairie of southwestern Louisiana. *Proceedings of the 17th North American Prairie Conference* 17:35–41.

Alvey, J., E. Brown, J. Byrnes, L. Culpepper, and B. Leggett. 2004. A comparison of field methods in magnetic gradiometer survey. *Mississippi Archaeology* 39:25–38.

Ament, R., A. P. Clevenger, O. Yu, and A. Hardy. 2008. An assessment of road impacts on wildlife populations in U.S. National Parks. *Environmental Management* 42:480–96.

Amorim, P. K., and M. A. Batalha. 2007. Soil-vegetation relationships in hyperseasonal cerrado, seasonal cerrado, and wet grassland in Emas National Park (central Brazil). *Acta Oecologica* 32:319–27.

Anderson, L. D. 1996. Guidelines for roadside revegetation to create wildlife habitat in northern Utah. Master's thesis, Utah State University.

Anderson, M. J. 2001. A new method for non-parametric multivariate analysis of variance. *Austral Ecology* 26:32–46.

Anderson, R. C. 1990. The historic role of fire in the North American grasslands. In *Fire in North American Tallgrass Prairies*, edited by S. L. Collins and L. L. Wallace, 8–18. Norman: University of Oklahoma Press.

———. 2006. Evolution and origin of the central grassland of North America: Climate, fire and mammalian grazers. *Journal of the Torrey Botanical Society* 133:626–47.

Anderson, R. C., J. S. Fralish, and J. M. Baskin, eds. 1999. *Savannas, Barrens, and Rock Outcrop Plant Communities of North America*. New York: Cambridge University Press.

Anderson, W., M. Casler, and B. Baldwin. 2008. *Improvement of Perennial Forage Species as Feedstock for Bioenergy*. USDA–Agricultural Research Service / University of Nebraska–Lincoln Faculty Paper 248.

ANHP (Alabama Natural Heritage Program). 2012. *The Rare, Threatened, and Endangered Plants and Animals of Alabama*. Alabama Natural Heritage Program, Auburn University.

Antrop, M. 2005. Why landscapes of the past are important for the future. *Landscape and Urban Planning* 70:21–34.

Apfelbaum, S. I. 2009. *Nature's Second Chance: Restoring the Ecology of Stone Prairie Farm*. Boston: Beacon Press.

Apfelbaum, S. I., and A. Haney. 2010. *Restoring Ecological Health to Your Land*. Washington, DC: Island Press.

Aranyosi, E. F. 1999. Wasteful advertising and variance reduction: Darwinian models for the significance of non-utilitarian architecture. *Journal of Anthropological Archaeology* 18:356–75.

Araujo, G. M., A. F. Amaral, E. M. Bruna, and H. L. Vasconcelos. 2013. Fire drives the reproductive responses of herbaceous plants in a Neotropical swamp. *Plant Ecology* 214:1479–84.

Archer, A. 1948. *Land Snails of the Genus* Stenotrema *in the Alabama Region*. Museum Paper 28. Tuscaloosa: Geological Survey of Alabama, University of Alabama.

Archer, S., D. S. Schimel, and E. A. Holland. 1995. Mechanisms of shrubland expansion: Land use, climate or CO_2? *Climate Change* 29:91–99.

Arkansas Geological Survey. 2012. Stratigraphic summary of the Mississippi Embayment and Gulf Coastal Plain. Accessed May 13, 2013. http://www.geology.ar.gov/geology/strat_missembayment_gulfplain.htm.

Arkansas Land Commissioner. 1999. *Arkansas: General Land Office Survey Notes 1815–Present*. 15 CD set. Little Rock: Arkansas Land Commissioner's Office.

Arkansas Natural Heritage Commission. 2013. Prairies, glades and barrens. Accessed November 6, 2013. http://www.naturalheritage.com.

Arkansas State Bureau of Immigration. 1888. *Facts concerning Arkansas*. Little Rock, AR: Woodruff Printing.

Arner, D. H. 1979. The use of fire in right-of-way maintenance. *Journal of Arboriculture* 5:93–96.

Arner, D. H., L. E. Cliburn, D. R. Thomas, and J. D. Manner. 1976. Use of fire, fertilizer and seed for rights-of-way maintenance in the southeastern United States. In *Proceedings of the First National Symposium on Environmental Concerns in Rights-of-Way Management, January 6–8, 1976*, edited by R. Tillman, 156–65. Mississippi State University.

Arner, D. H., and J. C. Jones. 2009. *Wildlife Habitat Management for Special Use Areas*. Mississippi State University Forest and Wildlife Research Center, and Mississippi Department of Wildlife, Fisheries, and Parks.

Arnold, M. S. 1991. *Colonial Arkansas, 1686–1804*. Fayetteville: University of Arkansas Press.

Asbjornsen, H., L. A. Brudvig, C. M. Mabry, C. W. Evans, and H. M. Karnitz. 2005. Defining reference information for restoring ecologically rare tallgrass oak savannas in the midwestern United States. *Journal of Forestry* 103:335–50.

Ashe, T. 1808. *Travels in America Performed in 1806*. London: William Sawyer.

Atkinson, J. R. 1979. A historic contact Indian settlement in Oktibbeha County, Mississippi. *Journal of Alabama Archaeology* 25:61–82.

———. 1987. The DeSoto Expedition through north Mississippi in 1540–41. *Mississippi Archaeology* 22:61–73.

———. 2004. *Splendid Land, Splendid People: The Chickasaw Indians to Removal*. Tuscaloosa: University of Alabama Press.

Atwill, L. 1992. Bow camp. *Field and Stream* 97:37–39.

Aviron, S., F. Herzog, I. Klaus, B. Schupbach, and P. Jeanneret. 2010. Effects of wildflower strip quality, quantity, and connectivity on butterfly diversity in a Swiss arable landscape. *Restoration Ecology* 19:500–508.

Bachand, R. R. 2001. *The American Prairie: Going, Going, Gone?* Boulder, CO: National Wildlife Federation, Rocky Mountain Natural Resource Center.

Ball, G. H. 1922 Variation in fresh-water mussels. *Ecology* 3:93–121.

Ball, J. 2000. The versatile Osage-orange. *American Forests* 106:60–63.

Ball, P. W., and A. A. Reznicek. 2002. *Carex*. *Flora of North America* 23:254–333.

Barbour, M. G., J. H. Burk, and W. D. Pitts. 1980. *Terrestrial Plant Ecology*. Menlo Park, CA: Benjamin/Cummings.

Barbour, P. J. 2006. Ecological and economic effects of field borders in row crop agriculture production systems in Mississippi. PhD diss., Mississippi State University.

Barden, L. S. 1997. Historic prairies in the Piedmont of North and South Carolina. *Natural Areas Journal* 17:149–52.

Barlow, C. 2001a. Anachronistic fruits and the ghosts who haunt them. *Arnoldia* 61:14–21.

———. 2001b. Ghost stories from the Ice Age. *Natural History* 110:62–67.

Barone, J. A. 2005a. The historical distribution of prairies in the Jackson Prairie belt and in western Mississippi. *Journal of the Mississippi Academy of Sciences* 50:144–48.

———. 2005b. Historical presence and distribution of prairies in the Black Belt of Mississippi and Alabama. *Castanea* 70:170–83.

Barone, J. A., J. W. Beck, M. B. Potter, S. R. Sneed, K. E. Stephenson, and E. J. Dollar. 2008. Distribution of canebrakes in 19th century Alabama. *Journal of the Alabama Academy of Science* 79:1–11.

Barone, J. A., and J. G. Hill. 2007. Herbaceous flora of blackland prairie remnants in Mississippi and western Alabama. *Castanea* 72: 226–34.

———. 2008. *Identification and Delineation of Black Belt Prairie Remnants along the Natchez Trace Parkway*. Report for the National Park Service, available at the library of the Mississippi Entomological Museum.

Barras, S. C., M. S. Carrara, R. A. Dolbeer, R. B. Chipman, and G. E. Bernhardt. 2000. Bird and small mammal use of mowed and unmowed vegetation at John F. Kennedy International Airport, 1998–1999. In *Proceedings of the Nineteenth Vertebrate Pest Conference, March 6–9, 2000, San Diego, California, USA*, 31–36.

Bartram, W. 1791. *Travels of William Bartram*. Edited by F. Harper. Reprint, New Haven, CT: Yale University Press, 1958.

Basinger, M. A., J. S. Huston, R. J. Gates, and P. A. Robertson. 1997. Vascular flora of Horseshoe Lake Conservation Area, Alexander County, Illinois. *Castanea* 62:82–99.

Batalha, M. A., M. V. Cianciaruso, I. A. Silva, and W. B. C. Delitti. 2005. Hyperseasonal cerrado, a new Brazilian vegetation form. *Brazilian Journal of Biology* 65:735–38.

Bates, J. D., R. F. Miller, and T. J. Svejcar. 2000. Understory dynamics in cut and uncut western juniper woodlands. *Journal of Range Management* 53:119–26.

Battaglia, M. A., R. J. Mitchell, P. P. Mou, and S. D. Pecot. 2003. Light transmittance estimates in a longleaf pine woodland. *Forest Science* 49:752–62.

Beckage, B., W. J. Platt, and L. J. Gross. 2009. Vegetation, fire, and feedbacks: A disturbance mediated model of savannas. *American Naturalist* 174:805–18.

Beckett, S., and M. S. Golden. 1982. Forest vegetation and vascular flora of Reed Brake Research Natural Area, Alabama. *Castanea* 47:368–92.

Belovsky, G. E., and J. B. Slade. 2000. Insect herbivory accelerates nutrient cycling and increases plant production. *Proceedings of the National Academy of Science* 97:14412–17.

Belue, T. F. 1996. *The Long Hunt: Death of the Buffalo East of the Mississippi*. Mechanicsburg, PA: Stackpole Books.

Benson, T. J., J. J. Dinsmore, and W. L. Hohman. 2007. Responses of plants and arthropods to burning and disking of riparian habitats. *Journal of Wildlife Management* 71:1949–57.

Berry, T., P. Beasley, and J. Clements. 2006. *Forgotten Expedition, 1804–1805: The Louisiana Purchase Journals of Dunbar and Hunter*. Baton Rouge: Louisiana State University Press.

Bettis, E. A., D. R. Muhs, H. M. Roberts, and A. G. Wintle. 2003. Last glacial loess in the conterminous USA. *Quaternary Science Reviews* 3:1907–46.

Betts, H. S. 1929. *The Strength of North American Woods*. Miscellaneous Publications No. 46. Washington, DC: US Department of Agriculture.

Bezanson, D. 2000. Natural vegetation types of Texas and their representation in conservation areas. University of Texas at Austin. Accessed May 2014. http://www.abisw.org/bezanson/.

Bierly, S. 2008. Terminal occupation community patterns at Lyon's Bluff (22OK520) in Oktibbeha County, Mississippi: Sedimentological, molluscan, artifactual, and geophysical evidence. Master's thesis, Mississippi State University.

Bisset, N. 2004. Restoration of a dry prairie: Methods and examples. In *Proceedings of the Florida Dry Prairie Conference*, edited by R. F. Noss and S. Singh. Sebring: Florida Center for Environmental Studies.

Blackwell, W. H., M. J. Powell, and G. Dukes. 1983. Fossil wood from Bayou Pierre and White Oak Creek, southwest-central Mississippi. *Mississippi Geology* 4:1–6.

Bohlen, P. J., S. Lynch, L. Shabman, M. Clark, S. Shukla, and H. Swain. 2009. Paying for environmental services from agricultural lands: An example from the northern Everglades. *Frontiers in Ecology and the Environment* 7:46–55.

Bond, T. B, C. D. Baumann, M. W. Lane, R. E. Thackston, and J. L. Bowman. 2005. Efficacy of herbicides to control Bermudagrass for enhancement of northern bobwhite habitat. *Proceedings of the Annual Conference of the Southeast Association of Fish and Wildland Agencies* 59:191–99.

Bond, W. J., and J. J. Midgely. 1995. Kill thy neighbour: An individualistic argument for the evolution of flammability. *Oikos* 73:79–85.

Bond, W. J., and B. van Wilgen. 1996. *Fire and Plants*. London: Chapman and Hall.

Boone County Historical and Railroad Society. 1998. *History of Boone County, Arkansas*. Paducah, KY: Turner.

Borsari, B. 2003. The coastal prairie habitat as a model to promote sustainability in the agricultural curriculum. In *Proceedings of the 18th North American Prairie Conference*, edited by S. Foré, 17–23. Kirksville, MO: Truman State University Press.

———. 2005. *Prairie as a Model for Sustainable Agriculture and the Paradox of Soil Fertility Management in Modern Farming Systems*. La Crosse, WI: USGS Upper Midwest Research Center.

———. 2011. Prairie garden of troubles. Accessed May 9, 2012. http://sciencecases.lib.buffalo.edu/cs/collection/detail.asp?case_id=570&id=570.

———. 2012. Curriculum framework for sustainability education. *Academic Exchange Quarterly* 16:74–78.

Borsari, B., K. Graetz, and J. Bernard. 2010. Learning about prairie restoration ecology in large, introductory biology courses. In *Proceedings of the 21st North American Prairie Conference*, edited by B. Borsari, N. Mundahl, L. Reuter, E. Peters, and P. Cochran, 199–204. Winona, MN: Winona State University.

Borsari, B., D. Johnson, W. Taylor, V. Kefeli, and S. Doherty. 2006. Conservation efforts and natural history of a prairie habitat at Jennings Environmental Education Center, Pennsylvania. In *Proceedings of the 19th North American Prairie Conference*, edited by D. Egan and J. A. Harrington, 149–57. Madison: University of Wisconsin–Madison.

Borsari, B., and M. Vidrine. 2005. Agriculture curricula in sustainability: An evolution across borders. *Journal of Sustainable Agriculture* 25:93–112.

Borsari, B., M. F. Vidrine, B. N. Mundahl, and M. Pastorek. 2012. The significance of micro-prairie reconstruction in urban environments. *Proceedings of the 23rd North American Prairie Conference* 23:70–77.

Boughton, E. H., P. J. Bohlen, and C. Steele. 2013. Season of fire and nutrient enrichment affect plant community dynamics in subtropical semi-natural grasslands released from agriculture. *Biological Conservation* 158:239–47.

Boughton, E. H., and R. K. Boughton. 2014. Modification by an invasive ecosystem engineer shifts a wet prairie to a monotypic stand. *Biological Invasions* 16:2105–14.

Boughton, E. H., P. F. Quintana-Ascencio, P. J. Bohlen, D. G. Jenkins, and R. Pickert. 2010. Land-use and isolation interact to affect wetland plant assemblages. *Ecography* 33: 461–70.

Bowles, M. L., and M. D. Jones. 2013. Repeated burning of eastern tallgrass prairies increases richness and diversity, stabilizing late successional vegetation. *Ecological Applications* 23:464–78.

Bradstock, R. A., M. Bedward, A. M. Gill, and J. S. Cohn. 2005. Which mosaic? A landscape ecological approach for evaluating interactions between fire regimes, habitat and animals. *Wildlife Research* 32:409–23.

Bragg, D. C. 2003. Natural presettlement features of the Ashley County, Arkansas area. *American Midland Naturalist* 149:1–20.

Bragg, D. C., and R. C. Weih Jr. 2007. Notable environmental features in some historical aerial photographs from Ashley County, Arkansas. *Journal of the Arkansas Academy of Science* 61:27–36.

Branson, D. H., A. Joern, and G. A. Sword. 2006. Sustainable management of insect herbivores in grassland ecosystems: New perspectives in grasshopper control. *Bioscience* 56:1–13.

Brennan, L. A. 1991. How can we reverse the northern bobwhite population decline? *Wildlife Society Bulletin* 19: 544–55.

Brennan, L. A., and W. P. Kuvlesky. 2005. North American grassland birds: An unfolding conservation crisis? *Journal of Wildlife Management* 69:1–13.

Brent, F. V. 1973. *Soil Survey of Oktibbeha County, Mississippi*. Washington, DC: Government Printing Office.

Brewer, J. S. 2011. Per capita community-level effects of an invasive grass, *Microstegium vimineum*, on vegetation in mesic forests in northern Mississippi (USA). *Biological Invasions* 3:701–15.

Bridges, E. L. 2006a. Landscape ecology of Florida dry prairie in the Kissimmee River region. In *Land of Fire and Water: The Florida Dry Prairie Ecosystem*, edited by R. F. Noss, 14–42. DeLeon Springs, FL: E. O. Painter.

———. 2006b. Historical accounts of vegetation in the Kissimmee River dry prairie landscape. In *Land of Fire and Water: The Florida Dry Prairie Ecosystem*, edited by R. F. Noss, 43–63. DeLeon Springs, FL: E. O. Painter.

Bridges, E. L., and S. L. Orzell. 1989. Longleaf pine communities of the West Gulf Coastal Plain. *Natural Areas Journal* 9:246–63.

———. 2005. *Vegetation Monitoring and Sampling Protocol for USDA-NRCS Wetland Reserve Program Sites at MacArthur Agro-Ecology Research Center, Highlands County, Florida*. Report submitted to Archbold Expeditions. FL.

———. 2013. *Vegetation Monitoring for USDA-NRCS Wetland Reserve Program Sites at MacArthur Agro-Ecology Research Center, Highlands County, Florida*. Report submitted to Archbold Expeditions, FL.

Briggs, J. M., G. A. Hoch, and L. C. Johnson. 2002a. Assessing the rate, mechanisms, and consequences of the conversion of tallgrass prairie to *Juniperus virginiana* forest. *Ecosystems* 5:578–86.

Briggs, J. M., and A. K. Knapp. 2001. Determinants of C-3 forb growth and production in a C-4 dominated grassland. *Plant Ecology* 152:93–100.

Briggs, J. M., A. K. Knapp, J. M. Blair, J. L. Heisler, G. A. Hoch, M. S. Lett, and J. K. McCarron. 2005. An ecosystem in transition: Causes and consequences of the conversion of mesic grassland to shrubland. *BioScience* 55:243–54.

Briggs, J. M., A. K. Knapp, and B. L. Brock. 2002b. Expansion of woody plants in tallgrass prairie: A fifteen year study of fire and fire-grazing interactions. *American Midland Naturalist* 147:287–94.

Brock, S. C., D. H. Arner, and D. E. Steffen. 1994. Seed yields of four moist-soil plants on Noxubee National Wildlife Refuge. *Proceedings of the Annual Conference of the Southeast Association of Fish and Wildlife Agencies* 48:38–47.

Brockway, D. G., K. W. Outcalt, B. L. Estes, and R. B. Rummer. 2009. Vegetation response to midstorey mulching and prescribed burning for wildfire hazard reduction and longleaf pine (*Pinus palustris* Mill.) ecosystem restoration. *Forestry* 82:299–314.

Brose, P., and D. D. Wade. 2002. Potential fire behavior in pine flatwood forests following three different fuel reduction techniques. *Forest Ecology and Management* 163: 71–84.

Brouillet, L., J. C. Semple, G. A. Allen, K. L. Chambers, and S. D. Sundberb. 2006. *Symphyotrichum*. *Flora of North America* 20:458–505.

Brown, J. R., and J. Carter. 1998. Spatial and temporal patterns of exotic shrub invasion in an Australian tropical grassland. *Landscape Ecology* 13:93–102.

Brown, L. E., K. Hillhouse, B. R. MacRoberts, and M. H. MacRoberts. 2002. The vascular flora of Windham Prairie, Polk County, east Texas. *Texas Journal of Science* 54: 227–40.

Brown, R. L. 2003. Paleoenvironment and biogeography of the Mississippi Black Belt. In *Blackland Prairies of the Gulf Coastal Plain: Nature, Culture, and Sustainability*, edited by E. Peacock and T. Schauwecker, 11–26. Tuscaloosa: University of Alabama Press.

Brown, R. N., and C. D. Sawyer. 2012. Plant species diversity of highway roadsides in southern New England. *Northeastern Naturalist* 19:25–42.

Brudvig, L. A., C. M. Mabry, J. R. Miller, and T. A. Walker. 2007. Evaluation of central North American prairie management based on species diversity, life form and individual species metrics. *Conservation Biology* 21:864–74.

Brye, K. R., and L. Moreno. 2006. Vegetation removal effects on soil quality in a native tallgrass prairie fragment in east-central Arkansas. *Natural Areas Journal* 26:94–100.

Brye, K. R., and C. P. West. 2004. Soil quality differences under native tallgrass prairie across a climosequence in Arkansas. *American Midland Naturalist* 152:214–30.

Buckner, J. L., and J. L. Landers. 1979. Fire and disking effects on herbaceous food plants and seed supplies. *Journal of Wildlife Management* 43:807–11.

Bugg, R. L, C. S. Brown, and J. H. Anderson. 1997. Restoring native perennial grasses to rural roadsides in Sacramento Valley of California: Establishment and evaluation. *Restoration Ecology* 5:214–28.

Burch, J. B. 1955. Some ecological factors of the soil affecting the distribution and abundance of land snails in eastern Virginia. *Nautilus* 69:62–69.

———. 1962. *How to Know the Eastern Land Snails*. Dubuque, IA: Wm. C. Brown.

Burke, M. J. W., and J. P. Grime. 1996. An experimental study of plant community invasibility. *Ecology* 77:776–90.

Burkhart, J. K. 2004. Vegetation response in field margins managed for northern bobwhite (*Colinus virginianus*) and negative impacts of Bermudagrass (*Cynodon dactylon*). Master's thesis, University of Georgia.

Bureau of Land Management General Land Office Records. Accessed May 20, 2013. http://www.glorecords.blm.gov.

Burger, L. W., D. McKenzie, R. Thackston, and S. J. DeMaso. 2006. The role of farm policy in achieving large-scale conservation: Bobwhite and buffers. *Wildlife Society Bulletin* 34:986–93.

Burgess, J. L., and C. W. Ely. 1907. *Soil Survey of Conway County, Arkansas*. Washington, DC: Field Operations of the Bureau of Soils.

Burnham, K. P., and D. R. Anderson. 2002. *Model Selection and Multimodel Inference: A Practical Information-Theoretic Approach*. 2nd ed. New York: Springer-Verlag.

Burton, J. D. 1973. *Osage-Orange*. Publication no. 0101–00315. Washington, DC: US Department of Agriculture.

Bush, J. K., and O. W. Van Auken. 1989. Soil resource levels and competition between a woody and herbaceous species. *Bulletin of the Torrey Botanical Club* 116:22–30.

Bushnell, D., Jr. 1927. *Drawings by A. DeBatz in Louisiana,*

1732–1735. Smithsonian Miscellaneous Collections 80, no. 5. Washington, DC: Smithsonian Institution.

Butcher, G. S., and D. K. Niven. 2007. Combining data from the Christmas Bird Count and the Breeding Bird Survey to determine the continental status and trends of North American birds. National Audubon Society, New York. Accessed August 2012. http://www.audubon.org/bird/stateofthebirds/CBID/content/Report.pdf.

Butler, T. J., and J. P. Muir. 2006. Coastal Bermudagrass (*Cynodon dactylon*) yield response to various herbicides. *Weed Technology* 20:95–100.

Cackowski, J. M., and J. L. Nasar. 2003. The restorative effects of roadside vegetation: Implications for automobile driver anger and frustration. *Environment and Behavior* 35: 736–51.

Caldwell, J. E. 2007. Clickers in the large classroom: Current research and best-practice tips. *CBE-Life Science Education* 6:9–20.

Call, R. E. 1891. *Annual Report of Geological Survey of Arkansas for 1889*. Vol. 2, *Geology of Crowley's Ridge*. Little Rock, AR: Woodruff.

Campbell, J. J. N., and W. R. Seymour Jr. 2011a. A review of the native vegetation types in the Black Belt of Mississippi and Alabama, with the suggested relationships to the catenas of soil series. *Journal of the Mississippi Academy of Sciences* 56:166–84.

———. 2011b. The vegetation of Pulliam Prairie, Chickasaw County, Mississippi: A significant remnant of pre-Columbian landscape in the Black Belt. *Journal of the Mississippi Academy of Sciences* 56:248–63.

———. 2012. The flora of Pulliam Prairie, Chickasaw County, Mississippi: A significant remnant of native vegetation in the Black Belt region. *Journal of the Mississippi Academy of Sciences* 57:160–95.

———. 2015. Botanical notes on the Black Belt region. III. http://bluegrasswoodland.com/Mississippi_Black_Belt.html.

Canham, C. D., A. C. Finzi, S. W. Pacala, and D. H. Burbank. 1994. Causes and consequences of resource heterogeneity in forests: Interspecific variation in light transmission by canopy trees. *Canadian Journal of Forest Research* 24:337–49.

Carr, S. C., K. M. Robertson, W. J. Platt, and R. K. Peet. 2009. A model of geographical, environmental and regional variation in vegetation composition of pyrogenic grasslands of Florida. *Journal of Biogeography* 36:1600–1612.

Carr, W. R. 2009. *An Annotated List of Peripheral and Disjunct Plant Species of Texas*. Austin: Nature Conservancy of Texas.

Chamberlain, D. E., R. J. Fuller, R. G. H. Bunce, J. C. Duckworth, and M. Shrubb. 2000. Changes in the abundance of farmland birds in relation to the timing of agricultural intensification in England and Wales. *Journal of Applied Ecology* 37:771–88.

Chase, J. M., M. A. Leibold, A. L. Downing, and J. B. Shurin. 2000. The effects of productivity, herbivory, and plant species turnover in grassland food webs. *Ecology* 81:2485–97.

Cherrill, A. J., and M. Begon. 1989. Predation on grasshoppers by spiders in sand dune grasslands. *Entomologia Experimentalis et Applicata* 50:225–31.

Chester, E. W., B. E. Wofford, D. Estes, and C. Bailey. 2009. *A Fifth Checklist of Tennessee Vascular Plants*. Fort Worth: Botanical Research Institute of Texas.

Cianciaruso, M. V., M. A. Batalha, and I. A. Silva. 2005. Seasonal variation of a hyperseasonal cerrado in Emas National Park, central Brazil. *Flora* 200:345–53.

Clark, H. L., G. J. Haley, E. J. Hebert, R. M. Hollier, and A. J. Roy. 1962. *Soil Survey of Acadia Parish, Louisiana*. US Department of Agriculture Soil Conservation Service in cooperation with Louisiana Agricultural Experimental Station.

Clark, J. W., Jr. 1969. Implications of land and fresh-water gastropods in archaeological sites. *Arkansas Academy of Science Proceedings* 23:38–52.

Clayton, L. A., V. J. Knight Jr., and E. C. Moore. 1993. *De Soto Chronicles: The Expedition of Hernando de Soto to North America in 1539–1543*. Vols. 1–2. Tuscaloosa: University of Alabama Press.

Collingwood, G. H. 1939. Osage-orange. *American Forests* 45:508–10.

Collins, O. B., F. E. Smeins, and D. H. Riskind. 1975. Plant communities of the blackland prairie of Texas. In *Prairie: A Multiple View*, edited by M. K. Wali, 75–88. Grand Forks: University of North Dakota Press.

Collins, S. L., and E. M. Steinauer. 1998. Disturbance, diversity and species interactions in tallgrass prairie. In *Grassland Dynamics: Long-Term Ecological Research in Tallgrass Prairie*, edited by A. K. Knapp, J. M. Briggs, D. C. Hartnett, and S. L. Collins, 140–58. New York: Oxford University Press.

Collins, S. L., and L. L. Wallace. 1990. *Fire in North American Tallgrass Prairies*. Norman: University of Oklahoma Press.

Cook, R. E., and J. C. Semple. 2004. A new name and a new combination in *Solidago* subsect. *Glomeruliflorae* (Asteraceae: Astereae). *Sida, Contributions to Botany* 21:221–24.

Cook, S. R. 2012. The Chickasaw Nation. Accessed June 2015. http://www.thechickasawvillages.com/ [containing several papers, review of literature and of unpublished original sources].

Cooke, C. W. 1939. Equivalence of the Gosport Sand to the Moodys Marl. *Journal of Paleontology* 13:337–40.

Correll, D. S., and M. C. Johnston. 1979. *Manual of the Vascular Plants of Texas*. Renner: Texas Research Foundation.

Crockett, D. 1987. *A Narrative of the Life of David Crockett of the State of Tennessee*. Lincoln, NE: Bison Books.

Crouch, V. E., and M. S. Golden. 1997. Floristics of a bottomland forest and adjacent uplands near the Tombigbee River, Choctaw County, Alabama. *Castanea* 62:219–38.

Cuming, F. 1810. *Cuming's Tour to the Western Country, 1807–1809*. Reprint, Carlisle, MA: Applewood Books, 2007.

Daar, S. 1994. Integrated approaches to roadside vegetation management. *IPM Practitioner* 16:1–8.

Dana, M. N., R. D. Kemery, and B. S. Boszor. 1996. *Wildflowers for Indiana Highways*. Report Publication FHWA/IN/JHRP-96/01.Washington, DC: US Department of Transportation, Federal Highway Administration.

———. 1997. *Wildflowers for Indiana Highways*. Joint Transportation Research Program, Paper 227. Purdue Libraries, Purdue e-Pubs Civil Engineering. https://docs.lib.purdue.edu/jtrp/227/.

Darby, W. 1818. *The Emigrant's Guide to the Western and Southwestern States and Territories*. New York: Kirk and Mercein.

Daubenmire, R. 1959. A canopy-coverage method of vegetational analysis. *Northwest Scientist* 33:43–64.

———. 1968. Ecology of fire in grasslands. *Advances in Ecological Research* 5:209–66.

Davis, J. D., S. D. Hendrix, D. M. Debinski, and C. J. Hemsley. 2008. Butterfly, bee and forb community composition and cross-taxon incongruence in tallgrass prairie fragments. *Journal of Insect Conservation* 12:69–79.

Davis, R. A. 1966. Revision of Lower Ordovician nomenclature in the upper Mississippi valley. *Journal of Geology* 74:361–65.

Davros, N. M., D. M. Debinski, K. F. Reeder, and W. L. Hohman. 2006. Butterflies and Continuous Conservation Reserve Program filter strips: Landscape considerations. *Wildlife Society Bulletin* 34:936–43.

DeBano, L. F., D. G. Neary, and P. F. Folliott. 1998. *Fire's Effects on Ecosystems*. New York: Wiley.

Denley, K. K., C. T. Bryson, and R. A. Stewart. 2002. Vascular flora of Yalobusha County, Mississippi. *Castanea* 67:402–15.

de Richarville, C. 1739. Report on the accounts furnished by Sr. de Richarville of the war against the Chickasaws. *Archives des Colonies*, Series C13C 4:202–5.

DeSelm, H., and N. Murdock. 1993. Grass dominated communities. In *Biodiversity of the Southeastern United States: Upland Terrestrial Communities*, edited by W. H. Martin, S. G. Boyce, and A. C. Echternacht, 87–142. New York: John Wiley and Sons.

Deyrup, M. 1996. Two new grasshoppers from relict uplands of Florida (Orthoptera: Acrididae). *Transactions of the American Entomological Society* 122:199–211.

Diamond, D. 1980. Remnant plant communities of the Fayette Prairie, Texas. Master's thesis, Texas A&M University.

———. 1993. Classification of the plant communities of Texas (series level). Unpublished manuscript. Austin: Texas Natural Heritage Program.

Diamond, D. D., and F. E. Smeins. 1993. Native plant communities of the blackland prairie. In *The Texas Blackland Prairie: Land, History, and Culture*, edited by M. R. Sharpless and J. C. Yelderman, 66–81. Waco, TX: Baylor University Program for Regional Studies.

Diboll, N. 1984. Mowing as an alternative to burning for control of cool season exotic grass in prairie grass plantings. In *Proceedings of the 9th North American Prairie Conference*. Moorhead, MN: Tri-College University Center for Environmental Studies.

Dickson, J. G., and T. B. Wigley. 2001. Managing forests for wildlife. In *Wildlife of Southern Forests: Habitat and Management*, edited by J. G. Dickson, 83–94. Blaine, WA: Hancock House.

Diggs, G. M., B. L. Lipscomb, and R. J. O'Kennon. 1999. *Shinners & Mahler's Illustrated Flora of North Central Texas*. Sida, Botanical Miscellany 16. Fort Worth: Botanical Research Institute of Texas.

Diggs, G. M., B. L. Lipscomb, M. D. Reed, and R. J. O'Kennon. 2006. *Illustrated Flora of East Texas*. Sida, Botanical Miscellany 26.

Diggs, G. M., and P. C. Schulze. 2003. Soil-dependent fire frequency: A new hypothesis for the distribution of prairies and woodlands in north central and east Texas. *Sida* 20:1139–53.

Dimbleby, G. 1967. *Plants and Archaeology*. London: Unwin Brothers.

Dixon, K. R., G. A. Feldhamer, J. E. Gates, and D. A. Harman. 1984. Activity of white-tailed deer along an interstate highway right-of-way. In *Proceedings of the Third International Symposium on Environmental Concerns in Rights-of-Way Management, February 15–18, 1982, San Diego,*

California, USA, edited by E. W. Colson and R. E. Tillman, 602–5. Mississippi State University.

Dollar, J. G. 2011. Responses of butterfly and forb communities to management of semi-natural grassland buffers. PhD diss., Mississippi State University.

Dollar, J. G., S. K. Riffell, and L. W. Burger Jr. 2013. Effects of managing semi-natural grassland buffers on butterflies. *Journal of Insect Conservation* 17:577–90.

Doxon, E. D., and J. P. Carroll. 2007. Vegetative and invertebrate community characteristics of Conservation Reserve Program fields relative to gamebirds in western Kansas. *American Midland Naturalist* 158:243–59.

Driscoll, D. A., D. B. Lindenmayer, A. F. Bennett, M. Bode, R. A. Bradstock, G. J. Cary, M. F. Clarke, et al. 2010. Fire management for biodiversity conservation: Key research questions and our capacity to answer them. *Biological Conservation* 143:1928–39.

Duncan, D. 2005. *Clickers in the Classroom: How to Enhance Science Teaching Using Classroom Response Systems.* New York: Addison Wesley and Benjamin Cummings.

Dye, D. H. 1993. Reconstruction of the de Soto expedition route in Arkanasas: The Mississippi Alluvial Plain. In *Expedition of Hernando de Soto West of the Mississippi, 1541–1543*, edited by G. A. Young and M. P. Hoffman, 36–57. Fayetteville: University of Arkansas Press.

Eades, D. C., D. Otte, M. M. Cigliano, and H. Braun. 2013. Orthoptera species file. Version 5.0/5.0. Accessed February 2013. http://Orthoptera.SpeciesFile.org.

Early, A. M. 2000. Winding Stair site. In *Forest Farmsteads: A Millennium of Human Occupation at Winding Stair in the Ouachita Mountains*, edited by A. M. Early, 69–92. Arkansas Archeological Survey Research Series 571. Fayetteville: Arkansas Archeological Survey.

Echols, S. L. 2007. Vascular flora of the blackland prairies and associated vegetation of central Georgia. Master's thesis, University of Georgia.

Echols, S. L., and W. Zomlefer. 2010. Vascular plant flora of the remnant blackland prairies in Oaky Woods Wildlife Management Area, Houston County, Georgia. *Castanea* 75:78–100.

Eckenwalder, J. E. 2009. *Ebenaceae. Flora of North America* 8:247–50.

Edmonds, M. 2001. Pleasures and pitfalls of written records. In *Historical Ecology Handbook: A Restorationist's Guide to Reference Ecosystems*, edited by D. Egan and E. A. Howell, 73–99. Washington, DC: Island Press.

Edwards, D. E., Jr. 1954. The physiography of Hinds County, Mississippi. Master's thesis, Mississippi State University.

Eldredge, I. F. 1911. Fire problems of the Florida National Forest. *Proceedings of the Society of American Foresters* 6:66–170.

Elsen, D., and R. Wieland. 2003. Rediscovery and management of prairie remnants of the Bienville National Forest, east-central Mississippi. In *Blackland Prairies of the Gulf Coastal Plain: Nature, Culture, and Sustainability*, edited by E. Peacock and T. Schauwecker, 239–45. Tuscaloosa: University of Alabama Press.

Emery, S. M., J. Uwimbabazi, and S. L. Flory. 2011. Fire intensity effects on seed germination of native and invasive eastern deciduous forest understory plants. *Forest Ecology and Management* 261:1401–8.

Emslie, S., and N. J. Czaplewski. 1999. Two new fossil eagles from the late Pliocene (late Blancan) of Florida and Arizona and their biogeographic implications. *Smithsonian Contributions to Paleobiology* 89:185–98.

Engber, E. A., and J. M. Varner. 2012. Reversing conifer encroachment with prescribed fire: Shifting mortality models toward restoration targets. *Restoration Ecology* 20:665–68.

Engle, D. M., and J. F. Stritzke. 1995. Fire behavior and fire effects on eastern redcedar in hardwood leaf-litter fires. *International Journal of Wildland Fire* 5:135–41.

Ervin, G. N., L. C. Majure, and J. T. Bried. 2006. Influence of long-term greentree reservoir impoundment on stand structure, species composition, and hydrophytic indicators. *Journal of the Torrey Botanical Society* 133:468–81.

Evans, E. W. 1984. Fire as a natural disturbance to grasshopper assemblages of tallgrass prairie. *Oikos* 43:9–16.

Evans, E. W., R. A. Rogers, and D. J. Opfermann. 1983. Sampling grasshoppers (Orthoptera: Acrididae) on burned and unburned tallgrass prairie: Night trapping vs. sweeping. *Environmental Entomology* 12:1449–54.

Everglades Headwaters National Wildlife Refuge proposal. 2010. Vero Beach, FL: Southeast Region, US Fish and Wildlife Service, Department of the Interior.

Eversull, L. G. 2005. The Twiggs Clay: Mineralogy, origin, and industrial properties of an Upper Eocene opaline claystone in the Coastal Plain Province of Georgia, U.S. PhD diss., Louisiana State University.

Faulkner, S., W. Barrow, B. Keeland, S. Walls, and D. Telesco. 2011. Effects of conservation practices on wetland ecosystem services in the Mississippi Alluvial Valley. *Ecological Applications* 21:S31–S48.

Featherstonhaugh, G. W. 1835. *Geological Report of an Examination Made in 1834 of the Elevated Country between the Missouri and Red Rivers.* Washington, DC: Gales and Seaton.

Feber, R. E., H. Smith, and D. W. MacDonald. 1996. The effects on butterfly abundance of the management of uncropped edges of arable fields. *Journal of Applied Ecology* 33:1191–205.

Fernald, M. L. 1950. *Gray's Manual of Botany*. 8th ed. New York: Van Nostrand.

Field, R. G., T. Gardiner, C. F. Mason, and J. Hill. 2005. Agri-environment schemes and butterflies: The utilization of 6 m grass margins. *Biodiversity and Conservation* 14:1969–76.

Fill, J. M., S. M. Welch, J. L. Waldron, and T. A. Mousseau. 2012. The reproductive response of an endemic bunchgrass indicates historical timing of a keystone process. *Ecosphere* 3:1–12.

Fitz, G. 1832. *Instructions for Surveying in the State of Mississippi*. R. Semple for the Surveyor General, Natchez, MS.

Flora of North America Editorial Committee. 1993–2010 (continuing). *Flora of North America North of Mexico*. New York: Oxford University Press [see also http://www.fna.org/].

Flora of North America Editorial Committee. 2002. *Flora of North America*. Vol. 23, *Magnoliophyta: Commelinidae (in Part): Cyperaceae*. New York: Oxford University Press.

Flores, D. L. 1984. *Southern Counterpart to Lewis and Clark*. Norman: University of Oklahoma Press.

Flynn, D. F. B., M. Gogol-Prokurat, T. Nogeire, N. Molinari, B. T. Richers, B. B. Lin, N. Simpson, et al. 2009. Loss of functional diversity under land use intensification across multiple taxa. *Ecology Letters* 12:22–33.

Folkerts, G. W. 1982. The Gulf Coast pitcher plant bogs [Mississippi, Alabama]. *American Scientist* 70:260–67.

Folkerts, G. W., M. A. Deyrup, and D. C. Sisson 1993. Arthropods associated with xeric longleaf pine habitats of the southeastern United States: A brief overview. In *The Longleaf Pine Ecosystem: Ecology, Restoration, and Management*, edited by S. M. Herman, 159–62. Proceedings of the Tall Timbers Fire Ecology Conference No.18. Tallahassee, FL.

Fonda, R. W. 2001. Burning characteristics of needles from eight pine species. *Forest Science* 47:390–96.

Fonda, R. W., L. A. Bellanger, and L. L. Burley. 1998. Burning characteristics of western conifer needles. *Northwest Science* 72:1–9.

Forbes, S. P. 1999. Community properties along biomass gradients in blackland prairies. Master's thesis, Mississippi State University.

Forman, R. T. T., and L. E. Alexander. 1998. Roads and their major ecological effects. *Annual Review of Ecology, Evolution, and Systematics* 29:207–31.

Forman, R. T. T., and R. D. Deblinger. 2000. Ecological road effect zone of a Massachusetts (USA) suburban highway. *Conservation Biology* 14:36–46.

Foster, B. L., K. Kindscher, G. R. Houseman, and C. A. Murphy. 2009. Effects of hay management and native species sowing on grassland community structure, biomass, and restoration. *Ecological Applications* 19:1884–96.

Foster, W. C. 1998. *La Salle Expedition to Texas: The Journal of Henri Joutel, 1684–1687*. Austin: Texas State Historical Association.

Foti, T. 1989. Blackland prairies of southwestern Arkansas. *Journal of the Arkansas Academy of Science* 43:23–28.

———. 1990. The vegetation of Saratoga Landing blackland prairie. *Journal of the Arkansas Academy of Science* 44: 40–43.

Foti, T., S. Simon, D. Zollner, and M. Hattenbach. 2003. Blackland prairie landscapes of southwestern Arkansas. In *Blackland Prairies of the Gulf Coastal Plain: Nature, Culture, and Sustainability*, edited by E. Peacock and T. Schauwecker, 94–109. Tuscaloosa: University of Alabama Press.

Fournié, J., and M. Chétail. 1984. Calcium dynamics in land gastropods. *American Zoologist* 24:857–70.

Fowler, A. C., R. L. Knight, and T. L. George. 1991. Effects of avian predation on grasshopper populations in North Dakota grasslands. *Ecology* 72:1775–81.

Frank, G. S., and W. J. Platt. 2006. Conservation and restoration of the *Pinus palustris* ecosystem. *Applied Vegetation Science* 9:7–10.

Freckmann, R. W., and M. G. Lelong. 2003. *Dichanthelium*. *Flora of North America* 25:406–50.

Freeman, C. C. 1998. Flora of Konza Prairie: A historical review and contemporary patterns. In *Grassland Dynamics: Long-Term Ecological Research in Tallgrass Prairie*, edited by A. K. Knapp, J. M. Briggs, D. C. Hartnett, and S. L. Collins, 69–80. New York: Oxford University Press.

Freeman, D. C., M. L. Brown, J. J. Duda, J. H. Graham, J. M. Emlen, A. J. Krzysik, H. Balback, et al. 2005. Leaf fluctuating asymmetry, soil disturbance and plant stress: A multiple year comparison using two herbs, *Ipomoea pandurata* and *Cnidoscolus stimulosus*. *Ecological Indicators* 5:85–95.

Freeman, J. E., and S. Jose. 2009. Role of herbicide in savanna restoration: Effects of shrub reduction treatments on the understory and overstory of a longleaf pine flatwoods. *Forest Ecology and Management* 257:978–86.

Friauf, J. J. 1953. An ecological study of the Dermaptera and Orthoptera of the Welaka area in northern Florida. *Ecological Monographs* 23:79–126.

Fritschle, J. A. 2008. Reconstructing historic ecotones using the Public Land Survey: The lost prairies of Redwood National Park. *Annals of the Association of American Geographers* 98:24–39.

Frost, C. C. 1993. Four centuries of changing landscape patterns in the longleaf pine ecosystem. In *The Longleaf Pine Ecosystem: Ecology, Restoration, and Management*, edited by S. M. Herman, 17–43. Proceedings of the Tall Timbers Fire Ecology Conference No.18. Tallahassee, FL.

Frost, C. C., J. Walker, and R. K. Peet. 1986. Fire-dependent savannas and prairies of the Southeast: Original extent, preservation status and management problems. In *Wilderness and Natural Areas in the Eastern United States: A Management Challenge*, edited by D. L. Kulhavey and R. N. Conner, 348–57. Nacogdoches, TX: Center for Applied Studies, School of Forestry, Stephen F. Austin State University.

Gagnon, P. R. 2009. Fire in floodplain forests in the southeastern USA: Insights from disturbance ecology of native bamboo. *Wetlands* 29:520–26.

Gagnon, P. R., H. A. Passmore, W. J. Platt, J. A. Myers, C. E. Paine, and K. E. Harms. 2010. Does pyrogenicity protect burning plants? *Ecology* 91:3481–86.

Galloway, P. 2000. Archaeology from the archives: The Chambers excavations at Lyon's Bluff, 1934–35. *Mississippi Archaeology* 35:23–90.

Gärdenfors, U. 1992. Effects of artificial liming on land snail populations. *Journal of Applied Ecology* 29:50–54.

Gardiner, E. S., and J. M. Oliver. 2004. Restoration of bottomland hardwood forests in the lower Mississippi Alluvial Valley, U.S.A. In *Restoration of Temperate and Boreal Forests*, edited by J. Stanturf and P. Madsen, 235–51. Boca Raton, FL: Lewis Publishers.

Gavin, D. 1843. *Journal of Major David Gavin's Horseback Ride, St. George to Mississippi and Return, 1843*. Typescript in Lauderdale County Department of Archives and History, Meridian, MS.

Gehring, J. L., and T. B. Bragg. 1992. Changes in prairie vegetation under eastern red cedar (*Juniperus virginiana* L.) in an eastern Nebraska bluestem prairie. *American Midland Naturalist* 128:209–17.

Gentry, J. L., G. P. Johnson, B. T. Baker, C. T. Witsell, and J. D. Ogle. 2013. *Atlas of the Vascular Plants of Arkansas*. Fayetteville: University of Arkansas Herbarium.

Geocaching.com. 2010. Glauconite in outcrops at the Suma River. Accessed September 2014. http://www.geocaching.com/geocache/GC35CT9_glauconite-in-outcrops-at-the-suma-river?guid=9d47e90f-dd12–4093-aace-0e41a60611cf.

George, R. J. 1987. Herbaceous flora of three Weches Formation outcrops in eastern Texas. Master's thesis, Stephen F. Austin State University.

George, R. J., and E. S. Nixon. 1990. The herbaceous flora of three Weches Formation outcrops in eastern Texas. *Sida* 14:117–27.

Gerstaecker, F. 1859. *Wild Sports in the Far West*. Boston: Crosby, Nicholls.

Gibson, D. J. 1988. Regeneration and fluctuations of tallgrass prairie vegetation in response to burning frequency. *Bulletin of the Torrey Botanical Club* 115:1–12.

Gibson, J. S. 1941. The Alabama Black Belt: Its geographic status. *Economic Geography* 17:1–23.

Glantz, S. A. 2002. *Primer of Biostatistics*. 5th ed. New York: McGraw-Hill.

Gleason, H. A., and A. Cronquist. 1991. *Manual of Vascular Plants of Northeastern United States and Adjacent Canada*. 2nd ed. Bronx: New York Botanical Garden.

Glitzenstein, J. S., D. R. Streng, G. L. Achtemeier, L. P. Naeher, and D. D. Wade. 2006. Fuels and fire behavior in chipped and unchipped plots: Implications for land management near the wildland/urban interface. *Forest Ecology and Management* 236:18–29.

Godfrey, C. L., C. R. Carter, and G. S. McKee. 1970. *Land Resource Areas of Texas*. Experiment Bulletin 1070. College Station: Texas Agricultural Extension Service.

Gordon, K. L., and J. B. Wiseman Jr. 1989. *Bienville National Forest Prairie Survey Final Report*. Museum Technical Report No. 7. Jackson: Mississippi Department of Wildlife, Fisheries, and Parks, Museum of Natural Science.

Gosse, P. H. 1993. *Letters from Alabama, Chiefly Relating to Natural History*. Tuscaloosa: University of Alabama Press.

Gould, F. W. 1962. *Texas Plants: A Checklist and Ecological Summary*. Texas Agricultural Experiment Station Leaflet 492. College Station: Texas A&M University Press.

———. 1975. *The Grasses of Texas*. College Station: Texas A&M University Press.

Govus, T. 2008. *Results from Field Verification of the University of Georgia's Ecological Systems Map for the Species at Risk Project, Part 1: Fort Benning, Robins Air Force Base and Fort Gordon, Georgia*. Prepared for the Nongame Conservation Section, Georgia Department of Natural Resources, Atlanta.

Gray, M. J., R. M. Kaminski, G. Weerakkody, B. D. Leopold, and K. C. Jensen. 1999. Aquatic invertebrate and plant responses following mechanical manipulations of moist-soil habitat. *Wildlife Society Bulletin* 27:770–79.

Gray, M. T. 2010. Use of General Land Office records and Geographical Information Systems for restoration of native

prairie patches in the Jackson Prairie region in Mississippi. Master's thesis, Mississippi State University.
Green, D., and D. Welker. 2003. Sustainable landscaping: The hidden impacts of gardens. US Environmental Protection Agency. Accessed May 25, 2011. https://archive.epa.gov/greenacres/web/pdf/smithsonian.pdf.
Greenfield, K. C., L. W. Burger, M. J. Chamberlain, and E. W. Kurzejeski. 2002. Vegetation management practices on Conservation Reserve Program fields to improve northern bobwhite habitat quality. *Wildlife Society Bulletin* 30: 527–38.
Greenfield, K. C., L. W. Burger, L. Golden, and P. Graham. 2005. *Light Disking to Enhance Early Successional Wildlife Habitat in Grasslands and Old Fields: Wildlife Benefits and Erosion Potential*. Natural Resources Conservation Service Technical Note 190–32. Washington, DC.
Greenfield, K. C., M. J. Chamberlain, L. W. Burger, and E. W. Kurzejeski. 2003. Effects of burning and disking in Conservation Reserve Program fields to improve habitat quality for northern bobwhite (*Colinus virginianus*). *American Midland Naturalist* 149:344–53.
Griffith, G., J. M. Omernik, and S. H. Azevedo. 1998. Ecoregions of Tennessee [Map poster]. Reston, VA: US Geological Survey. Accessed April 2014. ftp://ftp.epa.gov/wed/ecoregions/tn/tn_front.pdf.
Gruchy, J. P. 2007. An evaluation of field management practices to improve bobwhite habitat. Master's thesis, University of Tennessee.
Gruchy, J. P., C. A. Harper, and M. J. Gray. 2006. Methods for controlling woody invasion into CRP fields in Tennessee. In *Gamebird 2006: Quail VI and Perdix XII*, edited by S. B. Cederbaum, B. C. Faircloth, T. M. Terhune, J. J. Thompson, and J. P. Carroll, 315–21. Athens: University of Georgia, Warner School of Forestry and Natural Resources.
Guilfoyle, M. P. 2001. *Management of Bottomland Hardwood Forests for Nongame Bird Communities on Corps of Engineers Projects*. Ecosystem Management and Restoration Research Program Technical Notes Collection (ERDC TN-EMRRP-SI-21). Vicksburg, MS: US Army Engineer Research and Development Center.
Gustafson, D. J., D. J. Gibson, and D. L. Nickrent. 1999. Random amplified polymorphic DNA variation among remnant big bluestem (*Andropogon gerardii* Vitman) populations in Arkansas' Grand Prairie. *Molecular Ecology* 8:1693–701.
Gutzwiller, K. J., and S. K. Riffell. 2007. Using statistical models to study temporal dynamics of animal–landscape relations. In *Temporal Dimensions of Landscape Ecology: Wildlife Responses to Variable Resources*, edited by J. A. Bissonette and I. Storch, 93–118. New York: Springer.
Haag, W. R. 2012. *North American Freshwater Mussels: Natural History, Ecology, and Conservation*. Cambridge: Cambridge University Press.
Haag, W. R., and M. L. Warren Jr. 2008. Effects of severe drought on freshwater mussel assemblages. *Transactions of the American Fisheries Society* 137:1165–78.
Hale, S. L., S. K. Riffell, L. W. Burger, H. L. Adams, and J. G. Dollar. 2011. Fire ant responses to management of native grass field buffers. *American Midland Naturalist* 166:283–91.
Hall, T. F., and W. T. Penfound. 1943. Cypress-gum communities in the Blue Girth Swamp near Selma, Alabama. *Ecology* 24:208–17.
Hamilton, T. M. 1972. *Native American Bows*. York, PA: George Shumway.
Hamilton, W. J., Jr., and J. A. Pollack. 1961. The food of some lizards from Ft. Benning, Georgia. *Herpetologica* 17:99–106.
Hamm, J. 1989. *Bows and Arrows of the Native Americans: A Complete Step-by-Step Guide to Wooden Bows, Sinew-Backed Bows, Composite Bows, Strings, Arrows, and Quivers*. New York: Lyons and Burford.
———. 1992. Traditional choices. *Field and Stream* 97:24–25.
Hamrick, R., L. W. Burger Jr., J. C. Jones, and B. K. Strickland. 2007. *Native Warm-Season Grass Restoration in Mississippi*. Mississippi State University Extension Service Publication 2435.
Hanley, R. W. 1982. An analysis of the mollusk remains from site 22Cl814. In *Archaeological Investigations at the Yarborough Site (22Cl814), Clay County, Mississippi*, edited by C. Solis and R. Walling, 153–55. Report of Investigations 30. Tuscaloosa: Office of Archaeological Research, University of Alabama.
Hansen, W. D., B. F. Clardy, and S. Martin. 1997. Geologic map of the Dierks Quadrangle, Howard and Sevier Counties, Arkansas. Accessed May 13, 2013. http://www.geology.ar.gov/maps_pdf/geologic/24k_maps/Dierks.pdf.
Hardcastle, R. 1992. Osage flat bow. In *The Traditional Bowyer's Bible*, vol, 1, edited by J. Hamm, 131–48. New York: Lyons and Burford.
Hare, R. C. 1965. Contribution of bark to fire resistance of southern trees. *Journal of Forestry* 63:248–51.
Harper, C. A. 2007. Strategies for managing early succession habitat for wildlife. *Weed Technology* 21:932–37.
Harper, C. A., G. E. Bates, M. P. Hansbrough, M. J. Gudlin, J. P. Gruchy, and P. D. Keyser. 2007. *Native Warm-Season*

Grasses: Identification, Establishment and Management for Wildlife and Forage Production in the Mid-South. Knoxville: University of Tennessee Extension.

Harper, L. 1857. *Preliminary Report on the Geology and Agriculture of Mississippi*. Jackson, MS: E. Barksdale.

Harper, R. M. 1913a. A botanical cross-section of northern Mississippi, with notes on the influence of soil on vegetation. *Bulletin of the Torrey Botanical Club* 40:377–99.

———. 1913b. *Economic Botany of Alabama. Part 1. Geographical Report on Forests*. Monograph 8. Tuscaloosa: Geological Survey of Alabama.

———. 1914. Phytogeographical notes on the coastal plain of Arkansas. *Plant World* 17:36–48.

———. 1920. The limestone prairies of Wilcox County, Alabama. *Ecology* 1:198–208.

———. 1921. Geography of central Florida. In *Florida Geological Survey 13th Annual Report*, 71–307. Tallahassee: Florida Geological Survey.

———. 1927. Natural resources of southern Florida. In *Florida Geological Survey 18th Annual Report*, 27–206. Tallahassee: Florida Geological Survey.

———. 1943. *Forests of Alabama*. Monograph No. 10. Geological Survey of Alabama.

Harper-Lore, B. L., and M. Wilson. 2000. *Roadside Use of Native Plants*. 1st ed. Washington, DC: Island Press.

Harrington, J. A. 1995. Planning and implementation of a right-of-way native planting for Wisconsin Highway 51. In *Proceedings of the 14th Annual North American Prairie Conference: Prairie Biodiversity*, edited by D. C. Hartnett, 175–80. Manhattan: Kansas State University.

Hartnett, D. C. 1990. Size-dependent allocation of seed and vegetation reproduction in four clonal composites. *Oecologia* 84:254–59.

———. 1991. Effects of fire in tallgrass prairie on growth and reproduction of prairie coneflower (*Ratibida columnifera*: Asteraceae). *American Journal of Botany* 78:429–35.

Hartnett, D. C., and P. A. Fay. 1998. Plant populations: Patterns and processes. In *Grassland Dynamics: Long-Term Ecological Research in Tallgrass Prairie*, edited by A. K. Knapp, J. M. Briggs, D. C. Hartnett, and S. L. Collins, 81–100. New York: Oxford University Press.

Hawkins, T. S., D. A. Skojac, B. R. Lockhart, T. D. Leininger, M. S. Devall, and N. M. Schiff. 2009. Bottomland forests in the lower Mississippi Alluvial Valley associated with the endangered *Lindera melissifolia*. *Castanea* 74:105–13.

Hays, R. L., C. Summers, and W. Seitz. 1981. *Estimating Wildlife Habitat Variables*. Washington, DC: Office of Biological Services, Fish and Wildlife Service, US Department of Interior.

Heard, L. P., A. W. Allen, L. B. Best, S. J. Brady, W. Burger, A. J. Esser, E. Hackett, et al. 2000. *A Comprehensive Review of Farm Bill Contributions to Wildlife Conservation, 1985–2000*. Technical Report, USDA/NRCS/WHMI-2000. Washington, DC: Wildlife Habitat Management Institute, Natural Resources Conservation Service, US Department of Agriculture.

Heitmeyer, M. E., L. H. Fredrickson, and S. King. 2000. *An Evaluation of Ecosytem Restoration Options for the Grand Prairie Region of Arkansas*. Memphis, TN: US Army Corps of Engineers.

Henry, J. P. 1872. *Resources of the State of Arkansas with Description of Counties, Railroad, Mines, and the City of Little Rock, the Commercial, Manufacturing, Political and Railroad Center of the State*. 2nd ed. Little Rock, AR: Price and M'Clure.

Herreid, C. F. 1997. What is a case? Bringing to science education the established tool of law and medicine. *Journal of College Science Teaching* 27:92–94.

———. 2006. "Clicker" cases: Introducing case study teaching into large classrooms. *Journal of College Science Teaching* 36:43–47.

Hilgard, E. W. 1860. *Report on the Geology and Agriculture of the State of Mississippi*. Jackson: State of Mississippi.

Hill, J. G. (in prep). The grasshopper (Orthoptera: Acrididae) fauna of the Fall Line sandhills of Georgia, USA.

Hill, J. G. 2005. Disjunct distributions of *Pseudopomala brachyptera* and *Campylacantha olivacea* (Orthoptera: Acrididae) in blackland prairies of Mississippi. *Entomological News* 116:127–30.

———. 2006. Environmental variables affecting ant community composition in Mississippi's Black Belt Prairie and Flatwoods regions. Master's thesis, Mississippi State University.

———. 2007a. The grasshopper (Orthoptera: Romaleidae: Acrididae) fauna of Black Belt Prairie remnants in Mississippi and Alabama. *Journal of Orthoptera Research* 16:139–44.

———. 2007b. The grasshoppers (Orthoptera: Acrididae) of the Bibb County Glades, Alabama. *Marginalia Insecta* 2:1.

———. 2009. The grasshopper (Orthoptera: Acrididae) fauna of sand dunes along the Little Ohoopee River, Emanuel County, Georgia, USA. *Journal of Orthoptera Research* 18:29–35.

———. 2010. A new species of *Melanoplus* (Orthoptera: Acrididae: Melanoplinae) from the cedar glades of Tennessee, USA. *Journal of Orthoptera Research* 19:341–45.

———. 2012. Habitat associations of grasshoppers (Orthoptera: Acrididae) in the heterogeneous cedar glade environments of the Central Basin of Tennessee. *Journal of the Orthopterists' Society* 21:227–43.

———. 2014. Biological notes on some species in the *Melanoplus tribulus* species group (Orthoptera: Acrididae: Melanoplinae) with the description of two new species. *Transactions of the American Entomological Society* 140:1–16.

———. 2015. Revision and biogeography of the *Melanoplus scudderi* species group (Orthoptera: Acrididae) with a description of 21 new species and establishment of the *Carnegiei* and *Davisi* species group. *Transactions of the American Entomological Society* 141:252–350.

Hill, J. G., and R. L. Brown. 2010. The ant (Hymenoptera: Formicidae) fauna of Black Belt Prairie remnants in Alabama and Mississippi. *Southeastern Naturalist* 9:73–84.

Hill, J. G., and M. E. Dakin 2011. An annotated list of the grasshoppers (Orthoptera: Acrididae, Romaleidae) of the southeastern United States. *Mid-South Entomologist* 4: 39–48.

Hill, J. G., and J. Goddard. 2012. Medical and veterinary importance of grasshoppers, crickets, and katydids (Hexapoda: Orthoptera). *Journal of the Mississippi Academy of Science* 57:172–77.

Hill, J. G., and J. Seltzer. 2007. A note on additional plants found at Sixteenth Section (Osborn) Prairie. *Journal of the Mississippi Academy of Science* 52:295–97.

Hill, J. G., W. D. Stone, and J. L. Seltzer. 2009. Description of an upland oak-hickory forest in the Black Belt from Osborn Prairie, Oktibbeha County, Mississippi. *Journal of the Mississippi Academy of Science* 54:157–63.

Hill, J. G., K. L. Summerville, and R. L. Brown. 2008. Habitat associations of ant species (Hymenoptera: Formicidae) in a heterogeneous Mississippi landscape. *Environmental Entomology* 37:453–63.

Hill, K., and R. Horner. 2005. *Assessment of Alternatives in Roadside Vegetation Management*. Final Research Report, Agreement T2695, Task 67: Roadside Vegetation. Seattle: Washington State Transportation Commission.

Hill, R. T. 1888. *Neozoic Geology of Southwestern Arkansas*. Vol. 2 of *Annual Report of the Geological Survey of Arkansas for 1888*. Edited by John C. Branner. Little Rock, AR: Press Printing.

Hiscock, J. G., C. S. Thourot, and J. Zhang. 2003. Phosphorus budget analysis relating to land use for the northern Lake Okeechobee watershed, Florida. *Environmental Engineering* 21:63–74.

History of Benton, Washington, Carroll, Madison, Crawford, Franklin, and Sebastian Counties, Arkansas. 1889. Chicago: Goodspeed.

Hoadley, R. B. 1990. *Identifying Wood: Accurate Results with Simple Tools*. Newton, CT: Taunton Press.

Hodges, J. D. 1997. Development and ecology of bottomland hardwood sites. *Forest Ecology and Management* 90:117–25.

Hogue, S. H. 2000. Burial practices, mortality, and diet in east-central Mississippi: A case study from Oktibbeha County. *Southeastern Archaeology* 19:63–81.

———. 2003. Corn dogs and hush puppies: Diet and domestication at two Protohistoric farmsteads in Oktibbeha County, Mississippi. *Southeastern Archaeology* 22:185–95.

Hogue, S. H., and E. Peacock. 1995. Environmental and osteological analysis at the South Farm Site (22OK534), a Mississippian farmstead in Oktibbeha County, Mississippi. *Southeastern Archaeology* 14:31–45.

Holmes, W. C., A. E. Rushing, and J. R. Singhurst. 2005. Taxonomy and identification of *Isoetes* in Texas based on megaspore features. *Lundellia* 8:1–6.

Hoppe, K. A., P. L. Koch, R. W. Carlson, and S. D. Webb. 1999. Tracking mammoths and mastodons: Reconstruction of migratory behavior using strontium isotope ratios. *Geology* 27:439–42.

Hopwood, J. 2010. *Pollinators and Roadsides: Managing Roadsides for Bees and Butterflies*. Portland, OR: Xerces Society for Invertebrate Conservation.

Hornbach, D. J., V. J. Kurth, and M. C. Hove. 2010. Variation in freshwater mussel shell sculpture and shape along a river gradient. *American Midland Naturalist* 164:22–36.

Hotopp, K. P. 2002. Land snails and soil calcium in central Appalachian Mountain forest. *Southeastern Naturalist* 1:27–44.

Hough, R. B. 1924. *Handbook of the Trees in Northern States and Canada, East of the Rocky Mountains*. Lowville, NY: R. B. Hough.

Howard, W. E. 1950. Wildlife depredations on broadcast seedings of burned brushland. *Journal of Range Management* 3:291–98.

Howe, H. F. 1994. Response of early-flowering and late-flowering plants to fire season in experimental prairies. *Ecological Applications* 4:121–33.

Hubbell, T. H. 1932. *A Revision of the* Puer *Group of the North American Genus* Melanoplus, *with Remarks on the Taxonomic Value of the Concealed Male Genitalia in the Cyrtacanthacrinae (Orthoptera, Acrididae)*. Miscellaneous Publication 23.Ann Arbor: University of Michigan Museum of Zoology.

Hubricht, L. 1985. *The Distributions of the Native Land Mollusks of the Eastern United States*. Fieldiana, Zoology, New Series, no. 24.

Huddlestun, P. F., and J. H. Hetrick. 1986. *Upper Eocene Stratigraphy of Central and Eastern Georgia*. Atlanta: Georgia Geologic Survey, Environmental Protection Division, Department of Natural Resources.

Hudson, C. 1997. *Knights of Spain, Warriors of the Sun*. Athens: University of Georgia Press.

Huffman, J. M., and W. J. Platt. 2014. *Fire History of the Avon Park Air Force Range: Evidence from Tree-Rings*. Report submitted to Department of Defense, available at Avon Park Air Force Base.

Hughes, R. H. 1975. *The Native Vegetation in South Florida Related to Month of Burning*. Research Note SE-222. Asheville, NC: Southeastern Forest Experiment Station, US Forest Service.

Huijser, M. P., and A. P. Clevenger. 2006. Habitat and corridor function of rights-of-way. In *The Ecology of Transportation: Managing Mobility for the Environment*, edited by J. Davenport and J. L. Davenport, 233–54. Dordrecht, Netherlands: Springer.

Hulbert, L. C. 1986. Fire effects on tallgrass prairie. In *Proceedings of the 9th North American Prairie Conference*, edited by G. K. Clambey and R. H. Pemble, 138–42. Fargo: North Dakota State University.

———. 1988. Causes of fire effects in tallgrass prairie. *Ecology* 69:46–58.

Hulbert, L. C., and J. K. Wilson. 1983. Fire interval effects on flowering of grasses in Kansas bluestem prairie. In *Proceedings of the 7th North American Prairie Conference*, edited by C. L. Kucera, 225–57. Springfield: Southwest Missouri State University.

Huner, J., Jr. 1939. *Geology of the Caldwell and Winn Parishes*. Geological Bulletin No. 15. Baton Rouge: Department of Conservation, Louisiana Geological Survey.

Hunter, W. C., D. A. Buehler, R. A. Canterbury, J. L. Confer, and P. B. Hamel. 2001. Conservation of disturbance-dependent birds in eastern North America. *Wildlife Society Bulletin* 29:440–55.

Irving, R. S. 1980. Composition, production and management of eastern Arkansas prairies. In *Proceedings of the 7th North American Prairie Conference*, edited by C. L. Kucera, 281–86. Springfield: Southwest Missouri State University.

Irving, R. S., S. Brenholts, and T. Foti. 1980. Composition and net primary production of native prairies in eastern Arkansas. *American Midland Naturalist* 103:298–309.

Isbell, F., V. Calcagno, A. Hector, J. Connolly, W. S. Harpole, P. B. Reich, M. Scherer-Lorenzen, et al. 2011. High plant diversity is needed to maintain ecosystem services. *Nature* 477:99–202.

Jackson, J. F., D. C. Adams, and U. B. Jackson. 1999. Allometry of constitutive defense: A model and a comparative test with tree bark and fire regime. *American Naturalist* 153:614–32.

Jackson, L. L. 1999. Establishing tallgrass prairie on a grazed permanent pasture in the upper Midwest. *Restoration Ecology* 7:127–38.

Jamison, B. E., R. J. Robel, J. S. Pontius, and R. D. Applegate. 2002. Invertebrate biomass: Associations with lesser prairie-chicken habitat use and sand sagebrush density in southeastern Kansas. *Wildlife Society Bulletin* 30:517–26.

Jennings, J. D. 1947. Nutt's trip to the Chickasaw Country. *Journal of Mississippi History* 9:34–61.

Joern, A. 1979. Resource utilization and community structure in assemblages of arid grassland grasshoppers (Orthoptera: Acrididae). *Transactions of the American Entomological Society* 105:253–300.

———. 2005. Disturbance by fire frequency and bison grazing modulate grasshopper assemblages in tallgrass prairie. *Ecology* 86:861–73.

Johansson, L. J., K. Hall, H. C. Prentice, M. Ihse, T. Reitalu, M. T. Sykes, and M. Kindstrom. 2008. Semi-natural grassland continuity, long-term land-use change and plant species richness in an agricultural landscape in Oland, Sweden. *Landscape and Urban Planning* 84:200–211.

Johnson, A. M. 2000. *Best Practices Handbook on Roadside Vegetation Management*. Report No. Mn/DOT 2000–19. Minneapolis: Minnesota Technology Transfer (T2) / LTAP Program, Center for Transportation Studies, University of Minnesota.

Johnson, G. P. 2004. Lost and yet to be found: The snowy orchid (*Platanthera nivea*). *Native Orchid Conference Journal* 1:10–12.

Johnson, J. K. 1990. Cedar glades and protohistoric settlement: A reply to Peacock and Miller. *Mississippi Archaeology* 25:58–62.

———. 1997. Stone tools, politics, and the eighteenth-century Chickasaw in northeast Mississippi. *American Antiquity* 62:215–30.

———. 2000. The Chickasaws. In *Indians of the Greater Southeast: Historical Archaeology and Ethnohistory*, edited by B. G. McEwan, 85–121. Gainesville: University Press of Florida.

Johnson, J. K., H. K. Curry, J. R. Atkinson, and J. T. Sparks. 1984. *Final Report Cultural Resources Survey in the Line*

Creek Watershed, Chickasaw, Clay and Webster Counties, Mississippi. University: Center for Archaeological Research, University of Mississippi.

Johnson, J. K., P. K. Galloway, and W. Belokon 1989. Historic Chickasaw settlement patterns in Lee County, Mississippi: A first approximation. *Mississippi Archaeology* 24:45–52.

Johnson, J. K., and J. T. Sparks. 1986. Protohistoric settlement patterns in northeastern Mississippi. In *The Protohistoric Period in the Mid-South: 1500–1700*, edited by D. H. Dye and R. C. Brister, 64–87. Jackson: Mississippi Department of Archives and History.

Johnston, J. E., P. E. Heinrich, J. K. Lovelace, R. P. McColloh, and R. K. Zimmerman. 2000. *Stratigraphic Charts of Louisiana*. Baton Rouge: Louisiana Geological Survey, Louisiana State University.

Jonas, J. L., and A. Joern. 2007. Grasshopper (Orthoptera: Acrididae) communities respond to fire, bison grazing and weather in North American tallgrass prairie: A long-term study. *Oecologia* 153:699–711.

Jones, A. S., and E. G. Patton. 1966. Forest, "prairie" and soils in the Black Belt of Sumter County, Alabama, in 1832. *Ecology* 47:75–80.

Jones, E. M., and C. S. Waldrop. 1917. *Soil Survey of Chickasaw County, Mississippi*. Washington, DC: Government Printing Office.

Jones, J., D. S. Coggin, J. L. Cummins, and J. Hill. 2007. *Restoring and Managing Native Prairies: A Handbook for Mississippi Landowners*. Stoneville: Wildlife Mississippi.

Jones, R. L., C. L. Knight, and T. C. Majure 1997. *Endangered Mussels of Tombigbee River Tributaries: Tibbee Creek*. Museum Technical Report No. 49. Jackson: Mississippi Department of Wildlife, Fisheries, and Parks.

Jones, R. L., W. T. Slack, and P. D. Hartfield 2005. The freshwater mussels (Mollusca: Bivalvia: Unionidae) of Mississippi. *Southeastern Naturalist* 4:77–92.

Jones, S. B. 1971. A virgin prairie and a virgin loblolly pine stand in central Mississippi. *Castanea* 36: 223–25.

Jordan, T. G. 1977. Early northeast Texas and the evolution of western ranching. *Annals of the Association of American Geographers* 67: 66–87.

Jordan, W. R., III. 1997. Foreword. In *The Tallgrass Restoration Handbook for Prairies, Savannas, and Woodlands*, edited by S. Packard and C. F. Mutel. Washington, DC: Island Press.

Kahmen, S., P. Poschlod, and K. F. Schreiber. 2002. Conservation management of calcareous grasslands: Changes in plant species composition and response of functional traits during 25 years. *Biological Conservation* 104:319–28.

Kalmbacher, R. S., F. M. Pate, and F. G. Martin. 2002. Grazing evaluation of bahiagrass with *Evenia aeschynomene* and *Stylosanthes*. *Soil and Crop Science Society of Florida Proceedings* 61:10–15.

Kane, J. M., J. M. Varner, and J. K. Hiers. 2008. The burning characteristics of southeastern oaks: Discriminating fire facilitators from fire impeders. *Forest Ecology and Management* 256:2039–45.

Kane, J. M., J. M. Varner, E. E. Knapp, and R. F. Powers. 2010. Vegetation response to mechanical mastication in the northern Sierra. *Applied Vegetation Science* 13:207–20.

Karaman-Castro, V., and L. E. Urbatsch. 2006. *Boltonia*. *Flora of North America* 20:353–57.

Kartesz, J. T. 1999. A synonymized checklist and atlas with biological attributes for the vascular flora of the United States, Canada, and Greenland. In *Synthesis of the North American Flora*, version 1.0, edited by J. T. Kartesz and C.A. Meacham. Chapel Hill: North Carolina Botanical Garden.

———. 2012. BONAP's North American Plant Atlas (NAPA). Chapel Hill, NC: Biota of North America Program (BONAP). Accessed December 2014. http://bonap.net/napa.

Kaye, J. M. 1955. Certain aspects of the geology of Lowndes County, Mississippi. Master's thesis, Mississippi State University.

———. 1974. Pleistocene sediment and vertebrate fossil associations in the Mississippi Black Belt: A genetic approach. PhD diss., Louisiana State University.

Keith, E. 2007. Texas ladies' tresses orchid, *Sprianthes brevilabris*, rediscovered in Texas. *North American Native Orchid Journal* 13:113–14.

Kellison, R. C., and M. J. Young. 1997. The bottomland forest of the southern United States. *Forest Ecology and Management* 90:101–15.

Kilbride, K. M., and F. L. Paveglio. 1999. Integrated pest management to control reed canarygrass in seasonal wetlands of southeastern Washington. *Wildlife Society Bulletin* 27:292–97.

King, J. R., A. N. Andersen, and A. D. Cutter. 1998. Ants as bioindicators of habitat disturbance: Validation of the functional group model for Australia's humid tropics. *Biodiversity and Conservation* 7:1627–38.

King, J. R., and W. R. Tschinkel. 2006. Experimental evidence that the introduced fire ant, *Solenopsis invicta*, does not competitively suppress co-occuring ants in a disturbed habitat. *Journal of Animal Ecology* 75:1370–78.

———. 2008. Experimental evidence that human impacts

drive fire ant invasion and ecological change. *Proceedings of the National Academy of Science, USA* 105:20339–20343.

———. 2013a. Experimental evidence for weak effects of fire ants in a naturally invaded pine-savanna ecosystem in north Florida. *Ecological Entomology* 38:68–75.

———. 2013b. Fire ants are not drivers of biodiversity change: A response to Stuble et al. (2013). *Ecological Entomology* 38:543–45.

Klaus, N., and T. Patrick. 2002. *Draft Management Plan for the Blackland Prairies, Oaky Woods and Ocmulgee Wildlife Management Areas.* Atlanta: Nongame Wildlife and Natural Heritage Section, Georgia Department of Natural Resources.

Kline, V. M. 1997. Orchards of oak and a sea of grass. In *The Tallgrass Restoration Handbook for Prairies, Savannas, and Woodlands*, edited by S. Packard and C. F. Mutel, 3–22. Washington, DC: Island Press.

Knapp, A. K. 1984. Effect of fire in tallgrass prairie on seed production of *Vernonia baldwinii* Torr. (Compositae). *Southwestern Naturalist* 29:242–43.

Knapp, A. K., J. M. Briggs, S. L. Collins, S. R. Archer, M. S. Bret-Harte, B. E. Ewers, D. P. Peters, et al. 2008. Woody herb encroachment in North American grasslands: Shifts in growth form dominance rapidly alters control of ecosystem carbon inputs. *Global Change Biology* 14:615–23.

Knapp, A. K., J. M. Briggs, D. C. Hartnett, and S. L. Collins, eds. 1998. *Grassland Dynamics: Long-Term Ecological Research in Tallgrass Prairie.* New York: Oxford University Press.

Knapp, A. K., and L. C. Hulbert. 1986. Production, density, and height in flower stalks of three grasses in annually burned and unburned eastern Kansas tallgrass prairie: A four-year record. *Southwestern Naturalist* 31:235–41.

Knapp, A. K., and T. R. Seastedt. 1998. Grasslands, Konza Prairie, and long-term ecological research. In *Grassland Dynamics: Long-Term Ecological Research in Tallgrass Prairie*, edited by A. K. Knapp, J. M. Briggs, D. C. Hartnett, and S. L. Collins, 3–18. New York: Oxford University Press.

Knapp, E. E., J. M. Varner, M. D. Busse, C. N. Skinner, and C. J. Shestak. 2011. Behaviour and effects of prescribed fire in masticated fuel beds. *International Journal of Wildland Fire* 20:932–45.

Koschnick, T. J., W. T. Haller, and L. Glasgow. 2006. Documentation of landoltia (*Landoltia punctata*) resistance to diquat. *Weed Science* 54:615–19.

Kreye, J. K., J. M. Varner, J. K. Hiers, and J. Mola. 2013. Toward a mechanism for eastern North American forest mesophication: The role of litter drying. *Ecological Applications* 23:1976–86.

Kucera, C. L., and J. H. Ehrenreich. 1962. Some effects of annual burning on central Missouri prairie. *Ecology* 43: 334–36.

Kurtén, B., and J. M. Kaye. 1982. Late Quaternary Carnivora from the Black Belt, Mississippi. *Boreas* 11:47–52.

Kutschbach-Brohl, L., B. E. Washburn, G. E. Bernhardt, R. B. Chipman, and L. C. Francoeur. 2010. Arthropods of a semi-natural grassland in an urban environment: The John F. Kennedy International Airport, New York. *Journal of Insect Conservation* 14:347–58.

Ladd, D., and J. R. Thomas. 2015. Ecological checklist of the Missouri flora for floristic quality assessment. *Phytoneuron* 12:1–274.

Lapham, J. E. 1902. *Soil Survey of the Stuttgart Area, Arkansas* [Map]. Washington, DC: Field Operations of the Bureau of Soils, Department of Agriculture.

Laubin, R., and G. Laubin. 1980. *American Indian Archery.* Norman: University of Oklahoma Press.

Launchbaugh, J. L. 1955. Vegetational changes in the San Antonio prairies associated with grazing, retirement from grazing, and abandonment from cultivation. *Ecological Monographs* 25:39–57.

Lawless, P. J., J. M. Baskin, and C. C. Baskin. 2006. Xeric limestone prairies of the eastern United States: Review and synthesis. *Botanical Review* 72:235–72.

Lawson, E. R. 1990. *Juniperus virginiana* L. eastern redcedar. In *Conifers*, edited by R. M. Burns and B. H. Honkala, 131–40. Vol. 1 of *Silvics of North* America. USDA Forest Service Agricultural Handbook 654. Washington, DC.

Le Coeur, D., J. Baudry, B. Burel, and C. Thenail. 2002. Why and how we should study field boundary biodiversity in an agrarian landscape context. *Agriculture, Ecosystems and Environment* 89:23–40.

Leidolf, A. 1995. *A Floristic Survey of Rare and Endangered Plant Species of Oktibbeha County, Mississippi.* Jackson: Mississippi Natural Heritage Program.

Leidolf, A., and S. McDaniel. 1998. A floristic study of Black Prairie plant communities at Sixteen Section Prairie, Oktibbeha County, Mississippi. *Castanea* 63:51–62.

Leidolf, A., S. McDaniel, and T. Nuttle. 2002. The flora of Oktibbeha County, Mississippi. *Sida* 20:691–765.

Leopold, A. 1949. *A Sand County Almanac.* Warren, VT: Tamarack Press.

———. 1953. *Round River.* London: Oxford University Press.

Leps, J. 1999. Nutrient status, disturbance and competition:

An experimental test of relationships in a wet meadow. *Journal of Vegetation Science* 10:219–30.

Leps, J., and P. Smilauer. 2003. *Multivariate Analysis of Ecological Data Using CANOCO*. Cambridge: Cambridge University Press.

Lester, G. D., S. G. Sorenson, P. L. Faulkner, C. S. Reid, and I. E. Maxit. 2005. *Louisiana Comprehensive Wildlife Conservation Strategy*. Baton Rouge: Louisiana Department of Wildlife and Fisheries.

Li, M. H., J. R. Schutt, J. McFalls, E. K. Bardenhagen, C. Yong Sung, and L. Wheelock. 2008. *Successional Establishment, Mowing Response, and Erosion Control Characteristics of Roadside Vegetation in Texas*. Technical Report FHWA/TX-08/0-4949-1. Austin: Texas Department of Transportation Research and Technology Implementation Office.

Littell, R. C., G. A. Milliken, W. W. Stroup, R. D. Wolfinger, and O. Schabenberger. 2006. *SAS Systems for Mixed Models*. Cary, NC: SAS Institute.

Little, E. L., Jr. 1971. *Atlas of United States Trees*. Vol. 1, *Conifers and Important Hardwoods*. Washington, DC: US Forest Service.

Lo, E. Y. Y., S. Stefanovic, and T. A. Dickinson. 2007. Molecular reappraisal of relationships between *Crataegus* and *Mespilus* (Rosaceae, Pyreae)—two genera or one? *Systematic Botany* 32:596–616.

Logan, W. N. 1904. *The Geology of Oktibbeha County*. Geological and Industrial Survey of Mississippi, Report 1. Chicago: Rand McNally.

Lolley, T. 2000. Archaeology at the Lyon's Bluff site, a Mississippian and Protohistoric settlement in Oktibbeha County, Mississippi. *Mississippi Archaeology* 35:1–14.

Longleaf Alliance. The big picture. Accessed February 13, 2013. http://www.longleafalliance.org/.

Loughridge, R. H. 1884. Physical and agricultural features of the state of Arkansas. In *Report on Cotton Production in the United States; Also Embracing Agricultural and Physico-Geographical Descriptions of the Several Cotton States and of California. Part 1. Mississippi Valley and Southwestern States*, edited by E. Hilgard. Washington, DC: Government Printing Office.

Louisiana Department of Transportation and Development. 2000. Policy for roadside vegetation management. Louisiana Register: 1–60. Accessed August 6, 2013. http://wwwsp.dotd.la.gov/Inside_LaDOTD/Divisions/Engineering/Misc%20Documents/Policy%20For%20Roadside%20Vegetation%20Management.pdf.

Louv, R. 2005. *Last Child in the Woods: Saving our Children from Nature-Deficit Disorder*. Chapel Hill, NC: Algonguin Books.

———. 2011. *The Nature Principle: Reconnecting with Life in a Virtual Age*. Chapel Hill, NC: Algonguin Books.

Lovell, S. T., and W. C. Sullivan. 2006. Environmental benefits of conservation buffers in the United States: Evidence, promise, and open questions. *Agriculture, Ecosystems and Environment* 112:249–60.

Lowe, E. N., ed. 1913. *Forest Conditions of Mississippi*. Geological Survey Bulletin 11. Mississippi State University.

———. 1915. *Mississippi: Its Geology, Geography, Soils and Mineral Resources*. Jackson, MS: Tucker Printing House.

———. 1919. *Plants of Mississippi: A List of Flowering Plants and Ferns*. Geological Survey Bulletin No. 17. Mississippi State University.

Lundeberg, M. A., H. Kang, B. Wolter, R. Del Mas, N. Armstrong, B. Borsari, N. Boury, et al. 2011. Context matters: Increasing understanding with interactive clicker case studies. *Educational Technology, Research and Development* 59:645–71.

Lyman, R. L. 1994. *Vertebrate Taphonomy*. Cambridge: Cambridge University Press.

Lynch, P. 2012. Vascular flora and vegetation classification of the South Atlantic Coastal Plain Limestone Forest Association of central Georgia. Master's thesis, University of Georgia.

Lysne, S. J., K. E. Perez, K. M. Brown, R. L. Minton, and J. D. Sides. 2008. A review of freshwater gastropod conservation: Challenges and opportunities. *Journal of the North American Benthological Society* 27:463–70.

MacDonald, J. R. 1996. A survey of the flora of Monroe County, Mississippi. Master's thesis, Mississippi State University.

MacRoberts, B. R., and M. H. MacRoberts. 1995. Vascular flora of two calcareous prairie remnants on the Kisatchie National Forest, Louisiana. *Phytologia* 78:18–27.

———. 1996. The floristics of calcareous prairies on the Kisatchie National Forest, Louisiana. *Phytologia* 81:35–43.

———. 1997. Former distribution of prairies in northern Louisiana. *Phytologia* 82:316–25.

MacRoberts, B. R., M. H. MacRoberts, C. S. Reid, and P. L. Faulkner. 2009a. Vascular flora of a saline prairie in Winn Parish, Louisiana. *Journal of the Botanical Research Institute of Texas* 3:349–54.

———. 2009b. Vascular flora of Morse Clay prairies in northwestern Louisiana. *Journal of the Botanical Research Institute of Texas* 3:355–66.

MacRoberts, B. R., M. H. MacRoberts, and T. Witsell. 2011. Small scale vascular plant species richness in southwestern Arkansas blackland prairies. *Journal of the Botanical Research Institute of Texas* 5:743–51.

MacRoberts, M. H., and B. R. MacRoberts. 2002. West Gulf Coastal Plain prairies: A first approximation at a synthesis. In *Proceedings of the Third Eastern Native Grass Symposium*, edited by J. Randall and J. C. Burns, 5–18. Madison, WI: Omnipress.

MacRoberts, M. H., B. R. MacRoberts, and L. S. Jackson. 2003. Louisiana prairies. In *Blackland Prairies of the Gulf Coastal Plain: Nature, Culture, and Sustainability*, edited by E. Peacock and T. Schauwecker, 80–93. Tuscaloosa: University of Alabama Press.

Madison, L. A., T. G. Barnes, and J. D. Sole. 1995. Improving northern bobwhite brood rearing habitat in tall fescue dominated fields. *Proceedings of the Annual Conference of the Southwestern Association of Fish and Wildlife Agencies* 49:525–36.

———. 2001. Effectiveness of fire, disking, and herbicide to renovate tall fescue fields to northern bobwhite habitat. *Wildlife Society Bulletin* 29:706–12.

Maehr, D. S. 2007. Central Florida ranches are key to the black bear's future. *Florida Cattlemens' Magazine* 71: 90–96.

Maehr, D. S., and J. A. Cox. 1995. Landscape features and panthers in Florida. *Conservation Biology* 9:1008–19.

Mahler, W. E. 1987. *Leavenworthia texana* (Brassicaceae), a new species from Texas. *Sida* 12:239–42.

Main, M. B., and M. J. Barry. 2002. Influence of season of fire on flowering of wet prairie grasses in South Florida, USA. *Wetlands* 22:430–34.

Majure, L. C. 2007. The vascular flora of the Chunky River (Mississippi). *Journal of the Botanical Research Institute of Texas* 1:1179–202.

Majure, L. C., J. Hill, C. Doffitt, and T. C. Majure. 2011. The vascular flora of Lauderdale County, Mississippi, U.S.A. *Rhodora* 113:365–418.

MARIS (Mississippi Automated Resource Information System). 2009. County 10-meter digital elevation model. http://www.maris.state.ms.us.

Marks, P. L., and P. A. Harcombe. 1981. Forest vegetation of the Big Thicket, southeast Texas. *Ecological Monographs* 51:287–305.

Markwardt, D. 2005. Texas roadside wildflowers. *Native Plants Journal* 6:69–71.

Marshall, E. J. P., and A. C. Moonen. 2002. Field margins in northern Europe: Their functions and interactions with agriculture. *Agriculture, Ecosystems and Environment* 89: 5–21.

Marshall, R. 1977. Lyon's Bluff Site (22OK1) radiocarbon dated. *Journal of Alabama Archaeology* 23:53–57.

———. 1985a. Report on the MSU-UM summer dig at Lyon's Bluff site near Starkville. In *Anthology of Mississippi Archaeology, 1966–1979*, edited by P. Galloway, 60–61. Jackson: Mississippi Archaeological Association and Mississippi Department of Archives and History.

———. 1985b. Two radiocarbon dates for the Lyon's Bluff site (22-Ok-520). In *Anthology of Mississippi Archaeology, 1966–1979,* edited by P. Galloway, 93–94. Jackson: Mississippi Archaeological Association and Mississippi Department of Archives and History.

———. 1985c. Numerous burials found at Lyon's Bluff this summer. In *Anthology of Mississippi Archaeology, 1966–1979*, edited by P. Galloway, 61. Jackson: Mississippi Archaeological Association and Mississippi Department of Archives and History.

———. 1986a. The Protohistoric component at the Lyon's Bluff site complex, Oktibbeha County, Mississippi. In *Protohistoric Period in the Mid-South:1500–1700*, edited by D. H. Dye and R. C. Brister, 83–88. Archaeological Report 18. Jackson: Mississippi Department of Archives and History.

———. 1986b. Stylistic changes in the Mississippian house patterns at the Lyon's Bluff site, 22OK1, Oktibbeha County, Mississippi. *Journal of Alabama Archaeology* 32:25–38.

Martin, J. O., and E. P. Carr. 1903. *Soil Survey of Miller County, Arkansas*. Washington, DC: Field Operations of the Bureau of Soils, Department of Agriculture.

Maynard, E. E., and J. S. Brewer. 2013. Restoring perennial warm season grasses as a means of reversing mesophication of oak woodlands in northern Mississippi. *Restoration Ecology* 21:242–49.

McBryde, J. B. 1933. The vegetation and habitat factors of the Carrizo sands. *Ecological Monographs* 3:247–97.

McCook, L. M., and J. Kartesz. 2010. Preliminary checklist of Mississippi plants. Department of Biology, University of Mississippi, Oxford. Accessed August 2015. http://www.herbarium.olemiss.edu/checklist.html.

McCune, B., and J. B. Grace. 2002. *Analysis of Ecological Communities*. Gleneden Beach, OR: MjM Software Design.

McCune, B., and M. J. Mefford. 2011. *PC-ORD. Multivariate Analysis of Ecological Data*. Version 6. Gleneden Beach, OR: MjM Software.

McDonald, J. H. 2009. *Handbook of Biological Statistics*. 2nd

ed. Baltimore: University of Delaware Sparky House Publishing.
McGregor, S. W. 2000. *A Survey of Freshwater Mussels in Selected Tributaries of the Upper Tombigbee River System, Alabama and Mississippi, 1994–2000.* Tuscaloosa: Environmental Geology Division, Geological Survey of Alabama.
McGuire, W. W. 1834. On the prairies of Alabama. *American Journal of Science* 26:93–98.
McKee, J. J., and D. M. Cochran Jr. 2012. The role of landscape in the distribution of deer-vehicle collisions in south Mississippi. *Southeastern Geographer* 52:327–40.
McKinney, L. E., and N. H. Russell. 2002. Violaceae of the southeastern United States. *Castanea* 67:369–79.
McNaughton, S. J. 1993. Grasses and grazers, science and management. *Ecological Applications* 3:17–20.
McPhail, L. C. 1837. On the diseases that prevail in the southwestern parts of the United States—their causes, nature and treatment—a suite of observations. *American Medical Intelligencer* 1:201–3.
Meier, U. 2006. A note on the power of Fisher's least significant difference procedure. *Pharmaceutical Statistics* 5: 253–63.
Menges, E. S., W. G. Abrahamson, K. T. Givens, N. P. Gallo, and J. N. Layne. 1993. 20 years of vegetation change in 5 long-unburned Florida plant communities. *Journal of Vegetation Science* 4:375–86.
Menges, E. S., and D. R. Gordon. 2010. Should mechanical treatments and herbicides be used as fire surrogates to manage Florida's uplands? A review. *Florida Scientist* 73:147–74.
Menges, E. S., and P. F. Quintana-Ascencio. 2004. Population viability with fire in *Eryngium cuneifolium*: Deciphering a decade of demographic data. *Ecological Monographs* 74:79–99.
Menon, R., and M. M. Holland. 2012. Study of understory vegetation at the University of Mississippi field station in north Mississippi. *Castanea* 77:28–36.
Meyer, A. M. 1937. An ecological study of cedar glade invertebrates near Nashville, Tennessee. *Ecological Monographs* 7:403–43.
Michael, E. D., and C. J. Kosten. 1981. Use of different highway cover plantings by wildlife. In *Proceedings of the Second National Symposium on Environmental Concerns in Rights-of-Way Management, October 16–18, 1979, University of Michigan, Ann Arbor, Michigan, USA*, edited by R. E. Tillman, 50-1–50-7. Palo Alto, CA: Electric Power Research Institute.
Middleton, B. A. 2002. Nonequilibrium dynamics of sedge meadows grazed by cattle in southern Wisconsin. *Plant Ecology* 161:89–110.
Milko, L. V. N., M. Haddad, and S. L. Lance. 2012. Dispersal via stream corridors structures populations of the endangered St. Francis' satyr butterfly. *Journal of Insect Conservation* 16:263–73.
Miller, J. H., and K. V. Miller. 2005. *Forest Plants of the Southeast and their Wildlife Uses*. Athens: University of Georgia Press.
Mills, T. B. 1876. *A History of the North-Western Editorial Excursion to Arkansas*. Little Rock, AR: T. B. Mills.
Mink, J. N., W. C. Holmes, and J. R. Singhurst. 2010. *Penstemon oklahomensis* (Scrophulariaceae) in Texas. *Journal of the Botanical Research Institute of Texas* 4:471–72.
Mink, J. N., J. R. Singhurst, and W. C. Holmes. 2012. A new species of *Hymenoxys* (Asteraceae: Heliantheae) from Texas. *Novon* 22:56–59.
Minton, R. L., and K. E. Perez. 2005. A systematic checklist of the land snails of Louisiana. *Texas Journal of Science* 57:153–64.
———. 2010. Analysis of museum records highlights unprotected land snail diversity in Alabama. *American Malacological Bulletin* 28:91–95.
Mississippi Department of Wildlife, Fisheries, and Parks. 2005. *Mississippi's Comprehensive Wildlife Conservation Strategy, 2005–2015*. Jackson: Mississippi Department of Wildlife, Fisheries, and Parks.
Mississippi State University Department of Geosciences. 2010. Office of the Mississippi State Climatologist. Mississippi climate. Accessed August 22, 2013. http://www.geosciences.msstate.edu/scClimate.htm.
Mitchell, D. B., B. G. Lockaby, and E. F. Brantley. 2011. Influence of Chinese privet (*Ligustrum sinense*) on decomposition and nutrient availability in riparian forests. *Invasive Plant Science and Management* 4:437–47.
Mohlenbrock, R. H. 1986. *Guide to the Vascular Flora of Illinois*. Carbondale: Southern Illinois University Press.
———. 2011. *The Illustrated Flora of Illinois. Sedges: Carex.* 2nd ed. Carbondale: Southern Illinois University Press.
Mohr, C. T. 1901. *Plant Life of Alabama*. Geological Survey of Alabama, Monograph 5. Washington, DC: Government Printing Office.
Moleele, N. M., S. Ringrose, W. Matheson, and C. Vanderpost. 2002. More woody plants? The status of bush encroachment in Botswana's grazing areas. *Journal of Environmental Management* 64:3–11.
Moore, W. H., ed. 1985. *Geologic Map of Mississippi*. Originally compiled in 1969 by A. R. Bicker, and revised with

data submitted by Dr. E. E. Russell. Jackson, MS: Mercury Maps.

Moran, L. P. 1995. Soil and ecology remnants in the Jackson Prairie region of Mississippi. Master's thesis. Mississippi State University.

Moran, L. P., D. E. Pettry, and R. E. Switzer. 2003. Plant and soil interactions in prairie remnants of the Jackson Prairie region, Mississippi. In *Blackland Prairies of the Gulf Coastal Plain: Nature, Culture, and Sustainability*, edited by E. Peacock and T. Schauwecker, 146–63. Tuscaloosa: University of Alabama Press.

Moran, L. P., D. E. Pettry, R. E. Switzer, S. T. McDaniel, and R. G. Wieland. 1997. *Soils of Native Prairie Remnants in the Jackson Prairie Region of Mississippi*. Bulletin 1067. Mississippi State: Mississippi Agricultural and Forestry Experiment Station.

Morris, M. W., C. T. Bryson, and R. C. Warren. 1993. Rare vascular plants and associate plant communities from the Sand Creek Chalk Bluffs, Oktibbeha County, Mississippi. *Castanea* 58:250–59.

Morrison, J. L., and S. R. Humphrey. 2001. Conservation value of private lands for crested caracaras in Florida. *Conservation Biology* 15:675–84.

Morse, A. P. 1907. *Further Researches on North American Acridiidae*. Washington, DC: Carnegie Institution of Washington.

MNHP (Mississippi Natural Heritage Program). 1991. Inventory and restoration of Jackson Prairie openings, Bienville National Forest, in central Mississippi. Paper presented at the Third Annual Conference of the Society for Ecological Restoration, Orlando, FL.

———. 2006. *Special Plants Tracking and Watch Lists* [with updates by H. Sullivan, pers. comm.]. Jackson: Museum of Natural Science, Mississippi Department of Wildlife, Fisheries, and Parks.

Murphree, L. C., J. L. Anderson, R. M. Ferguson, M. C. Garber, M. G. Martin, K. H. Miller, L. H. Nichols, et al. 1966. *Soil Survey of Monroe County, Mississippi*. Publication Series 1961, No. 37. USDA Soil Conservation Service, in cooperation with the Mississippi Agricultural and Forestry Experiment Station.

Murphree, L. C., J. S. Huddleston, and L. H. Nichols. 1974. *Soil Survey of Chickasaw County, Mississippi*. USDA Soil Conservation Service, in cooperation with the Mississippi Agricultural and Forestry Experiment Station.

Murphree, L. C., and K. H. Miller. 1976. *Soil Survey of Clay County, Mississippi*. USDA Soil Conservation Service, in cooperation with the Mississippi Agricultural and Forestry Experiment Station.

Murphy, K. E., J. K. Vidrine, D. R. McDaniel, and C. L. Godfrey. 1986. Soil survey of St. Landry Parish, Louisiana. Baton Rouge: US Department of Agriculture Soil Conservation Service in cooperation with Louisiana Agricultural Experiment Station and Louisiana State Soil and Water Conservation Committee.

Murphy, T. R., D. L. Colvin, R. Dickens, J. W. Everest, D. Hall, and L. B. McCarty. 2007. *Weeds of Southern Turfgrasses*. IFAS Extension. Gainesville: University Press of Florida.

Nairne, T. 1708. *Nairne's Muskhogean Journals: The 1708 Expedition to the Mississippi River*. Reprint, 1988, Jackson: University Press of Mississippi.

National Climatic Data Center. 2013. Annual climatogical summary text. Verona Experimental Station. Accessed January 2014. http://www.ncdc.noaa.gov.

Native American Seed. 2011. Junction, Texas. Accessed March 10, 2011, and July 31, 2013. http://www.seedsource.com/.

Natural Fibers Information Center. 1987. *The Climate of Texas Counties*. Austin: Natural Fibers Information Center, University of Texas at Austin, in cooperation with Office of State Climatologist, Texas A&M University, College Station.

Nature Conservancy. 2003. *The West Gulf Coastal Plain Ecoregional Conservation Plan*. San Antonio, TX: West Gulf Coastal Plain Planning Team, Nature Conservancy.

———. 2006. *Conservation by Design: A Strategic Framework for Mission Success*. Arlington, VA.

———. 2009. *The Conservancy's Programs and Places in Arkansas* [Map]. http://www.nature.org/wherewework/northamerica/states/arkansas/projectprofiles/.

NatureServe. 2004. *International Ecological Classification Standard: Terrestrial Ecological Classification. National Forests of Northern Mississippi (Delta, Holly Springs, Tombigbee)*. Final Report. Arlington, VA: NatureServe.

———. 2012. National vegetation classification. http://www.natureserve.org.

———. 2012–2013. NatureServe Explorer, an online encyclopedia of life. http://explorer.natureserve.org.

Nelson, J. C. 1997. Presettlement vegetation patterns along the 5th principal meridian, Missouri Territory, 1815. *American Midland Naturalist* 137:79–94.

Nemec, K. 1996. *White Bladderpod Conservation Efforts*. Houston, TX: Ecological Services, Clear Lake Field Office, US Fish and Wildlife Service.

Nesom, G. L. 2010. Notes on *Fraxinus profunda* (Oleaceae). *Phytoneuron* 32:1–6.

Niemiller, L. M., R. G. Reynolds, B. M. Glorioso, J. Spiess, and B. T. Miller. 2011. Herpetofauna of the cedar and associated habitats of the inner Central Basin of Tennessee. *Herpetological Conservation and Biology* 6:127–41.

NOAA (National Oceanic and Atmospheric Administration). 2012. Office of Coast Survey. *Map Showing the Distribution of the Slave Population of the Southern States of the United States—1860.* Accessed September 3, 2012. http://historicalcharts.noaa.gov/historicals/preview/image/CWSLAVE.

Noss, R. F. 2006. Introduction to the Florida "dry" prairie. In *Land of Fire and Water: The Florida Dry Prairie Ecosystem*, edited by R. F. Noss, 1–2. DeLeon Springs, FL: E. O. Painter.

———. 2013. *Forgotten Grasslands of the South: Natural History and Conservation.* Washington, DC: Island Press.

Noss, R. F., A. Cooperrider, and R. S. Schlickeisen. 1994. *Saving Nature's Legacy: Protecting and Restoring Biodiversity.* Washington, DC: Island Press.

Noss, R. F., E. T. LaRoe III, and J. M. Scott. 1995. *Endangered Ecosystems of the United States: A Preliminary Assessment of Loss and Degradation.* Biological Report 28. Washington, DC: National Biological Service, US Department of Interior.

Noss, R. F., W. J. Platt, B. A. Sorrie, A. S. Weakley, D. B. Means, J. Costanza, and R. K. Peet. 2015. How global biodiversity hotspots may go unrecognized: Lessons from the North American coastal plain. *Diversity and Distributions* 21:236–44.

Nowacki, G. J., and M. D. Abrams. 2008. The demise of fire and "mesophication" of forests in the eastern United States. *Bioscience* 58:123–38.

Nuttall, T. 1821. *A Journal of Travels in the Arkansas Territory during the Year 1819, with Occasional Observations on the Manners of the Aborigines.* Philadelphia: T. W. Palmer. Reprint, 1980, Norman: University of Oklahoma Press.

O'Kennon, R. J., and G. L. Nesom. 2001. *Liatris aestivalis* (Compositae). *Sida* 19:768.

Oksanen, J. 2012. Unconstrained ordination: Tutorial with R and vegan. Accessed September 10, 2013. http://cc.oulu.fi/~jarioksa/opetus/metodi/sessio1res.pdf.

Oliver, C. D., E. C. Burkhardt, and D. A. Skojac. 2005. The increasing scarcity of red oaks in Mississippi River floodplain forests: Influence of the residual overstory. *Forest Ecology and Management* 210:393–414.

Olson, D. M., and F. L. Wackers. 2007. Management of field margins to maximize multiple ecological services. *Journal of Applied Ecology* 44:13–21.

Ormeño, E., B. Céspedes, I. A. Sánchez, A. Velasco-García, J. M. Moreno, C. Fernandez, and V. Baldy. 2009. The relationship between terpenes and flammability of leaf litter. *Forest Ecology and Management* 257:471–82.

Ormsbee, P., F. A. Bazzaz, and W. R. Boggess. 1976. Physiological ecology of *Juniperus virginiana* in old fields. *Oecologia* 23:75–82.

Orr, W. D. 1992. *Ecological Literacy: Education and Transition to a Postmodern World.* New York: State University of New York Press.

Ortmann, A. E. 1920. Correlation of shape and station in freshwater mussels (Naiades). *Proceedings of the American Philosophical Society* 59:268–312.

Orzell, S. L., and E. L. Bridges. 1999. Dry prairie. In *The Ecosystem*, 10–66. Vol. 2 of *Multi-Species Recovery Plan for the Threatened and Endangered Species of South Florida.* Technical/Agency Draft. Atlanta: US Fish and Wildlife Service.

———. 2006a. Floristic composition of the south-central Florida dry prairie landscape. In *Land of Fire and Water: The Florida Dry Prairie Ecosystem*, edited by R. F. Noss, 64–99. DeLeon Springs, FL: E. O. Painter.

———. 2006b. Species composition and environmental characteristics of Florida dry prairies from the Kissimmee River region of south-central Florida. In *Land of Fire and Water: The Florida Dry Prairie Ecosystem*, edited by R. F. Noss, 100–135. DeLeon Springs, FL: E. O. Painter.

———. 2006c. Floristic composition and species richness of subtropical seasonally wet *Muhlenbergia sericea* prairies in portions of central and south Florida. In *Land of Fire and Water: The Florida Dry Prairie Ecosystem*, edited by R. F. Noss, 136–75. DeLeon Springs, FL: E. O. Painter.

Ott, R. L., and M. Longnecker. 2010. *An Introduction to Statistical Methods and Data Analysis.* 6th ed. Belmont, CA: Brooks/Cole.

Otte, D. 1981. *The North American Grasshoppers.* Vol. 1, *Acrididae: Gomphocerinae and Acridinae.* Cambridge, MA: Harvard University Press.

———. 2012. Eighty new species of *Melanoplus* from the United States (Acrididae: Melanoplinae). *Transactions of the American Entomological Society* 138:73–167.

———. 2014. *Aptenopedes* and two related genera in Florida with a description of fourteen new species (Acrididae: Melanoplinae). *Transactions of the American Entomological Society* 140:245–92.

Ovenden, G. N., A. R. H. Swash, and D. Smallshire. 2008. Agri-environment schemes and their contribution to the conservation of biodiversity in England. *Journal of Applied Ecology* 35:955–60.
OVPD (Oklahoma Vascular Plants Database). 2013. Accessed November 12, 2013. http://www.oklahomaplantdatabase.org/.
Packard, S., and C. F. Mutel, eds. 1997. *The Tallgrass Restoration Handbook for Prairies, Savannas, and Woodlands.* Washington, DC: Island Press.
Panshin, A. J., and C. de Zeeuw. 1980. *Textbook of Wood Technology.* New York: McGraw-Hill.
Panzer, R. 1988. Managing prairie remnants for insect conservation. *Natural Areas Journal* 8:83–90.
———. 2003. Importance of in situ survival, recolonization, and habitat gaps in the postfire recovery of fire-sensitive prairie insect species. *Natural Areas Journal* 23:14–21.
Panzer, R., K. Gnaedinger, and G. Derkovitz. 2010. Prevalence and status of conservative prairie and sand savanna insects in the Chicago Wilderness Region. *Natural Areas Journal* 30:73–81.
Panzer, R., and M. W. Schwartz. 1998. Effectiveness of a vegetation-based approach to insect conservation. *Conservation Biology* 12:693–702.
Parr, C. L., and A. N. Anderson. 2006. Patch mosaic burning for biodiversity conservation: A critique of the pyrodiversity paradigm. *Conservation Biology* 20:1610–19.
Parr, C. L., H. G. Robertson, H. C. Biggs, and S. L. Chown. 2004. Response of African savanna ants to long-term fire regimes. *Journal of Applied Ecology* 41:630–42.
Paveglio, F. L., and K. M. Kilbride. 2000. Response of vegetation to control of reed canarygrass in seasonally managed wetlands of southwestern Washington. *Wildlife Society Bulletin* 28:730–40.
Peacock, E. n.d. Molluscan remains from 22OK1147. Manuscript on file, Cobb Institute of Archaeology, Mississippi State University.
———. 1992. Some additional notes on forest reconstruction in the Black Belt. *Mississippi Archaeology* 27:1–18.
———. 1993. Reconstructing the Black Belt environment using leaf impressions in daub. *Southeastern Archaeology* 12:148–54.
———. 1998. Fresh-water mussels as indicators of prehistoric human environmental impact in the southeastern United States. 2 vols. PhD thesis, University of Sheffield.
———. 2000. Assessing bias in archaeological shell assemblages. *Journal of Field Archaeology* 27:183–96.
———. 2002. Shellfish use during the Woodland Period in the Middle South. In *The Woodland Southeast*, edited by D. G. Anderson and R. Mainfort, 444–60. Tuscaloosa: University of Alabama Press.
———. 2012. Archaeological freshwater mussel remains and their use in the conservation of an imperiled fauna. In *Conservation Biology and Applied Zooarchaeology*, edited by S. Wolverton and R. L. Lyman, 42–67. Tucson: University of Arizona Press.
Peacock, E., and J. Gerber. 2008. Using land snails and freshwater mussels to chart human transformation of the landscape: An example from northeast Mississippi, U.S.A. In *Case Studies in Environmental Archaeology*, 2nd ed., edited by E. J. Reitz, C. M. Scarry, and S. J. Scudder, 123–41. New York: Springer Science Business Media.
Peacock, E., and S. H. Hogue. 2005. A new series of absolute dates from Lyon's Bluff (22OK520), east-central Mississippi. *Southeastern Archaeology* 24:46–58.
———. 2008. A qualification concerning Beck et al.'s (2006) mention of a possble DeSoto-era artifact from the Lyon's Bluff site (22OK520), Mississippi. Southeastern Archaeological Conference.
Peacock, E., C. Jenkins, P. F. Jacobs, and J. Greenleaf. 2011. *Archaeology and Biogeography of Prehistoric Freshwater Mussel Shell in Mississippi.* British Archaeological Reports, International Series 2297. Oxford, UK: Archaeopress.
Peacock, E., S. W. McGregor, and A. A. Dumas. 2014. Behavioral, environmental, and applied aspects of molluscan assemblages from the lower Tombigbee River, Alabama, USA. In *Trends and Traditions in Southeastern Zooarchaeology*, edited by T. M. Peres, 186–212. Gainesville: University Press of Florida.
Peacock, E., and R. Melsheimer. 2003. Terrestrial gastropods from archaeological contexts in the Black Belt province of Mississippi. In *Blackland Prairies of the Gulf Coastal Plain: Nature, Culture, and Sustainability*, edited by E. Peacock and T. J. Schauwecker, 27–47. Tuscaloosa: University of Alabama Press.
Peacock, E., and W. F. Miller. 1990. Protohistoric settlement patterns in northeast Mississippi and the cedar glade hypothesis. *Mississippi Archaeology* 25:45–57.
Peacock, E. R., A. Palmer, Y. Xia, W. Bacon-Schulte, B. Carlock, and J. Smith. 2010. Chemical sourcing of a prehistoric freshwater shell artifact using laser ablation-inductively coupled plasma-mass spectrometry. *Archaeology of Eastern North America* 38:91–99.
Peacock, E., and J. Rafferty. 2013. Bet-hedging model as an explanatory framework for the evolution of mound building in the southeastern United States. In *Beyond Barrows: Current Research on the Structuration and Perception of*

the Prehistoric Landscape through Monuments, edited by D. Fontijn and A. J. Louwen. Leiden, Netherlands: Sidestone Press.

Peacock, E., J. Rafferty, and S. H. Hogue. 2005. Land snails, artifacts, and faunal remains: Understanding site formation processes at prehistoric/protohistoric sites in the southeastern United States. In *Archaeomalacology: Molluscs in Former Environments of Human Behavior*, edited by D. Bar-Yosef, 6–17. Oxford, UK: Oxbow Books.

Peacock, E., C. R. Randklev, S. Wolverton, R. A. Palmer, and S. Zaleski. 2012. The "cultural filter," human transport of mussel shell, and the applied potential of zooarchaeological data. *Ecological Applications* 22:1446–59.

Peacock, E., and M. C. Reese. 2003. A comparison of three methods of paleoenvironmental analysis at an archaeological site on the Mississippi Black Prairie. In *Blackland Prairies of the Gulf Coastal Plain: Nature, Culture, and Sustainability*, edited by E. Peacock and T. J. Schauwecker, 64–79. Tuscaloosa: University of Alabama Press.

Peacock, E., J. Rodgers, K. Bruce, and J. Gray. 2008. Assessing the pre-modern tree cover of the Ackerman Unit, Tombigbee National Forest, North Central Hills, Mississippi, using GLO survey notes and archaeological data. *Southeastern Naturalist* 7:245–66.

Peacock, E., and T. Schauwecker, eds. 2003a. *Blackland Prairies of the Gulf Coastal Plain: Nature, Culture, and Sustainability*. Tuscaloosa: University of Alabama Press.

———. 2003b. Introduction: The nature, culture, and sustainability of blackland prairies. In *Blackland Prairies of the Gulf Coastal Plain: Nature, Culture, and Sustainability*, edited by E. Peacock and T. J. Schauwecker, 1–7. Tuscaloosa: University of Alabama Press.

Peacock, E., and J. Seltzer. 2008. A comparison of multiple proxy data sets for paleoenvironmental conditions as derived from freshwater bivalve (Unionid) shell. *Journal of Archaeological Science* 35:2557–65.

Peattie, D. C. 1953. *A Natural History of Western Trees*. Boston: Houghton Mifflin.

Peck, J. E. 2010. *Multivariate Analysis for Community Ecologists: Step-by-Step Using PC-ORD*. Gleneden Beach, OR: MjM Software Design.

Perez, K. E., and R. L. Minton 2008. Practical applications for systematics and taxonomy in North American freshwater gastropod conservation. *Journal of the North American Benthological Society* 27:471–83.

Peterjohn, B. G. 2003. Agricultural landscapes: Can they support healthy bird populations as well as farm products? *Auk* 120:14–19.

Peterjohn, B. G., and J. R. Sauer. 1999. Population status of North American grassland birds from the North American Breeding Bird Survey, 1966–1996. *Studies in Avian Biology* 19:27–44.

Peterson, N. J. 1983. The effects of fire, litter, and ash on flowering in *Andropogon gerardii*. In *Proceedings of the 8th North American Prairie Conference*, edited by R. Brewer, 21–24. Kalamazoo: Western Michigan University.

Pettry, D. E. 1977. *Status of Mississippi Soil Surveys*. Information Sheet 1276. Mississippi State: Mississippi Agricultural and Forestry Experiment Station.

Pettry, D. E., and R. E. Switzer. 1993. *Expansive Soils of Mississippi*. Bulletin 1067. Mississippi State: Mississippi Agricultural and Forestry Experiment Station.

Phipps, J. B. 1990. *Mespilus canescens*, a new rosaceous endemic from Arkansas. *Systematic Botany* 15:144–49.

Pilsbry, H. A. 1939–1948. *Land Mollusca of North America (North of Mexico)*. Monograph No. 3, vols. 1 and 2. Philadelphia: Academy of Natural Sciences of Philadelphia.

Platt, S. G., and C. G. Brantley. 1997. Canebrakes: An ecological and historical perspective. *Castanea* 62:8–21.

Platt, W. J. 1999. Southeastern pine savannahs. In *Savannahs, Barrens, and Rock Outcrop Plant Communities of North America*, edited by R. C. Anderson, J. S. Fralish, and J. M. Baskin, 23–51. Cambridge: Cambridge University Press.

Platt, W. J., G. W. Evans, and M. M. Davis. 1988. Effects of fire season on flowering of forbs and woody herbs in longleaf pine forests. *Oecologia* 76:353–63.

Platt, W. J., J. M. Huffman, M. G. Slocum, and B. Beckage. 2006. Fire regimes and trees in Florida dry prairie landscapes. In *Land of Fire and Water: The Florida Dry Prairie Ecosystem*, edited by R. F. Noss, 64–99. DeLeon Springs, FL: E. O. Painter.

Platt, W. J., S. L. Orzell, and M. G. Slocum. 2015. Seasonality of fire weather strongly influences fire regimes in South Florida savanna-grassland landscapes. *PLoS One* 10:e0116952.

Poole, J. M., W. R. Carr, D. Price, and J. R. Singhurst. 2007. *A Field Guide to the Rare Plants of Texas*. College Station: Texas A&M University Press.

Pope, S. T. 1962. *Bows and Arrows*. 3rd ed. Berkeley: University of California Press.

Posner, A. W. 2012. Bird, plant, and herpetofaunal associations in cove forests and pine plantations in Mississippi. Master's thesis, Mississippi State University.

Post, L. C., ed. 1969. *Louisiana as It Is: A Geographical and Topographical Description of the State 1873*. By Samuel H. Lockett. Baton Rouge: Louisiana State University Press.

Preszler, R. W., A. Dawe, C. B. Shuster, and M. Shuster. 2007. Assessment of the effects of student response systems on student learning and attitudes over a broad range of biology courses. *CBE-Life Science Education* 6:29–41.

Quarterman, E. 1950. Major plant communities of Tennessee cedar glades. *Ecology* 31:234–54.

Radford, A. E., H. E. Ahles, and C. R. Bell. 1968. *Manual of the Vascular Flora of the Carolinas*. Chapel Hill: University of North Carolina Press.

Rafferty, J. 1995. A seriation of Historic period aboriginal pottery from northeast Mississippi. *Journal of Alabama Archaeology* 41:180–207.

———. 2001. Determining duration at prehistoric sites: Short-term sedentary settlement at Josey Farm, NE Mississippi. *Journal of Field Archaeology* 28:347–66.

———. 2003. Prehistoric settlement patterning on the Mississippi Black Prairie. In *Blackland Prairies of the Gulf Coastal Plain*, edited by E. Peacock and T. Schauwecker, 167–93. Tuscaloosa: University of Alabama Press.

Rafferty, J., T. R. James, K. McMahon, J. Alvey, and E. Peacock. 2003. Geophysical evidence bearing on the community plan at a Mississippian mound site, Oktibbeha County, Mississippi. Poster presented at the 60th Southeastern Archaeological Conference, Charlotte, NC.

Rafferty, J., and E. Peacock. 2008. The spread of shell tempering in the Mississippi Black Prairie. *Southeastern Archaeology* 27:253–64.

Rafferty, M. D. 1996. *Rude Pursuits and Rugged Peaks: Schoolcraft's Ozark Journal, 1818–1819*. Fayetteville: University of Arkansas Press.

Rankin, H. T. 1974. *Black Belt Prairie: Montgomery County, Alabama, and Vicinity*. Alabama Agricultural Experiment Station Bulletin 454. Auburn, AL: Auburn University.

Rankin, H. T., and D. E. Davis. 1971. Woody vegetation in the Black Belt Prairie of Montgomery County, Alabama, 1845–46. *Ecology* 52:716–19.

Ransom, M. D., C. W. Rice, T. C. Todd, and W. A. Wehmueller. 1998. Soils and soil biota. In *Grassland Dynamics: Long-Term Ecological Research in Tallgrass Prairie*, edited by A. K. Knapp, J. M. Briggs, D. C. Hartnett, and S. L. Collins, 48–68. New York: Oxford University Press.

Rausher, M. D., and F. J. Odendaal. 1987. Switching and the pattern of host use by *Battus philenor* butterflies. *Ecology* 68:869–77.

R Core Team. 2012. *R: A Language and Environment for Statistical Computing*. Version 2.14.1. Vienna, Austria: R Foundation for Statistical Computing.

Redmon, J. 2010. State OKs $28.7 million for Oaky Woods. *Atlanta Journal-Constitution*. Accessed May 13, 2013. http://www.ajc.com/news/news/local/state-oks-287-million-for-oaky-woods/nQnsr/.

Reeder, K. F., D. M. Debinski, and B. J. Danielson. 2005. Factors affecting butterfly use of filter strips in midwestern USA. *Agriculture, Ecosystems and Environment* 109:40–47.

Reese, M. C. 2000. Analyses of various methods used to quantify historical plant assemblages in east-central Mississippi. Master's thesis, Mississippi State University.

Reid, C. M., P. L. Faulkner, M. H. MacRoberts, B. H. MacRoberts, and M. Bordelon. 2010. Vascular flora and edaphic characteristics of saline prairies in Louisiana. *Journal of the Botanical Research Institute of Texas* 4:357–79.

Reitalu, T., L. J. Johansson, M. T. Sykes, K. Hall, and H. C. Prentice. 2010. History matters: Village distances, grazing and grassland species diversity. *Journal of Applied Ecology* 47:1216–24.

Resende, I. L. M., L. J. Chaves, and J. Â. Rizzo. 2013. Floristic and phytosociological analysis of palm swamps in the central part of the Brazilian savanna. *Acta Botanica Brasilica* 27:205–25.

Rhoades, C., T. Barnes, and B. Washburn. 2002. Prescribed fire and herbicide effects on soil processes during barrens restoration. *Restoration Ecology* 10:656–64.

Robbins, L. E., and R. L. Myers. 1992. *Seasonal Effects of Prescribed Burning in Florida: A Review*. Miscellaneous Publications No. 8. Tallahassee, FL: Tall Timbers Research.

Robertson, K. M., and T. L. Hmielowski. 2014. Effects of fire frequency and season on resprouting of woody plants in southeastern US pine grassland communities. *Oecologia* 174:765–76.

Robertson, P. A., G. T. Weaver, and J. A. Cavanaugh. 1978. Vegetation and tree species patterns near the northern terminus of the southern floodplain forest. *Ecological Monographs* 48:249–67.

Roberts-Pichette, P., and L. Gillespie. 1999. *Terrestrial Vegetation Biodiversity Monitoring Protocols*. EMAN Occasional Paper Series, Report No. 9. Burlington, Ontario: Ecological Monitoring Coordinating Office.

Robinson, R. A., and W. J. Sutherland. 2002. Post-war changes in arable farming and biodiversity in Great Britain. *Journal of Applied Ecology* 39:157–76.

Robison, N. D. 1983. Archeological records of naiad mussels along the Tennessee-Tombigbee waterway. In *Report of Freshwater Mussels Workshop, 26–27 October, 1982*, edited by A. C. Miller, 115–29. Vicksburg, MS: Waterways Experiment Station, US Army Corps of Engineers.

Rollins, R. C. 1963. The evolution and systematics of *Leavenworthia* (Cruciferae). *Contributions from the Gray Herbarium* 192:3–198.

Romans, B. 1775. *Concise Natural History of East and West Florida*. Facsimile of the first edition, 1962. Gainesville: University Press of Florida.

Rossen, J., and J. Olson. 1985. The controlled carbonization and archaeological analysis of the southeastern U.S. wood charcoals. *Journal of Field Archaeology* 12:445–56.

Rostlund, E. 1957. The myth of a natural prairie belt in Alabama: An interpretation of historical records. *Annals of the Association of American Geographers* 47:392–411.

Rudolph, J. L. 1983. Lamar period exploitation of aquatic resources in the middle Oconee River valley. *Early Georgia* 11:86–103.

Russell, K. N., H. Ikerd, and S. Droege. 2005. The potential conservation value of unmowed powerline strips for native bees. *Biological Conservation* 124:133–48.

Ryan, K. C., E. E. Knapp, and J. M. Varner. 2013. Prescribed fire in North American forests and woodlands: History, current practice, and challenges. *Frontiers in Ecology and the Environment* 11:e15–e24.

Sampson, F., and F. Knopf. 1994. Prairie conservation in North America. *BioScience* 44:418–21.

Sargent, C. S. 1905. *Manual of the Trees of North America*. 1st ed. Boston: Houghton Mifflin.

SAS Institute. 2007. *SAS/STAT 9.2 User's Guide*. Cary, NC: SAS Institute.

Saucier, R. T. 1994. *Geomorphology and Quaternary Geologic History of the Lower Mississippi Valley*. Vol. 1. Vicksburg, MS: Waterways Experiment Station, US Army Corps of Engineers.

Sauer, J. R., J. E. Hines, and J. Fallon. 2008. *The North American Breeding Bird Survey, Results and Analysis 1966–2007*. Version 5.15.2008. Laurel, MA: USGS Patuxent Wildlife Research Center.

Scarff, F. R., and M. Westoby. 2006. Leaf litter flammability in some semi-arid Australian woodlands. *Functional Ecology* 20:745–52.

Schambach, F. F. 2003. Osage-orange bows, Indian horses, and the Blackland Prairie of northeastern Texas. In *Blackland Prairies of the Gulf Coastal Plain: Nature, Culture, and Sustainability*, edited by E. Peacock and T. Schawacker, 212–36. Tuscaloosa: University of Alabama Press.

Scharf, E. A. 2010. Archaeology, land use, pollen and restoration in the Yazoo Basin (Mississippi, USA). *Vegetation History and Archaeobotany* 19:159–75.

Schauwecker, T. J. 1996. A comparison of blackland prairie relicts in Arkansas and Mississippi. Master's thesis, Mississippi State University.

Schauwecker, T., R. Brzuszek, B. Cooke, and K. Grala. 2011. Historical forest patterns and the analysis of site-scale forest gradients for natural area management. *Natural Areas Journal* 31:43–50.

Schauwecker, T., and J. MacDonald. 2003. Blackland prairie plant communities of northeast Mississippi: Composition, threatened species and assemblage response to disturbance. In *Blackland Prairies of the Gulf Coastal Plain: Nature, Culture, and Sustainability*, edited by E. Peacock and T. Schauwecker, 246–53. Tuscaloosa: University of Alabama Press.

Schlesinger, W. H., J. F. Reynolds, G. L. Cunningham, L. F. Huenneke, W. M. Jarrell, R. A. Viriginia, and W. G. Whitford. 1990. Biological feedbacks in global desertification. *Science* 247:1043–48.

Schmitz, D. W., C. L. Wax, and E. Peacock. 2003. Water-resource controls on human habitation in the Black Prairie of north-central Mississippi. In *Blackland Prairies of the Gulf Coastal Plain: Nature, Culture, and Sustainability*, edited by E. Peacock and T. Schauwecker, 194–211. Tuscaloosa: University of Alabama Press.

Scholes, R. J., and S. R. Archer. 1997. Tree-grass interactions in savannas. *Annual Review of Ecology and Systematics* 28:517–44.

Schotz, A., and M. Barbour. 2009. *Ecological Assessment and Terrestrial Vertebrate Surveys for Black Belt Prairies in Alabama*. Montogmery: Alabama Department of Conservation and Natural Resources.

Schramm, P. 1990. Prairie restoration: A twenty-year perspective on establishment and management. In *Recapturing a Vanishing Heritage: Proceedings of the Twelfth North American Prairie Conference*, edited by D. D. Smith and C. A. Jacobs, 169–78. Cedar Falls: University of Northern Iowa.

Schuster, M. F., and S. McDaniel. 1973. Vegetative analysis of a black prairie relict site near Aliceville, Alabama. *Journal of the Mississippi Academy of Science* 19:153–58.

Scudder, S. H. 1878. The Florida Orthoptera collected by Mr. J. H. Comstock. *Proceedings of the Boston Society of Natural History* 19:80–94.

Segars, W. I. 1987. *Soils of Georgia*. Cooperative Extension Service, Agronomy 2–2, Bulletin 662. Athens: University of Georgia College of Agriculture.

Segelquist, C. A., and W. E. Green. 1968. Deer food yields in four Ozarks forest types. *Journal of Wildlife Management* 32:330–37.

Seigel, A. B., and J. L. Lockwood. 2010. How increasing levels of private land enrollment in conservation agreements affect the population viability of grassland birds. *Biodiversity Conservation* 19:2343–57.

Seltzer, J. L. 2007. Determining the presence of cultural bias in wood charcoal from Lyon's Bluff (22OK520), a mound center in Oktibbeha County, Mississippi. Master's thesis, Mississippi State University.

Seltzer, J. L., and E. Peacock. 2011. Determining the seasonality of Mississippian house construction via the analysis of plant impressions in daub: New data from the Lyon's Bluff site (22OK520), Oktibbeha County, Mississippi. *Southeastern Archaeology* 30:123–26.

Semple, J. C., and R. E. Cook. 2006. *Solidago*. *Flora of North America* 20:107–66.

Shackford, J. A., and S. J. Folmsbee. 1973. A narrative of the life of David Crockett of the State of Tennessee. Facsimile edition. Knoxville: University of Tennessee Press.

Shea, A. 1998. *Floral Remains in Archaeological Data Recovery at the McKnight Site (22co560) Coahoma County, Mississippi*. Edited by S. Chapman and R. Walling. Memphis, TN: Panamerican Consultants.

Shelby, C. A., M. K. Pieper, D. E. Owen, T. J. Freeman, A. C. Wright, and V. E. Barnes. 1998. *Texarkana Sheet*. 1:250,000 geology map. Sellards Memorial Edition.

Shelford, V. E. 1974. *The Ecology of North America*. 7th ed. Urbana: University of Illinois Press.

Siemann, E., D. Tilman, J. Haarstad, and M. Ritchie. 1998. Experimental test of the dependence of arthropod diversity on plant diversity. *American Midland Naturalist* 152: 738–50.

Simmons, T. R. 1968. The vascular plants of the Kilgore Hills, Clay County, Mississippi. Master's thesis, Mississippi State University.

Simond, F. W., and T. C. Hopkins. 1891. *Geology of Benton County*. Vol. 2 of *Annual Report of the Geological Survey of Arkansas, 1891*. Fayetteville, AR.

Singhurst, J. R. 2010. *Blackland Prairie Limestone Glade Ephemeral Pool*. Ecological Association Comprehensive Report. NatureServe Explorer. Accessed November 8, 2013. http://www.natureserve.org/explorer.

Singhurst, J. R., M. White, J. N. Mink, and W. C. Holmes. 2011. *Castilleja coccinea* (Orobanchaceae): New to Texas. *Phytoneuron* 32:1–3.

Skeldon, M. A., M. A. Vadeboncoeur, S. P. Hamburg, and J. D. Blum 2007. Terrestrial gastropod responses to an ecosystem-level calcium manipulation in a northern hardwood forest. *Canadian Journal of Zoology* 85:994–1007.

Skojac, D. A., C. T. Bryson, and C. H. Walker. 2007. Noteworthy collections from the Yazoo-Mississippi Delta region of Mississippi. *Journal of the Botanical Research Institute of Texas* 1:769–75.

Slocum, M. G., B. Beckage, W. J. Platt, S. L. Orzell, and W. Taylor. 2010a. Effect of climate on wildfire size: A cross scale analysis. *Ecosystems* 13:828–40.

Slocum, M. G., and S. L. Orzell. 2013. A structural equation model analysis of relationships among ENSO, seasonal descriptors and wildfires. *PLoS ONE* 8:e75946.

Slocum, M. G., W. J. Platt, B. Beckage, S. L. Orzell, and W. Taylor. 2010b. Accurate quantification of seasonal rainfall and associated climate-wildfire relationships. *Journal of Applied Meteorology and Climatology* 49:2559–73.

Slocum, M. G., W. J. Platt, and H. C. Cooley. 2003. Effects of differences in prescribed fire regimes on patchiness and intensity of fires in subtropical savannas of Everglades National Park, Florida. *Restoration Ecology* 11:91–102.

Small, R. K. 1933. *Manual of the Southeastern Flora*. Chapel Hill: University of North Carolina Press.

Smeins, F. E., and D. D. Diamond. 1983. Remnant grasslands of the Fayette Prairie, Texas. *American Midland Naturalist* 110:1–13.

Smith, D. 1992. Tallgrass prairie settlement: Prelude to demise of the tallgrass ecosystem. In *Proceedings of the 12th North American Prairie Conference*, edited by D. D. Smith and C. A. Jacobs, 195–99. Cedar Falls: University of Northern Iowa.

———. 2012. Restoring a national treasure: Investment for the future. In *Proceedings of the 22nd North American Prairie Conference*, edited by D. Williams, B. Butler, and D. Smith, 32–237. Cedar Falls: University of Northern Iowa.

Smith, E. A. 1883. *Geological Survey of Alabama*. Montgomery, AL: W. D. Brown.

Smith, J. L., and J. V. Perino. 1981. Osage-orange (*Maclura pomifera*): History and economic uses. *Economic Botany* 35:24–41.

Smith, L. M. 1995. *State Rare Plant Communities in Louisiana*. Baton Rouge: Louisiana Natural Heritage Program.

Smith, L. M., N. W. Gilmore, R. P. Martin, and G. D. Lester. 1989. *Keiffer Prairie/Forest Complex: A Research Report and Preliminary Management Plan*. Baton Rouge: Louisiana Natural Heritage Program.

Smith, M. T. 1991. Indians of Mississippi: 1540–1700. In *Native, European, and African Cultures in Mississippi, 1500–1800*, edited by P. Galloway, 31–42. Jackson: Mississippi Department of Archives and History.

Smith, S. G. 2001. Taxonomic innovations in North American *Eleocharis* [CYPERACEAE]. *Novon* 11:247.

Smith, S. G., J. J. Bruhl, M. S. González-Elizondo, and F. J. Menapace. 2002. *Eleocharis*. *Flora of North America* 23: 60–120.

Soil Survey Staff. 2003. *Soil Survey of Vernon Parish, Louisiana*. Washington, DC: Natural Resources Conservation Service, US Department of Agriculture.

———. 2013. Official soil series descriptions. USDA Natural

Resources Conservation Service. Accessed May 13, 2013. https://soilseries.sc.egov.usda.gov/.

Sorrie, B. A. 2011. *A Field Guide to Wildflowers of the Sandhills Region: North Carolina, South Carolina, and Georgia.* Chapel Hill: University of North Carolina Press.

Spasojevic, M. J., R. J. Aicher, G. R. Koch, E. S. Marquardt, N. Mirotchnick, T. G. Troxler, and S. L. Collins. 2010. Fire and grazing in a mesic tallgrass prairie: Impacts on plant species and functional traits. *Ecology* 91:1651–59.

Speed, J. D. M., E. J. Cooper, I. S. Jonsdottir, R. Van Der Wal, and S. J. Woodin. 2010. Plant community properties predict vegetation resilience to herbivore disturbance in the Arctic. *Journal of Ecology* 98:1002–13.

Squitier, J. M., and J. L. Capinera. 2002. Habitat associations of Florida grasshoppers (Orthoptera: Acrididae). *Florida Entomologist* 85:235–44.

Stanturf, J. A., S. H. Schoenholtz, C. J. Schweitzer, and J. P. Shepard. 2001. Achieving restoration success: Myths in bottomland hardwood forests. *Restoration Ecology* 9:139–200.

Stephenson, K. 2011. Distribution of grasslands in 19th century Florida. *American Midland Naturalist* 165:50–59.

Stephenson, L. W., and A. F. Crider. 1916. Geology and ground waters of northeastern Arkansas. Washington, DC: Government Printing Office.

Stiles, J. H., and R. H. Jones. 1998. Distribution of the red imported fire ant, *Solenopsis invicta*, in road and powerline habitats. *Landscape Ecology* 13:335–46.

Stromayer, K. A., R. J. Warren, A. S. Johnson, P. E. Hale, C. L. Rogers, and C. L. Tucker. 1998. Chinese privet and the feeding ecology of white-tailed deer: The role of an exotic plant. *Journal of Wildlife Management* 61:1321–29.

Stuble, K. L, L. D. Chick, M. A. Rodriguez-Cabal, J. P. Lessard, and N. J. Sanders. 2013. Fire ants are drivers of biodiversity loss: A reply to King and Tschinkel (2013). *Ecological Entomology* 38:540–42.

Svedarsky, W. D., M. A. Kuchenreuther, G. J. Cuomo, P. Buesseler, H. Moechnig, and A. Singh. 2002. *A Landowner's Guide to Prairie Management in Minnesota*. Crookston: Northwest Research and Outreach Center, University of Minnesota.

Svejcar, T. J. 1990. Response of *Andropogon gerardii* to fire in the tallgrass prairie. In *Fire in North American Tallgrass Prairie*, edited by S. L. Collins and L. L. Wallace, 19–27. Norman: University of Oklahoma Press.

Swain, H. M. 1998. Archbold Biological Station and the MacArthur Agro-Ecology Research Center. *Bulletin of the Ecological Society of America* 79:114–20.

Swain, H. M., P. J. Bohlen, K. L. Campbell, L. O. Lollis, and A. D. Steinman. 2007. Integrated ecological and economic analysis of ranch management systems: An example from south central Florida. *Rangeland Ecology and Management* 60:1–11.

Swain, H. M, E. H. Boughton, P. J. Bohlen, and L. O. Lollis. 2013. Trade-offs and ecosystem services and disservices on a Florida Range. *Rangeland* 35:75–87.

Swanton, J. R. 1911. *Indian Tribes of the Lower Mississippi Valley and Adjacent Coast of the Gulf of Mexico*. Smithsonian Institution Bureau of American Ethnology, Bulletin 43. Washington, DC: Government Printing Office.

———. 1969. *The Indians of the Southeastern United States*. Smithsonian Institution Bureau of American Ethnology, Bulletin 137. Grosse Point, MI: Scholarly Press.

Symstad, A. J., E. Siemann, and J. Haarstad. 2000. An experimental test of the effect of plant functional group diversity on arthropod diversity. *Oikos* 89:243–53.

Tallamy, D. W. 2009. *Bringing Nature Home: How You Can Sustain Wildlife with Native Plants*. Updated/expanded version. Portland, OR: Timber Press.

ter Braak, C. J. F. 1990. *CANOCO—a FORTRAN Program for Canonical Community Ordination by (Partial) (Detrended) (Canonical) Correspondence Analysis and Redundancy Analysis*. Version 3.10. Ithaca, NY: Microcomputer Power.

ter Braak, C. J. F., and P. Smilauer. 2002. *CANOCO Reference Manual and CanoDraw for Windows User's Guide: Software for Canonical Community Ordination*. Version 4.5. Ithaca, NY: Microcomputer Power.

Tevesz, M. J. S., and J. G. Carter 1980. Environmental relationship of shell form and structure of Unionacean bivalves. In *Skeletal Growth in Aquatic Organisms*, edited by D. C. Rhoads and R. A. Lutz, 295–332. New York: Plenum Press.

Thiers, B. 2012 [continuously updated]. Index Herbariorum: A global directory of public herbaria and associated staff. New York Botanical Garden's Virtual Herbarium. http://sweetgum.nybg.org/ih/.

Timme, S. L. 2007. *Wildflowers of Mississippi*. Jackson: University Press of Mississippi.

Tobe, J. D., K. C. Burks, R. W. Cantrell, M. A. Garland, M. E. Sweeley, D. W. Hall, P. Wallace, et al. 1998. *Florida Wetland Plants: An Identification Manual*. Tallahassee: Florida Department of Environmental Protection.

Tower, J. A. 1961. The changing Black Belt—a geographical review. *Journal of the Alabama Academy of Science* 32: 479–85.

Towne, E. G. 1995. Influence of fire frequency and burning date on the proportion of reproductive tillers in big blue-

stem and Indian grass. In *Proceedings of the 14th Annual North American Prairie Conference*, edited by D. C. Hartnett, 75–78. Manhattan: Kansas State University.
Towne, E. G., and A. K. Knapp. 1996. Biomass and density responses in tallgrass prairie legumes to annual fire and topographic position. *American Journal of Botany* 83: 175–79.
———. 2003. Vegetation dynamics from annually burning tallgrass prairie in different seasons. *Journal of Range Management* 56:185–92.
Transportation Research Board. 2005. *Integrated Roadside Vegetation Management: A Synthesis of Highway Practice*. Washington, DC: National Cooperative Highway Research Program.
Tschinkel, W. S. 2006. *The Fire Ants*. Cambridge, MA: Belknap Press of Harvard University Press.
Turner, B. L., H. Nichols, G. Denny, and O. Doron. 2003. *Atlas of the Vascular Plants of Texas*. Vol. 1, *Dicots*; Vol. 2, *Ferns, Gymnosperms, Monocots*. Sida, Botanical Miscellany 24.
USDA (US Department of Agriculture). 2004. *Practice CP33 Habitat Buffers for Upland Wildlife*. Farm Service Agency, Notice CRP-479. Washington, DC.
———. 2006. *US General Soil Map (STATSGO2)*. Soil Survey Staff, Natural Resources Conservation Service. Accessed April 1, 2012. https://www.nrcs.usda.gov/wps/portal/nrcs/detail/soils/survey/geo/?cid=nrcs142p2_053629.
———. 2009. *Soil Survey of San Augustine and Sabine County*. Temple, TX: Natural Resources Conservation Service.
———. 2010–2013. PLANTS Database. Natural Resources Conservation Service. http://plants.usda.gov/.
———. 2012a. *Soil Surveys for Counties of the U.S.A.* Natural Resources Conservation Service. Accessed May 13, 2013. http://websoilsurvey.nrcs.usda.gov/app/WebSoilSurvey.aspx.
———. 2012b. *Official Soil Series Descriptions (OSDs)*. Natural Resources Conservation Service. Accessed November 30, 2017. *https://www.nrcs.usda.gov/wps/portal/nrcs/detail/soils/survey/class/?cid=nrcs142p2_053587*.
———. 2012c. *Fire Effects Information System (FEIS)*. Rocky Mountain Research Station, Fire Sciences Laboratory (Producer), US Forest Service. Accessed November 30, 2017. https://www.fs.fed.us/rmrs/tools/fire-effects-information-system-feis.
US Department of Transportation. 2011. How the highway beautification act became a law. Federal Highway Administration. Accessed August 6, 2013. https://www.fhwa.dot.gov/infrastructure/beauty.cfm.
US Environmental Protection Agency. 2012. Greenacres: Landscaping with native plants. Great Lakes National Program Office, Chicago, IL. Accessed July 31, 2013. https://archive.epa.gov/greenacres/web/html/index.html.
USFWS (US Fish and Wildlife Service). 1992. *White Bladderpod (*Lesquerella pallida*) Recovery Plan*. Albuquerque, NM.
———. 2011. Proposed rules: Endangered and threatened wildlife and plants; 12-month finding on a petition to list *Calopogon oklahomensis* as threatened or endangered. *Federal Register* 76, no. 192 (October 4, 2011). https://www.gpo.gov/fdsys/pkg/FR-2011-10-04/pdf/2011-25530.pdf.
———. 2012. Proposed rules. *Federal Register* 77, no. 176 (September 11, 2012).
Vanatta, E. S., B. D. Gilbert, E. B. Watson, and A. H. Meyer. 1914. *Soil Survey of Ashley County, Arkansas*. Washington, DC: Government Printing Office.
Van Bruggen, T. 1992. *Wildflowers, Grasses and Other Plants of the Northern Plains and Black Hills*. 4th ed. Rapid City, SD: Fenske Printing.
van der Schalie, H. 1939. *Medionidus mcglameriae*, a new naiad from the Tombigbee River, with notes on other naiads of that drainage. *Occasional Papers of the Museum of Zoology* 407:5–7.
Van der Valk, A. G. 2007. Development of post-disturbance vegetation in prairie wetlands. In *Plant Disturbance Ecology: The Process and the Response*, edited by E. A. Johnson and K. Miyanishi, 341–70. Amsterdam, Netherlands: Elsevier Press.
Van Lear, D. H., W. D. Carroll, P. R. Kapeluck, and R. Johnson. 2005. History and restoration of the longleaf pine-grassland ecosystem: Implications for species at risk. *Forest Ecology and Management* 211:150–65.
Vasconcelos, H. L., M. F. Leite, J. M. S. Vilhena, A. P. Lima, and W. E. Magnusson. 2008. Ant diversity in an Amazonian savanna: Relationship with vegetation structure, disturbance by fire and dominant ant. *Austral Ecology* 33:221–31.
Veldman, J. W., W. B. Mattingly, and L. A. Brudvig. 2013. Understory plant communities and the functional distinction between savanna trees, forest trees, and pines. *Ecology* 94:424–34.
Vickery, P. D., and P. W. Dunwiddie, eds. 1997. *Grasslands of Northeastern North America: Ecology and Conservation of Native and Agricultural Landscapes*. Lincoln: Massachusetts Audubon Society.
Vidrine, M. F. 2010. *The Cajun Prairie: A Natural History*. Eunice, LA: Self-published.
Vidrine, M. F., C. M. Allen, B. Borsari, L. Allain, and S. R.

Johnson. 2001a. The Cajun Prairie Restoration Project. In *Proceedings of the Seventeenth North American Prairie Conference*, edited by N. P. Bernstein and L. J. Ostrander, 151–54. Mason City: North Iowa Area Community College.

Vidrine, M. F., C. M. Allen, and W. R. Fontenot. 1995. *A Cajun Prairie Restoration Journal: 1988–1995*. Eunice, LA: Gail Q. Vidrine Collectables.

Vidrine, M. F., B. Borsari, J. Cordes, M. McClendon, and C. Allen. 1999. Landscaping with prairie in southwestern Louisiana: The Cajun Prairie and the natural landscaping paradigm. *Proceedings of the Louisiana Academy of Sciences* 62:49.

Vidrine, M. F., and T. Hillman, 2010. The Cajun Prairie: A natural history. Cajun Prairie Habitat Restoration Society. http://cajunprairie.org.

Vidrine, M. F., G. Vidrine, D. Vidrine, and C. Vidrine. 2001b. The Cajun Prairie Gardens: Demonstrations of sustainable landscaping in southwestern Louisiana. In *Proceedings of the 18th North American Prairie Conference*. Kirksville, MO: Truman State University.

Vogel, J. A., R. R. Koford, and D. M. Debinski. 2010. Direct and indirect response of tallgrass prairie butterflies to prescribed burning. *Journal of Insect Conservation* 14:663–77.

Vogl, R. J. 1973. Fire in the southeastern grasslands. In *Proceedings of the 12th Tall Timbers Fire Ecology Conference*, 175–97. Tallahassee, FL: Tall Timbers Research Station.

Wackerman, A. E. 1929. Why prairies in Arkansas and Louisiana. *Journal of Forestry* 27:726–34.

Waggoner, J., S. A. Clark, K. E. Perez, and C. Lydeard. 2006. A survey of terrestrial gastropods of the Sipsey Wilderness (Bankhead National Forest), Alabama. *Southeastern Naturalist* 5:57–68.

Wailes, B. L. C. 1854. *Report on the Agriculture and Geology of Mississippi*. Mississippi State Legislature. Jackson, MS: E. Barksdale, State Printer.

Walker, B. 1928. *The Terrestrial Shell-Bearing Mollusca of Alabama*. Miscellaneous Publication No. 18. Ann Arbor: Museum of Zoology, University of Michigan.

Ward, D. B. 2006. Keys to the flora of Florida–14, *Viola* (Violaceae). *Phytologia* 83:242–52.

Waselkov, G. A., B. M. Wood, and J. M. Herbert. 1982. *Colonization and Conquest: The 1980 Archaeological Excavations at Fort Toulouse and Fort Jackson, Alabama*. Archaeological Monograph 4. Montgomery, AL: Auburn University.

Watson, G. E. 1979. *Big Thicket Plant Ecology: An Introduction*. Big Thicket Museum Publication Series, No. 5. Saratoga, TX.

Watters, G. T. 2000. Freshwater mussels and water quality: A review of the effects of hydrologic and instream habitat alterations. In *Proceedings of the First Freshwater Mollusk Conservation Society Symposium, 1999, Chattanooga, Tennessee*, edited by R. A. Tankersley, D. I. Warmolts, G. T. Watters, B. J. Armitage, P. D. Johnson, and R. S. Butler, 261–74. Columbus: Ohio Biological Survey.

Watts, A. C., and G. W. Tanner. 2003. Fire and roller chopping have varying effects on dry prairie plant species (Florida). *Ecological Restoration* 22:229–30.

———. 2006. Restoration of dry prairie using fire and roller chopping. In *Land of Fire and Water: The Florida Dry Prairie Ecosystem*, edited by R. F. Noss, 225–29. DeLeon Springs, FL: E. O. Painter.

Weakley, A. S. 2012–2015. Flora of the southern and Mid-Atlantic states. Chapel Hill: University of North Carolina Herbarium. Working draft at http://www.herbarium.unc.edu/flora.htm.

Weaver, J. E. 1968. *Prairie Plants and Their Environment: A Fifty-Year Study in the Midwest*. Lincoln: University of Nebraska Press.

Weaver, J. E., and T. J. Fitzpatrick. 1934. The prairie. *Ecological Monographs* 4:109–295.

Webb, D. H., H. R. DeSelm, and M. Denniss. 1997. Studies on the prairie barrens of northwestern Alabama. *Castanea* 62:173–84.

Webb, R. H., H. G. Wilshire, and M. A. Henry. 1983. Natural recovery of soils and vegetation following human disturbance. In *Environmental Effects of Off-Road Vehicles: Impacts and Management in Arid Regions*, edited by R. H. Webb and H. G. Wilshire, 279–302. New York: Springer-Verlag.

Webb, S. D. 1977. A history of savanna vertebrates in the new world. Part I: North America. *Annual Review of Ecology and Systematics* 8:355–80.

———. 1990. Historical biogeography. In *Ecosystems of Florida*, edited by R. L. Myers and J. J. Ewel, 70–100. Orlando: University of Central Florida Press.

Webb, W. M., and F. S. Guthery. 1983. Response of wildlife food plants to spring discing of mesquite rangeland in northwest Texas. *Journal of Range Management* 36:351–53.

Weiher, E., S. Forbs, T. Schauwecker, and J. B. Grace. 2004. Multivariate control of plant species richness and community biomass in blackland prairies. *Oikos* 106:151–57.

Whelan, R. J. 1995. *The Ecology of Fire*. Cambridge Studies in Ecology. Cambridge: Cambridge University Press.

Wherry, E. T. 1955. *The Genus* Phlox. Philadelphia: Morris Arboretum.

White, H. L., W. C. Holmes, and J. R. Singhurst. 1998. *Sedum pulchellum* (Crassulaceae) in Texas. *Sida* 18:623–26.

White, M., and G. L. Nesom. 2012. Two intersectional hybrids in *Liatris* (Asteraceae) from east Texas. *Phytoneuron* 21:1–6.

White, P. S., and J. L. Walker. 1997. Approximating nature's variation: Selecting and using reference information in restoration ecology. *Restoration Ecology* 5:338–49.

White, P. S., and R. D. White. 1996. Old-growth oak and oak hickory forests. In *Eastern Old-Growth Forests: Prospects for Rediscovery and Recovery*, edited by M. B. Davis, 178–79. Washington, DC: Island Press.

Whitney, G. G. 1986. Relation of Michigan's presettlement pine forests to substrate and disturbance history. *Ecology* 67:1548–59.

Whitney, G. G., and J. P. DeCant. 2001. Government land office surveys and other early land surveys. In *Historical Ecology Handbook: A Restorationist's Guide to Reference Ecosystems*, edited by D. Egan and E. A. Howell, 147–72. Washington, DC: Island Press.

Whittemore, A. T. 1997. *Ranunculus*. *Flora of North America* 3:88–117.

Wieland, R. G. 2000. Ecology and vegetation of LeFleur's Bluff State Park, Jackson, Mississippi. *Journal of the Mississippi Academy of Sciences* 45:150–84.

Wilder, H. J., and C. F. Shaw. 1906. *Soil Survey of the Fayetteville Area, Arkansas*. Field Operations of the Bureau of Soils.

Willard, R., J. Morin, and O. Tang. 2010. *Assessment of Alternatives in Vegetation Management at the Edge of Pavement: Washington State Department of Transportation, Pavement Edge Vegetation Management*. Final Report, WA-RD 736.1. Olympia, WA.

Willcox, E. V., G. W. Tanner, W. M. Giuliano, and R. McSorley. 2010. Avian community response to grazing intensity on monoculture and mixed Florida pastures. *Rangeland Ecology and Management* 63:203–22.

Williams, J. D., S. L. H. Fuller, and R. Grace 1992. Effects of impoundment on freshwater mussels (Mollusca: Bivalvia: Unionidae) in the main channel of the Black Warrior and Tombigbee Rivers in western Alabama. *Bulletin of the Alabama Museum of Natural History* 13:1–10.

Williams, S. 1954. An archeological study of the Mississippian culture in southeast Missouri. PhD diss., Yale University.

Williamson, S., L. W. Burger, S. Demarais, and M. Chamberlain. 2002. Effects of northern bobwhite management practices on red imported fire ants. In *Quail V: Proceedings of the 5th National Quail Symposium*, edited by S. J. DeMaso, W. P. Kuvlesky, F. Hernandez, and M. E. Berger, 151–55. Austin: Texas Parks and Wildlife Department.

Wipff, J. K., and R. A. Thompson. 2003. *Urochloa*. *Flora of North America* 23:492–507.

Wiygul, S., K. Krans, R. Brown, and V. Maddox. 2003. Restoration of a prairie remnant in the Black Belt of Mississippi. In *Blackland Prairies of the Gulf Coastal Plain: Nature, Culture, and Sustainability*, edited by E. Peacock and T. Schauwecker, 254–61. Tuscaloosa: University of Alabama Press.

Woods, A. J., T. L. Foti, S. S. Chapman, J. M. Omernik, J. A. Wise, E. O. Murray, W. L. Prior, et al. 2004. *Ecoregions of Arkansas*. Reston, VA: US Geological Survey.

Woods, A. J., J. M. Omernik, D. R. Butler, J. G. Ford, J. E. Henley, B. W. Hoagland, D. S. Arndt, et al. 2005. *Ecoregions of Oklahoma* [Map poster]. Reston, VA: US Geological Survey. ftp://newftp.epa.gov/EPADataCommons/ORD/Ecoregions/ok/ok_eco_pg.pdf.

Woods, J. C. 1967. *Soil Survey of Houston and Peach Counties, Georgia*. Washington, DC: US Department of Agriculture, Soil Conservation Service.

Worthen, E. L. 1909. *Soil Survey of Clay County, Mississippi*. Washington, DC: Government Printing Office.

Wright, V., T. Hopkins, and S. Smith. 2005. Grasshoppers of Konza Prairie. Accessed December 2012. http://keep.konza.k-state.edu/prairieecology/grasshoppers.html.

Yager, L. Y., D. L. Miller, and J. C. Jones. 2011. Woody shrubs as a barrier to invasion by cogongrass (*Imperata cylindrica*). *Invasive Plant Science and Management* 4:207–11.

Yatskievytch, G. 2006a. *Steyermark's Flora of Missouri*. 2nd ed. Vol. 1. St. Louis: Missouri Botanical Garden Press.

———. 2006b. *Steyermark's Flora of Missouri*. 2nd ed. Vol. 2. St. Louis: Missouri Botanical Garden Press.

Young, S., and V. Claassen. 2007. *Evaluating Alternative Methods for Vegetation Control and Maintenance along Roadsides: Study II*. Research Technical Agreement # 65A0137. California Department of Transportation and University of California, Davis.

Zedler, P. H. 2007. Fire effects on grasslands. In *Plant Disturbance Ecology: The Process and the Response*, edited by E. A. Johnson and K. Miyanishi, 397–440. Amsterdam, Netherlands: Elsevier Press.

Zollner, D., S. Simon, and T. Foti. 2003. A plant community classification for Arkansas's blackland prairie ecosystem. In *Blackland Prairies of the Gulf Coastal Plain: Nature, Culture, and Sustainability*, edited by E. Peacock and T. Schauwecker, 110–45. Tuscaloosa: University of Alabama Press.

Contributors

Charles M. Allen is a researcher and wildlife scientist associated with the Center for Environmental Management of Military Lands and a professor of biology, retired from the University of Louisiana at Monroe. He has researched and published scientific papers and books on the Louisiana flora, especially prairies and grasses.

Brent T. Baker is a botanist for the Arkansas Natural Heritage Commission, where he works to identify and protect rare species. He has 15 years of experience studying and working with the flora of Arkansas, both in and out of the field, has served on the Arkansas Vascular Flora Committee, and in 2013 was an editor of the *Atlas of the Vascular Plants of Arkansas*. He is also an active member and past president of the Arkansas Native Plant Society.

John A. Barone is a professor in the Department of Biology at Columbus State University, Columbus, Georgia. A plant ecologist, he has studied the history, diversity, and structure of blackland prairies for 15 years.

Gail Barton is a landscape consultant and garden writer in Meridian, Mississippi. She has studied and propagated blackland prairie plants in Mississippi and Alabama for many years. Barton has been instrumental in establishing a one-acre prairie garden at the University of West Alabama. She is currently contracted to write detailed plant descriptions for New Moon Nursery, a wholesale grower of prairie plants and wildflowers.

Patrick J. Bohlen is the director of landscape and natural resources and the arboretum, and a professor of biology at the University of Central Florida. His current research interests include ecosystem ecology, urban ecology, ecology and valuation of urban green space, and wetland services.

Bruno Borsari is a professor in the Department of Biology and land steward at Winona State University, Winona, Minnesota. An agro-ecologist, he has studied and restored prairies for 23 years.

Elizabeth H. Boughton is a research biologist and the program director of agro-ecology at Archbold Biological Station, a nonprofit research and education organization located in the heart of Florida. For the last decade, Elizabeth's research has focused on how fire, grazing, and flooding shape the unique plant communities in wetlands and grasslands of the northern Everglades.

Edwin L. Bridges is an independent botanical and ecological consultant specializing in providing educational and research services to government agencies and nonprofit organizations. He has over 40 years of extensive experience in plant taxonomy and ecology, plant community ecology, effects of land management activities on rare species and ecosystems, and floristic and ecological field research. His research experience includes every southeastern state, with emphasis on Florida and the Longleaf Pine region of the Outer Coastal Plain.

Nicholas Brown is a wildlife lakes specialist for the Minnesota Department of Natural Resources in Fergus Falls, Minnesota. He currently manages shallow lakes and wetlands in northwestern Minnesota for the benefit of wildlife habitat.

L. Wes Burger Jr. is a Dale H. Arner Professor of Wildlife Ecology and Management in the Department of Wildlife, Fisheries, and Aquaculture at Mississippi State University. Over the past 20 years, Wes has conducted a body of research addressing wildlife response to conservation practices in working landscapes. His research interests include restoration of native grasslands and fire-dependent southern pine-grasslands. His research focus includes northern bobwhite, grassland songbirds, and pine-grassland bird species. Wes has been a frequent cooperator with USDA FSA, NRCS, and APHIS. He led the NRCS Bobwhite Restoration Project and is currently coordinating the National CP33 Monitoring program.

J. J. N. Campbell is from London, England, but since 1976 he has lived in Kentucky, where he received a PhD from the University of Kentucky in 1980. He has been involved in research on the flora and vegetation of that state, and he worked for the

Nature Conservancy from 1987 to 2007; see bluegrasswoodland.com for details. He occasionally wanders into adjacent states extending this research, and he maintains an interest in comparisons with Europe and Asia.

Jolie G. Dollar is currently the Pacific Island Region agroforestry consultant for the nonprofit Xerces Society for Invertebrate Conservation. For more than 10 years, she has studied the conservation of beneficial insects within agricultural and agroforestry systems.

S. Lee Echols is a conservation biologist for the North American Land Trust. His research interests include southeastern floristics and vegetation classification, blackland prairies, and private land conservation.

Edward D. Entsminger is a research associate in the Department of Sustainable Bioproducts/Forest and Wildlife Research Center at Mississippi State University. He has three professional certifications as an Associate Wildlife Biologist (AWB), Associate Fisheries Professional (FP-A), and a Wetland Professional in Training (WPIT). He is a wildlife and fisheries biologist/ecologist and has studied prairies, wetlands, and forested environments for plant and animal species reclamation and diversity.

Thomas L. Foti holds an MS in botany (ecology) from the University of Arkansas. He is the retired chief of research and plant community ecology at the Arkansas Natural Heritage Commission. He developed the first map and description of the ecoregions (natural divisions) of Arkansas and has participated in subsequent revisions and updates. He has been studying and working to protect remnants of Arkansas prairies since the 1960s.

Toby Gray is a research associate with the Gulf Coastal Plains and Ozarks Landscape Conservation Cooperative at the Mississippi State Department of Geosciences.

John W. Guyton is the director of Mississippi State Extension's Arthropod Adventures Program, which manages the world's oldest Entomology and Plant Camp, the Arthropod Zoo, BugFest, and related 4-H programs. He is also involved with the Entomology Society of America's Education and Outreach Committee, the Mississippi Native Plant Society, and the Mississippi Environmental Education Alliance and edits the *Gloworm* newsletter.

JoVonn G. Hill is an assistant research professor with the Mississippi Entomological Museum in the Department of Biochemistry, Molecular Biology, Entomology, and Plant Pathology at Mississippi State University. He is interested in the biodiversity, biogeography, ecology, and natural history of the southeastern United States and is an expert in the taxonomy and ecology of the ant and grasshopper fauna of the Southeast.

Raymond B. Iglay is a research associate in the Department of Wildlife, Fisheries, and Aquaculture at Mississippi State University. He is an applied research biologist with interests ranging from community ecology to individual-based modeling. When not in the office, he manages a small business helping landowners achieve their wildlife conservation goals.

Jeanne C. Jones is a wildlife science professor in the Department of Wildlife, Fisheries, and Aquaculture at Mississippi State University (MSU), where she has conducted research and taught courses in wildlife conservation and restoration for more than 30 years. She is a Certified Wildlife Biologist and Fellow of the Wildlife Society (TWS) and holds officer positions in TWS working groups. In 2009, she received the Outstanding Woman Faculty Award at MSU from the Presidential Committee on the Status of Women.

Lisa McInnis has an undergraduate degree and a master's degree in biology, and a PhD in forestry. Her applied research interests relate to fire in southeastern forests and reintroduction of fire for prairie restoration. Lisa has been working in natural resource management for the National Park Service for 13 years.

Reed F. Noss is Provost's Distinguished Research Professor at the University of Central Florida. He studies biogeography, fire ecology, vegetation science, conservation biology, and many other subjects. He has more than 300 publications, among them seven books, including *Forgotten Grasslands of the South: Natural History and Conservation.*

Steve L. Orzell studies the native flora and plant communities of the southeastern United States, with emphasis on peninsular Florida. His current research focuses primarily on fire-climate-vegetation relationships and fire adaptations of the flora in subtropical and tropical savannas and grasslands. Additionally, he is interested in documenting and describing the unknown plant biodiversity of peninsular Florida.

Marc G. Pastorek is a horticulturist and amateur prairie ecologist specializing in developing prairie landscapes by way of locally sourced seed. He consults, designs, and builds projects in eastern Texas, Louisiana, Mississippi, and Alabama. He and his wife, Candi, have owned and operated their consulting firm, Pastorek Habitats, LLC, since 1998.

Evan Peacock is a professor of anthropology at Mississippi State University. His research is centered on long-term human/nature interactions, including documenting and quantifying prehistoric human environmental impacts. He specializes in the analysis of molluscan remains, especially freshwater mussel and land snail shells.

Samuel K. Riffell was an associate professor of wildlife and undergraduate coordinator in the Department of Wildlife, Fisheries, and Aquaculture at Mississippi State University. His interests included wildlife conservation in agricultural landscapes, road ecology, landscape ecology, ornithology, and dragonfly biology. Deceased.

Timothy J. Schauwecker is the coordinator of the Landscape Contracting and Management program in the Department of Landscape Architecture at Mississippi State University. His teaching focuses on basic ecology, project management, and sustainable practices. His research interests are biodiversity management, water management, and watershed planning.

Jennifer L. Seltzer is a research associate at the Mississippi Entomological Museum. She is responsible for the curation of the Macrolepidoptera, bark beetle surveys, and various collection projects to further enhance the museum collection. Other areas of interest include charcoal identification from southeastern archaeological sites and the impact of humans on prehistoric landscapes.

W. R. "Randy" Seymour Jr. is an avid naturalist and proprietor of Riders Mill Farms, a working farm and wildlife sanctuary in Hart County, Kentucky, near Mammoth Cave National Park. In 1997, he published *Wildflowers of Mammoth Cave National Park*, a guide to species of the whole region. With his son, he founded Roundstone Native Seed, Inc., which is now a leading producer of native grass and wildflower seed for the eastern United States.

Jason R. Singhurst is a botanist and plant ecologist with the Texas Parks and Wildlife Department in Austin. He has conducted field-oriented research on the rare plants and vegetation ecology of Texas for the past 20 years. In 2008, Jason coauthored *Rare Plants of Texas*.

Reggie E. Thackston received his bachelor's and master's degrees in forest resources from the University of Georgia School of Forest Resources in 1976 and 1978, respectively. He is a TWS Certified Wildlife Biologist and a member of the Society of American Foresters. He currently serves as the Private Lands program manager of the Wildlife Resources Division, Georgia Department of Natural Resources.

James W. Tomberlin is a wildlife biologist with North Carolina's Wildlife Resources Commission. During his seven and a half years working with Georgia's Wildlife Resources Division, he focused on bobwhite quail management, pine savanna restoration, and Farm Bill policy and implementation at local, state, and national levels.

J. Morgan Varner is an assistant professor in the Department of Forest Resources and Environmental Conservation at Virginia Tech in Blacksburg. He is a fire ecologist focused on the interactions among fire, trees, and herbs in the southeastern and western United States.

Malcolm F. Vidrine is a retired professor of biology at Louisiana State University, Eunice. As an evolutionary ecologist, he restores prairies and studies natural landscaping. Two twenty-first-century books, *The Cajun Prairie: A Natural History* and *Mites of Freshwater Mollusks*, summarize much of his life's work.

Matt White teaches American history at Paris Junior College in northeastern Texas. He has spent many hours searching for lost prairie remnants, which culminated in the book *Prairie Time: A Blackland Portrait*. An avid birder and amateur botanist, he has published several journal articles about both subjects as well as contributions to the 2002 *Birds of Northeast Texas*.

C. Theo Witsell is a botanist and ecologist for the Arkansas Natural Heritage Commission and a research associate at the Botanical Research Institute of Texas. He was an editor of the *Atlas of the Vascular Plants of Arkansas* and has authored numerous articles and technical publications on the flora of the southeastern United States—especially of Arkansas, where he works to identify and protect rare species and high-quality natural areas.

Wendy B. Zomlefer is a professor in the Department of Plant Biology at the University of Georgia and curator of the herbarium there. Her research focus is on general floristics of the Southeast and the systematics of petaloid monocot groups. She is also well known for her book *Guide to Flowering Plant Families*, which she both wrote and illustrated.

Index